Living in Fear
The Francoist Genocide of Spain
1936-1949

An appalling humanitarian catastrophe

Enhanced Second English Edition of
FRANCISCO MORENO GÓMEZ's
La Victoria Sangrienta
by
MAGDALENA GORRELL JAÉN

SWEETSPIRE LITERATURE
— MANAGEMENT —

CONTENTS

APPENDIX

FOREWARD TO THE NEW ENGLISH EDITION OF *VICTORIA SANGRIENTA*

Francisco Moreno Gómez

With this new English edition of my 2014 book, *La Victoria Sangrienta*, now published as *Living in Fear. The Francoist Genocide of Spain, 1939-1949,* we wish to draw attention to the great unknown humanitarian catastrophe that Francoism created in Spain. An appalling humanitarian catastrophe without parallel, which almost continues plunged into oblivion, not only in Spain itself, but also worldwide. When the World War II Allies, the victors in 1945, decided not to interfere in Spain against Franco's fascist dictatorship, the policy of forgetfulness and impunity was buried and forever sealed in silence.

The aftermath of the 1936 event that in Spain was a military coup in all respects, was a terrible civil war. The violent clash that pitted the supporters of the coup against those whose who stood for the legal government of the Republic did not end with Franco's bloody victory in 1939. It was followed by a widespread, violent dictatorship that lasted almost 40 years, in full view of and the total indifference of the western democratic world.

The Axis powers of Germany and Italy, Nazis and fascists everywhere, the model adopted by the Spanish military and the Spanish right-wing, were overwhelming in favour of Franco and the supporters of the coup whom they actively promoted. As if this were an apparent rehearsal of Hitler's program for the rest of Europe, the Axis sent troops, bombers, tanks and military supplies. The legitimate government of the Spanish Republic only had recourse to limited international aid, namely the thousands of heroic volunteers from fifty countries who travelled to democratic Spain to assist in the country's defence against fascism.

From the onset, with the outbreak of violence the leaders of the military coup started a horrific wave of summary executions by firing squad of more than 100,000 civilian Republicans of all ages: city and country supporters, farmhands and workers of all sorts. This first massacre, introductory to the Francoist regime, was followed throughout the post-war period by the slaughter of another 40,000 individuals, the flight into exile of more than 450,000 Spaniards, and countless other calamities.

Nowadays, in Spain, there is not a minimum consensus regarding the memory of the war and the dictatorship. There is no generally accepted official or public memory about the serious violations of human rights by Francoists. Spain appears divided by inimical interpretations of the common past because the heirs of the winners still refuse to ever acknowledge anything regarding the memories of the defeated. These latter-day successors to Franco continue to justify the military coup and to praise Francoism, even as they pay lip-service to the country's present-day claim to democracy.

The general true knowledge of the crimes of Francoism in Spain, in my case, based on the data for the province of Cordoba, have slowly becoming known since the death of the dictator in 1975, some 85 years ago. Very slowly and with great difficulty, because the conservative political parties, which in their day supported Francoism, have been systematically opposed to any initiative to make the truth known to the Spanish people. These conservative parties are the ones guilty for today's continuing ignorance of the Francoist criminality and the reality of the victims, in their many forms, especially those that are hidden in anonymous mass graves.

The memorialist movement that began in 2001 has had a major impact on the critical review of the country under Francoism, even more so than the work of academic historians. The public work of the different fora and memorialist associations has grown in importance in the last few years, particularly thanks to the most positive support of the several UN organizations dedicated to human rights. Furthermore, in recent years there has been a steady public interest in and support of the motto of the memorialist associations, that is, continuing search for truth, justice and reparation, with a view to ensure that this history does not repeat itself.

In order for the truth to be known in Spain, the following have been particularly helpful: 1) The work of territorial historiographic monographs (provincial and local) of a group of dedicated historians, as well as quite a few books by survivors relating their memories; 2) The work of several volunteer Hispanists who have published extremely well grounded works (Paul Preston, Ian Gibson, Gabriel

Jackson, Helen Graham, etc.); 3) The great investigative work of the Associations and Groups for the Democratic Historic Memory who have been very active since the beginning of the 21st century, especially as regards the exhumation of mass graves.

The spark that initiated the memorialist phenomenon appears to be the exhumation of thirteen bodies in a mass grave in Priaranza del Bierzo (León), in October 2000. In August 2002, the phenomenon of the disappeared in Spain and the beginning of the opening of mass graves was drawn to the attention of the UN High Commission for Human Rights. Most important, it was the third generation of those who suffered during the civil war (grandchildren and other descendants), who decided to break with the disremembering and to honour the victims; to dig up the graves and bury with dignity, the remains of those assassinated and buried in anonymous mass graves. By 2014, 2,382 mass graves had been identified throughout Spain, from 280 of which 6,000 cadavers were exhumed.

In Cordoba capital, there is a Platform for the Commission of Truth, a Forum for the Historic Memory and a Memory Association regarding those assassinated during the war in the capital alone, entitled 'Let us Cry' or 'Association of the 4,000'. Today, in Cordoba province, there are ten additional Historic Memory associations or fora, as in Cordoba capital, Aguilar, Castro del Río, Belmez and La Carlota.

Elsewhere in Spain, there have been numerous memorialist activities. In Catalonia, they created the Democratic Memorial of Catalonia; in Seville, the All the Names Project; in Galicia another project, 'Names and Voices', in 2006. In 2004, the Universidad Complutense of Madrid created the 20th Century Chair of Historic Memory. Also in 2004, the Spanish Government created the Interministerial Commission for the study of the situation of the Victims of the Civil war and Francoism.

This brings us to the first historic memory legislation, Law 52/2007. Judge Baltazar Garzón's famous October 16 2008 ruling triggered an uproar as he declared the initiation of proceedings for 'illegal detention and disappearance' against 35 Francoist leaders, beginning with Franco himself. His daring to do so cost him dearly, as he was disbarred. In Spain, where the conservative political parties and centres of power, such as the judiciary, are desperately attempting to continue a blanket of silence, forgetfulness and impunity regarding so much bloodshed, all mention of Francoist crimes continues to be taboo.

On the international front, many organizations have taken positions respecting the Spanish memorialist movement. In the UN there was the International Convention for the Protection of All People with Forced Disappearances of 29 December 2006. The alarming, forced forgetfulness regarding disregard of human rights during the civil war, led the Nizkor Team, in 2004, to publish a Report on the Impunity of the Francoist crimes. Amnesty International published a Report of its own in 2006. A first trip to explore the democratic situation of Spain regarding these victims occurred between May and July 2012

In 2013, the UN sent the Working Group on Forced Disappearances to Spain. Its Final Report dated 2-7-2014, told democratic Spain that it must comply with 43 demands to repair the past, without forgetting the case of the 30,960 children lost or stolen by the Francoists. Pablo Geiff, UN Special Relator for the Promotion of Truth, Justice, Reparation and Guarantees of no Repetition regarding the victims of Francoism, visited Spain in January 2014. The final report of his visit was published in July 2014.

The social democratic Spanish government reacted positively and finally, 19 October 2022, it approved the second and definitive Law of Democratic Memory that declared that the 1936 military coup was illegal, the sentences of the Francoist courts were declared null and void, regulated the right of all the victims to the truth, and made noteworthy declarations such as the following:

> "All the laws of the Spanish Government, including Amnesty Law 46/1977 of 15 October, shall be interpreted and applied in keeping with the principles of International Humanitarian Law, according to which war crimes, those of lèse humanity, genocide and torture shall be considered unprescribably and not amnestied"

As a Spanish law, it is like a miracle, an infinite reparation.

Still, the Spanish society, schooled for so many years in the strictures of dismemory and forgetfulness, is generally apathetic and indifference to any mention of Francoism, as our people appear almost exclusively captured by the *carpe diem* of today's consumer society. Nevertheless, things are slowly changing. In the political plane, the social democrats have become aware that this silence and forgetfulness attitude has enabled the right-wing to take over all reports of a transition with their motto: "Do not meddle with the past".

I close this report with the news that today, as I write, the remains of five women who had been murdered in 1937, were exhumed in Palma de Mallorca January 28, 2023. A fitting headline to draw the attention of all the world.

"THE REMAINS OF PICORNELL AND THE ROGES DEL MOLINAR ARE RETURNED TO THEIR FAMILIES.

A total of 540 persons took part this Saturday in rendering homage and returning the remains of Aurora Picornell and the Roges del Molinar, an act that was celebrated in the Balearic Conservatory of Music. This was attended by the second Vice-Presidency of the Government and Ministry of Labour and Social Economy, Yolanda Díaz, the minister of Equality, Irene Montero, the president of the Government, Francina Armengol, the vice-president of the Autonomic Council, Juan Pedro Yllanes, and the Mayor of Palma, among other authorities.

Speaking on behalf of the Government, Yolanda Díez, said:

There can be no peace nor future without memory, as nothing solid can be built on forgetfulness. For this reason, this Saturday we are in Palma, to pay tribute and return the valiant women who fought against fascism to their families, because the democratic memory is an act of love for one's country, as well as a lesson, so that society, especially its young people, do not remain indifferent when facing hate".

Major German & Italian Attacks
against the Republicans
during the Spanish Civil War (1936-1939).

Main Rebel centres

Main Republican centres

Land battles

Naval battles

Bombed cities

Concentration camps

Massacres

Refugee camps

WHAT THE WORLD NEEDS TO KNOW ABOUT THE CRIMES OF FRANCOISM.

Francisco Moreno Gómez

The Francoist repression was much more than the firing squads. The punishments emanated from the world of the prisons, an infernal realm from which all the methods employed for punishing and disciplining the vanquished, the so-called "multi-repression", stemmed. The unequivocal mission of the entire Francoist world, under the direction of the State and its operatives, was to castigate the vanquished, where the prison realm provided a whole range of very numerous and very complex repressive measures.

Franco's governing project had nothing in common with the Republican penal system and that which occurred in the Republican zone. The difference between the two was abysmal, which is why we need to examine the Francoist system from a scientific, historical point of view.

A. To begin, the Francoist prison realm was principally noted for its ceremonial executions by firing squad of an estimated 40,000 Republicans during the post-war period. (This does not include the 100,000 individuals executed during the war, statistics that are generally supported by the generation of Spanish historians who have studied this matter during the last 30 years.) According to the personal witness declarations of the archivists employed by the First Territorial Military Archives in Madrid,[1] 300,000 individuals are recorded as offenders. In our book *Victimas de la guerra civil*,[2] based on data from two-thirds of the Spanish provinces, we estimated that 140,000 individuals were liquidated under Franco. Most specialist authors on this subject refer to this as the great massacre.[3]

---

[1] *Archivo Militar Territorial Primero* (Madrid) - First Territorial Military Archives.

[2] *Victims of the civil war,* Santos, Juliá (Coord.), Temas de Hoy, Madrid, 1999. p. 411.

[3] Francisco Espinosa Maestre (Coord.), *Violencia roja y azul,* op. cit., p. 78. Reports 130,199 Republican victims; in the last few years this number has been corrected upwards several times.

The great problem of calculating the Republican losses is the existence of the *desaparecidos* [disappeared], the majority of whom are not documented or for whom there are no records. Judge Baltazar Garzón, in his 16 October 2008 decision, reported that these totalled 114,226.[4] There is, however, a mini story involving this total, considering that it was released by the press in *El País*, 17 October 2008, when the study was still incomplete. In his document "Report on the Francoist repression"[5] that F. Espinosa presented at Judge Garzón's Court, the minimum number of disappeared was calculated as 129,472. Still, the "precise" number of mortal victims of Francoism is still impossible to calculate because of the following:

a. almost daily removals from and executions by firing squads in the concentration camps (Castuera, La Granjuela, San Marcos de León, Uclés monastery, Ronda, San Cristóbal…).

b. inexact, although considerable, numbers of prisoners who died in the forced labour detention camps.

c. thousands of Spaniards who died in the French concentration camps during the first three months of their flight into exile;[6]

d. several hundred prisoners who died in the North African forced labour camps; (in both these cases, there was a massive chaos resulting from the number of refugees fleeing Franco's terror.)

e. approximately 6,000 Spaniards interned and abandoned in Nazi concentration camps.

f. almost 15,000 individuals incarcerated in Francoist jails (an exact number is lacking) who died from hunger and privations, especially in 1941. Current data from a dozen prisons and the number of victims is already estimated at more than 5,000.

g. 3,500 members of the Republican Resistance killed in the countryside.

---

[4] United Nations document: Human Rights Council. Report of the Working Group on **Enforced** or **Involuntary** Disappearances, Mission to Spain 23-30 Sept 2013, Addendum 2 July 2014. https:/digitallibrary.un.org

[5] *Informe sobre la represión franquista*, F. Espinosa.,

[6] Francisco Moreno Gómez, *Trincheras de la República, 1937-1939*, El Páramo, Córdoba, 2013, p. 560. Antonio Vilanova reports on 14.672 deaths the first three months in exile in the South of France, in his book *Los olvidados. Los exiliados españoles en la segunda guerra mundial*, Ruedo Ibérico, Paris, 1969, p. 10, he cites "official" French sources. Geneviève Dreyfus-Armand quotes the same data in *El exilio de los republicanos españoles en Francia*, Crítica, Barcelona, 2000, p. 53.

h. over 1,500 civilians eliminated by the Guardia Civil on the grounds of the "Law of Fugitives" during the period that I call the "triennial of terror" (1947-1949). In Cordoba province alone, 160 are known to have been shot in roadside ditches, especially in 1948 (at the time that the Universal Declaration of Human Rights was being approved in the UN), and a similar number killed in this manner in other locations.

This genocide continued. In 1949, when in August the Geneva Convention was being signed, hundreds of Republicans were executed without the benefit of a trial or court hearing in the Seville mountains and elsewhere in the country. In 1950, for example, the *paseo*[7] was applied without pity in Nerja (Málaga, Granada, etc.).

There are no details about this human catastrophe. Furthermore, when one talks about "martial law ending on 7 April 1948" as being the end of the repression, one is making a great mistake because that very year, and all during the following year, even more persons were being executed all over Spain, victims of the so-called "Law of Fugitives".

B. One often hears speak of the victorious Francoist government's fraudulent pronouncement, that "all those whose hands are not sullied by blood, have nothing to fear". Today, this declaration has been discredited by all students of the repression and the number of individuals who fell into the trap. It was a trick designed to catch the unwary within the Spanish borders and to hoodwink the international community by appearing that Franco's government was respecting the terms of the 1929 Geneva Convention. Peter Anderson has wisely explained this question.[8] The said Convention forbade all condemnation for "rebellion" or "political crimes" in

---

[7] *"The ride"*. At night, armed trucks of Francoists would show up at suspected republican supporters' homes, knock on the door and take the men, saying they would be "going for a ride". They were then taken to a prison where they were held until the next day, when most were executed by firing squad without further ado. Also used to apply the Law of Fugitives.

[8] Peter Anderson, "Francisco Franco, ¿Criminal de guerra?", in *Hispania Nova*, number 10, 2012. Also, Peter Anderson & M. A. del Arco Blanco (Eds.), *Lidiando con el pasado. Represión y memoria de la guerra civil y del franquismo*, Comares, Granada, 2014, and "Escándalo y diplomacia. La utilización de los consejos de guerra para mantener la represión franquista durante la guerra civil", pp. 83-100.

civil war, and only accepted convictions for "ordinary crimes". Consequently, Franco's "bloody hands" declaration was a ruse to cover up his crimes and deceive the international community, on the grounds that "delinquents only", not political prisoners, "were being executed". This fallacious statement was exceedingly popular and is still a frequently quoted figure of speech by Spanish conservatives, who make spectacular declarations.

In point of fact, the great repression was not created to punish individuals guilty of breaches of common law, but for political reasons such as simply appearing to be sympathetic to a left-wing political party, belonging to a trade union, the Frente Popular party, being a Mason, and so forth .[9],[10] This is where we find the famous political repression law against "Free Masonry and Communism" of 1 March 1940.

An example of the arbitrary mature of that type of "justice" is that which occurred to Antonio Varo Granados, from Pozoblanco (Cordoba). Resident in Madrid in 1936 at the time of the events in his village, he was condemned to the "common garrotte" in 1940 on the grounds of the following "crime of blood". He was accused of the fact that at the beginning of the *Glorious Event*, he was surprised in Madrid, quite a distance from his town. Furthermore, even though he held the destiny of his town in his hands, he stayed in Madrid rather than travel home to prevent the commission of violations and outrages.[11] Varo Granados escaped being put to death thanks to a Francoist captain whom he had hidden in his house in Madrid during the civil war. An example of the bad luck of a vanquished individual whose life depended on the winner's courts.

C. Another often mentioned fallacy is that the repression was only applied "**at certain levels of responsibility in the left-wing parties...**" and that "**each case was examined on its own**". Here

---

[9] The writer, Ferrán Fontana Grau (Tarragona) was shot by firing squad on the grounds that he was the author and Director of several plays based on and promoting revolutionary tendencies, which were presented in Reus theatres; that he also wrote poems on the same subject and with the same purpose, such as the one entitled "Song from the Rear" that was also sung at the theatres. He was executed 8 August 1939. *Pilatos,* J. Subirats Piñana, *1939-1941* and *Prisión de Tarragona,* F. Pablo Iglesas, Madrid, 1993.

[10] Further details can be found in *Damnatio Memoriae,* Vol. I, pp. 168 and following.

[11] Case number 26.454/1939, Pozoblanco Court, 22 April 1940.

we must make the following quite clear: the Francoist repression was first directed against the system (the II Democratic Republic) and then against its elite and against the socio-political bases that sustained the system. The firing ranges executed people of all kinds and conditions, both politically significant and insignificant. We must not forget that the Francoist repression, among other aspects, was also a class repression against "shirtless persons who wore espadrilles". The plan was to decimate members of the left wing, at the very least, so that they would be unable to raise their heads for decades. That is what Manuel Díaz Criado, the Seville butcher, meant when he declared "thirty years from now, there will be no living soul here".[12] In the Cortes, the Spanish Parliament, José Calvo Sotelo praised the "product" of a good slaughter with his statement that the 40,000 executions of the Commune ensured sixty years of social peace.[13] It is manifestly clear that the objective was that of a political, not penal, repression.

It is not true that the cases were examined on a case-by-case basis as, until mid 1940, the courts-martial were collective. In Madrid, for example, the prisoners were taken to be judged in trucks, to the Las Salesas where they were put on show to the public. One hour was not long enough to read out the list of the accused. Their appointed lawyers, who did not even know them, limited their defence to ask the Court for "clemency". Collective trial is how we describe the extremely swift court martial of the *Three Roses* which began in Madrid in August 1939. In fact, these young women were not "bloodstained", they were simple militants of the JSU [Socialist Youth]. Undoubtedly, it was a political trial. An example of the perfect reconstruction of a collective court martial can be seen in Benito Zambrano's 2011 movie, *La voz dormida*.

I must insist on the fallacy that those who were executed by firing squad were guilty of "crimes of blood". We know that at the beginning of 1938, there were 106,822 republican prisoners in Franco's hands, graded by the Concentration and Prison

[12] Antonio Bahamonde, *Un año con Queipo de Llano*, Espuela de Plata, Seville, 2005, p. 159.
[13] Manuel Álvaro Dueñas, "La legitimación de la represión franquista", in Mirta Núñez Díaz-Balart, *La gran represión. Los años de plomo del franquismo*, Flor del Viento, Madrid, 2009, p. 101.

Camps Inspectorate[14] as belonging to four groups: a) indifferent or apathetic); b) unconvinced; c) disgruntled; and d) presumed authors of crimes. The latter represented 2.13% of the inmates.[15] Thus, little more than two percent of the multitude of Franco's prisoners could be accused of having "bloodstained hands" or some other excess. As it happened, few were "bloodstained"; the overwhelming majority were political.

D. We cannot also study the Francoist repression without examining another crucial and terrible subject, the systematic practice of torture. Very few historians address this, although all the prisoners, without exception, suffered while they were being arraigned. This was such a generalized and brutal torture, that more than a few did not survive in the hands of the torturers. In this matter the oral testimony and dozen extremely valuable diaries are fundamental, most of which have now come to light with the current move towards democracy. Torture was "a fundamental principle" of Francoism, as it is with all dictatorships and tyrannical governments. Simultaneously with the torture of men, there was the *humiliation of women* by *shaving their heads, forcing them to drink castor oil*, and *stripping them naked* before parading them in public.

  Also noteworthy is the *widespread wave of terror* following continuous house arrests and raids that terrorized not only the partners of those who were apprehended and the inmates of the prisons themselves, and poured into the streets, affecting relatives, neighbours, friends and so forth. The Francoists well knew that the execution of an individual not only spread terror inside the detention centres, as it did so outside, into the everyday lives of the people.

E. Prisons and penal detention centres also were a source of battalions of forced labour sent out into the streets, something that already existed during the war but was much more prevalent in the post-war

---

[14] *Inspección de Campos de Concentración y Prisoneros* - Concentration and Prison Camps Inspectorate.

[15] Antonio D. López Rodríguez, *Cruz, bandera y caudillo. El campo de concentración de Castuera*, CEDER-La Serena, Badajoz, 2006, p. 72.

period. They had various names such as Disciplinary Battalions of Worker Soldiers, Militarised Prison Colonies, Penal Assignment Centres, etc.[16] All these were governed by an extremely severe internal regime with a high mortality rate.

There were forced labour camps in the Centre-South of Spain, under the Republic. Totana (Murcia) Labour Camp (which is what it was called) was established by an Order from the Minister of Justice (Juan García Oliver) of 28 December 1936. It was in the ancient convent of the Capuchins and surrounding land. Another Order of 11 January 1937 created the Corps of Labour Camps Supervisors (something that was unconceivable in the Franco zone, where the camps were supervised by falangists, the military and Francoist supporters). For example, the Order states:

"Article 8. Every guard is required to inform his superior when an inmate appears to be ill... or who requires some special care; the Director is required to ensure that the Doctor examines the said inmate, without delay. Article 9. The guards shall not use any unnecessary force when dealing with the inmates," And further ahead: *"An inmate who during the day has worked regularly with good behaviour, shall be paid a bonus of 50 centavos."*

There were few more than a thousand inmates in Totana. (Under Law, there could be no more than two thousand inmates in a camp, nor anyone more than 60 years old.) The above clearly shows an unmatched difference as compared to Franco's forced workers.

---

[16] The breakdown of the terrible theme of forced, or slave, labour can be seen in the work of Santos Juliá (coord.), *Víctimas de la guerra civil,* Temas de Hoy, Madrid, 1999, 2004 edition, pp. 277 and following. Also in Francisco Moreno Gómez, *La victoria sangrienta, 1939-1945,* cit., pp. 519 and following; Fernando Mendiola and Edurne Beaumont's documentary, *Desafectos. Esclavos de Franco en el Pirineo,* Eguzki Bideoak, Memoriaren Bideoak, 30', 2007; Mariano Agudo and Eduardo Montero's documentary, *Presos del silencio,* Intermedia Producciones, Seville, 2004; Rafael Torres, *Los esclavos de Franco,* Oberón, Madrid, 2000; Isaías Lafuente, *Esclavos por la patria,* Temas de Hoy, Madrid, 2002. This was always "forced" labour. The only volunteers were criminal offenders who accepted to "convert their sentence through work", something totally from francoist forced labour and only represented some 5% of all inmates.
Another note: No Disciplinary Battalions of Worker Soldiers were created during the war; only during the post-war period. Only Labour Battalions were created during the war. Also, nobody was sentenced to "forced labour" as an inmate's being sent to forced labour was determined after he had been sentenced. It is also false to believe that resort to forced labour was much more limited than in other dictatorships: in fascist western Europe, only Hitler enslaved more people than Franco.

Furthermore, under Franco there were more than 100 such camps in Spain, while under the Republic, there were no more than ten.[17] In addition to the Totana Republican forced labour camp, the SIM (Servicio de Información Militar) managed another 6 camps in Catalonia: Tivissa, Concabella, Ornells de Na Gaia, etc. Although there were some deaths under the Law of Fugitives, it would be out of the question to pretend that one cannot compare the Francoist and the republican camps.[18]

F. Going forward, the realm of the prison was the perfect greenhouse for the *ideological repression* by the prison chaplains especially, as it provided the force to impose the ideology of the winners in the minds of the vanquished inmates. This repression was extended to their families as inmates who had had a civil wedding, were forced to be married by the church before their wives were allowed to visit their imprisoned husbands. Just one of the many ways that the Francoists used to exert pressure the minds of the vanquished, in an attempt to exterminate their republican, lay or class thoughts. In addition to the role of mass slaughter, Raphael Lemkin cites this as the essence of his concept of genocide: to take the thoughts of the vanquished by force, thus compelling them to accept the ideology (national-Catholicism) of the winners.[19]

Another widespread form of Francoist repression was the *destruction of Republican families* by executing the fathers, imprisoning the mothers, and sending the children to orphanages, an organized plan to ensure that the ideology of the parents was not transmitted to their children. Thousands of families were scattered or broken up as some members were murdered, others fled in exile or took to the hills, or simply disappeared.

---

[17] Payne-Palacios' remarkable theories regarding the forced labour camps were based on the work of Julius Ruiz, an expert in undervaluing the Francoist repression and hyper valuing the Republic, "Work and Don't Lose Hope: Republican Forced Labour Camps during the Spanish War", in *Contemporary European History*, 18, 4, 2009, pp. 419-441.

[18] Francesc Badía is the source of data on the Republican Forced Labour Camps, in *Els Camps de Treball a Catalunya durant la Guerra Civil, 1936-1939*, Abadía de Montserrat, Barcelona, 2001. This book suffers from an excess of religious pathos which has too much influence on its content.

[19] Raphael Lemkin, *El dominio del Eje en la Europa ocupada*, Prometeo Libros, Buenos Aires, 2008. *Rule In Occupied Europe: Laws of Occupation, Analysis of Government, Proposals for Redress*. First published in 1944. General Editor Joseph Perkovich.

Yet another harrowing repressive measure with a widespread impact was the theft and disappearance of children. When families were destroyed, unattached or unsettled children were placed in religious orphanages. The same occurred to children whose mothers were in prison and from whom they were taken by force. These so-called "orphaned" children were then used to feed irregular adoptions. Several UN organizations have estimated that more than 30,960[20] children were stolen in this way by Francoists.

Another harrowing image is that in Franco, the dictator's, mind, punishing children during the post-war period by sending them to prison, placing them in orphanages, making them disappear by erasing their identities, and otherwise harming them with ill-treatment and starvation, was undisputed.[21]

The issue of the "theft of children" was again addressed, more recently, by the *Fiscalia General* [Attorney General] in Circular 2/2012, by which this was proscribed as unlawful and as a major crime under the Law.

---

[20] The Francoist legal basis for the theft or disappearance of children, resides first in Order of 30 March 1940, that prohibited children older than 3 years to remain with their mothers in prison. Next, Law of 4 December 1941, allows the authorities to change the names of destitute children. Absolutely required reading is Ricard Vinyes, Montse Armengou and Ricard Belis' wo, *Los niños perdidos del franquismo,* Plaza y Janés, Barcelona, 2002, and the viewing of the documentary with the same name and date produced by Televisión de Cataluña. This data about the catastrophic persecution of children is recorded in UN document Report of the Working Group on *Enforced* or *Involuntary* Disappearances, Mission to Spain 23-30 Sept 2013, published by the UN Human Rights Council 2 July 2004.

Regarding disappeared adults, the said preliminary report estimates these as 114.226, an incomplete number that was recorded in the Minutes of ex-Judge Garzón's Court. To this number, we need to add those executed on the order of the courts-martial, although these are not considered as actually "disappeared".

[21] One of the children who suffered, Ernesto Caballero Castillo, published a book, Vivir *con memoria,* El Páramo, Cordoba, 2011, in which he describes many of the things that happened to him, in detail.

INTRODUCTION TO THE HISTORY OF THE CIVIL WAR.

An ongoing Pact of Silence.

The Spanish civil war ended almost 85 years ago, and bells celebrating Franco's victory pealed all over the land. This was no Surrender at Breda nor Convention of Vergara.[22] There was no thought of any form of reconciliation. Spain was swept by a whirlwind of repression and revenge, the precursor of a humanitarian catastrophe without bounds as both during and after the war, the perpetrators of the 1936 military coup massacred the best of Spain. Supporters of the insurgent Franco and his military only represented half of the Spanish population; the other half, excluded and repressed at the most productive stage of their lives, barely managed to survive at the feet of a totalitarian New State that swiftly moved to subjugate them as it sank its teeth into a vanquished people that was unable to flee its grasp.[23]

Today, our society knows little of these events as the impact of that victory has continued to this day to be filtered by an enduring right-wing policy of enforced silence and memory destruction so as to prevent History from attributing any blame on the aggressors of the past. The so-called Pact of Silence is a misnomer, as there was no such pact. What remains is the *desmemoria*, the planned eradication of every memory of that period, because memory itself is accusative. From the end of the civil war to the death of Franco in 1975, including the so-called Period of Transition towards democracy during the 1950s, in schools and universities there was and continues to be no detailed or even superficial teaching of the implications of the 1936 military coup and its consequences. A consequence of the persistence of so many iniquities from the past, nowhere in any field of knowledge except that of the Spanish civil war, do we find such a high degree of academic discord, such confusion in society, so many enduring

---

[22] Two notable historical military treaties in which the Spanish victors were notably magnanimous.

[23] Of those who managed to escape, hundreds of thousands of Spaniards were lost to their country when they went into exile. Liberty for those who fled to France and were interned there, would be short-lived when they were sent back to Spain.

myths and fallacies, so many ideological influences engulfing the historical reality.

The worst of all are the myths and the fallacies used to substantiate an uncompromising objectivity in the most ideological formats possible: an ideology of theoretical neutrality; an ideology that follows the theory of equidistance; an ideology that promotes the theory of equivalency, that 'everyone was the same'; an ideology that considers that there are first class victims (present-day victims of ETA and terrorists - 858 and 191) and second class victims (executed by Francoism, 140,000)... Of even greater concern today is the fanatic dogma, the intractable beliefs of those who refuse to listen to reason, of the crazy men who have never said they were sorry. That is how it all begins: the creation of the most absolute confusion.

As Rogelio López Cuenca said: "(...) The end result is an embellishment of History, a mythology of the past; an utterly artificial reinterpretation of past events, assembled and beautified as if this were a product that one wishes to sell on the market(...)".[24]

Ángel Viñas has also drawn attention to the same mindset, to the tendency of some sectors to sugar-coat the past and try to sell us a pig in a poke: "If in Hungary or Slovakia, also European Union member states, we note with some concern of instances of whitewashing the Fascist past (...) In Spain, we must not fail to react to the actions of some untrustworthy academics and half-baked, shameless members of the media, lest we suffer that which occurred in Chile where in all seriousness, the official position is to gloss over General Pinochet's dictatorship by describing it as just a military regime."[25]

In Spain, the fiends responsible for covering up the fascist-totalitarian-dictatorial past have been running loose for some time as they gloss over 'the darkest page of Franco's dictatorship that historiographers have attempted and are attempting to conceal.' Reig Tapia further stated (as did Judge Garzón) that July 18 1936 "marked the beginning of a crime against humanity, whereby Franco and the regime that he fathered were much more criminal than General Pinochet in Chile or Slobodan Milosevic in Serbia (...) How can one

---

[24] *Tesis: Carretera de Almería (1ª Parte)*, (Thesis: Road to Almería – Part I) Internet documentary (available on You Tube).

[25] Ángel Viñas (Ed.), *En el combate por la historia* (In the fight for History). Pasado & Presente, Barcelona, 2012, p. 24.

explain that Pinochet and Milosevic are considered guilty of genocide and that the whole world trembles at their crimes against humanity, whilst at the same time there are those who are tearing their hair out in irritation because Franco is being considered in like manner?"[26]

This and much more form the Iberian pyramid of socio-political contradictions and the mountain of vested interests regarding the Spanish civil war. Although many books and papers have been written on the Francoist repression and these have circulated amongst the minority and a certain elite, the 'new history', the history of the Democracy, has not reached the bulk of the Spanish population. As a result, the wide field of that which is called 'public opinion' continues to reflect a false understanding of the history of Francoism that has remained intact, well and truly wrapped up and unravelled to this day, despite any number of recent studies. Perhaps, if the television media were to have helped us after the death of Franco, at least by airing a modicum of important documentaries or interviews with individuals who had a lot to say, Francoism might have been unmasked as was the 'Jewish case'. In the matter of the 'Spanish case' it has not been so. Spanish mass media (especially television) tirelessly persist in presenting the public with a grand design that might feed their stomachs but not our intelligence. In the case of historic topics – 'our case' – it is not a matter of official censorship but of a self-censorship that has existed since the death of the Dictator. The reporter instinctively knows which topics well-received and which ones are not, so he is content to ignore 'that which must remain unspoken', without being ordered by anyone to do so. This being so, what has happened had to happen: the false Francoist history continues to be disseminated with impunity and nothing can be done about that. Worse still, the major mass media are in the hands of the many and varied right-wing groups, precisely those for whom Francoism is a sacred creed to be safeguarded at all costs.

Today, well into the 21st century, it has become increasingly difficult to present the Spanish people with a responsible study of Francoist crimes - genocide, crime against humanity, or war crimes - however you wish to call them. In forty-five years of democracy, the

---

[26] Alberto Reig, "La pervivencia de los mitos Francoistas" (The survival of the Francoist myths), p. 912.

history of the Second Republic and its destruction by the 1936 military coup has been written against the political tide. Even greater have been the difficulties in unravelling the history of the great Francoist repression during the war and in the post-war period.

More than half the Spanish population have been force-fed a simplistic conservatism and the evident socio-political so-called pact of silence that existed during the period of transition; they are not ready to sit back calmly and receive the historic truth. They show no interest in learning of the humanitarian catastrophe for which Francoism was responsible. No one knows, or wants to know, the full truth of what they were taught (or mis-taught) at school or the misinformation they get through the media.

> "I have not heard anyone say that we should forget the Holocaust, forget the 'train of death' that went to Auschwitz, forget Pinochet (...) But in Spain, we had to draw a thick veil, forget all our relatives, forget the sufferings and the anguish, and all the rest. Here, I know not why, we are supposed to forget everything, to erase it all and turn over a new page; we are not even supposed to seek those responsible, and they even are against our attempts to obtain closure [exhumation and identification of the dead]."

So spoke Clara González in 2003, whose four uncles lay in the Piedrafita de Babia (León) mass grave.[7] Clara's aunt, Isabel González, one of those supremely distinguished Spanish women and whose two brothers also lay in that grave, commented in words worthy of a philosopher:

> "What was the purpose of all of this? What good has come from killing these people? What has been the good of these deaths and the deaths of so many others?"[27]

---

[27] Montse Armengou and Ricard Belis, in the documentary entitled *Las fosas del silencio* (The mass graves of silence). Televisión de Catalunya, 2003, 30'. The subject is the disinternment of 7 bodies in the Piedrafita de Baia mass grave, among which, both González brothers, Clara's uncles.

So why the forced oblivion? Just compare the attention paid and the official concern with the recognition of victims of today – the 858 victims of ETA and 191 victims of the 11 March terrorist attack, which is both fair and necessary, with the total amnesia, the *damnatio memoriae*, of the psychological support to the relatives of yesterday's victims who still seek closure? Surviving families such as Clara González's who will never forget:

> "The Fal Rafael Sánchez Ferlosio, Interview in *EL PAÍS*, 22 May 2007.angistas celebrated their killing those who lie in the mass grave by forcing my mother [Isabel González, whose brothers had just been shot], who was known to be a good cook, to prepare them a meal, and my aunt Asunción and another of my mother's sisters-in-law to play the tambourine and entertain them whilst they feasted on some of the family's lambs they had also killed."[28]

The history of Spain is truly extremely complicated. Little tolerance is to be expected in the general socio-political climate through which this work hopes to open a way forward. Teaching the Spanish people their 20th century history is difficult as it is a two-fold problem: those who do not know on the one hand and those who do not want to know on the other. Although people may talk about the many cases of genocide throughout history, they never speak of the Francoist genocide, mention of which has been vetoed until today by its perpetrators. Despite this, as we historians try to reconstruct these events, we are striving to create a public record. The history is there, and the deeds are there.

The following work was governed by the historian's three fundamental principles: truth, accuracy and documentation. Since 1978, I have engaged in reconstructing the details of the great Francoist repression, first in the city of Córdoba, as it was applied throughout the province of Córdoba as a whole. On these pages, I have set down, in black and white, the results of all that I have researched, that I have

---

[28] Ibid. The author's interviews with Asunción Álvarez, Isabel González and her niece, Clara González.

obtained from written sources, that I have been told first-hand by victims, from witness accounts. You could say that this is my narration of everything that I have seen and heard in the voices, the files and the faces of the victims and their families.

First, however, I need to address the great labyrinth of present-day hostility towards the history of the Francoist repression: the fanaticism of those who do not wish to know (as well as those who throw stones at historians for several reasons). I totally agree with Sánchez Ferlosio who said that "You cannot convince anybody of anything." [29] Perhaps one might convince 22nd century readers, when through information and culture, the Spanish people have espoused the reality of the facts.

To begin with, Spanish historians of the civil war need to accept that they have a problem with the political right-wing's extremely conservative stance that has been handed down without interruption by the Francoists. The intractability of the Right against any research into the history of the civil war inspired an ad hoc publication led by Ángel Viñas: *En el combate por la historia* (2002).[30] The purpose of this closing of ranks is not difficult to detect: it is a question of preventing people from knowing exactly what occurred under Franco, a project to destroy all memory of that time, to throw more soil over the graves of the victims and to wipe out more than half a century of Spanish history. The Spanish right-wing (political Right – entrepreneurs, financial institutions; social, judiciary, military, ecclesiastic, mediatic and academic Right) are determined to erase the recent past: *Delenda est historia*. There are a great many right-wingers and moreover, they are at the heart of the present Government, which is why historians must investigate and write against a tide of disapproval. What is surprising is that the political Left is also failing to make the grade. Spanish social-democracy has been seriously negligent when it comes to assuming the historical truth.

Unlike some European right-wing groups, the Spanish Right lacks the minimum antifascist traditions of the French so-called civilized Right, inherited from Charles de Gaulle. The Spanish Right not only lacks an anti-fascist tradition as it also lacks a democratic tradition. Put to the test during half a century of thraldom and travesty under

---

[29] Rafael Sánchez Ferlosio, Interview in *EL PAÍS*, 22 May 2007.
[30] Ángel Viñas. *In the fight for History*. 2002

the self-styled parties of change, when a true democracy worthy of its name arrived in Spain for the first time in 1931, followers of the Spanish Right (especially in the military barracks, casinos and church vestries) dedicated themselves to boycotting the Second Republic until they were able to demolish it following the 1936 military coup. When that democracy was restored in 1977, the Right imposed conditions of impunity, self-amnesty and forgetfulness of the past, creating Francoist gallows under which a perplexed, weak and disoriented Left was forced to march. Today, the Spanish right-wing groups ignore the international organizations that are demanding compensation for a past defiled by 140,000 murdered or 'disappeared' individuals. They take no notice of the mechanisms of so-called 'transitional justice' or 'universal justice' under International Law, as endorsed by several United Nations bodies. During the past year (Fall-Winter, 2013-2014), no less than three UN bodies called Spain to task for neglecting the issue of the disappeared and for neglecting to pay due attention to the victims and/or create a Committee of Truth. On each occasion, the governing right-wing party has mocked these bodies, replying to their comments with a jingle praising the 'model' transition and the 'reconciliation' falsehoods, curtly shooing them away like so many pesky flies.

In addition to the Report from the UN Commission against Torture (November 2009), there are other reports that need to be mentioned. Following a week-long visit to Spain, the UN Working Group on Enforced or Involuntary Disappearances published a Preliminary Report September 30, 2013.[31] The Working Group reported that, in Spain, there had been grave and widespread violations of human rights during the civil war, citing provisional figures of 114,226 disappeared and 30,960 children stolen under Francoism ('systematic sequestration of children'). Commenting on the lack of links and communication between victims' groups and the state authorities and the 'lack of any national plan for the search for disappeared persons,' the Working Group declared that it was "a matter of urgency that the Government begin a search for the truth, and in particular, make of the establishment of the fate and whereabouts of

---

[31] UN Committee on Enforced Disappearances, Working Group report on Spain, A/HRC/27/49/Add.1.

the disappeared persons, an immediate priority." The Working Group also noted that mapping of the mass graves was not yet complete and that there remained "other important challenges," including "the lack of any law on access to information and the difficulty in accessing the archives", among others. The politicians of the ruling PP party coldly showed the UN commissioners the door. Is there anyone brave enough to put Rajoy to work searching for victims of Francoism in the fields, along the roads and in the ditches of Spain? On October 13, 2013, a mass beatification of 522 Francoist 'martyrs', was attended by Government ministers, yet today when one tries to obtain recognition for Republican 'saints', whenever the same civil authorities can throw stones and garbage at these, they do so.

On 3 February 2014, at the end of his official visit to Spain, Pablo de Greiff, UN Special Rapporteur on the promotion of truth, justice, reparation and guarantees of non-recurrence, spoke of his findings. He said that the victims and their associations with whom he had been in contact felt that they were insufficiently recognized and listened to. A primary target of his statement was the Amnesty Law of October 1977, that he said was a breach of international conventions to which Spain was a party, such as the International Civil and Political Rights Pact, whose Article 2.3 prohibits amnesty for serious violations of human rights, signed by Spain on 28 September 1976 and ratified 27 April 1977, before the enactment of the Amnesty Law that, in Spain, is taken as the law that puts an end to the matter[32]. (I would add, like similar laws in South America, almost all of which have been abolished, albeit not in Spain. In other words, the so-called 'model transition' in Spain was no model anywhere.) De Greiff clarified that although the Amnesty Law suspended penal responsibility, what it could not shelve was an investigation of the acts, at the very least. He showed his concern for the Government's failure to update the map of the mass graves of the disappeared in Spain (the figure still stands at 2,382 mass graves found, containing some 45,000 individual remains). As regards the pillar of truth, in reality, there has never been an official policy in this respect. The archives continue to hold dossiers classified as confidential on the grounds of the right to privacy, documents that

---

[32] *Lei de punto final.*

are inaccessible to the international entities that wish to consult them on the grounds of the right to the truth.

De Greiff continues, listing a whole series of incongruences in Spain: the Historic Memory Law did not rescind the sentences handed down by the Francoist courts, contrary to that which was done in Germany, for example, nor has there ever been any mention of restoring the private property that was seized by Francoists. Francoist signs and symbols continue to be displayed all over the country. He added that they had received ambiguous information about the way that the civil war and the dictatorship were taught in schools. Lastly, among other points, he refers to the undermining of the legislation that governs the Spanish courts' application of International Law. (Here, I should mention something that I have referred to elsewhere, and that is that the Spanish courts are schooled on the fringes of modern International Law, and that they function within a kind of judicial autocracy clearly inherited from Francoism, totally secluded within the country's borders from the outside world.)

In defence of those judges who have not lost a sense of what is right and proper, Joaquín Bosch, Judges for Democracy speaker, deserves a special mention for the article that he published on the Internet, entitled 'Ten things you should know about the crimes of the Franco regime.'[133] Despite this, one reads the most incredible claptrap about the victims of Francoism, such as the events in the Provincial Court of Córdoba, in the Dorado Luque Case, when the Court of Appeals rejected Appeal 355/2006 against the 2nd Court sentence (3.651/2006 of 11 August) disallowing any official criminal responsibility. The Court justified its ruling for the sake of reconciliation, citing an agreement of a Parliamentary Committee dated 20 November 2002, devoid of any force of law, according to which one should avoid any kind of initiative that might "reopen old wounds or stir civil confrontation". To make matters worse, also according to the Court, "Argentina was a military coup; in Spain, it was a war." The sentence further contained such an amount of legal nonsense, barroom chatter and topics and fallacies about the civil war, that it appears that no one was able to comprehend the gravity

---

[33] Joaquín Bosch, "Las diez cosas que deberíass aber sobre los crímenes del Francoismo." *Diario Público.es,* Madrid, 19-10-2013.

of exactly that which is going on in the Spanish legal system today, which astounds foreign observers.

It is no secret that the Spanish right-wing experienced democracy without even barely absorbing any of it. The right-wing has never condemned Francoism, two out of three right-wingers continue to defend the 18th July 1936 insurrection, they voted against the Law of Historic Memory and, when the right-wing was recently elected to the Government, they left it without a budget, they continue daily to tarnish the memory of the Second Republic, they have forever erased its name from the Transition and the Constitution, they persist in the slander with which the Francoists demonized the Second Republic... I have never forgotten a statement I heard at a meeting in Huesca: "The French Right, from de Gaulle onwards, has always maintained an antifascist tradition, contrary to the Spanish Right, which has never adopted such a position because it is a readaptation of Francoism."

Given such a background one cannot be surprised that the anti-memory movement has erupted so aggressively in Spain. (In the Fall of 2013, I was stunned to hear of a public jumble sale of Francoist memorabilia in Quijorna, Madrid, under the approving eyes of the local Lady Mayor.) In early February 2014, *Fachas*[34] destroyed several monuments that had been erected to fallen Republicans on the Ebro battlefields: the monument to the 43rd Division, the one dedicated to General Líster... In Cantabria, Fachas destroyed a monument to the guerrillas, the monument that Jesús de Cos cared so much for. In Fuente Palmera (Córdoba), they decapitated the monument to Captain Ximeno.

A Galician politician recently declared that "if Republicans were executed, it was because they must have done something". At about the same time, a high-ranking member of the Partido Popular (PP), Rafael Hernando, referred to the Second Republic as "the Regime that ended up with a million dead" and soon afterwards astounded everyone by declaring, on television, that "some people only think of disinterring their relatives when there are subsidies on offer". The PP speaks out daily in open contempt for the victims of Francoism. Another PP leader, Jaime Mayor Oreja, speaks of the "the

---

[34] Colloquial expression used to describe right-wingers of every size and shape.

extraordinary tranquillity of life under Francoism". Matters are no better nowadays. April 22nd there is a celebration in Burgos in honour of General Yagüe, the butcher of Badajoz, in outright contempt for the law, with the profanity of despots and the obstinacy of fanatics acting without impunity.

AD LIMINA

Combativeness, propaganda, political activism and ideology
"Language as a weapon for mass destruction"
by
Mirta Núñez Diaz-Balart
**Chair and Professor of 20th Century Historic Memory
Universidad Complutense de Madrid**

F rancisco 'Paco' Moreno Gómez has, with notable tenacity, investigated the Francoist repression in Córdoba, the Andalusian province that Franco used to evaluate the efficacy of his program to eradicate the Republican half of Spain who opposed his insurrection. The figures are clear, as Paco Moreno points out: four thousand Republicans slaughtered in the immediate post-war period in Cordoba city alone, as compared to a handful of victims amongst those who supported the military coup. An appalling proportion that did not temper the violence that the rebels had earlier demonstrated in the territory they occupied during the civil war. In Córdoba province, a strong class system that opposed the aristocracy and wealthy landowners and the expectations of the bulk of the working class helps explain why the vengeance of the Falangist upper class fell so heavily upon the Republicans who struggled for a more egalitarian social order.

Even before the fall of the Republican government, the rebels began taking over the entire administrative apparatus of the government in the territories they had conquered, first through extra-judicial executive orders, then by rule of law under edicts from Franco's self-proclaimed government in Burgos directed at creating a new social order for the country. Where there had been collective management systems, the new lords of the manor and the *Señoritos* bore down on these latter-day vassals with a re-creation of a medieval form of servitude. The new serfs were not only manual laborers and shepherds, but teachers, doctors, lawyers and numerous other individuals who believed in the democratic ideals of the Republic, even though they might not wear the espadrilles that traditionally shod the feet of the working and farming classes.

With his meticulous investigation, Moreno Gómez demonstrates how the blood bath was programmed from the onset by the military coup led by Generals Franco and Mola, as another way of dominating the people. Part of their plan was to undo everything that had been achieved during the Republic's brief tenure, beginning with the laws themselves. 'Combativeness, propaganda, political activism and ideology' were placed at the service of Spain's *Army of Salvation* and its partners. With this expression, the author draws attention to yet another example of all that he calls 'the use of language as a weapon for mass destruction', the forerunner of a manner of thought that still exists to this day.

Moreno Gomes, a professor for many years until his recent retirement, reminds us of the role of the Church in using virtual machetes to hack a path across a society that was beginning to become secular. A Church that returned to rites and formalities that harked back to the Inquisition as a means of purifying society. Nor did the Church spare the children as it used them as key players for establishing the new National Catholicism, a pillar of the new regime. The author reports eyewitness statements that are at the same time both sarcastic and tragic, such as when Ernesto Caballero recalls how whenever a childhood friend swore as they worked, he would look around fearfully, waiting to be struck by the bolt of lightning with which he had been threatened by the nuns who employed the boys. Few priests and nuns did not fit that mould. Paco Moreno recalls the words of one of the few Republican village priests, Marino Ayerra, who was able to escape into exile and who defined those times as 'a new Middle Ages that spreads its dark wings over the homes'. Even more striking is his reflection about the militarized and clerical life of those times: "Why did they want the Fascist Party when they already had the Church?"

Prior to examining the events of the immediate post-war, the 'Bloody Victory', he examines the quagmire of indecision by which the National Defence Council (CND), entrusted with negotiating a peaceful surrender, failed to save an immense territory containing half a million troops. In the South of the country, by the end of Franco's infamous 'Victory Walk', army units with thousands of combatants had become an equal number of prisoners.

One satisfaction that one gets from this book is how its author speaks loudly and clearly about the public historiographic debates of today, putting all who question the existence of the genocide or attempt to sugar-coat the repression, into their place.

Paco Moreno, whose books are examples of thorough research into the violent actions of the fascists who rebelled against the democratic order, does not shirk from using a magnifying glass to divulge, in some detail, the atrocities that were committed. Disgusted with the Church's ongoing syrupy disavowal of the events through its supposed compassion for the dramatic occurrences of the past, Moreno reports cases such as those of the children of the Mayor of Villanueva de Córdoba, executed by firing squad, who were forced to beg in order to survive whilst their mother went into hiding after she was stripped naked and forcibly subjected to the castor oil and shaven head treatment. The author does not mince his words as he states that this is an example of the insurgents' intent to shatter the moral fortitude of the defeated, or 'disaffected' as they were called. For most, survival was their sole objective. The author does not shrink from shedding light on the innumerable dead from starvation, within and without the prisons and so many other 'rod and thwack' complexes such as the many concentration camps.

His aside about economic matters is especially relevant. As he says, insufficient importance has been given to the Francoist authorities' abolition of the Republican currency, which forced widespread penury upon the defeated. Still, when all was said and done, this was just yet another feature of the programmed pillaging, first during the war through extra-judicial seizures and appropriations and, when it was almost over, the legalization of the spoils of war rule under the so-called Law of Political Responsibilities. This law was a weapon that would make ghostly appearances, years after the accused had been executed, adding to the misery of their families.

Franco and his fellow fascists were intent at creating a New State where society kneels before the sword and the cross. All means for attaining this goal were acceptable, including weapons of mass psychological destructions such as free rein given to snitches and informers, the use of offensive nicknames and insults in order to convince the persecuted of their own guilt, in addition to the press such as the Cordovan newspaper *Azul* which voluntary collaborated with this mission.

In conclusion, Paco Moreno asks whether the plethora of crosses erected at the Valley of the Fallen and similar shrines and that have populated cities and towns all over Spain from the moment they were occupied by the rebels, are not yet another feature of the anger that the government encouraged against the repressed. These and other public memorials, raised where they are continuously visible to all, remain as symbols dedicated to ensuring the permanency of a memory – the victors' memory. Thanks to Paco Moreno's scientifically led research, we now have a first-class weapon with which to work against the victors' enduring desire to obliterate all recollection of those who perished and to dilute the pain and injustice suffered in silence by the defeated, despite multiple latter-day Francoist attempts to erase their memories and re-write History.

I

FORTY YEARS OF OPPRESSION
AND RETALIATION
LIFE AND PROPERTY. SPOILS OF WAR

Franco's 'victory walk'. Half of spain in chains. Early features of the fascist victory. Return of the defeated. Humiliation and confiscation.

> *"... There is nothing worse than the pairing of the barracks mentality with that of the sacristy, since traditional uncouth Spanish Catholicism can hardly be considered Christian..."*
> **Unamuno**
> Last public speech
> Salamanca, 12 October 1936

Franco's Victory Walk

Franco's *Paseo de la Victoria* heralded the end of the civil war March 26, 1939, a so-called walk as there was no fighting, no significant offensive action of any kind by the rebel forces. Throughout history, all wars have ended with a decisive battle, except for the Spanish civil war. Historically, this may never have been the case, but in Spain it now happened. This was not the fault of the fighting population but of the professional military leaders and their fellow Coryphaei whose defeatism at the end sold out the Republic as did the rebel Nationalists who rose against it in 1936. At the end of the conflict, the eternally clear-thinking Juan Negrín, last Loyalist Premier of Spain who presided over the defeat of the Republican forces, bade farewell to Juan Simeón Vidarte, a fellow politician, with the following words: *"Go without fear, because there is no instance in history when an army of more than half a million men has surrendered without a* fight."[1]

Franco set March 25 as the deadline for the *Consejo Nacional de Defesa* – CND[1] to surrender the Republican Air Force. However,

---

[1] National Defence Council, a military junta under Juan Negrín, formed in order to negotiate a peace deal.

as some of the planes had flown out of the country and several pilots delayed meeting these demands, not a single plane landed in Burgos on that date. Major Segismundo Casado Lopez who had meanwhile assumed command of the Republican Army in an attempt to negotiate a ceasefire, called Franco frantically, imploring, whining, and promising that the Air Force would be handed over the next day. Franco, ignoring these histrionics, declared that as certain, actually nonexistent, negotiations had failed, nothing but an unconditional surrender was acceptable, and he was ordering his Army to begin the Paseo de la Victoria. Casado was told that every town and village in the country should fly a white flag if they wished to avoid the consequences of continued action by Franco's air force and artillery.

As promised, Franco's Victory Walk began at dawn March 26. The Republic was no more; it had been stabbed to death by treason and Franco, with his self-appointed Army, proceeded to take possession of a no longer existent entity. Beginning in the South, the Francoist Nationalist Army crossed the unmanned lines in the region of Peñarroya. The Moroccan Army, under General Yagüe (responsible for the slaughter at Badajoz), set off in the west near Cordoba and the Andalusian Army, under General Muñoz Castellanos in the region of Espiel and the port of Calatraveño. These troops were joined by the Cordoba Army Corps under General Borbón and the Army of Extremadura led by General Soláns. The army marched in the region of Hinojosa and El Viso towards Santa Eufemia and Almadén, and on the right, towards Pozoblanco and Villanueva de Córdoba. The frontlines had totally disappeared. Companies, brigades, and entire divisions abandoned their weapons and wandered all over the place as the Republican troops attempted to make their way back to their hometowns. This was an extraordinary phenomenon of the en masse disbanding of an entire army. White flags flew from every town and village tower as far as the eye could see. Resignation and anxiety reigned in constricted chests. When Franco heard of the 'sightseeing' nature of the walk on that day, he promptly ordered another wiping-up operation throughout the province of Toledo, towards Mora.

The same day, the Moroccan Army marched through the towns of Hinojosa del Duque, Belalcázar, Fuente La Lancha, Villanueva del Duque, Villaralto and El Viso. The 24th Division, under the command of Colonel Rodríguez de la Herranz and Lieutenant-Colonel Manuel

Vázquez Sastre entered Belalcázar. Although the majority of Republican troops in this sector had fallen back towards Santa Eufemia, at the end of the day a light column of Francoist troops reached this village and accepted the surrender of 5,000 soldiers in the region. It is said that some troops shot their rifles upwards into the air as a symbolic act of defiance, before throwing down their weapons.

A long walk home

The Andalusian Army continued its march from the small port of Calatraveño during the morning, entering the towns of Alcaracejos, Añora, Dos Torres and ... Pozoblanco! The Nationalists were finally able to enter Pozoblanco, but this time without a fight.[2] Here, they captured all the equipment, supplies and files belonging to the Republican 8th Army Corps and took 3,100 prisoners. The *Los Jubiles*, civilian freedom fighters who formed an illustrious guerrilla corps that fought alongside the 88th Republican Army Corps, retreated into the Cordoba hills, taking with them many weapons, including mortars, the army gave them, declaring "We shall not surrender, nor shall we leave Spain."

All the regiment's units were ordered to retreat to Puertollano and all the senior officers, to Torrecampo. Ildefonso Castro who had been appointed commander of the 8th Army Corps during the last days of the war, left with Major Emiliano Mascaraque Castillo[ii], another 8th Army Corps senior officer, intending to go to Ciudad Real in hopes of getting away during the first few days. They spent the night under a heavy snowstorm and arrived in Puertollano by car on the 27th, where they fell into the hands of the Falangistas and were imprisoned. A few days later they were transferred to Ocaña penitentiary.

The National Defense Council that appeared to have been created with the sole purpose of saving half of the defeated Republic from being hunted down, both civilian and military, proved to have been a colossal failure. The Nationalist Air Force was making its last bombing attack on Pozoblanco, whilst in Villanueva de Córdoba, it was a case of every man for himself. Several leftists had been arrested by Falangistas near the San Rafael flour factory where they were taken.

---

[2]   The medieval city of Pozoblanco remained fiercely loyal to the Republic throughout the civil war, defeating several attacks by General Queipo de Llano's Nationalist troops in March 1937.

Rafael Rodriguez, nicknamed *Tres Cuartos* and who had had a bit too much wine to drink, boasted, audaciously, that he was the Mayor of the town. Claiming that he had to go to the toilet, he escaped through the back door and joined his family in Villanueva de Córdoba.

On March 23, already, Lieutenant Lorenzo Cepas Rico[3] a communist from Villanueva de Córdoba, desperately tried to save himself from the approaching enemy and to obtain help to rescue Villanueva's Communist Municipal Authorities[4] who had meanwhile been jailed by the Nationalists. After obtaining a safe-conduct pass to travel from Puertollano to Villanueva, he immediately went to Pozoblanco to see the 8th Army Corps' quartermaster, but the officer in charge of transportation refused to give him some trucks. He returned to Villanueva and obtained help from the 114th Brigade in the form of 35 soldiers and their lieutenant and two trucks. The evening of March 26, they immediately left the Las Navas farmhouse where they were staying, to attack the Villanueva jail where his comrades were being held. They entered the town at dawn but before they arrived at the jailhouse, they ran into Vilches' wife who told them that the 8th Army Corps'quartermaster had already been there and had released the men who then left on the road to Conquista.

They drove to Conquista where they found a train packed with people that was going nowhere. They drove around town all day without finding José Caballero and the others. Hearing that the Nationalists had entered Villanueva and were advancing towards Cardeña, and afraid that their retreat might be cut off, they drove out of town down the airport runway as far the railway station and from there, to Puertollano. In the town square, they ran into two fellow communists from Villanueva who refused to join them, claiming that a new front was about to be created in the Sierra Morena. They continued on the road to Cartagena.

---

[3] Lieutenant Rico was interned in the North African concentration camps and only returned to Villanueva with the Democracy where he slowly recovered his health and enjoyed a well-deserved respite from so much misfortune. Had Moreno Gómez just decided to speculate on what Rico might have related when he returned home on a stretcher and analyzed his memories from a distance as others do, they would have been lost to History as later, it would have been too late. It is remarkable how in these days, so many other historians and universities yawn with boredom at any mention of this kind of research.

[4] Gabino Cabrera, Madero, Bartolomé Caballero, José 'Carnes', Vilches and Francisco Sánchez Muñoz, among others.

In Damiel, the Officer in Charge, a doctor who was preparing to surrender his command to the Nationalists, refused to grant them a safe-conduct pass and some gasoline. Another communist from Villanueva, Francisco Copado, tried to intercede for us but was turned down and urged them to get away from there as quickly as they could. Lieutenant Rico decided to act and, drawing his revolver, obtained the safe-conduct pass and the gasoline at gunpoint. In Manzanares, a company of Nationalist *Guardias de Assalto*[5] barred the way, again forcing them to resort to gunpoint – revolvers and machine guns. In Albacete, both trucks broke down but fortunately they were able to join a small convoy led by an Infantry Lieutenant. In Murcia, another armed control point, this time accompanied by some civilian Falangistas. They slowed down, pretended to stop, and stepped on the gas at the last moment and drove off at full speed.

Finally, they arrived at the port of Cartagena, on the ssoutheastern coast, where they hoped to board the *Campido*, an oil tanker that was waiting to take as many as it could to safety. After some difficulty getting safe-conduct passes for their group of some 50, they managed to get on board just as the ship was raising its anchor. Lieutenant Rico was the last one to board and not too soon. As they sailed off, they could see Falangista flags flying all over Cartagena.

Everyone set off in an attempt to go home, some went north, some went south, east, west shedding bitter tears as they bade farewell to good companions through hardships and suffering. Everyone from the same town or region got together to form groups for the walk home. Lieutenant Carbonero, three officers from his company and some countrymen began walking and did not stop until they reached Salvañete, where they received a hunk of bread each, only stopping in Cañete to rest at 4 p.m. and eat some of the bread and a little salt cod they had brought with them. White flags flew in every village they passed and everywhere, dirty, ragged, limping, exhausted and starving men with backpacks were going from door to door begging for something to eat.

Nightfall and they continued walking, forming a line with the strongest in front and the weakest at the end, as they fell behind, until they reached Pajaroncillo. Just before they entered the town,

---

[5]   An elite armed urban police force, similar to Pretorian Guards.

they bought a flock of sheep from some shepherds so that when they reached the town, the people could cook them some food. They expected to run into Nationalist troops at any moment and had no idea how they would be received. They found a haystack and slept there until 5 a.m. March 30, when they continued to Carboneras, hoping to find a train for Cuenca.

When they arrived at the station, soaking wet and very tired, they found it full of soldiers, so they had settled in a goods wagon. People were already saying that the Nationalists had arrived in the town, but none were visible. The next day, Lieutenant Carbonero and his companions decided to continue to walk as far as Cuenca. Five or six kilometers before Fuentes, it began to pour and the snow heavily. Finally, they arrived in Fuentes where they were warmly received at the first house they knocked.

Dawn April 1 was splendid, and they continued to walk to Cuenca, when they met the first soldiers sporting red and yellow Nationalist armbands and Nationalist *Guardia Civil*[6] at the station closest to Cuenca. Several trucks laden with soldiers overtook them and, as they passed, greeted them with the Falangist salute.[7] When they finally arrived at a control post, a civil guard and several soldiers stopped them and asked for their documents. As they had none, they were told to stand at the side of the road where others were already waiting. At 2 p.m., a guard arrived, ordered everyone to fall in and then marched them across all of Cuenca, to the Seminary. Feeling more cheerful, they had no idea what awaited them.

After waiting more than three hours at the entrance to the Seminary, they fell into groups of 100 and went in. They were hungry but found out that they would not be fed. April 2 came and went and they had not a bite to eat all day. That night, a few groups of 20 got some food but it quickly ran out. Again, they went to bed without eating, suffering a gnawing hunger. Finally, April 4, at noon, they were fed some lentils and a piece of a bread roll that had been cut in four.

And so, the days passed until the officers in charge began taking statements, after which some were given leave to go home. Those who

---

[6]   Civil Guard. Spanish paramilitary police.
[7]   The Francoist salute, identical to the Nazi salute, where the right arm is raised outstretched and the had is open flat, palm facing downwards, often referred to by Spaniards as a "salute in the Roman manner," i.e., like one of Caesar's centurions.

had been officers were taken elsewhere, some say to Zaragoza. The rest left in small groups until about 900 of the approximately 2,000 they had numbered at the beginning, were left.

April 24 they were told that those who remained were to be taken to a concentration camp in Corunna, Galicia. When the train on which they travelled arrived ear Ponferrada, a small mining town, the workers and the women looked at them with pity and gave them some bread. They got off the train at Santiago de Compostela and started walking. Outside the station, the bulk of the local population were waiting for them. Some women cried and others gave them a bit to eat when the officer was not looking. They marched on until they reached a village call Labacolla, the site of the concentration camp.[iii]

It is through accounts such as these that Moreno Gómez presents us with a vivid understanding of what the 'Black Week' of victory meant for defeated Spain. Casimiro Jabonero's sufferings were those of all who returned home after the defeat. When they arrived at their destination, almost none had time to embrace their families as they were immediately interned. For many, their future was a case of *Vae victis!*

Meanwhile, March 26 the Nationalists pressed forth with their military success as Falangistas, militiamen, legionnaires and Moroccan troops trekked across the north of Cordoba province countryside, hunting their quarry like so many birds of prey. The Francoist military issued a Press Release on that day, stating that entire *Rojo*[8] battalions with their Commanders had raised the white flag of surrender to Franco's forces; 10,000 being the number of prisoners and individuals who turned themselves in.

From Pozoblanco, troops of the 60th Francoist Division, under Lieutenant-Colonel Aguilera, arrived in Añora. Others from the 115gh Division entered Dos Torres, where Aguilera was confirmed as military commander, a position he held until some days later he was replaced by Juan Benítez Tatay, the militia commander who ordered the first blood bath.

Azul, the Francoist newspaper of Cordoba, published numerous illustrations of the Fascist parade through the semi-destroyed streets

---

[8] *Rojo*, or red, meaning all Republicans or members of any leftist party who supported the legal Democratic regime and therefore, were not loyal to Franco's *Causa General* manifesto or to his regime.

of Pozoblanco lined with locals saluting in the Roman manner, and multiple photographs of an open-air mass celebrated to purify the town from the so-called *Red filth*.[9] Priests as distinguished participants in the front rows, mobilizing the masses in aid of National Catholicism.[9]

Francoist Air Force Operational Reports[iv] also say something about the events of March 26:

at 4 p.m.:
- [Republican] cars and trucks at km. 36 on the road from Alcaracejos to Pozoblanco. – Seven Rojo trucks from this village to Villanueva de Córdoba and an ambulance going the other way. Our troops at 1 km. to the east of Villaralto. No traffic of any kind on the road from El Viso to Pozoblanco; likewise, regarding the villages of Añora, Dos Torres and Pedroche.

and again at 5 p.m.:
- El Viso, 2 kms to the north, has been occupied and overtaken. Our troops are 1.5 kms south of Pozoblanco and advancing. Our troops are visible in Hinojosa del Duque. 30 trucks, apparently belonging to the Rojos, are travelling in this direction on the road from Santa Eufemia to Almadén.

The Victory Walk continued on March 27. At the same time that the Nationalist troops reached the village of Almadén, a detachment from the Andalusia Army Corps (the 40th Division, under Colonel Badía) set off from Pozoblanco. At noon, these troops entered the regional capital, Villanueva de Córdoba.

That morning, the last Republican leaders had rushed to apply urgent safety or salvation measures to protect the mostly communist population from the cruel persecution that threatened them.

Laura Contreras[v], a teacher at the Villaviciosa School, and Maria Josefa López ran to get the keys to the Communist Party Headquarters where they burned the lists of members in the files. They then went to the Antifascist Women's Centre where other comrades had done the same. The Villanueva communist leaders that had been freed from jail were waiting for them outside Conquista. Bartolomé Nieto and

---

[9] *Azul*, Cordoba, April 1 & 2 1939.

Laura were told to go to Puertollano and then to Ciudad Real. They set off on foot but when they passed a river near Puertollano, Laura was so exhausted that she felt like killing herself.

In Puertollano, they got a train to Valencia that was so filthy they became infested with lice. During the entire trip, Laura saw many people commit suicide by jumping onto the tracks. At Valencia railway station, they met a communist leader from Villanueva de Córdoba, Gabino Cabrera, who was thinking of going undercover in Valencia. They decided to go back to their hometowns and hide nearby in the Fuencaliente Sierra mountains. In the end, they did not dare to do so, returning that night to Villanueva and taking refuge in Gabino's house. Later, Laura walked along across the fields and handed herself in to the authorities in Villaviciosa.

These were not isolated experiences. The Republican half of Spain was overwhelmed with similar fears and exploits. Everywhere, there were signs of what was to come.

At 2 p.m. on March 27, after an airplane inspection and confirmation that a white flag had been hoisted in the town tower, Nationalist troops entered Villanueva de Córdoba on the road from Pozoblanco and on the road from Obejo. There were several columns from the 60th Division under Colonel Baturone, a Falangista deputation from López Tienda and half a brigade of regular troops, among others. They had left Pozoblanco at 10 A.M. and halted in Los Barreros, to confirm that there was no opposition to their advance. The Moroccans marched on one side of the road and Baturone's Infantry, on the other side. As soon as Baturone arrived, he set up his headquarters in Emilio Reina's house on Calle Herradores. Soon afterwards, General Queipo de Llano himself arrived, accompanied by Commander Ampliato, head of the Secret Service; they left a few hours later as they had another appointment in Almadén. It was a cold Spring day and had snowed. Captain Ignacio Pizarro, of the Guardia Civil, was appointed military commander of the town. He set up his headquarters and the Military Court at the renamed *Plaza del Generalissimo*, number 9.[10] That day, 750 Republican military prisoners and their commanding officer, Lieutenant Domingo Muñoz Sánchez, surrendered in Villanueva, as did a communications company, a squadron of foot soldiers and a Workers Brigade.

---

[10] This square, renamed in honour of Franco, lives on in the nightmares of the families of victims of the reprisals who were taken to those headquarters where they lost their skins, quite literally.

Francoist Air Force Operational Reports for March 27: beginning at 10:30 a.m., reads:

- Our troops can be seen at km. 87 on the road from Pozoblanco to Villanueva de Córdoba. Nothing is moving in this village. In Pedroche, 200 metres from the town, on the Pozoblanco road, there are two large caliber guns to which a white flag is attached. There is no change to the lines in Santa Eufemia; a column is marching towards there on the Las Pilillas road. There is no movement at all for 4 kms from km. 101 onwards. Before Almadén, there are groups of militias waving white flags making their way there and to Almandines. There are 15 carriages at the railway station that crosses the road to Almadén. There are 90 carriages and two engines at the Almadénejos railway station.

at 11 a.m.:

- The village of Pedroche is occupied as are 2 kms beyond it; troops continue to march towards Torrecampo; troops are also advancing from Dos Torres towards El Guijo. Troops SW from Santa Eufemia are advancing to the NE. A column of tanks and trucks are advancing on the Pozoblanco road towards Villanueva de Córdoba; they are some 8 kms from that village and we note many people appearing peaceful. All the farms are flying white flags and so is the village of Alamillo.

at 12 noon:

- Cover given to the 4th and 5th Squadrons that are bombing the Almadénejos railway station.[11] The bombing over, we reconnoitred the front, noting a column of trucks advancing on the road to Almadén, in the Santa Eufemia sector. To the east of that road our troops are advancing, in such a way that we presume they are not finding any resistance. Another column, further east, is advancing towards the river Guadalmez. Our troops have occupied the villages of Villanueva de Córdoba and Pedroche. The population in the village of Torrecampo are waving white flags and the people in the neighbourhood are waiting for the arrival of our troops.

---

[11] Author. Presumably attacking the 90 aforementioned railway carriages.

at 6:30 p.m.., the last report of the day:

- We recognize Adamuz, still not occupied by our troops from the Villafranca sector, who remain in their positions. Near the port, our troops are marching on the road from Villanueva de Córdoba to Adamuz, towards the latter. A column is leaving Villanueva on the road to Venta de Cardeña; the vanguard is about 1 km from there. Conquista appears deserted. Our troops occupy both Torrecampo and San Benito. Our troops from Santa Eufemia have occupied Alamillo and another column, coming down from the mountains, has overrun it to the east.

Even as the Francoist *Regulares*[12] entered Villanueva de Córdoba on the last day of the war, there was no stopping the looting and raping. In other words, the violent and repressive practices of the Nationalist troops as a whole, most especially the Moroccan troops, occurred everywhere during the occupation of every town and village without exception, at the beginning, during and at the end of the war. A tragic example of this occurred in Villanueva, March 27 1939, when the Moroccan troops, marching down the road to Obejo, passed the La Atalayuela farm, stopped and raped a woman named Catalina Maestre whilst they held her husband at gunpoint.[vi] Near the village, they killed a man who had no idea of what was going on, Pedro Capitán Moreno, and they threw his body into a well where he was only discovered some three weeks later. The Arabs were apparently trying to steal his livestock. For some other reason Moreno Gómez was unable to ascertain, two Moroccans were shot and buried next to the washing tanks, near the beginning of the road to Cardeña.

In another incident, a family from Almodóvar del Rio – men, women, and children – had taken refuge on a farm near La Charquita, a couple of kilometres from that village. Two Moroccan soldiers went there with obvious evil intent. One of the men saw them coming and hid; the other man, Miguel Claus Salado, did not, either because he could not hide in time or because he stayed to protect his four children. The Arabs arrived and held him at gunpoint whilst they raped his aunt and another woman. As the Arabs left, they shot Miguel Claus and he died in his sister-in-law's arms.[vii] Other eyewitnesses attested that the Nationalist troops spent those

---

[12] Professional or career soldiers. Not conscripts.

11

days looting, not just the towns and villages, but all the surrounding farms as well. Others arrived at the La Alcarria farm, which belonged to Luna Gómez Rodríguez, and took everything they could lay their hands on, including her sewing scissors.[viii]

The victorious troops continued their advance during the afternoon of March 27. At 5:30 p.m., troops of the 102nd and 112 Divisions, under Colonel Castejón (the butcher of Puente Genil), entered Cardeña. Elsewhere, the 105th Cadiz Battalion under Commander Luis Gómez entered Torrecampo, and we know from another Air Force report that they were greeted b an excited crowd saluting with raised right arms and shouting *Viva Franco!*

March 28, Nationalist troops marched from the area of the Guadalquivir River and Villafranca towards Adamuz, 'cleaning up' the hills from the Guadalmellato River in Adamuz. (A photo in the Cordoba *Azul* newspaper shows the prisoners being marched down one of the streets.) Many of the regular army soldiers whose looting and misdeeds mentioned earlier, converged that day in Villanueva de Córdoba.

Conquista was the last village in the province of Cordoba to be visited by such 'illustrious' conquerors, was occupied March 28 by an Infantry unit under the command of Second Lieutenant Valderrramas. These troops also marched through Azuel, whilst other units in the neighbouring province of Jaén, 450 Nationalist prisoners captured by the Republican army during the last battle in January, the battle of Cordoba-Extramadura, were freed in San Benito. [ix]

General Yagüe arrived in Almadén, which had been occupied March 27 and his troops entered Puertollano March 29. Republican Artillery Major Blanco Pedraza, previously an enthusiastic *Casadista* [13] from Villanueva de Córdoba, was entrusted with negotiating the surrender of the town. At the last minute, his assistant José Arévalo Toril, stripped him piece by piece of his insignia as the Nationalist troops arrived at the railway station in carriages Blanco himself had ordered. Even though Blanco, when negotiating the surrender, had begged that only Spanish troops should occupy Puertollano, Yagüe

---

[13] *Casadistas*. Supporters of Sigesmundo Casado, a socialist Army officer, who as a member of the National Defense Committee (CND), advocated negotiating a peaceful surrender with Franco in an attempt to avoid reprisals. Francom, who was only interested in an unconditional surrender, made false promises, ignoring Casado and the CND.

sent the train back full of Moroccan troops. So much for Franco's promises to the Casadistas.

The Republican commander of the Extremadura Army Corps, General Escobar (later executed) and a senior officer, Enrique Ruiz-Fernells, were captured in Almadén. By then, there were more than 60,000 military prisoners from the region of Cordoba, Almadén and Puertollano. Added to these, were hundreds of civilians considered hostile to Franco, the so-called 'disaffected' or *Rojos*, being crammed into every kind of building, warehouses, yards, convents and especially the Puertollano bullring.

The truth is that the war was not over for anybody. For some, the enormous task of victory and vengeance, persecution and vigilance, coercion, anger, had just begun. For others, hunger, overcrowding, torture, death, the destruction of their homes, exile, illness and despair, awaited them. The long, dark night of Francoism was just beginning, impelled by something much greater than vengeance alone: the planned extermination of an entire people. There can be no sweetening of the pill or denying of the facts. The fog of times past cannot erase the crimes against humanity. The reality was what it was, as Herbert R. Southworth wrote: "You were right, gentlemen, it was a crusade, but the cross was a swastika."[x]

The documents and photographs of these tragic days cannot be ignored nor must be forgotten. One well-known, albeit unsourced, photograph shows a line of haggard, defeated soldiers marching under white flags of surrender. On April 1, *ABC* of Seville printed photographs of Queipo de Llano, the Andalusia and Extremadura war criminal, proudly walking down the streets of Almadén. The same day, the paper published an image of thousands of prisoners being marched to a concentration camp, along the road towards Peñarroya-Pueblonuevo. According to Carlos Menéndez,[xi] one of the prisoners who walked from Alcaracejos to Pueblonuevo, people everywhere lined the streets shouting *Viva Franco!* as they raised their right arms in the Roman salute. How many Falangistas came out during those and the coming days?

Several months later, October 3, the Bishop of Cordoba, Adolfo Pérez Muñoz, appears in photos saluting with his right arm raised. To this day, the Spanish Catholic Church still gives little or no indication that it may be critical of, nor does it express its regret at what happened then. The Church remained unwilling to as for forgiveness when faced with its scandalous identification with fascism. In September 1971, a synod of bishops and priests proposed that the Church publicly ask for forgiveness on the grounds that in those days they did not know how to be true 'ministers of reconciliation'. The motion was rejected because it was not approved by the statutory two-thirds majority of those present.

Republican prisoners in the concentration camp at Villalba, Tarragona, August 23, 1938. Biblioteca Digital Hispánica. BNE, CC BY-NC-SA.

Half of Spain in chains

Prisoners, thousands of prisoner s everywhere, half Spain was under arrest. That was Franco's Spain: boundless humiliation and wire fencing. It never occurred to the victorious *Nuevo Estado*[14] that there could have been any form of reconciliation – a concept that has been discussed to exhaustion until it has become meaningless. In fact, there was no reconciliation, nor could it ever have been achieved from the onset: the plan was to *exterminate* a people, the reason why the military coup was conceived and launched. Without an extermination, the military coup lost its purpose.

---

[14] Franco's New State.

The three parties to the coup (found in military barracks, the casinos[15] and church sacristies) demanded a plan that would 'exterminate, cleanse, disinfect and massacre' all the 'Godless' everyone who was 'anti-Spain', every Freemason, Marxist, Anarchist and all the promoters of laicism, modernity and 'dissolutionary' ideas. The Church considered that all non-Catholic ideas, ostensibly those of the Republican democracy, were dissolutionary. On the other hand, the unholy tripartite did not consider the Nazi, Falangista or Francoist ideas, nor the Roman salute that the hierarchy of the times exhibited so vigorously, could be considered to be the slightest dissolutionary.

The number of prisoners in Francoist hands was beyond belief and unique in the History of Spain. Added to the aforementioned 60,000 prisoners, were the troops captured at the Caceres-Badajoz-Toledo front.

Units of the regular Army, such as the one in which Corporal José Pérez Navarrete, a conscript from Pozoblanco served, were left without direction when the war ended.[xii] His unit was sent to the village of Alcaudete de la Jara, where they remained until March 28. Confused, the soldiers stood down and everyone was sent on his own way. Corporal Navarette and several comrades reported the next day to the so-called Nationalist Army, in the village of Espinoso del Rey, where they went from the frying pan into the fire.

About 4 p.m. on the last day of the month, they were informed that they belonged to the Rojo Army and that there were some 7,000 of them. Without their having had anything to eat, they were herded onto trucks, driven across hills and mountains, places totally unknown to them, until at 2 a.m. the next day, they arrived at Casa de la Jaeña. More dead than alive, they were led at dawn to a nearby farm that was surrounded by wire. They were stacked there like espadrilles on shelves and there they stayed four days without food or water.

Looking like skeletons, they were then taken to a nearby farm and told that as 'Marxist prisoners' they would be placed in a concentration camp. They were fed very little food every 24 hours and they lost all hope of their ever returning to their homes. April 21, they were issued travel passes and released to make their way to the Calera railway station and home by train.

---

[15] Not to be confused with gambling casinos. Generally, gentlemen's clubs frequented by the wealthy and upper classes of society, often described as *señoritos*.

This description of the La Jaeña concentration camp at Aldeanueva de Barbarroya, Toledo, has not previously been mentioned in studies of this period. Located in the middle of the fields, Republican prisoners were kept totally isolated from the neighbouring villages. All the officers in the camp were quickly separated from the rank and file, then taken to Talavera, where they were summarily executed. The railway station mentioned in the above account is the one at Calera y Chozas, Toledo.

All over Spain, never-ending lines of exhausted prisoners, wearied by despair and hunger, were marched along country roads and down village streets; more than 40,000 in the Centre of the country and a similar number on the Levante[16] front. There is no denying the numbers. Nevertheless, when Nagrín told Juan Simeón Vidarre that never before in History had an army of half a million men surrendered without a fight, he did not say whether this number included those who had already fled to France or not.

At least two to three hundred thousand Republican combatant fell into Francoist hands. When you add the astronomical number of civilians who were wandering all over Spain and were being arrested right, left and centre during those days, as well as the great many regular soldiers and other combatants who managed to make their way home only to be immediately imprisoned, the total number of prisoners goes through the roof. Thus, the expression 'half Spain in chains' or better still, the title of a recent television documentary: *Spain, an immense prison.*[xiii]

Never in the History of Spain has there ever been such a humanitarian catastrophe. Under Franco, all historical statistics were outstripped, everywhere and in every possible scenario. That is why when Moreno Gómez speaks of the concept of crimes against humanity, genocide or extermination, he is not exaggerating the reality, contrary to the efforts by today's heirs of the victors who persist in belittling, or declaring off-limits, the mention, let alone the study of the repression that under Franco can be compared to that under Hitler and Stalin.

---

[16] *Levante* is the name by which the eastern region of Spain, along the Mediterranean coast, is known.

The astonishing number of prisoners in Cordoba province alone, were sent mainly to three partially destroyed villages enclosed by ditches and wire fencing. Valsequillo, L Granjuela and Los Blásquez. There were others in the region of Peñarroya-Pueblonuevo. After the fall of the La Serena pocket (Summer 1938), many more were captured in Fuenteobejuna, in Cordoba capital and elsewhere in the province. According to Lopez Rodriguez, the first lot of prisoners taken at the La Serena pocket totalling exactly 6,280, we not sent to the Castuera concentration camp – it was not yet completed – but to Campillo and Guareña and then, to camps in Mérida (3,605), Fuenteobejuna (481), Cordoba capital (2,194) and Almendralejo.[xiv]

The camps in Valsequillo, La Granjela and Los Blásquez began to fill up after the defeat following the Republic's last battle at the beginning of February 1939, the battle of Cordoba-Estremadura, that resulted in almost 6,500 prisoners. Another 700 were interned in Almendralejo and Fuenteobejuna as well as La Isla in Huelva. The Nationalist army expected to capture such an enormous umber of prisoners that in March 1939, the military headquarters began rushing to complete the Castuera concentration camp in Badajoz, which was not complete until the day on which war ended, when there was a new flood of prisoners. López Rodriguez further reports that in the immediate post-war period, an additional 7,500 were imprisoned in Valsequillo, 8,153 in La Granuela and 1,342 in Los Blázquez, although these numbers rose and fell from one day to another.

The last days of the war, many combatants from the area north of Cordoba fled towards Puertollano and Ciudad Real, as well as to other locations in La Mancha, such as Almadén, where they were eventually captured. A small concentration campo in Chillón held 750 prisoners and another one in Almadenejos, over 1,000.

Extra-judicial violence was the norm during the Victory Walk. The unmitigated account of a war crime, by a soldier of the 41st Republican Division in Extremadura, Albino Garrido, has recently come to light, been recorded and can be listened to on the Internet.[xv]

"On March 28, 1939, once we had laid down our weapons, the officers of the 41st Division decided that a committee should be sent to contact the Nationalist troops that were entering Extremadura from the

17

north. The men entrusted with negotiating this delicate 'surrender' were a Captain (a physician, head of the 66th Brigade Hospital), a Militia Lieutenant and several other soldiers, of which I was one. They presented themselves to Lieutenant Colonel Francisco Adame Triana, commander of the 2nd Regiment of the 19th Nationalist Division, who was arriving from north of Badajoz.

Taking the Captain and the Lieutenant aside, Colonel Adame asked the Captain why he had not gone over to the Nationalists. He replied that it was difficult, that he was engaged in medical treatments and that he was born in Murcia. Next, Adame asked the Militia Lieutenant which academy he had attended. He replied that he had not attended any military academy, that he was a member of the Militia.

'Ah so', replied the officer: 'Therefore, you are a Lieutenant of the People's Army.' Without another word, Adame turned to the Nationalists and called out: 'Aim! Fire! That is how we do justice in Nationalist Spain.'

The bodies of both negotiators were left where they fell. I and the other members of the group were sent to Pantano de Cijara concentration camp, near Casstilblanco, and later to Castuera."

There we many tragedies in the manner by which the Spanish Falangistas handled their victory. Moreno Gómez was able to record the details of a few such crimes during the first days, but the majority, almost all, remain forever unknown. One largely forgotten crime devastated the Republican Navy when March 5, 1939, Admiral Miguel Buiza Fernández Palacios fled the naval base in the port of Cartagena that was under heavy attack, with a view to saving as much of the fleet as he could. He made his way to French Algeria which he knew was tantamount to handing the fleet over to Franco as a few days earlier the French Government had recognized the Burgos Government as the only legal Spanish government. Denied entrance to Oran, the French authorities directed the fleet to the Tunisian port of Bizert (a French protectorate) where it arrived March 11 and was interned by the French.

Except for a few sailors left on board the several ships on guard duty, the remainder of the crews asked for political asylum and were interned in a nearby concentration camp. When Admiral Francisco Moreno arrived in Bizerte April 14 to take command of the now Nationalist Navy, the French authorities polled the Republican crew. Of the 4,000 naval personnel, 2,350 chose to return to Spain, almost all Senior Staff Officers and a great many crew, especially machinists. Unfortunately, as the 'Hounds of Victory' were baying for blood, Admiral Moreno ordered the summary slaughter of 50 Republican sailors who had decided to return to Spain. The remainder were disembarked in Cadiz, where they were sent to Rota and other concentration camps in that province.

Early features of the Fascist victory

Falangista flags fluttered victoriously all over Cordoba province. A first report after the end of the March 1939 Victory Walk, was of a burst of looting by the victors, a precursor of the unbridled seizure of the spoils of war that would soon loom over the defeated. No sooner did the Nationalist troops enter the towns and villages, that they attacked the military stores whose guards had disappeared. In Pozoblanco, the victorious soldiers did not wait a minute before plundering the goods. They even left an amount to be shared amongst the 'new' merchants.[17] In Villanueva de Córdoba, the contents of the great La Alpujarra warehouse, as in many other towns, were seized.

The most important confiscation in the region was the recently harvested grain and all the livestock kept on Cabañeros farm, belonging to some 40 collective farms in Los Pedroches, March 26. Days later, Nationalists seized a large number of animals also belonging to Los Pedroches collective farmers that were being kept at E Yegüero farm in Cardeña. Bartolomé Cabrera Peralbo, a member of the UGT trade union, who was responsible for caring for the livestock, was totally helpless to prevent the seizure.

---

[17] Falangistas and other friendly civilians who had seized premises belonging to defeated Republican shop owners and merchants.

The Nationalist military commander of Cardeña summoned Peralbo and ordered him to hand over all the weapons on the property, reminding him that he would be responsible for everything there was, even if he had to go and get it himself. Peralbo replied that there was nobody there who controlled anything, that they had had some two hundred suckling pigs, each weighing about 50 pounds, but that two days previously soldiers had shot them and taken them away. The goats that they had disappeared, and of the breeding rabbits, only the metal cages remained. The soldiers had taken all their food store. They opened the wine store and spent all day removing bottles and demijohns, saying that it was wine for the Administrative Officer's wine cellar.[xvi]

When the soldiers arrived at the Villanueva collective farms in Loma del Caballero, Manuel Bustos Badia[xvii] tells how everyone became afraid and went abandon the animals. Two Falangista henchmen, Freco 'El Tirador' and Mariano 'El Gitano' arrived and, on Torrico's orders, took all that was left.

As the Nationalist troops took a town or village, they seized many types of collective ventures, both agricultural and livestock, that they then handed over to traditional gentlemen farmers, primarily major landowners who had returned to the land after having absented themselves from Cordoba and Seville for the past three years. They brought with them a new rural fascism and the return of the old feudal ways and customs of deferential day labourers that had prevailed under the monarchy.

In Cordoba capital, right-wingers and municipal authorities could not contain their joy when March 28, Eduardo Valera Valverde, the new Governor of Cordoba, inflamed with the spirit of Frncoism, addressed the people of Cordoba in celebration of the new *Duce* or *Caudillo*'s occupation of Madrid. His speech was published in its entirety in the local newspaper, *ABC*.[xviii]

At the beginning of April, the Falangista Cordoba authorities took symbolic possession of the towns and villages in the north of the province (Los Pedroches district), by making the rounds of the region with a retinue comprising of Governor Valera; the provincial head of the Falange (FET and JONS),[18] Fernando Fernandez de Córdoba;

---

[18] FALANGE ESPAÑOLA DE LAS JUNTAS DE OFENSIVE NACIONAL-

the president of the Provincial Appellate Court, José Aguilar; and the Inspector of Health, Luís Nájera, among others.

Days before, March 28, Los Pedroches was visited by the Social Welfare caravan bringing some snacks and light supplies for the civil population. When they arrived at Alcaracejos, they found the town deserted and partly destroyed. They then went on to other towns, such as Almadén. There were quite a few such expeditions during the first few days of the conquest, directed at gaining some sympathy from the unfriendly and apprehensive civil population.

An example of this occurred in November, when Nationalist units approached Madrid with some trucks full of melons and watermelons for populace and with the statue of Saint Rafael they had taken from Cordoba to preside over a celebratory open-air mass planned for San Bernardo square.[19] Madrid, however, was still resisting so the Nationalists were forced to cancel the mass and return the statue to Cordoba, under a pouring rain, leaving the melons and watermelons to rot.

The last few days of March 1939 were bitterly cold – it snowed around Los Pedroches – as Nationalists prepared for the 'Holy Week of the Victory' (April 1, Palm Sunday, to April 7, Holy Friday). House searches, arrests, first interrogations, beatings and floggings, and the other usual Francoist practices were the norm everywhere, clearly part of a planned, not 'improvised' shedding of blood, as some have written without justification.

Later, Moreno Gómez shows how Franco liberally applied the so-called extra-judicial *Lei das Fugas*[20] throughout the recently

---

SINDICALISTA. Union of the Committees of the National Syndicalist Offensive in Spain, a coalition of right-wing fascist parties in Spain, the sole legal party under Franco.

[19] The statue of Saint Rafael, patron saint of Cordoba, is jealously protected by the Cordobans as a symbol of their leadership and power over the centuries, of which they are so proud. Taking it to Madrid to preside over a Fascist mass was an insult to Republicans in both cities as it was an affirmation of Francoist power over the nation.

[20] Law of Escapes. Prisoners were taken out for 'walks' from which they never returned. These walks were known as *paseos*, the tongue-in-cheek name given to the extra-legal execution of prisoners or captives. Guards, leading prisoners from one jail to another, would drop back a distance and on the pretence that the prisoners ahead were 'escaping', shoot them in the back. Another practice was for members of a firing squad to loosen their captive' bonds and remove their blindfolds, then turn around pretending to give them a chance to get away. As the captives fled, they were shot by the whole firing squad. Anyone who survived received a 'mercy shot' in the back of the head. This in total violation of the Geneva Convention, which Franco in any case, did not recognize.

conquered Centre-South region during the months of April and May, as part of these fascist Easter Victory celebrations. Again, this was the affirmation of the alliance of the sword and the cross: two faces of the same coin. Gunpowder and incense, as Hilari Raguer wrote.[xix]

The clergy resumed the splendour of the beginning of the century, and they strutted down the streets in front of their congregations, entertaining cheers, hymns and a multitude of celebratory acts: *Te Dwum laudamus*, alleluias and thanksgivings. During those dramatic early days, crowds gathered on the streets primarily for military or religious reasons, attracted by the parades or processions or a combination of both.

Men clothed in black cassocks, with wide-brimmed hats and flowing capes returned, with unusual force, to the social and political centre stage after three years' absence. Falange flags hung side-by-side with religious banners. Open-air masses were celebrated everywhere, always with a strong military presence, resounding with the hymns and *Vivas!* proper to the moment, embellished with all the symbols of a Fascist state soaked in holy water, contrary to the pagan versions seen in Germany and Italy. This was, without a shadow of a doubt, the most fanatic Holy Week in Spanish history.

Today it is difficult to find eyewitness descriptions for the many kinds of memories of that period. Certainly, in Coroba, where on Palm Sunday the Church organized a picturesque pilgrimage to Fuensanta Church to give thanks for the victory. In Alsasua, Navarra, however, we do have the account of Father Marino Ayerra, one of the few priests who rejected the military coup and the ensuing slaughter.[xx] In his book on his experience in Alsasua, Father Ayerra tells of a great manifestation that the Nationalists organized in Alsasua to celebrate Franco's victory, to the overwhelming joy of a few and the stifled concentrated pain of almost all the population , already subjugated and again bound in chains. It is his description that a new Middle Age had spread its dark wings over their homes.

> "In Alsasua, as everywhere, the hills echoed with pealing bells and the sound of fireworks, calling all the neighbouring villagers to attend. A military band led the triumphant march, beginning at Military Headquarters then into the parish church where a thanksgiving *Salve Regina* was sung. The cortège then

went up and down the streets of the town until it arrived at the church of the Capuchin monks where the *Salve Regina* was again sung. The procession then returned to the starting point – Military Headquarters – where the Military Commander gave a closing speech."

When the procession arrived at the parish church, a few steps in front of Military Headquarters, Father Ayerra felt compelled to say a few words to everyone who was present:

"By all means, *pax vobis, pax vobis*, for all. Let us celebrate the triumph of Christ and not the triumph of the Passion. Let this be our motto, therefore, not the *vae victis*, the woe to the vanquished of usual fights between brothers (...) There is no need for me to remind you of the nobility of the Spanish and Christian knight who, when facing his defeated enemy, raised his sword and saluted him. Honour and glory to the victor and honour to the vanquished!"

around the village. Ahead, members of the public sang Nationalist and Militia hymns. Behind, military, civil and religious authorities followed. As the procession arrived at the church of the Capuchin monks, a tall, imposing Capuchin monk came out under a rain cloak. At the end of the hymn to the Virgin Mary, he turned towards the crowd and in a loud, bombastic voice, addressed the assembly:

"The God of the Armies has triumphed once again. Does this mean that the war is over? Absolutely not! The life of Man is a permanent state of war against the enemies of God, subjugated, yes, but rebellious, always ready to rise again, to mutiny... No! The war is not over. We are always at war... To you, yes. *Pax vobis...* But to the others! *Non est pax inpiis!* No peace of any kind for the Ungodly!

The Capuchin monk's speech was a brutal response to Mariano Ayerra and in tune with what every member of the Spanish clergy, and the victors as a whole, were preaching at the moment. There was to be no pardon. This was the voice of the exterminators, at its most evident and brutal. The fact of the matter was that 99.9% of all Spanish clergy identified themselves with the Capuchin monk who so strongly criticized Father Ayerra's pacifying words. On this day, no other cleric in victorious Spain dared speak out in such a manner or express such an opinion, just this humble Basque parish priest in Alsasua in the Catholic province of Navarra.

In the other half of Spain, the defeated population cried in secret, leaving their children bewildered as to their overwhelming sadness. As one older gentleman said, one of the saddest memories he had of the end of the war, when he was only eight years old, was hearing his grandmother cry all night as trucks of victorious soldiers passed their house at the entrance to Tarancón, Cuenca, all night.[xxi]

The first Francoist town councils, or *Ayuntamientos*, were quickly created during Summer 1919 and the first Mayors bore surnames belonging to the great landed gentry. In Pozoblanco, the first Mayor was Antonio 'El Niño' Herrero Martos, a wealthy rancher. In Villanueva, it was the no less wealthy land mogul Antonio Casimiro. Afterwards, some were replaced by medium-sized landowners.

The new town councils were set up as was the norm in a militarized state, under the aegis of the military authority from whom all legitimacy issued. Captain Casas Ochoa of the military judiciary was responsible for the mountain villages. The new councillors 'swore upon their honour' to faithfully carry out their respective functions with zeal, austerity and energy, taking inspiration from Nationalist Spain and General Franco's regulations, then sent telegrams of congratulations to Franco, Queipo de Llano and other Nationalist leaders.

The first proposal every new council unanimously approved was to purchase portraits of General Franco and José Antonio Primo de Rivera, the dictator, and a crucifix. This was immediately followed by the decision to proceed with the utmost urgency to purge the Town Council of Republicans, banish them from the town and replace the entire staff, from the porter to the highest-ranking employee. In each council, this extensive, huge and almost unprecedented purge would be supervised by one of the new councillors who was appointed to manage the disciplinary proceedings.

Consultation of the Minutes of Cordoba province town councils, immediately draws attention to the avalanche of claims from Falangistas for municipal jobs, based on allegations of being ex-prisoners, ex-combatants, relatives of *caídos*[21], war wounded and other similar entitlements. At the same time, these job seekers attempted to add to the 'merit' of their claim by pointing out how cruelly they had dealt with the defeated when doing house searches and detaining individuals and detailing the first beatings they gave in improvised prisons, military headquarters or Nationalist barracks. The new national sport became to have a couple of drinks, go down to the town jail and beat up some prisoners.

Positions in the Municipal Police were filled by the most unsavoury individuals. Never before had these guards enjoyed such extra-judicial power and given such a carte blanche, as in the first post-war period. So much so, that in La Rambla, the first depositions from defeated returnees were taken by the head of the Municipal Police. In Villanueva de Córdoba, the head of the Municipal Police, Bartolomé 'Berenguer' Cepas Díaz, became one of the nightmares of the entire neighbourhood and especially of those deemed to be 'disaffected'. This was also true of others such as Vicente 'Salado' Muñoz Fernández (whose brothers were among the caídos) and Diego 'El Chunga' member of the recently created Military Police, under the tragically infamous SIPM[22] Lieutenant, Leopoldo Mena. According to the Municipal Minutes of July 14, 1939, when applying for the job El Chunga alleged in his favour that he was both a veteran and a wounded soldier.[xxii] Another infamous Municipal Policeman was Emilio 'El del Lunar' Santofimia because of his crazy methods. It is interesting that he, as well as the jobseeker before him, were both defectors from the left. Many of the newly employed also had prior criminal records, such as Berenguer who had served as a guard with Torrico and had been jailed before the war for having killed a hunter in the countryside.

Within the context of post-war rural Fascism, rural policemen also played a significant role because of their past connections with

---

[21]   *Caídos* or 'fallen'. Generic designation for everyone who was killed while serving in the Nationalist Army or died for the Francoist cause. Hence, the *Los Caídos* national monument outside Madrid.

[22]   SIPM – *Serviçio de Información y Policía Militar* – Military Intelligence and Police Service.

the large landholdings. They actually became material executors of the repression and the punishments as they helped the Guardia Civil. This was the murky world of informers who cropped up like poison mushrooms all over the place, including the capital, that Moreno Gómez had studied since 1985, a topic that was recently addressed by Oscar Rodríguez in his work in the phenomenon of the informers.[xxiii]

The victorious town councils, in addition to their major role in the flowchart of the repression (gathering and supplying information, classifying individuals, etc.), were also responsible for managing the pensions allocated to widows or relatives of municipal employees who had been employed since Summer 1936. When relatives of employees who had been purged by the Francoists dared to ask for any back pay, the managers denied these requests that they deemed inadmissible.

As a whole, the first view of the victory was of a country festooned with the paraphernalia of public acts, military homages and parades, blessings of flags and victors, processions and calls for revenge, and clergy and military speaking with a single voice, during a month of April crazed by triumphalist paroxysms. Even Francesc Cambó whom nobody could suspect of leftist sympathies exclaimed in amazement that 'Spain had become a permanent celebration of victorious power, a shameless exhibition of revenge and vendettas.'

Adding insult to injury on the anniversary of the declaration of the Second Republic, April 14, Cordoba's Town Council organized a homage to the eccentric Falangista General Millás Astray, consisting of the gift of a one-metre-tall image of the Sacred Heart of Jesus handed to him by the Major of the provincial capital, José Maria Verastegui. 'A roué in penitent's clothing,' as the poet Antonio Machado might have described him.[23]

April also witnessed the *delirium tremens* of the victors and Franco's triumphal walk around the capitals of Andalusia. Saturday, April 15, Franco arrived in Seville, the parting shot of a magnificent tournée in praise of his base – his 'loyal' multitude. The other multitude would be sent to the dungeons. The central act of the tour was the great victory parade that was celebrated in Seville, Sunday 16, a key date in the Regime's first steps because as the victors paraded, Vatican

---

[23] Antonio Machado. Eminent Spanish poet and writer and Republican intellectual (1879-1939).

Radio transmitted Pope Pius XII's message of blessing to Franco and the heroes of the victory.

The April 16 celebrations in Seville were a Nationalist military highpoint. The magnificent parade, down Avenida de Mayo, was led by General Yagüe (the butcher of Badajoz) at the head of the Moroccan Army Corps. Next came the Extremadura Army Corps, led by General Soláns, followed by the flagbearers of the *Falange Española*[24] and the *Requeté*,[25] led by Colonel Redondo; the Granada Army Corps with its commander, General González Espinosa; and the Cordoba Army Corps under General Borbón. These were followed by the Andalusia Army Corps led by General Muñoz Castellanos and Colonels Castejón (the butcher of Puente Genil) and Maturone (the butcher of Palma del Rio). Next came other units, Artillery, Air Force, etc. Taking the salute, Franco's slight silhouette, accompanied by Queipo de Llano (who would not be 'Viceroy' of Andalusia for much longer), the Minister of Government, Serrano Súñez[26], and assorted Sevillian members of the hierarchy, who stood with their arms raised in the Falangista salute as the bells of the cathedral pealed.

Despite all the pomp and circumstance, the Seville parade was but a pale image of the magnificent May 19 Victory Parade in Madrid, with its march-past of 120,000 soldiers, Moroccans, legionnaires, Guardia Civil and an 'entire cortège of nights and paladins', a remarkable display of military force that Ruben Dário might have chosen for his poem, "Triumphal March".[xxiv]

Back in Andalusia, Franco was cosseted and revered for three days in Seville before continuing his triumphal tour across the region. At the same time the Law of Fugitives began to be applied without distinction throughout the villages and across the fields of the Cordoba hill, against the first defeated individuals captured by the Francoist troops.

On the afternoon of April 18, pealing bells announced the dictator's arrival in San Fernando (where dozens upon dozens of sailors had been executed) and in Cadiz capital; April 19, Malaga (where 4,000 victims have been recently exhumed from mass graves in San Rafael cemetery alone); April 20, Granada (where García Lorca lay

---

[24] *Falange* for short. Coalition of right-wing fascist parties in Spain.
[25] *Requeté*. Paramilitary Carlist Militia, part of Franco's coalition during the civil war.
[26] Franco's brother-in-law.

in an unmarked grave in Viznar). At noon, April 21, the victorious cortège made a symbolic visit to the La Cabeza Sanctuary before entering Jaén, the city Queipo de Llano had bombed. That same day, at 5:30 p.m., Franco was acclaimed in Tendilla Square in Cordoba capital by a huge crowd of Falangistas and right-wingers, both local and many others brought in from surrounding villages, arms raised in the Falangista salute. The Tendillas shook with the noise of the Falangistas, who were hoarse from so much shouting.

Franco, who would stand impassively on the bloody pedestal of half a million dead and another half million exiled, thousands of homes demolished, and thousands of families destroyed and a democracy in shreds, was acclaimed in Cordoba by his followers with the customary hymns and *Vivas!* The provincial authorities of Cordoba had earlier gathered to welcome him in Villa del Rio, where the dictator first set foot in Cordoba province. The Mayor of Cordoba and General Borbón waited on the riverbank in Alcolea to receive him. Cordoba, the martyred city, was about to become the private hunting grounds for the military governor Colonel Ciriaco Cascajo Ruíz, Bruno Ibáñez Gávez – *Don Bruno*, Luíz Zurdo, Reverend Ildefonso Hidalgo and friar Jacinto de Chucena.

Franco's high-pitched voice[27] rang out shrilly from the balcony of the Góngora Institute, the humiliated seat of the intelligentsia, whose Director and Professor, the distinguished Antonio Jaén Morente, was at that moment somewhere on the Pacific Ocean on his way to exile in Ecuador. Next to Franco, General Queipo de Llano referred to 'criminal Marxist bombing' despite the fact that it was he who ordered the bombing of Jaén on April 1, 1937, killing 155 people in a single day, more than all the casualties of the Government bombing of Nationalist positions in the capital throughout the entire war. Following the bath of the crowd, the dictator went for refreshment to the mansion of the Duke of Hornachuelos, José de La Lastra y Hoces, who was taking up his position as a Military Prosecutor at summary court martials.

After Seville, April 24, Franco returned to Burgos, the seat of his government until October 18 when Franco moved into the El Pardo

---

[27] Franco, considered to be the most inarticulate of all the fascist leaders of the time because he was unable to utter eight sentences in a row, spoke in a high-pitched and choppy manner.

royal palace in Madrid. These celebrations marked the end of the wartime Francoist government of the city and province the Nationalists had occupied during the war, as military officials were replaced by other high-ranking officers whose experience and inclination was apparently more suited to the methods of the repressive government that followed the surrender of the Republican forces. Those were tough times for justice, for common sense, for cool heads.

Queipo de Llano's days as Lord of Lives and Properties were numbered. Franco relieved him as Commander in Chief in July and on the twentieth of that same month, appointed General Andrés Saliquet Commander of the II Region in his stead, on the pretext that the butcher of Andalusia and Badajoz should enjoy a 'well-deserved rest'. His farewell was set for July 18, the date chosen by the *ABC* newspaper of Seville to honour his retirement by publishing an insufferable poem by Luis Pérez Solero which had been earlier dedicated to General la Casa González Byass.

> *'What a seed*
> *the General has sown*
> *in the city of Seville!*
> *How he has converted the evil*
> *that suffocated the people*
> *to well-being and affection!*
>
> *Today the child*
> *who had been so forgotten,*
> *shall enjoy entertainment and joy*
> *thanks to Don Gonzalo.*
> *Make the child laugh;*
> *a child who laughs is not bad!*
>
> *When he becomes a citizen*
> *In this new-born Spain,*
> *he will remember what*
> *General Queipo de Llano is doing today,*
> *which is to give bread with one hand*
> *and justice, with the other.'*

Colonel Ciriaco Cascajo remained for a while as military governor of Cordoba although despite his 'achievements' as the butcher of several thousand individuals in the capital, among other nefarious deeds, he had to wait a long time to reap the reward of a promotion to General, which he only received February 28, 1939. The servile nonsense began immediately afterwards when a public subscription was raised to buy him his General's sash. This subscription was considered closed much later, in October, after 24,000 pesetas had been raised. In addition to a villa in the heart of Cordoba capital, a gift from Cordoba right-wingers in 1936, he received many other gifts. Everywhere, people sang the Colonel-General's praise and generosity, his 'blood accomplishments and his role in the military coup.

The *ABC* of Cordoba, a city of beggars, orphans, hunger, imprisoned and misery, celebrated the events of July 18 1939 by publishing a passage from a particularly pathetic speech of Cascajo's in which he extolled the innovations he claimed the new Regime had introduced to Cordoba during the three years since the beginning of the war: soup kitchens for the needy, a profusion of Social Welfare dining rooms for children and adults, a special kitchen for diabetics, a charity dining room equal to the finest in Spain; a sanatorium to treat tuberculosis such as was not thought of during the many years when it could have been built; three groups of cheap houses for labourers and workers where 200 families could live in comfort and health; a model children's crèche; and in addition to all this, the unlimited amount of help given to ex-combatants and to the families of those who died in battle...

Meanwhile, the last few months of the war brought several changes to the military and civil leadership of Cordoba. Brigade General Francisco Formoso Blanco replaced Cascajo as military governor November 23. The civil governor, Valera Valverde, was replaced earlier, August 30, by another career military officer, Artillery Captain Joaquín Cárdenas Lavaneras. The Mayor, José María Verastegui, was replaced November 6 by the well-known Cordoba politician, José Tomás Valverde Castilla who, having trained under the Primo de Rivera dictatorship, was a staunch supporter of Calvo Sotelo in 1936. His term of office was brief, as he resigned as Mayor on November 28 to become Civil Governor of Seville.

The position of Mayor of Cordoba was handed on to Manuel Sarazá Murcia, who had served as Mayor earlier during the tragic days of 1936 and during the 'Terror of Don Bruno', the leading war criminal and genocide in this provincial capital city.[28]

In the end, the first view of the victory was that of a country enveloped by an intense wave of patriotic euphoria, Falangista parades, open-air masses widespread detentions and accusations, executions and starvation in the jails. On the other hand, festivals promoting social welfare, long queues for any kind of food and public subscriptions of all kids, such as one for the reconstruction of the La Cabeza Sanctuary. Every town and village was busy erecting monuments and crosses to the Nationalist *caídos*. In Cordoba capital, a cross to the fallen was placed on Malamuerta Tower, which was odd because the only fallen in Cordoba were 4,000 Republicans who were assassinated.

In towns all over the province, such as Fernán Núñez, victory parties were celebrated with barrels of wine, fireworks and marching bands, regardless of the cost of these festivities during this time of extreme poverty and suffering.

---

[28] Bruno Ibánez Gálvez, Lieutenant-Colonel of the Guardia Civil appointed by Queipo de Llano as Head of Public Order in Cordoba and who, with his reign of terror during his short tenure in office, was responsible for no less than four thousand executions in Cordoba and numerous imprisoned or disappeared. Upon his departure to a new posting, 'Don Bruno' stated that he could claim no merit for his achievement – he simply limited himself to signing the lists that were presented to him by his deputies and by the clergy.

RETURN OF THE DEFEATED.
HUMILIATION AND CONFISCATIONS.

"The absolute secrecy surrounding the repression during forty interminable years, together with the cynical declarations of pro-Francoist leaders as they deny the most evident realities, still leave seeds of doubt in some minds regarding the dramatic truth."
Sócrates Gómez, *Tiempo de Historia,*
January 1980.

Democratic Spain falls into the clutches of the fascist victors.

No sooner had Franco declared victory that there was no doubt as to what awaited the return of the defeated to their hometowns. The Spanish countryside became peopled by terrified individuals, itinerant families, an indescribable coming and going of the population, some searching for a safe haven, others walking non-stop, without respite, towards the homes they had long ago abandoned. The few working vehicles, trucks and trains were crammed with thousands of Spaniards; women, old people and children walked for miles to the limit of their endurance, painting an unbelievable panorama of pain, fear, dark forebodings – a human catastrophe that is difficult to reconstruct today given the very few historians who are willing to undertake this task.

One of the victorious government's more astute stratagems were the general instructions encouraging all the defeated, combatants or not, to return to their home tows (where they could be more easily hunted down). The only breath of hope that the defeated clung to was the fabrication that victorious Francoists spread everywhere: 'He whose hands are not sullied with blood has nothing to fear'. The truth would soon become apparent as the ruse was visible to all. The fact is that this never was a tacit, unspoken fallacy. Franco himself openly used the 'hands not stained with blood' sophism in his bare-faced statements to the world. Franco, who wanted to appear to be complying with the 1929 Geneva Convention, which only permitted enemies to be punished for 'ordinary crimes' and not for 'political transgressions', not even for 'rebelling' as this was not contemplated as a crime in cases of civil war. He wanted to show the world that he was pursuing criminals and not politically motivated individuals. Yet another example of how Franco manipulated the people and used language as a weapon of mass destruction. Hence, Paul Preston's sound judgement in his book on Franco: *The great manipulator.*[xxv]

The month of April 1939 saw a constant coming and going of trucks laden with soldiers and endless columns of defeated combatants on foot, guarded by the victors, railway stations packed with people, both civilians and ex-combatants downloaded from some goods wagons to be loaded onto others towards multiple destinations. All the defeated avoided going through town centres because of the open animosity of the victorious Francoists and the new authorities, eager to punish every Republican they could lay their hands on. Ex-combatants particularly suffered the ill treatment meted out in the towns by the same people who had previously respected them when they were the ones with guns. It was the unhappy humiliating return for both civilian refugees and ex-combatants who had been surprised by the end of the war in distant battle fronts and were considered pariahs everywhere they went. The pain of the return, worse than the difficult moments of the earlier evacuation months, was instinctive: a search for the natural sanctuary of one's fatherland.[xxvi]

Still, the defeat was so overwhelming, the victors so vindictive and ruthless, that for the defeated there no longer was any kind of fatherland, neither large nor small.

Moreno Gómez remarked that although there have been defeats throughout history, he could not recall a single one where there was so much planned revenge, not just against ex-combatants but worse, against the entire civilian population that did not support the victorious regime. Even at the end of World War II in Europe, although there were reprisals against the leading defeated elite, none against the civilian population.

When the ex-combatants returned to their hometowns and their families, they were faced with many tragic situations that break one's heart, such as when José Manuel Matencio who, with a companion, took almost a month to get home, hiding during the day and walking at night. That night, his young sons hearing a noise in the kitchen got out of bed and went to see what it was. There, they saw two men whose unshod feet were bandaged with bloody rags. One, whom they did not recognize, was crying because they did not want him to touch them.[xxvii]

It was not just ex-combatants who were returning home, but also the thousands of families who had earlier fled the fighting and taken refuge in Republican-held territory. Elisa Carillo's family returned to Posadas by train from Argamasilla de Calatarava, Ciudad Real. As they got off the train, the village schoolmaster, a Falangista, recognized her father and called out to the Guardia Civil: "Fine fish we caught tonight! For now, handcuff him!" They quickly took him away, first to Cordoba jail then to jail in Palencia. From there, the family was notified that he had died 'of a heart attack'.[xxviii]

Returning families had harrowing experiences, as was the case of Francisca Adame's family that had also taken refuge in Ciudad Real and was returning to their home in Posadas where her father felt they would be safe from the authorities who were looking for him. They got as far as a little village, Miguel Turra, 4 kms away, where they joined other refugees. That night the war ended.

At daybreak, a Guardia Civil and a priest came accompanied by a woman crying out: 'Long live Christ the King!' and a truck full of Moroccan soldiers. They were sent to a house for refugees where they remained several days with nothing but lentils to eat. As the villagers passed on their way to open air mass that was to be held in the village hall, they yelled: "Make the refuges come out! Make them come out!" Francisca was chosen to come out and they shouted *Cara al sol! Cara al sol!* But she did not know what they meant.[29] Finally, arriving by train in Posadas, they were greeted by a couple of Guardia Civiles shouting: "Refugees over; refugees, here!"

Francisca's uncle who was waiting for them took the family to her grandmother's house, but her father and all officers of rank were taken aside. They were told that a ship was coming for them but when they got to the port of Alicante, no ship had arrived. They were taken to a cinema and then to a prison camp. Half of the men died in that camp and the remainder were shuttled from one camp to another – San Fernando Fort, Santa Barbara Castle, Elche, until they were brought to Cordoba to be tried by a military court.[xxix]

---

[29] *Cara al sol* (Face to the sun). The title of the Spanish Falange party anthem.

Rafael García Contreras described his father's return to Pedro Abad in April 1939 because his father believed the Caudillo's words that no one whose hands were not sullied with blood could return home. They left by train in a cattle wagon from Rus, Ciudad Real, and made their way to their house to find that only the walls and the street door remained standing. His father threw his kit bag to the ground and hugged his mother and sister who were crying.

There was no time for anything else. Behind them stood two Guardia Civiles who took his father away to their barracks where he received beating after beating and was tortured, which left him with a large wound in his back. One day, Rafael's aunt carried him in her arms so that he could see his father as he was taken from the barracks to the jail. He was dripping blood all over his body. November 20, after his wounds had healed, the Guardias took him back to the village and shot him against the Pedro Abad cemetery wall.[xxx]

Many people made their way back on foot, walking all or part of the way: Bartolomé Marín, from La Rambla, found himself in Valencia at the end of the war and he walked back as far as Jaén where he obtained a safe-conduct pass in Ubeda A truck took him to Bujalance. From there, he again walked to La Rambla.[xxxi] Others did not choose to return to their hometowns: many people of little political importance remained in Madrid, Valencia and other towns, working at whatever they could, always keeping their thoughts to themselves.

After the July 25, 1936 slaughter of so-called anarchists in the village of Ferán Núñez at the hands of the Spanish Foreign Legion,[30] refugees from the violence, and there were a great many, escaped along the usual route for these country villages. After briefly wandering through fields and farms (El Alcaparro, etc.), they ended up making their way to Espejo and Castro del Río, Bujalance (to join the Andalusia-Extremadura Republican forces), and eventually to villages in Jaén and also to Manzanares, another CNT-National Workers Union sanctuary. Arcángel Bedmar Gonzálvez says that 42 of them were lost forever in battle.

---

[30] Fighting on the Nationalist side, the Spanish Foreign Legion had a reputation for brutality that in the early months of the civil war, was used to intimidate and terrorize the opponents of Franco's uprising, particularly anyone professing to be communist or socialist or a member of a trade union, all whom Franco labelled anarchists.

The refugees who survived returned under appalling conditions, packed in all kinds of transport, dirty and dejected, dragging their few belongings. Their condition was so bad, that April 17, 1939, the head of the Provincial Board of Health informed its branch in Fernán Núñez of the high incidence of scabies and other parasitic afflictions of the skin among those who were returning from the Red Zone, ordering it to take appropriate measures to remedy the situation.[xxxii]

At least 1,500 people – 440 women, 271 men and 688 ex-combatants and an indeterminate number of children under the age of 16 – returned to Fernán Núñez. Many others were unable to return in the early days, either because they were interned in concentration camps, jails or forced labour battalions, all over Spain. It was a dramatic number of returnees for a village of only 11,000 inhabitants. To these figures, we need to add the numbers for all the other towns and villages in Spain to get an idea of the indescribable extent of the internal migration.

Bedmar adds that, upon their return, many discovered that their relatives had either taken their own lives or had died during their extended absence, as occurred to Andrés Osuna Sánchez who had fled and been imprisoned after the war, returned to find that his wife, Ana Creso Rosal, aged 45, had hung herself in January 1937.

Furthermore, when the refugees returned to their homes, they usually discovered that their homes had been ransacked by Falangistas who had seized all their furniture and household goods. They could not complain. They had to suffer in silence and go out and beg for references from members of the regime (persons of 'recognized solvency'), in order to obtain the release of their fathers, husbands, sons or brothers from the concentration camps or temporary jails in which they had been interned. Sadly, these references were often denied them.

As far as acts of humiliation are concerned, the mothers of twelve neighbours in the village of Santa Cruz who had been assassinated during the first days of the coup, were put to sweeping the town square. What a far cry from what we call 'psychological support for victims' that is so widely offered victims of violence today, or the so-called 'request for forgiveness to victims' families'.

In those days there was only one kind of therapy for relatives of the victims: humiliation, degradation, castor oil and shaven heads. In the same village of Santa Cruz, Falangistas shaved the head of an elderly lady named Benilde, a cousin of the Mayor's, stripped her naked and forced her to drink castor oil, just because she dared to ask for the return of a bed. She died a few days later. When the relatives of Juan José Gómez Gálvez (four children and their sick mother) returned from Torredelcampo, Jaén, where they had taken refuge, they found their house occupied by a Falangista. He finally gave them a room in which to live, but he continued to occupy the building as if he owned it. Such was the so-called behaviour beyond reproach of these persons 'of good standing' and of 'recognized solvency'.

The punishment and extermination program was terrible and unbearable. The victorious Right, traditional *Defender of the Family*, had no qualms in destroying thousands upon thousands of homes and families all over Spain. When victory was declared, the Francoist population began to mistreat all the refugees they could find. They kicked the refugees out of their homes or any shelter they had been able to find, and left them to fend for themselves outdoors, without the slightest sign of the charity or love for one's fellow man that was so vigorously preached by Catholics. For example, many individuals and families who had fled Montilla in July 1936 returned during that tragic month of April 1939. The Gómez Márquez family and others who had taken refuge in Jabalquinto, Jaén, and were housed in the local barracks, were forcibly ejected by a shrieking crowd of Catholic Francoist women. They spent three days outdoors without any shelter and at the mercy of the elements, until they were able to get on a filthy goods train that returned them to Montilla.[xxxiii] The Ruiz Morales family took four days to arrive on a goods train from Murcia. Like these, thousands more painful examples.

There are notable examples of the great unofficial confiscation or economic oppression (plundering and looting, to call a spade a spade) of Montilla that affected the people much more than the later official economic repression under the Law of Political Responsibility. To be a Falangista or member of the local militia was very profitable in those days. Since 1936, the Francoist town council in Montilla had assumed the extra-legal management of property belonging to all those who had earlier fled the hostilities. A local Bureau was set up to make an inventory of their assets and property; when the refugees returned, they found that their homes had already been plundered by the Falangistas and all their furniture and chattel gone. As regards urban or rural real estate, most of these properties were forever lost to their lawful owners. The Bureau expropriated ex-Mayor Manuel Sánchez' house on Calle Ciprés and a bar belonging to his family. Socialist Gregorio Sánchez lost his house on Calle Las Prietas and some land, although in the case of the land, thanks to an influential relative, he was able to recover it after some lengthy negotiations. Josefa Martínez, who had been widowed during the war and had no means of living when she returned, appealed to the Municipal Police for the return of a vineyard she had rented in 1936. In reply, she was beaten and called a whore, with total disregard for the fact that a young child accompanied her. For many years after the war, the municipality seized the crops that the family of Rafael García Espejo who had gone into exile, grew on his farm, El Barrizal.

Although some returnees arrived in Montilla on foot, the majority did so in 'third class carriages', to use Antonio Machado's expression. Guardia Civiles and armed Falangistas were posted on station platforms to hunt and capture. The most appreciated prizes were leftist leaders and senior and administrative officers of the recently defeated Republican Army. These would be led with cheers and shouts of delight to the Nationalist Army Headquarters on Calle Ancha or to the Francoist Guardia Civil barracks where they were subjected to a first interrogation, the first beatings and humiliations.

Many ex-combatants returned by means of safe-conduct passes from the innumerable concentration camps all over Spain. If they did not, the Falangistas would go looking for them. A committee of guards and Falangistas eagerly did the rounds of the Alicante jails and camps on fishing expeditions. This was how they were able to capture and bring back Juan Cordoba Zafra and Manuel Alcaide Aguilar, commanders of the Republican Army, the last democratic Mayor of Montilla, Manuel Sánchez Ruíz, and city councillor Francisco Merino Delgado whose tragic fate will be described further ahead.

Returned men had to comply with two requirements as soon as they arrived. First, they had to go at once to the Guardia Civil and present themselves for a first interrogation (for some unknown reason, the first interrogations in La Rambla were overseen by the head of the Municipal Police). That done, almost all were sent to the improvised jail that had been set up for this purpose, which in

Fernán Núñez was the village cinema which began by housing some 200 prisoners. Again, according to Bedmar, Fernán Núñez prison guards amused themselves by beating the prisoners with rods. One day, in front of all the prisoners, they beat three by breaking a chair over their heads: Amor Jiménez, Pedro Antúnez and Antonio Naranjo, When the prisoners were taken from the jail to the barracks to make further statements, the Guardia Civil beat them again. On other occasions, prisoners would be taken out of the jail and set to clean the streets. That and many other humiliations, acts of degradation and aggravation, all calculated to make life unbearable for the defeated.

In his study of Lucena,[xxxiv]Arcángel Bedmar records 27 prisoners who in July 1940, were transferred to the Montilla jail and later, to Cordoba. In the latter prison, 4 of the almost 1,000 who died from starvation in the 1940s were from Lucena. It would appear that the final number of Lucena prisoners was not higher because of the 132 executions involved in the 'total cleansing' of the town in 1936. In April 19939, a concentration camp in Lucena held 305 prisoners, not necessarily all from Lucena. On the other hand, amongst the arbitrary actions of the victors, there is the case of Antonio Fuillerat Carrasco, a socialist from the village of Jauja who travelled to Lucena from Manzanares concentration camp in April, was court-martialled and curiously, absolved. When he returned to Jauja, he was beaten several times by the Falangistas. According to his grandson Rafael, "weakened by the beatings, upsetting experiences and a disease that he contracted when harvesting rice, he died young, aged 41 years".[xxxv]

At least eight ex-combatants returned to the tiny village of Montemayor during that black Spring. Antonio Jiménez Marín, who like others had left to fight for the Republic, ended up in exile in France in Barcarés concentration camp. June 1939, he decided to return to Spain, turned himself in in Montemayor, was tried in Cordoba and sentenced to 30 years in prison. José Carmona Aguilar, a Lieutenant in the Militia, was sent to Barcarés, returned, and was sent to the Lacolla Workers Brigade. Other Montemayor Barcarés inmates included José Luque Solano whose father obtained his release, and Rafael Moreno Llamas and Antonio Nadales Luque who were also released.

Applications for the release of prisoners were usually successful, but only insofar as they were to swell the ranks of the Forced Labour

Brigades or Disciplinary Worker Soldiers Brigades, such as Brigade 51 of Oyarzun (Guipúzcoa), where Franco's New Spain received Antonio Nadales. Although José Francisco Luque Moreno was also released thanks to his family's efforts, he was unable to return from France until June 1940.[xxxvi] In Spain, he was warmly welcomed at Miranda del Ebro concentration camp by Ángel Carmona Jiménez, eminent Montemayor school teacher who had had a brilliant career during the war at the heart of the FETE-UGT[31] labour union. A friend of Modoaldo Garrido's, before the war he had taught at the famous Academia Espinar in Cordoba. Lastly, among the residents that Montemayor lost into exile, special note is made of the great combatant, Alejandro Cabello Sánchez. Seriously wounded in the war, he ended up as a brilliant university professor in Hungary and in Cuba. If Franco's military coup could cause such damage in tiny Montemayor, imagine its total impact on every town in Spain.

When the defeated returned to the hamlet of Rute, they were greeted by the sounds of the victors' marching bands. However, no sooner did the menfolk arrive, after thousands of trials and complicated ways and by-ways, they had to follow the usual procedure: a first deposition before the Guardia Civil and a first report about those considered to be hostile to the regime, the so-called disaffected. Those who did not immediately land in jail were not free: they had to report every evening at 8 p.m. at the doors to the Guardia Civil barracks, answer the rollcall and sing *Cara al Sol*.[xxxvii] The Fascist population delighted in watching the 'red rabble' bite the dust. Sundays, they had to attend the noon mass at Santa Catarina church so that they could be purified of 'dissolutionary ideas'. At the end of the war, Las Palomas anise factory was used as a temporary prison to house some 80 prisoners who were put to work repairing the Caracabuey road.

This sample of the sufferings of the Cordoba country folk would not be complete without mention of the case of the defiant city of Baena, the subject of an extensive study by Arcángel Bedmar.[xxxviii] The Military Historical Archives hold a file for this town of great conflicting passions, listing the names of prisoners from the Red Zone calculated as at least 2,174. This indication of the great many persons

---

[31] FEDERACIÓN DE TRABAJADORES DE LA ENSEÑANZA-UGT. Federation of Teaching Professionals – General Union of Workers.

forcibly displaced from the villages following the military coup, is yet another example of this authentic humanitarian catastrophe. As usual, the returnees were required to present themselves to the Guardia Civil barracks or to Army Headquarters, for a first interrogatory deposition.

In Baena, the prisoners were interned in the jail on Plaza Vieja and in Tercia jail, together with prisoners from Albendin, Valenzuela and Luque. In February 1940, they were transferred to the Santa Maria de Scala Coeli convent in Castro del Rio that held 1,500 prisoners. In October of the same year, prisoners from all over the province were packed into Cordoba's New Prison, where there was a real extermination. Prisoners were not held for long periods of time in any of the great Francoist prisons, but often moved from one to another, the cynically so-called 'penitentiary tourism'. One can follow the tracks of these unfortunate men from every town in Cordoba province, in and out of Franco's prisons all over Spain.

In Baena, as everywhere else, the prisoners were systematically beaten for the amusement of Falangistas, either in the prison itself during interrogations or in the barracks during a summary proceeding. Without a doubt, the Francoist Guardia Civiles were the great specialists in beatings and torture. Bedmar describes the agony of a prisoner who survived being interrogated and being held for four hours on his knees, a garrotte around his neck, whilst they beat him on his back. A member of the Zarabanda family, whom they had hung by his arms while he was tortured, was unable to withstand the pain and later died in the hospital. José Tarifa Gálvez, who drafted an appeal to the Cordoba Judge Advocate in July 1940, explained that "The [accused's] entire deposition is totally false. He was forced to say this in the Guardia Civil barracks whilst they beat him with a rod". Such cases of unbearable torture led to many suicides, as we shall later see.

No less than 172 of the defeated from Baena were sent to forced labour in the Workers Brigades or the Disciplinary Worker Soldiers Brigades (Los Pastores, Algeciras; Los Barrios, Cádiz; Melilla; Tetuán; at the Labacolla airfield where there were many from Albedín, etc.). At least two from Baena died in the Melilla or San Roque Brigades. Thirteen from Baena died from starvation in Cordoba jail. These included: Antonio Romero León (his brother Juan was a victim of the Nazis in Guse, 1941), Antonio Arroyo León (his son Antonio Arroyo

Rodríguez committed suicide in Tremp jail in 1939 and his other son, Manuel Arroyo Rodríguez, was executed in 1943).

There are numerous examples of how entire working-class families were destroyed time and again by the so-called Defenders of the Family who took communion daily and beat their chests in repentance for their sins. It was cynicism and cruelty raised to the nth degree, a thorough repression at multiple levels, one that too many historians are still not taking seriously, engrossed as they are in refuting undeniable facts. No matter how one looks at it, this as a programmed, systematic extermination.

When they could, the imprisoned wrote heart-rending letters to their family, thanking them for their efforts to send them clothes and food whenever they could, as without them, they felt it would be impossible for them to go on. Similar letters to the one that 43-year-old Joaquin Moreno Muñoz, imprisoned in Cordoba, sent his family February 1942:

> "My dear parents and brothers... Mother, I received everything that you say you sent me, as you do not know how well I have been and that I am better from the weakness that I had. If you can continue to come or send me clothes and what food you can, I will be better off. If you cannot, it will be impossible for me to go on... You have no idea how pleased a prisoner is to know that he is remembered, more so in the circumstances in which I find myself. Nobody knows that better than he who has to endure these situations and how they devour you. Tomás of my heart: do whatever you can, both for the sake of my health and to relieve me of this unbearable burden..."

Worse was to happen to him one month later, March 10. He was taken to San Rafael cemetery in Cordoba and executed.[32]

A look at the National Institute of Statistics records for the 1930-1940 decade in Baena shows that whilst the number of widowers rose 16.3%, which was in agreement with the population growth rate, the

---

[32] Copies of this and many other letters from prisoners are reproduced elsewhere.

number of widows rose 46.15% (588 in all), witness to the extent of the extermination of the men from that town, executed or from acts of war.[xxxix]

There were hundreds of thousands such tragedies all over Spain, as we are now learning of in a great many recent televised documentaries dedicated to recovering the historic memory of Spain and to countering the prevalent historically correct discourse on the basis of the truth as related by survivors' eyewitness reports. A frequent complaint referred to the prison chaplains whom most considered to be nasty and malicious. They said the chaplain would tell them that they were evil and that they were redeeming their sins by doing forced labour. Still, they kept quiet and did their best to raise each other's spirits and never lose their dignity.

> "In one way or another they wanted to exterminate you, either by firing squad, by starvation or forced labour... The priest was evil, very evil, he was a Francoist. He would tell us that we were redeeming our sins with our work. We had to listen to all of this and keep quiet. We were the evil ones... But we never lost our dignity; whenever we could, we raised each other's spirits... They were unable to destroy our morale, they couldn't.[xl]

José Espejo Ruiz and Jesús Maria Romero Ruíz recently published two interesting studies of the effects of the Francoist repression.[xli] The latter's work is particularly important as it gives details of the first depositions of the 583 returnees who were interrogated when they returned to La Rambla in 1939: 297 interrogated by the Guardia Civil; 286 interrogated by the head of the Municipal Police. A number of women and under-age children must be added to that total. Although sone of these may be duplicates, the list is an eloquent sample of the great many people in the South of Spain who fled the rebels in 1936 and the great many who were forced to return in humiliation in 1939 and were subjected to the strictures of the Francoist extermination program.

We can draw many conclusions from these interrogations. More than a thousand residents of La Rambla fled to the Republican zone upon the outbreak of the civil war. Entire families and many workers from a same farm frequently formed groups who walked or rode animals together. One returnee who was interrogated stated that he was part of a group of 150 farm workers on Mina de las Puertas farm who, with their families, set off on foot towards Espejo, taking 20 horses with them, walking all night until they reached Castro del Rio. The route was similar for everyone: they first went to Espejo which was packed with a multitude of country folk. One interrogation revealed a curious fact that in order for them to be given food in Espejo, they had to be carrying a rifle. Then, onwards to Castro del Rio, Bujalance. From there, they split up: some to villages in Jaén province; others, to Vila del Río, Montoro and Andújar. Many went on towards Ciudad Real, Manzanares, and quite a few reached Madrid. Some enlisted as volunteers in the several Republican militia regiments, whereas others worked in orchards, made charcoal or laboured in the El Centenillo mines whilst they waited to be called up by the Republican draft.

The massive displacement of country folk towards loyal villages was not always totally voluntary. Tenant farmers were reluctant to abandon their farms, orchards, melon patches and crops, or their farming implements. Some did leave voluntarily, primarily because they knew that if they stayed, they would be imprisoned. Modesto García, a volunteer runaway, stated that when the rebel movement broke out on August 16, 1936, he was in a field watching his father's goats. His brother came and told him that a municipal policeman had stopped by the house with orders for him to present himself to the Guardia Civil barracks. He decided to go to the Red (Republican) Zone, and he left that same night for Castro del Rio.

The interrogations further revealed just how active patrols of armed militia were when it came to recruiting farm workers who would not otherwise have left of their own accord:[xlii]

- Antonio Álvarez declared that on December 22, 1936, two paid militia arrived one night at Zueros, a charcoal-workers' shanty, and told those who were there to go with them as life was good in the Red Zone. So, they all did: four men and two families, arriving in Linares where they continued to work as coalmen.

- Alfonso Lucena said that when the Rojo forces retreated from the Torres Cabrera railway station, a Red Cavalry Patrol dragged several of them towards the unliberated zone, to Andújar and Marmolejo.

- Antonio Osuna described another case regarding coal workers, this time from Las Pilas farm, where four militiamen arrived on horseback and took him and his family to Espejo.

- José Pérez stated that he left August 1, 1936, with his mother and one of his sisters, abandoning a melon field that he had in Pobletes.

- Antonio Romero declared that whist he was working on Miguelo Farm, two armed militia appeared and forced hi and three others to follow them to Montoro, October 23, 1936.

- Juan Toledano tells how he was working one night on a melon patch he shared fifty-fifty with Pastoseco, when a patrol appeared, none of which were known to him, and forced him to go to Espejo, Andújar.

- Tentecarretas, near Montalbán, when a patrol arrived including El Confitero, El Tropillo and others he knew, ordering them go to Espejo. However, he was unsupportive of their cause so in January 1939 he went over to the Nationalists who received him, not in a hotel as he expected, but in the La Aurora (Malaga) concentration camp.

- Cristóbal Villegas declared that finding himself in a watermelon patch that he and his father worked in Los Arenales, La Rambla, a patrol of ten to twelve armed men appeared on horseback and forced them to go with them.

- Another individual interrogated tells how when he was travelling from Fernán Núñez along the road to La Rambla, he met a group of men whom he did not know because they were not from his village, who forced him to follow them to Espejo.

Residents of La Rambla whose lives, homes and farms had been destroyed as a result of the military coup, enlisted in great numbers in the Jaén, Ciudad Real and Madrid militia units (at the end of 1936 these became Mixed Brigades). A noteworthy example is the men who went to Madrid to enlist in the Cordoba Volunteer Battalion (about twenty) that had been created by the distinguished Member of Parliament Antonio Jaén Morente, under the command of his nephew, Antonio Jaén Romero[33]. A large part of this battalion was later incorporated in the 103rd Militia Battalion (MB) and another part, in the 40th MB. Elsewhere, many enlisted in the so-called Torres Battalion, from Valdepeñas, commanded by Carlos Cornejo, from that town, and later incorporated in the 2nd MB. Others joined the Montilla Company that soon became the Second Battalion of Jaén, later the 92nd MB. Many from La Rambla enlisted in several other battalions of Jaén Militia, including the famous Garcés Battalion, later to become the 73rd MB. In the latter, however, the great majority of recruits came from Montilla and Espejo and, especially, from Villanueva de Córdoba, the birthplace of this famous unit. Everyone who returned from France ended up in one of the concentration camps in the North. In two or three cases, the returnees were immediately sent to Workers Brigades.

Most who returned to La Rambla in April 1939 and the following months had previously been issued safe-conduct passes and recently released from concentration camps all over Spain. Others had passes issued by Workers Brigades, town military headquarters, and some by Falangista mayor or leaders. One is astounded by the great may concentration camps in Spain through which the wretched ex-combatants of La Rambla passed before they were released: no less than 50, located in the north, south, east and west of the country. Considering that this was the case of a single village, just imagine the incredible hodgepodge of prisoners from all over the country who were enduring the chaos and suffering of this humanitarian catastrophe without equal in the history of Spain.

---

[33] Maternal grandfather and cousin of Magdalena Gorrell Jaén, author of this translation.

Table 1.
List of Francoist concentration camps housing ±500 prisoners from La Rambla village in 1939

Campo de concentración de Alcalá de Henares
Campo de concentración de Albatera, Alicante
Campo de concentración de Alcoy, Alicante
Campo de concentración del Castillo de Santa Bárbara, Alicante (2)
Campo de concentración de la Plaza de Toros de Alicante
Campo de concentración de Almería
Campo de concentración de Teruel, Aragón (2)
Campo de concentración de Aranda de Duero
Campo de concentración de Guardiola, Barcelona
Campo de concentración de Horta de San Jorge, Barcelona (3)
Campo de concentración de Deusto, Bilbao (2)
Campo de concentración de Los Escolapios, Bilbao
Campo de concentración de Cáceres
Campo de concentración de Puerto Real, Cádiz (2)
Campo de concentración de Rota, Cádiz
Campo de concentración de la Plaza de Toros de Castellón
Campo de concentración de Ciudad Real
Campo de concentración de Daimiel, Ciudad Real (2)
Campo de concentración de Manzanares, Ciudad Real (4)
Campo de concentración de Puertollano, Ciudad Real
Campo de concentración de La Granjuela, Córdoba (3)
Campo de concentración de Valsequillo, Córdoba
Campo de concentración de Benalúa de Guadiel, Granada
Campo de concentración de Bucos, Granada
Campo de concentración de Padul, Granada (5)
Campo de concentración de Pinos Puente, Granada
Campo de concentración de la Isla de Saltés, Huelva
Campo de concentración de San Juan del Puerto, Huelva (3)
Campo de concentración de La Especia (?)
Campo de concentración de León
Campo de concentración de Lérida
Campo de concentración de La Aurora, Málaga (2)
Campo de concentración de Málaga (2
Campo de concentración de Miranda de Ebro (5)
Campo de concentración de Ronda, Málaga
Campo de concentración de Murguía, País Basco
Campo de concentración de Camposancos, Pontevedra (2)
Campo de concentración de Corbán, Santander (2)
Campo de concentración de Los Carmelitas, Tarragona
Campo de concentración de Reus, Tarragona
Campo de concentración de Manel, Valencia
Campo de concentración de Porta Coeli, Valencia (8)
Campo de concentración de Orduña, Vizcaya
Campo de concentración de Zamora
Campo de concentración de Toro, Zamora

Source: Jesús Maria Romero Ruiz. *Recuperación de la memoria histórica de La Rambla*. La Rambla, City Council, 2010.

Note: The above numbers in brackets refer to prisoners identified in a specific camp. In 1938, more than 170,000 defeated Republicans were imprisoned in

Francoist concentration camps all over Spain. In 1939, this number rose from between 367,000 to 500,000 inmates in more than 180 temporary and permanent concentration camps. In 1946, 137 forced labor and concentration camps were still operational, housing 30,000 political prisoners. The last concentration camp, in Mirana del Ebro, was closed in 1947.

In every town and village, the first examination of all the returnees required the interrogator to do the following: verify the safe-conduct pass; summarize the declarant's passage through the Republican zone; list the military units to which he belonged; record the names of fellow countrymen with whom he was in contact in the other zone and provide the following information regarding them: what bloody deeds does he know they committed, what does he know about them, which ones died in battle, which ones got married and where, which ones held a rank of some kind or were political commissars, and so forth.

Consequently, the massive volume of information from the interrogations provided the Francoists with more information about the Republican zone than the Republicans themselves. On that basis, they were able to target their persecution directly to their prey and tailor its harshness to their victims.

One ex-combatant who returned to Villanueva de Córdoba was Militia Captain Pedro Torralbo Gómez, who had walked there from Jaén. According to his son José's account, when his father left Cazorla he went to Jaén. Uunable to find his family, he walked to his village. It took him several days to do so because he had to keep hiding in the hills. He hid in the well-known Calceras market garden near the railway station whose owner was brave enough to send a message to his wife. She met Pedro in the orchard, bringing him civilian Jaénclothes so that he could change out of uniform. Pedro Torralbo had intended to take to the hills and join Julián Caballelro's guerrillas, but as he was preparing to take refuge on El Minguillo farm where one of his uncles was the manager, he was spotted by Juan Grande, the forest ranger, who turned him in. So began Pedro Torralbo's calvary: Villanueva jail, relentless torture, Burgos prison and finally, a firing squad in Cordoba June 3, 1941.[xliii]

The Cordoba hills, where the Los Pedroches township was still smoking from the last shots from the trenches, resounded with the cunning proclamations of a devious *Caudillo intent on capturing Rojos.*[xliv]

IF YOUR HANDS ARE NOT STAINED WITH
COMMON CRIMES, COME.
FRANCO OFFERS YOU PEACE,
WORK, BREAD AND JUSTICE.

IF YOU HAVE NOT COMMITTED ANY CRIMES,
YOU HAVE NOTHING TO FEAR.
NATIONALIST SPAIN IS FAIR AND GENEROUS.

NATIONALIST SPAIN PROTECTS THE PRISONER
WHO HAS COMMITED NO CRIMES.

NATIONALIST SPAIN OFFERS YOU BREAD AND A
PARDON, WHAT ARE YOU WAITING FOR?
WE OFFER YOU CAUDILLO FRANCO'S GENEROSITY.

All lies. A ruse for internal consumption and to feed the hypocritical international image of a 'magnanimous' Caudillo who only killed common criminals and no one who might be guilty of political transgressions. Western Europe believed or appeared to believe it. Such are the eternal mechanisms of the *real politik*. Franco, with his wily proclamations, was like a fox: he attempted to captivate the chickens with promises of generosity and good will. In the hundreds of personal interviews conducted over many years, survivors repeatedly mentioned how they had fallen for Franco's lies and their great disillusionment when the torture and death sentences began to rain upon them.

In a letter to Moreno Gómez, Pedro Donaire, who fought in Cordoba and in Madrid before enlisting in the elite urban police force, described how he was taken in by the victors' subterfuge and false promises for returnees. He had ended the war in Barcelona and fled to France where the French badly treated him, and as he had had no news of his wife and children, parents and brothers and sisters for almost three years, he decided to return to Spain. He knew that his hands were not sullied with blood and believed that if he did go back, he would be welcomed with open arms by the Francoists. Together with his commanding officer, several Lieutenant Colonels and other

Republican military commanders, they decided to return to their Fatherland.

After they handed themselves in, they were taken to San Sebastian and then to Burgos where they were assigned a Military Judge. Transferred in a cattle wagon on a goods train to Barcelona, they were briefly court-martialled. One of the Colonels who was sent with them, Torres Iglesias, was immediately executed. The others were sent to a concentration camp where, as prisoners of military significance, they were candidates for the death penalty.[34] So much for the value of the Caudillo's word of honour.[xlv]

In actual fact, although the percentage of individuals implicated in 'blood' crimes was minimal as few were officially recorded as such, in more than half of Spain, the number of persons murdered by Franco's regime can be counted in the thousands in Navarra, Soria, Zamora, Galicia, Baleares, Canary Islands, and so forth. As other authors have said, the matter of 'crimes' and 'blood' was nothing more than double-talk employed by Francoists to draw the wool over the eyes of the credulous, both within and outside Spain. What the regime intended was not to eliminate inexistent 'common criminals' but instead to eradicate the very social foundation of the Republic – its authorities, leaders and politically aware citizens. It intended to wipe out all ethereal manifestations of revolution, no matter how far-fetched, as an excuse for abolishing the democratic Republican regime and all that it stood for: modernity, reformism and laity. Once that had been achieved, the regime would proceed to build a New totalitarian State on the blood-soaked land, a New Spain, one that was National Catholic or National Unionist, and to restore the privileges enjoyed in the past by the military, the Church and the bourgeoisie, especially the major agrarian landowners.

Earlier, Moreno Gómez described the flood of prisoners in the north of Cordoba province at the end of March 1939, following the

---

[34] Illustrates the remarkable naivety even of Republican Military Police commanders and officers who returned from France apparently unaware that all the officers of the defeated Army were interned sin concentration camps where they were classified as group C prisoners and whose military significance alone ensured that they were candidates for the death penalty. Not to mention the fourth group of prisoners, lesser ranks classified as group D, who were accused of crimes for which only 2 or 3% were sent directly to the firing squad, and those classified as group B (disaffected of no political significance), who were also summarily shot.

beginning of the 'Victory Walk'. We saw how part of the 8th Army Corps of the Republic fled towards Almadén-Puertollano-Ciudad Real. Another part filled the countless jails that were created in the towns and villages and concentration camps in the Cordoba hills: Valsequillo, La Granjuela, Los Blázquez, Cerro Muriano and a few more. A very few scattered vehicles were able to escape at full speed across La Mancha province, towards Alicante, only to be captured by Italian troops and interned in the bullring, in Santa Barbara castle, or in Albatera concentration camp.

The first days of the victory, Francoist Military Police spread throughout the recently conquered towns and villages accompanied by its Information Service (SIPM) that had under its command the 'Cleansing' or 'Information' Committee, setting up regional headquarters in the larger towns and branches in the smaller villages. In the north of Cordoba province, the Military Police Headquarters were in Pozoblanco. In Villanueva de Córdoba, there was a branch commanded by the terrifying Lieutenant Leopoldo Mena, notorious in the village as the SIMP Lieutenant who killed several individuals by beating them with rods.

In those days, the Military Police and its Information Committee were assisted by a host of unscrupulous individuals that the victory drew from the sewers of society and whose names live on in the *vox populi*. In Pozoblanco, these included: Francisco *El Muleta* Ballesteros (strongman for Antonio Herrero, a wealthy landowner), Juan Andrés *El Pichón* Dueñas (who ended up throwing himself under a train), the security guard *Guijuelo*, and the infamous Teodoro Válero (deputy to Lieutenant Pepirillo), among others. Particularly hated in Cordoba capital was the disreputable Vélasco. He, like so many others, had a miserable end. The oppressors used this scum and then cast them aside.

On March 27, 1939, the SIPM began to pack the prisoners into the most inhospitable places in Villanueva de Córdoba: the anti-aircraft shelter that had been dug in the main square, Pepe Barrón's factory yard, Romo's house on Calle Herradores and Ángel Díaz' fenced-in yard outside town, next to the Ramírez factory. The latter, a temporary holding camp, held several hundred men prisoner without any provisions for several days, except for that which some families were able to bring. The convoys for La Granjuela and for Castuera in Badajoz province left mainly from here.

The permanent, albeit improvised, jails were located in the Municipal Depot, Juan Herrero's house on Calle Conquista and the Fuente Vieja Schools.

In Pozoblanco in May 1939, the jail at number 40, Calle Dr. Rodríguez Blanco was used first as a prison for women and later until the end of the year, as a prison for both sexes. The main prison was the Prisión del Partido, which according to the attending physician, Juan Redondo Muñoz, housed 990 prisoners in 1939. In 1940, they numbered 313 (many had been sent to the capital), and 336 in 1941, according to his records for the number of vaccinations against typhus and smallpox. His notes also mention that between 1939 and 1940, 105 Republicans were shot in Pozoblanco, an extremely low and inexact number particularly as Dr. Redondo appears not to have made note of the 100 prisoners who 'passed' between April and May 1939.[xlvi]

Then began the reprisals. In Villanueva de Córdoba, on April 19, Diego *El Chunga* took Manuel Cruz *El Pichaco* prisoner at his house; the next morning, he appeared to have been shot, along with another five individuals, under the extra-legal Law of Fugitives that in those days was applied at full speed, right, left and centre, in the villages of Los Pedroches district. This was a first 'shock of the violence' that was to come. Not the 'improvised repression' that Sánchez Marroyo[xlvii] mistakenly called it, because nothing that occurred under Francoism after July 18 was improvised; everything had been planned, very carefully planned indeed. Villaralteño family, Francisco Rubio, who had returned sick from the front. El Chunga dragged him out of bed and had him taken to jail on a stretcher. El Chunga also arrested Juan Gómez Luna, brother-in-law to the Villaralteño sisters. All those arrested were executed.[35] Later, El Chunga and his cohorts went for Fructuoso Prieto, who had been identified as a communist and was hiding in his house. They threatened his wife until he came out of

---

[35] Visit by Moreno Gómez to the Villaralteño family home on Calle Torrecampo in Villanueva during the 1980s, where he met with two despondent women in black (all the men in the family had been executed after the war), who told him this story. Both have since died, taking with them the few eyewitness reports that the author was able to save, when it was still possible to do so, despite all the supposed facilities for researchers during the transition period. The silent despair of those two women was the scar they bore of the genocide.

hiding and gave himself up. They arrested socialist Cecilio Ruíz at the farm where he was working. These hatchet men then roamed the streets threatening the wives of Rojos. El Chunga went to Calle Lepanto, to the house that had belonged to the Communist Member of Parliament, Adriano Romero (imprisoned in Ciudad Real) and threatened to shoot his sister Dolores if she dared to show herself at the front door.[xlviii] He then walked right up to Francisca Cabrera in the street, shouting: "Where are your brothers? They are going to die with their boots on." Her brother Gabino Cabrera, a Captain in the 73rd Militia Battalion, was beaten to death a few days later.

The police, the Guardia Civil or the Falange in every town and village in the province waited for the returnees (both those who were travelling without documents and those who had safe-conduct passes from the concentration camps). They waited for them at the railway station and at the bus stops and took then directly to jail without giving them time to see their families, On the day that Fernando Carrión, ex-Mayor of Peñarroya for the Popular Front, arrived, he found the military police waiting for him at the railway station. He was nevertheless able to get away and see his children to whom he said goodbye because 'it was the last time they would meet'. He later handed himself in to the police, was sentenced to death, but managed to survive.

The well-known Joaquim Gómez Tienda, nicknamed *Transio,* a trade union leader from Baena and an altruistic and messianic person, symbol of the agrarian utopia in the South attempted to take refuge in Úbeda (Jaén). He was soon discovered by Falangist Lieutenant Mariano Ariza who ordered him to present himself in Baena. So, he did.. No sooner did he arrive that he was tied up. He asked that he be allowed to see his parents whom he had not seen for three years. His request was refused. The day he was captured was celebrated by the local fascist population, notably by a schoolteacher Fernando *El Carlista*, who organized a children's demonstration calling for his death. He was executed June 22.

The ex-Mayor of Fuente Obejuna, Agustín León, tried to hide in the village of Albó (Almeria). Four months later, he was informed upon by local Falangistas and taken to his hometown where he was executed November 6 1939, under the attentive eye of the bloodthirsty SIPM Lieutenant Flores.

From these and other eyewitness accounts, Moreno Gómez was able to reconstruct some interesting facts that do not appear in the written summaries of the official 'interviews', painful accounts that so clearly represent the tragic inner history of the towns during the operation for the hunt and capture of Rojos.

Undoubtedly, at the end of 1939, many returning defeated were encouraged to return home by Franco's promise of a pardon to all whose hands were not stained with blood, and that somehow, they would be able to return to their pre-war life. When Rafael Bedmar Guerrero arrived in Puente Genil, he saw to his surprise that the Guardia Civil was controlling all the exits from the railway station and that everyone coming from 'the other' Spain was being arrested. Aware that his appearance had greatly changed during the past three years, he calmly walked over to one of the guards who did not recognize him and asked for directions to the Town Hall. He walked down the road to his without raising suspicion.

Arriving home, nobody expected him, not his mother who broke into tears as for the past three years she did not know whether he was dead or alive. She told him that his brother Manuel had died fighting on the Jarama front and set about preparing hi a meal. Rafael had not yet finished eating when two Guardia Civiles knocked at the door. Without a bye-your-leave, they entered the house and arrested him and took him their Headquarters. So much for the promised pardon.[xlix]

A brother of Adriano Romero's, Member of Parliament Antonio Romero, walked back to Villanueva de Córdoba from the Granada front over several exhausting days. As he approached the village, he heard of the cruel treatment that the defeated were being subjected to and decided to hide on a farm. Several months later, September 24, 1939, he undertook a daring cross-country journey on foot to France. He was incredibly lucky as he crossed the border November 8. He was still alive not long ago when he told Moreno Gómez his story. Like his, there are numerous unforgettable accounts of the hardships endured by the thousands of ex-combatants who travelled along roads and cross country during those months of 1939, as they made their way back to their hometowns or attempted to hide as well as they could.

Extra-judicial economic repression.
Widespread pillaging and plunder.

Now follows a brief overview of the outrageous phenomenon that was the extra-judicial widespread direct seizure of belongings and the pillaging that the victors engaged in from the first day of the war, against the property and assets of everyone whom they classified as *disaffected*. The volume of this widespread theft is not recorded, obviously because direct theft, like torture, is never recorded. Although the recent claims of the Historic Memory Movement have been directed at the damage done to the lives of the victims, they are forgetting the damage to the property and the economies of those very same victims. It is very well to clamour for 'truth, justice and reparation', but we must not forget to call for the recovery and restitution of the victims' material property. The latter not only lost their lives, they and their families also lost their property and their savings and were reduced to begging, to suffering in abject poverty.

In cases of transitional justice (or in the absence of any such justice, as in Spain), account is also taken of the mechanisms of restitution, as we have seen in South Africa or in Germany, where the Courts ruled against enterprises that became rich from slave labour. In Spain, however, there has been an absolute impunity regarding the perpetrators of those crimes, the authors of the mass thefts and those who became rich by employing slave labour. The reality of the widespread plundering conducted by the authors of the military coup and the victors can only be very loosely inventoried through the eyewitness accounts of the victims or their relatives. That is what Moreno Gómez proposed to attempt to do next.

The legion of those who persist in refuting the events that occurred under Franco and who continue to infest Spanish society and some academic circles, must know that the crime of genocide, as defined by Raphael Lemkin, does not refer solely to the crime of exterminating human groups (blood crimes) but to many other crimes as well, such as "actions taken to ruin the economic existence of the members of a community."[l] In his famous work, *Axis Rule*, Lemkin lists eight methods of genocide: "political, social, cultural, economic, biologic, physical, religious and moral."[li] Of these, Franco committed six of them in full, including the economic genocide of his political opponents.

Lemkin further goes into great detail to explain his concept of genocide: as a "coordinated plan of different actions intended to provoke the disintegration of the political and social institutions, culture, national feeling, religion and separate economic existence and the destruction of personal safety, liberty, health, dignity, as well as the life, of any individuals who belong to those groups". A perfect description of the great Francoist repression. Also of great concern to Lemkin was the impact of acts of genocide against life, liberty and property: "the three pillars of the devastation that Franco wrought, with malice aforethought, upon those he defeated.

Lemkin's expectations were shattered at the Paris Convention for the Prevention and Sanction of Genocide under United Nations resolution of 9 December 1948 (*Resolution 160 A.III*). Only 'ethical and religious' criteria were approved. France at least tried to include 'opinions' as a criterion. Following pressure from the USSR, the 'political' criterion was eliminated. Pressure within the USA delayed this country's signing the Convention until 1988. It is a known fact that genocide is the work of states, which is why so many took care not to 'get their fingers burnt'. That which was approved, was approved, but it is Lemkin's thoughts and doctrine that interest us, as well as everything that he wanted to teach the world in terms of 'universal justice' regarding intra border massacres and acts of 'barbarism'. Therefore, to Spanish negationists of today, Moreno Gómez says: "If you care to look at the history of the civil war and its aftermath seen through Lemkin's eyes, you will find it increasingly difficult to continue to deny the Francoist crimes of genocide. Furthermore, according to Gregory H. Stanton, denial *per se*, is also a form of genocide, both the contemporary denial of the facts and the denial after the facts."[lii]

Today's historians and promoters of the Historic Memory of Spain now have to do some basic field work. Curiously, oral testimonies are the only documents that can put us on the track of the Francoist plundering as we cannot expect to find this information among existing written documentation. José Francisco Luque Moreno, a historian from Montemayor, has contributed some very interesting data regarding the direct confiscation. This looting that we call 'unofficial', 'direct' or 'unregulated' extra-judicial economic repression, was no more than theft by the fistful, an amassing of perceived spoils of war.

An example of this are the several residents of Montemayor who returned to their homes in 1939 to discover that all their household belongings, carts and livestock had been plundered.[liii]

One of so many similar cases occurred in the village of Castrillo de la Vega, Burgos, after José Brizuela's father was assassinated and his family was forced to hand over 400 kilos of the wheat they had harvested that summer to the same people, he told Mirta Núñez, who murdered his father July 29. His 10-year-old brother and himself, aged only 7, had helped in that harvest.[liv]

The humiliation of the returned, the ransacking of their homes and the pillaging of their belongings was not a sporadic event. It was widespread practice and those who fought it often suffered the consequences. Francisco Poyatos, a Cordoba attorney, kept written records of some cases of this pillaging in the immediate post-war period, such as this one in Adamuz when he was consulted by a neighbour from the Cordoba hills, to no avail.

- 'I have a very handsome mule. The local boss insists that I must sell it to him. I flatly refused and he replied with threats. Can he force me to sell it to him?'
- "Not at all, I said. That gentleman cannot seize your mule."

A few weeks later, the attorney heard that the neighbour who consulted him had been executed.[lv]

All legal niceties went by the board when local Falangistas in a town cast their greedy eyes on a disaffected person's goods and property. Spring 1941, Manuel Cañuelo and Juan Medina, two leftist scum in El Viso, Cordoba, decided that they would confiscate a shepherd's hunting dog. They followed this up by taking his sheep and his house. The shepherd, aware that harassment and beatings at the barracks were liberally applied to anyone who objected to such acts, and that individuals were freely arrested on the slightest allegation of plotting with the guerrillas in the hills, decided that he had no choice but do just that. He grabbed his 16-year-old son and fled to the hills. This son, José Murillo, became the famous guerrilla *Comandante Ríos*.

Musé Murillo Alegre. His father, was captured in 1946; he hanged himself in Cordoba prison.[lvi]

Not even the clergy were exempt from temptation as was the case of a priest who lived in the same building in Madrid as Dr. Joaquim Sama Naharro who, when he was allowed out on parole from prison, was punished and sent 250 kms distant. The director of the prison in Cordoba, a very shrewd man, told Dr. Sama that it appeared that someone was coveting his apartment. How right he was. The greedy priest denounced Dr. Sama to the authorities and gave them some very damaging information regarding him and his family. That was all it took.[lvii]

In Cordoba, Dr. Sama Naharro gave Moreno Gómez an example of how even the family of a disaffected person who had been executed was not exempt from the threat of a seizure. When Manuel Fernández Contreras, a barber who was a member of the Izquierda Republicana was executed in Pozoblanco, the local fascist seized his house. Fortunately, however, as the house was in his wife's name, she was later able to get it back.

The same was not the case of Bartolomé Cabrera Peralbo, who told Moreno Gómez that one day his family was left with only the clothes on their backs after their house was ransacked and emptied, and they were kicked out. They said that they looted the house because Bartolomé himself had seized it. One day, he ran into Juan Félix *El Pichón* Dueñas who was wearing Bartolomés trousers and his fur jacket... His wife desperately tried to get them to return her stocking machine and her daughter Maria's sewing machine, but although she was assured they would get them back, nothing happened. Juanito Garcia told Bartolomé's wife that he would talk to Antonio Bautista and if he agreed her machines would be returned to her. Nothing happened.[lviii]

On other occasions, rivalry between village tradesmen, more than greed, appeared to play a part. Francisco Pino, the head of the Falange in Villanueva del Rey, Cordoba, seized his rival's bakery. Soon afterwards, the other village baker who also owned a flour factory, Francisco Vizuelte, appeared shot, apparently in order to get rid of the competition.

The total lack of compassion for the families of the executed was notable and widespread. It mattered not that a widow and her children might find themselves without a roof over their heads after her property was seized. José Torralbo Rico recalls when in 1939, after his father Militia Captain Pedro *Quadrado* Torralbo Gómez was executed in Cordoba, his mother left their home in Jaén where they lived while he was on military duty, to return to her parents' house in Villanueva de Córdoba. They left in a hurry, leaving her household goods behind. José's mother managed to come to an arrangement with someone she knew who owned a truck and who agreed to help her bring back everything she had left. When they got to the house, it was empty. It had been thoroughly looted and not a handkerchief was left. Beds, clothes, blankets... they took it all. The family was not only left without anything to eat, they did not even have clothes to wear, nor beds to sleep in, nor blankets to keep them warm. From then onwards, they slept on a pile of straw in his grandparent's attic.[lix]

The epidemic of seizure and plundering got totally out of control as every possible event or act on the part of a disaffected family was an opportunity for graft, not to say, the outright theft of goods and property. In Pedroche, a family appealed to the head of the Falange for a reference so they could get their son out of a concentration camp. The Falangista agreed but only on the condition that they hand over a pig they had fattened for slaughter.[lx] Then there was the case of Manuel Cañete Esteban, a land worker and treasurer of the Workers Centre in La Rambla who, with two other prisoners, was taken away in a truck on a *paseo* on 6 September. In desperation, he managed to jump out of the truck and run off, but he was caught and shot in an olive grove. The Falange seized a plot of land he owned in Montalban: his family was unable to get it back.

Beatriz Blancas Pino from Adamuz, sister to the famous guerilla leader nicknamed *Veneno*, told how after her father's death, *Señoritos* from the village entered her mother's house and raided it of everything of value – several silk Chinese shawls, her father's dress suits and other valuables they said now belonged to them, including properties she had inherited from her grandmother – a three-storied house, olive groves and fields, goats, and sheep. When she went to pay her taxes hoping to keep her property, they threatened to arrest her. She lost everything.[lxi]

Like the above tales of the pillaging and ransacking, too many others such as those of Juan *El Escribano* from Villanueva, Matías

Romero Badía, also from Villanueva de Córdoba, Juan A. Velasco Diez from El Saucejo, Josefa de la Fuente from Cabezas de San Juan and the poet Juan Ramón Jiménez, in Madrid, and so very many more, remain to be told. Sadly, as time passes, few of the older survivors are still with us and too many of their children are still afraid of talking of their families' experiences.

Thanks to the birth of the Movement for the Historic Memory of Spain and its multiple branches throughout the country, more and more people are coming out with their memories. When Moreno Gómez attended a Historic Memory Conference in Seville, November 24, 2012, he was able to obtain numerous first-hand depositions of unrelenting suffering and theft.

A veteran of these conferences, Francisca Adame said: "The post-war period was much worse than the war. The books do not tell what we went through, you had to have lived it. I do not want to die without getting it all off my chest."

A television documentary filmed in several small villages of Navarra was clearly directed at calling attention to the extra-judicial economic damage suffered by Republicans and from which the authors of the coup reaped many benefits. The documentary also contains very clear oral testimonies of the suffering in Lesaka, Peralta, Murchante, Lerin, Larraga, Sartaguda, Carcar, Castejón and Burgí, first-hand depositions of great historical value. All are heart-breaking, but none quite so much as the following ones:[36]

- "The Guardia Civil came for my father, but he had already fled. A few days later they returned at night with a truck and took my mother who had a baby in arms, and until today..."
- "They came to my house at 2 a.m. and ordered us to open the door, if not, they would break it down. My sister said that she also wanted to know what they were doing to my father. 'Right you are, get dressed!' She got dressed and went with them. We have not seen either of them since."

---

[36] Oral testimonies recorded in the above documentary: Fernando Gurrea (Lerin), Gloria Villafranca (Peralta), Josefine Campos (Peralta), Pedro Lantz (Lesaka), Josefina Lamberro (Larraga), José Jarauta (Murchante), Justino Berrozpe (idem), Antonio Barros (idem), Félix Moreno (Sartaguda), Julio Sesma (idem), Paz Moreno (idem), Laureano Socarro (Carcar) Josefina Jiménez (Castejón) and Vicente Lacasia (Burgí).

- "The gang went around the town, leading women whose heads they had shaved, making them shout 'Long live the whores that we are!' They took the children out of the schools so they could watch and applaud. They shaved the head of one woman whose husband was imprisoned and forced her and her young children to drink castor oil. My mother had to go begging and we had to go to the Social Welfare office and beg the soldiers for bread".

The same documentary mentions the frequent cases of forced labour in Falangista homes, villages and country houses: "People who were considered to be Rojos, who had escaped from 'the truck'[37] were housed with right-wing families and on farms... 'You will go here, you will go there'... where they were forced to work for no pay, harvesting and threshing the grain... When leftists were sent to work at the airfield, they did so for free" ...

From Barcelona, Moreno Gómez received Antonio Reseco's report on the tragic fate of his maternal grandfather and the family's possessions, in 1940.[lxii] Juan Rivera. Who owned a brickworks in Belmez, spent the war in the Republican zone because he had been elected Deputy Mayor by one of the Republican parties. When he returned to Belmez after the victory and was arrested, he was charged with being a 'communist', to be eliminated and informed that 'all his properties should be given to deserving individuals in the village'. Juan was court-martialled in Fuente Obejuna and executed in Hinojosa del Duque May 25, 1940. He was also sentenced under the Law of Political Responsibilities and his family was fined 100 pesetas.

As a rule, all working families' homes, especially those who during the war had gone to the Republican zone, were looted by Francoists who seized their furniture, household goods and other belongings, as this book repeatedly describes. Faced with all these accounts, it is incomprehensible how today, those who like to sweeten the pill when speaking of the Franco Regime, can continue to spout their shameful, false version of the events, without ever having come into contact with victims or survivors of the great plunder.

---

[37] Referring o, the *paseo* truck on which Falangistas loaded the men they captured and took away, usually to prison and often summarily executed under the Law of Fugitives.

To make matters worse, if possible, on top of it all the Church and its defenders were always present to 'bless' the booty and lend their support to Francoists and Falangistas, totally in disregard of the Sixth Commandment *Thou shalt not kill*. They also ignored the Eighth Commandment *Thou shalt not steal*. The Ninth Commandment *Thou shalt not bear false witness against your neighbour*, also fell by the wayside of course. None of the Francoist courts-martial, especially, paid heed to the latter. The Tenth Commandment *Thou shalt not covet your neighbour's house, your neighbour's wife, nor his male servant, nor his female servant, nor his ox, nor his donkey, nor anything that is you neighbour's*, fared no better. 'National Catholicism and the Ten Commandments': what a ground-breaking subject for a Ph. D, thesis this would be.

There can be no doubt that the victors saw themselves as legitimate proprietors of lives and property. Quite literally. Consultation of the residency registers was the order of the day. Pairs of Falangistas, municipal police or the usual opportunists, felt free to knock at every door. When someone was taken away under arrest, the looters entered his house, taking everything they wanted: pots and pans, food, beds and mattresses. When the first executed individuals fell wearing new clothes or new shoes, the members of the firing squad took their victims' clothes and shoes. When they met one of their victims' widows in the village square, they would jeer at her: 'Just you wait. Tomorrow you will suffer the same fate as your bastard husband!'.[38]

Confiscation of Property Registers.

From July 17, 1936, onwards, the perpetrators of the military coup and the victors "unofficially" began an impressive amount of extra-judicial direct confiscation, the out-and out theft of property belonging to Republican families. This was undoubtedly much greater than the "official" wartime plundering through the confiscation of Property Registers[39] and the post-war application of the Law of Political Responsibilities. Nonetheless, this was a three-pronged economic cataclysm: simple looting and ransacking by opportunists,

---

[38] Tragic anecdote based on real facts, often repeated in Cordoba villages.
[39] List of Republican individuals and corporate owners of property.

expropriations to finance the consolidation of the New State and the regime's intent to apply an economic punishment to the defeated to force them to succumb.

When addressing the matter of Franco's economic annihilation of his enemies, we are again reminded of Raphael Lemkin, the great authority on crimes against humanity, who listed 'actions undertaken to ruin the economic existence of the members of a community' as one of the eight types of genocide.[lxiii] The Confiscation of Property Registers, however, was not just created for repressive purposes, but for another reason: to help finance the cost of the war.

Nowadays, there is a greater penchant for studying only the regulations and the legislation, instead of supplementing this knowledge by going into the field and obtaining first-hand accounts. Of note is something that is evident: that the great looting cataclysm, from the first day of the uprising, began long before the official rules for it were laid down. In other words, the rules were published as a form of retrospect benediction. To begin with, General Queipo de Llano's extra-judicial edicts in Andalusia were published well before the normative edicts were drafted in Burgos. Queipo's Edict Number 13 (August 18, 1936, instituted the 'Confiscation of property belonging to instigators of violence, propagandists and rebels'. Days earlier, August 10, 1936, the Civil Government of Cordoba published a 'List of Organizations' (union, artistic, pharmaceutical, recreational, etc.) 'that are voluntarily disbanding themselves and putting their property at the disposal of the Army of Spanish Salvation', as the Governor sarcastically put it.[lxiv]

Those were the days of uncontrolled plundering and endless 'patriotic' subscriptions that Moreno Gómez described in his 2008 book.[lxv] It was not just the humble families of the disaffected who suffered, but also the Francoist regime's loyal supporters who were continuously called to donate gold and jewels, make multiple patriotic subscriptions, adopt the one single dish meal a week and the day 'without dessert' regimes, and so forth. An outburst of calls for donations that were unsupportable for some and ruinous for others. Those were the days when the Nationalist Military were the protagonists of the great robbery and many of its leaders made their fortune during those tragic years. Just where did all those donations of gold and jewels, monies and seized belonging go?

With Edict number 34 of September 11, 1936, Queipo de Llano ordered that the money and property obtained in the recently occupied towns and villages must be handed over to the military authorities. Property belonging to individuals who were considered liable under Franco's Edict of War, was to be pre-emptively seized by the military authorities. Consequently, as the rebellious troops advanced, their successes in the field resulted in the thorough looting of homes abandoned by fugitive Republicans. Here again, Queipo de Llano acted extra-judicially in advance of the Burgos edict that was only published as National Defence Council Decree number 108 on September 13.

The entire 'official' Francoist pillaging setup derives from this Decree 108, according to which the seizure of property applied to all the belongings of the Republican political parties and of trade unions in general (who owned a substantial amount of real estate) and most especially to all disaffected persons or entities opposed to the Movement. Management of the plundered goods was entrusted to the military officers involved in the revolt and who were granted full powers over lives and property. In fact, Decree 108 set the official stamp on this means of financing the war, the building of the New State and the enrichening the many individuals. Decree 108 was later amended and expanded as Decree-Law of January 10, 1937, which created the Central and the Provincial Committees for the Administration of Seized Property.

The self-styled 'Viceroy' of Andalusia, General Queipo de Llano, did not cease his flood of edicts for expropriations. Edict 29 of 2 September ordered the local Guardia Civil commanders to post a list of the names of the 'instigators of the present rebellion', with a view to freezing their bank accounts. In Edicts 49 of November 5 and 57 of December 29, Queipo attempted to be as specific as possible regarding the targets of his pillaging[lxvi] by referring to militancy, sophistry, political and ideological activism, in other words, the entire gamut of Republican action.

José Francisco Luque Moreno has studied some interesting aspects of the great theft in the village of Montemayor under the so-called Confiscation of Property edicts, where the seizure of property also included the revocation and holding back of all kinds of payments owed to the individuals on the register. For example, in September

1939, at least 20 mules were seized in the village, with the approval of the village Military Command and the provincial Agricultural Recovery Committee and handed over to the town council who took them to Mingo Hijo farm where they were sold at auction. There was the case of Gabriel Gómez Marín, arrested at the beginning of the municipal detentions, who lost the contents of his house in Calle Feria, the grain he had harvested and his land. He was able to recover part of his property during the post-war period thanks to some influential family contacts.[lxvii]

The Confiscation of Property edicts in Montemayor affected no less than 428 individuals (many more than the Law of Political Responsibilities that in this village was only applied to 71 persons). From December 1936 onwards, the local military commander was ordered to take charge of all the material goods belonging to individuals on the register: beds, tables, chairs, mattresses, clothes, sundry belongings and sewing machines, as well as all the livestock. Large warehouses were built to store the booty. The little that escaped storage was sold at public auction. Palma Garrido, a private tutor who owned a tobacconist's shop, provided one of hundreds of examples of this expropriation of property: his name was placed on the confiscation list, his shop was seized by local authorities and given to the widow of a Guardia Civil.

An August 21 telegram from the Civil Governor ordered the Mayor and the Commander of the Guardia Civil in Montemayor to seize the livestock on the farms that had been split up under the Agricultural Reform. October 8, the same Civil Governor asked for a list of all seizable rural properties. 33 orders for seizure were issued. At the end of the war, when the returnees arrived in Montemayor only very few of the legal owners were able to recover their property which they did with great difficulty and over a long period of time.[lxviii]

The case of Natividad Rodrigo, daughter of a couple from Burgos, become known. They had been taken in a raid of 18 persons who were said to have 'disappeared', leaving behind three children, including Natividad who was 5 years old at the time. May 1937, her father, Restituto Rodrigo was ordered to appear before the Confiscation of Property Court, although he had been dead for eight months. The procedure for legitimizing the looting of his property had begun: first, he was given ten days to appear before the Court. Meanwhile, the authorities took out information regarding his political behaviour from the mayor, the municipal judge and the parish priest. The reports were inevitably negative: 'he was a leftist', 'he participated in a First of May rally', 'he voted for the Popular Front at the last election', 'he expressed himself against the Movement and he was considered to be a member of the Socialist Party'. The next step was the valuation of the property, which in his and his wife's case consisted of one house, several farms, a market garden and all kinds of household goods, valued at 1,279 pesetas[40]. All the property was embargoed in 1939 and given to a relative who had meanwhile become a member of the Falange.[lxix]

These were the usual procedures for the confiscation of property by the aforementioned Central Committee for the Administration of Confiscated Property (Decree-Law of 10 January 1937) and its Provincial Committees, presided by the Civil Governor, with an army officer as prosecuting judge. In this way, Republicans not only lost their lives or their freedom, but they also lost their belongings that went to swell the pockets of the victors.

Rich pickings were the bank accounts that were confiscated or frozen. According to Central Committee data, between July 1936 and March 1938, in 15 provinces alone, the Francoists filed 75,000 indictments against persons whose bank accounts were ordered to be attached. Another 100,000 indictments were expected from another 20 provinces.[lxx] Equally valuable was the real estate that was seized from Republican political parties and syndicates that usually ended up in the hands of the Falange (more than 33 buildings in Cordoba province).[lxxi] The management of the entire setup was always left to the military.

This mass confiscation of real estate and other assets, farms, industries and businesses, was very profitable for the military raptors. In Toledo, for example, half of the rural property was confiscated.[lxxii] Many lowlifes supported the new regime solely because it enabled them to steal businesses, trade, property and personal belongings, even if killing was involved, as in Fernán Núñez, Cordoba, where Fernando

---

[40] Following Franco's devaluation of the Republican currency, in 1939, 1,280 PTA were worth USD 136 (1 USD = 9.41 PTA). The relative worth, if this were wealth held in 2023, is USD 14,724.05. Downloaded from "Purchasing Power Today of a US Dollar Transaction in the Past," MeasuringWorth, 2023. URL: www.measuringworth.com/ppowerus/.

Valle Luque and his wife Maria Antonia Jiménez were murdered August 16, 1936, for their shop. This wartime chapter about the wanton confiscations is just one that illustrates how under Francoism, military coup, Crusade and armed robbery all come together as one.

According to Barragán Moriana, in Cordoba province, Montemayor was the village that suffered the most from the Confiscation of Property Edict, which he reports as 410 indictments, although Luque Moreno raises this to 428. This was followed by La Rambla, with 300; Fernán Núñez, with 286; and Espiel, 262.16 Cordoba villages suffered more than 100 indictments each.[lxxiii]

The Cordoba Provincial Committee for Confiscations was presided by the Civil Governor, Valera Valverde, a reactionary who had an intriguing history during the Republic, as he was implicated in the 1932 failed uprising against the government led by José Sanjurjo. The military prosecutors were Clemente Heras, Pedro Herrera, Antonio J. Rueda Roldán and Luís Estrada Pérez.

Antonio Barragán Moriana unearthed the early confiscation proceedings against several distinguished Cordoba Republicans, such as Antonio Jaén Morente, Member of Parliament for the Popular Front, prosecuted with a devilish rhetoric by Pedro Herrera on November 2, 1936: '...resident in this capital and cursed son of the same, Antonio Jaén Morente, resident at Calle Juan de Mena number 6....' Also indicted was Antonio Jaén's personal secretary, the physician Mariano Moya Fernández, who accompanied Antonio Jaén when he was appointed Ambassador to Lima (Peru) in 1933. Mariano Moya was saved by a miracle, despite having been indicted and banished to Burgos in 1936. Judge Enrique Poole Escat, was another of those prosecuted because they were 'friends of Jaén Morente and frequented his office, as well as followers of the politics of the Republican Left'. The list of those indicted under the Confiscation of Property edict was later expanded to include the elite of Cordoba Republicans, upper middle class, middle class and notables, attorneys, physicians, professors, political leaders and so on, such as Rafael Castejón (also banished from Cordoba in 1936), Juan Díaz del Moral and many others.[lxxiv]

When the <u>Law of Political Responsibilities</u> was published February 9, 1939, all the indictments under the Confiscation of Property edict pending at the end of the war were transferred to the jurisdiction of the new law and would continue to be prosecuted until the LPR was repealed December 12 1966.

The grand finale of the victory celebrations was illuminated by these repressive procedures. More than ever, from the end of the war onwards, the victorious Francoists would assume their ownership of lives and property and implement a new repression, a form of multi-repression that ensured that the defeated would be hassled from all sides. In addition to the aforementioned wartime economic repression and the one that was expected during the post-war, life became insufferable for the defeated who were surrounded by a tremendous vacuum. The outlook for those who belonged to the so-called other half of Spain was so bad that all who could, emigrated.

The end of the war also marked the creation of yet another instrument with which Francoists would humiliate the Spanish people: the mortification that was forced upon the prisoners who had been detained en masse and who packed the several concentration camps in Cordoba and Castuera, and their families who had to seek references of good behaviour in order to obtain the safe-conduct pass without which a prisoner would not be set free. Each reference had to be signed by at least two right-wing 'guarantors' of a prisoner's character.

The safe-conduct pass was compulsory for everyone who left his home for a walk, who travelled or even left the camp to go and work in the neighbouring fields. Once, when in a fit of rage, the Falange representative confiscated the inmates' passes, harvesters and olive growers could not go out into the fields and these local labourers lost the few days of casual work they had.

Furthermore, even the simplest bureaucratic act required the presentation of a *good conduct certificate*, signed by the parish priest, a representative of the local Falange and the municipal police or local Guardia Civil post, the three power brokers of National Catholicism - the National Syndicalist Revolution that swept the country. In other words, unadulterated social control. Francoists believed that half of the population was a suspicious mass that had to be quarantined. Not for days, but for years.

There now arose another major problem, one faced by the young men who were being conscripted into the military service and who had to obtain a favourable classification from the Local Recruitment Boards whose primary mission was to define the recruits' character in light of the Glorious National Movement. After a usually cursory examination, the young men were classified as either loyal or disaffected. The latter were forced to purge themselves of all leftist tendencies by serving in one of the *Disciplinary Battalions of Worker Soldiers*. The Enlistment Boards comprised a President (the Mayor), a representative of the Falange, another ex-captive, a representative of the military authorities, and two manual labourers. The supreme authority was always military. Not only had the New Spain been converted into one huge prison, it had now also become one huge military barracks.

In addition to the usual general harassment, the defeated population were now faced with the drama of how to survive in a clearly hostile environment, which led to multiple desperate situations (suicide, emigration, begging). Those whose names had been placed on a blacklist were forbidden to work and were excluded from everything. There was no place for the defeated in Franco's Spain. All of this formed the kaleidoscope of the multi-repression, a form of genocide that was not simply limited to bloodshed, killing and physical destruction, but especially directed at propelling half a country, a complete social class and an entire defeated sector into the eye of a doctrinal hurricane.

Multi-repression consisted of much more: of eliminating almost all due process of law, all chances of survival, the right to food and the right to work; of depriving the defeated of their freedom and throwing them into jail, denying them their Fatherland and forcing them into exile; of abolishing the rights of parents over their children, besieging the defeated on all sides and destroying every glimmer of hope. From the moment the military uprising began July 17, 1936, ethnic cleansing and extermination in Spain could be summarized as the exclusion of half the Spanish population, not just physically but in every other possible way, solely because of the peoples' political ideas and social standing.

Hunger as an Instrument of Oppression and Genocide.

Returning to the doctrine of Raphael Lemkin, the father of the theory of genocide, and his formulate of the condemnation of barbarous crimes against humanity (i.e., genocide), in which, among the several criminal features to which he draws attention, he includes 'actions to ruin the economic survival of the oppressed'.[lxxv] He later added 'economic genocide'[lxxvi] and drew attention to the elemental rights to 'life, liberty and property'.[lxxvii] If Moreno Gómez repeatedly refers to Lemkin, it is because nowadays negationists obstinately continue to refuse to admit to the several aspects of Francoist criminality. Furthermore, when it comes to introducing extreme hunger as a feature of the defeated's existence, something more sinister becomes apparent. Hunger was an example of modern fascist methods for obliterating an individual's personality, a prime instrument of psychological and ideological destruction.

From time immemorial, oppressors have known than one of the most effective instruments for breaking down a people's moral integrity, perhaps more than taking away their freedom or their life, is to reduce them to extreme poverty and destitution. A person who finds himself totally insolvent automatically becomes bereft of all ideology and in unable to think of anything else other than his sustenance. Extreme poverty is an instrument for humiliating and controlling people. The collected letters from the great poet, Miguel Hernández, subjected to starvation in jail, only mention food, especially in 1941. He writes begging letters to all and sundry, he only thinks of food and never refers to either poetry or literature that were his passion.[lxxviii]

Following the Nationalist military victory, the defeated learned what hunger truly was, in its most literal meaning, as we shall see later in Chapter 6, when Moreno Gómez addresses the lethal hunger in Francoist prisons. Meanwhile, Francoists became the owners of lives and property as they forced their victory on the population with successive extra-judicial bouts of confiscation and pillaging.

There is a terrible economic reality that studies of the post-war period have paid little attention to: the abolishment of the Republican currency and its extremely serious economic consequences. With victory, Franco enacted the worst of all possible penalties: with a single stroke of his pen, the population was left penniless. No less than 13,251 million pesetas of Republican currency, plus another 10,350 million on deposit at banks, were declared null and void. A sum total of 23,607 million pesetas was no longer legal tender.[lxxix] In one fell swoop, the defeated were left without a centavo and left to beg. This marked the beginning of the abject poverty and indescribable hunger that, aggravated by many other adverse circumstances, would torture the defeated population.[41]

The repressive methods of absolute impoverishment were widely practiced by European fascists during the 1930s and 1940s. Suffice to recall the skeletal images of the Nazi victims, transported by machines like so many bales of hay, to the mass graves. These skeletal victims of Hitler and of Franco remain a permanent burden on History's conscience, if one accepts that History has a conscience. For those who do not, was penury just yet another stratagem like so many others, all that was done to further the interests (*real politik*) of the powers-that-be? It is interesting to note that the tragic year of 1941 is notorious as the year when the greatest number of persons died from starvation, both in Nazi concentration camps and in Franco's prisons. The Jewish case and the Spanish case (the one that nobody talks about) coincided in 1941. During this year, hunger as an instrument of oppression and genocide caused inconceivable suffering in Spain and in Germany. Nazis and Francoists, fraternally united by a common denominator, *mutatis mutandis*.

In addition to the genocide in Francoist prisons, hunger became one of the horses of the apocalypse that Franco released upon all the defeated in Spain. It was not enough for the dictator to defeat his opponents; they also had to bite the dust, emaciated by hunger, yet another link in the chain of humiliation to ensure that his opponents would not raise their heads 'for thirty years'.

The repressive poverty began with *the total confiscation of property* (direct seizures, sentences for the confiscation of property and the Law of Political Responsibilities), followed by the *absolute exclusion from employment* whereby the defeated were prohibited from accepting any public job, from applying for teaching positions, from owning any kind of businesses (considered spoils of war by the victors), not forgetting the *blacklists* by which anyone who had not humiliated

---

[41] In order to give the reader an idea of the values in terms of the 1939 PTA and the extent of the economic consequence, the 1939 PTA- USD value is given at $1 = 9.41 PTA. If the 23,607 million pesetas were converted as wealth held, in 2023 the USD value would be $271,606 million. Downloaded 3 Jan 2023 from "Purchasing Power Today of a US Dollar Transaction in the Past," MeasuringWorth, 2023. URL: www.measuringworth. com/ppowerus/.

himself sufficiently according to the strictures of National Catholicism, could not even obtain work as a day labourer.

Considerable research needs to be done to quantify the total number of dead from starvation in Franco's Spain. Moreno Gómez reports how every town and village in Cordoba province, without exception, has lists of persons who 'died from starvation' in varying numbers. Half a dozen here, a dozen or more there, depending on the location, during the entire post-war period and especially 1945-1946, the 'years of hunger'. He further regrets that when he was doing field work in Cordoba in 1978 and 1979, he only took note of the number of violent deaths, neglecting these other side effects of the repression.

Moreno Gómez delves into the memories from his own childhood that remind him of those tragic days, as do others who come from country towns. He remembers the many poor who begged on the streets in every town, a phenomenon that only began to disappear during the 1960s. He remembers the many people who roamed the fields searching for acorns and all kinds of oddly edible greens – hedge mustard, wild spinach, purslane, thistles, cress, leeks, nettles, dandelions... and vegetables – turnips, radishes, beans, vetch... in an attempt to stave their hunger. All of this was forbidden by the Guardia Civil who posted guards at the entrances to the villages. They searched everyone who arrived, confiscated their sacks, bags or back-packs and gave them a beating. Those victims were the 'itinerant carpet sellers'[42] of yesteryear, society's poor that governments like to exhibit to 'show that there exists a rule of law.

Moreno Gómez recalls his childhood in Villanueva de Córdoba:

"We who come from country villages, I repeat, have the good fortune of possessing memories, memories that are surely alien to city people. I still have my ration card, dated 1948. I am not trying to teach anybody anything, more so because my personal memory dates from the 1950s and not the terrible 1940s.

---

[42] The expression 'itinerant carpet sellers' alludes to the droves of North African indigent travellers who went from door to door in Southern Europe during the 1960s and 70s, selling cheap homemade carpets as a way of making a living.

I remember that my family, of modest means, never had desserts after meals. A box of oranges was taken to the farms on days when the pigs were slaughtered. The rest of the time there was no dessert other than what we could gather from our orchard, almost always only in summer (plums, figs or watermelons). In the Fall, we had quince. Other fruit, such as bananas, were an inaccessible luxury. We sometimes sold old tools to be able to buy some bananas.

What can I say about milk? Milk in cartons was an invention of the 1960s. In my home, when the cow or the goat gave birth, we had milk; if not, which was most of the year, there was none. More peculiar still was the matter of coffee, which people today will remember was known as 'good coffee' and that in my house was only purchased on the days of the pig slaughter or when the sewing woman came. The rest of the year we drank 'coffee' made from toasted barley and was nothing more than black, hot water During the war and the post-war years, everybody drank barley coffee for breakfast, the same as in the prisons. It is hard to imagine this today, these days of a consumer society.

If this is what I can dredge from my childhood memories of the 1950s, almost during the 'fat cow' days of the Regime, what must it have been like immediately after the war? If a child of the 1950s can remember the pain of that terrible austerity, even when his family fortunately had some land and an orchard, what kind of miserable existence was it like for all those who had not even the smallest plot? I remember often being told of people who died from starvation, their bellies swollen from eating boiled greens without olive oil.

If I have mentioned these snippets of memory, it is because I sometimes read dissertations written by young people of today on the daily life under Francoism and they sound like travel postcards from Lapland, not fascist Spain."

A recent eyewitness account of the first years of the 1940s and halfway through the decade, describes the hunger in the villages of Cordoba as a public disaster, at least among the people who were further from the Regime and as such, did not share in the victory pie. During the 1940s, the Campiña countryside of Cordoba, on the left bank of the Guadalquivir, with single crop grain farms, had very few leftovers and even less to trade, as opposed to the hillsides north of Cordoba, with their meadows, oak groves, abundance of orchards and vegetable gardens, green fields and fruit. Because of this, during that decade, families from villages in the Campiña would walk for days until they reached villages in the hills, and knock on the doors of huts and farms, begging for food. "Around 1946, a family, possibly from Villa del Rio, arrived at the old Venta Velasco, Villanueva de Córdoba, farm, and knocked at the swineherd's hut where Petra Romero Huertas lived. The family of strangers included the father and a young woman of some eighteen years of age, thin as a rake and almost naked as she only wore a simple housecoat. Petra Romero asked them in and gave the young woman some underwear and other clothing. The young woman's appearance was extremely pitiful. They left, saying that they were heading for Conquista. Petra Romero asked them to stop by the hut on their return. A few days later, they did, but the father was alone. His daughter had died, and he had buried her in Conquista. I went to the Civil Registry but could find no record of her death. Perhaps it was not recorded before the family returned to its home in Campiña." [lxxx] That is how one died of starvation in Franco's Spain." Ana Tudela reports that in the province of Jaén, the infant mortality rate in 1942 was as high as 30% because of the chronic hunger and malnutrition.[lxxxi]

Many villages set up hostels, ramshackle buildings that housed dozens and dozens of destitute families who tried to live as they best could, combing the countryside or begging for alms. Villanueva de Córdoba sheltered a great many poor families in Los Bretes (a group of schools) during the 1940s. The villagers considered 'the groupies' as they called them, outcasts.

Franco even invested in concentration camps for those who had been banished from towns and for beggars. The years of 1941-1942 (the years of the Spanish Auschwitz) saw the creation of Las Arenas (La Algaba, Seville) concentration camp, that housed some 300 prisoners employed in digging a canal between La Algaba and Guillena and other slave labour, of which 144 died of starvation. They were not shot. They simply starved to death.[lxxxii]

Forcing the defeated to endure the most absolute poverty was a major operation to ensure that defeated Spain should be left with empty stomachs, because when a person is starving, he thinks of nothing else except food. With hunger as a tool, Franco could eliminate all aspirations, abolish all ideals, obliterate all political thoughts. All this, embellished with a cynical paternalism reinforced by total humiliation, leads human beings to believe that their very survival is a gift from the tyrant. The humiliated thank the dictator, not just for the crust of bread, but also because he has allowed them to live. The life of the defeated becomes the victor's act of charity. When state-programmed violence, fear and poverty appear, it is Fascism rearing its ugly head.

This was a campaign to harass and humble the defeated, to force them to pass under the Francoist gallows of charity and alms. The Falange and the Church behaved as if they were the maestros of that terrible out-of-tune orchestra, through Social Welfare, Catholic Action, Delegate Committees for the Imprisoned, etc. At the same time that they dispensed charity, they increased their power over the Rojos. For all of these reasons, the great economic repression, expressed as poverty and destitution, gave the victors magnificent results in terms of the social control of the defeated.[lxxxiii]

We are reminded of the great misery of the people, when we recall the story of the ex-Mayor of Villanueva de Córdoba's children, forced to beg in the streets without their mother knowing, whilst their father, Julián Caballero, fought the rebels as a guerrilla leader in the Sierra Morena. Just this image of the impoverished family of a man who had played such an important role during the Second Republic was sufficient to fill all the local National Catholic leaders with delight. Destitute families North of Cordoba invented a solution to help them care for their children: they fostered them out to neighbouring farms where they would tend the livestock in exchange for food and perhaps, also some clothing and shoes.

There was a counterpoint to the people's miserable existence during the post-war that was as picturesque as it was dramatic: the *estraperlo*, or black market, which in turn was a consequence of the rationing. A black marketer might operate either as a wholesaler or as a small retailer. The poor frequently went to the latter as their only salvation. That rationing could not provide sufficient food for everyone is demonstrated by the fact that the consumption of black-market staples such as sugar, vegetables and olive oil increased three-fold during that period. As expected, this activity was prohibited under an October 23, 1941, law, with punishments that went as far as the death penalty. As usual, the truth of the matter was that it was the small retailers and rarely, the wealthy wholesalers, who bore the brunt of the repression. On October 27, 1941, a military court was set up in each province to prosecute crimes of hoarding, hiding and sale of products at exorbitant prices, and some black marketers were condemned to three months' service in Forced Labour Battalions.

In October 1940 the government created the Tax Inspectorate, to seek out black market crimes and hoarding, but it was unsuccessful. People would store goods with which to speculate and so would merchants, so that when the tax inspectors came by there was a mad rush to cover up or hide products. Even so, the inspectors themselves could be easily bought.

A ministerial May 14, 1939, Decree, created the ration card that was first given to the family as a whole. The purpose was to enable families to acquire rationed staples at a set price, which they could do only at the specific shop to which they had been assigned.[lxxxiv] It became rapidly evident that there were insufficient rations to meet the demand, which gave rise to the quaint method of a black market where people paid exorbitant prices for their goods. A kilo of sugar that cost 1.90 pesetas when rationed, could cost 20 pesetas on the black market. A litre of olive oil costing 3.75 pesetas, could cost 30 pesetas.

What happened the marketer? The needy became black marketers who traded in a very small way, constantly pursued by the Guardia Civil. Men and women would take advantage of short train runs to go and find goods to sell. On their return, before they reached the stations where the Guardia Civil was always waiting, they would toss their sacks out of the windows and later go back and pick them up. If they were caught by the Guardia Civil, not only would they lose their goods, but they would be given a good beating. It frequently happened that the wives of Guardia Civiles would keep part of the goods their husbands seized and do some black-market trading of their own with persons in their confidence. The municipal police also participated in this curious trade as they would sometimes come to an agreement with the black marketers and share goods that could be seized.

The following sign was posted in one of the cemeteries: 'If you are not a black marketer, Falangista or clergyman, I'll be waiting for you here this winter'. From the heights of power, the Regime's leaders also lined their pockets.

Tobacco was rationed in June 1940 and quickly entered the black market. Moreno Gómez remembers watching his father sow tobacco in the vegetable patch, next to the pimentos, and then dry it in a room on the farm. There was a special ration card for tobacco, for males. There was another ration card for bread. Milk became a luxury good to such an extent that <u>Degree of June 16 1943</u> banned the consumption of milk in cafés, bars and restaurants as if it were a pharmaceutical product. No wonder that the prison staff stole milk that was destined for the Cordoba Provincial Prison by the jug, as they did other foodstuffs. At the end of the day, the prisoners were only left with turnips.

An <u>April 5, 1943, Decree</u> officially created the individual ration card, although that had been in existence since the previous year, according to a newspaper's sarcastic message to hungry Spain: 'The individual ration card will prevent the wastage of three million rations'.[lxxxv] At this time, there were 27 million ration cards in Spain, one for each member of the population. And so it went for the wretched country, resigned to bread and onions, until the end of May 1952 when rationing was abolished and the market for staples was liberalized.

In 1941, the Cordoba Civil Government published the regulations and prices for rationed goods almost every week in the press. Black market goods arrived in Cordoba mainly by train, hidden in the locomotives of the trains that arrived at dawn, by the engine drivers who, generally speaking, collaborated with the black marketers. They knew that many poor people, war widows and proscribed individuals of all kinds, could only survive thanks to this trade. Before the trains entered Cordoba, the engine drivers would throw the goods outside the windows for the black marketers who were waiting at previously agreed spots to collect them (Las Margaritas level crossing, La Residencia bridge in Cercadilla, Viaducto, Santos Pintados, etc.). The main station itself was strictly controlled by the Guardia Civil who climbed on the train and searched it thoroughly before passengers were allowed off. Those who were caught with food were arrested and taken to the Tax Inspectorate where their goods were seized, they were fined and mistreated (head shaving and doses of castor oil were the order of the day). Although all this could provide excellent subject matter for today's investigative cinema, Francoism and all its cesspits are mostly ignored by Spanish film makers. There is still a lot of fear.

In Cordoba capital, an individual who hunted black marketers was Moreno, a municipal policeman who specialized in catching those who brought goods in by train, and who sowed panic in the queues for rationed goods by confiscating women's ration cards for the slightest reason. Likewise, Ballesteros, a corporal in the Guardia de Assalto. Mothers with their children queued since dawn, taking charcoal burners with them to keep warm. Ballesteros would appear, confiscate ration cards from whoever he felt like and use them for himself. If a woman refused to hand hers over to him, he would beat her to a pulp; if a merchant protested, Ballelseros would threaten to close his shop. He would be even harsher when dealing with female black marketers: he would arrest them, beat them, fine them and most especially, confiscate everything they had, which he kept for himself. On other occasions, if a woman wished to avoid complications, he would force her to choose between giving him half of her goods or have sex with him. If she refused, he would have her arrested, shave her head, strip her naked, purge her with castor oil and send her down the street. He particularly liked to have sex with minors, for which he was disciplined.[lxxxvi] Another Guardia Civil in Cordoba, nicknamed *El Dino*, terrorized female black marketers whom he took to the La Calahorra post, purged them with castor oil and took their goods. He applied all sorts of torture to the men.

In the middle of this Francoist *Patio de Monipodio,* or conglomerate of persons with unlawful intentions, there appeared a grotesque 'benefactor' in the person of Ramón Risueño Catalán, Civil Governor and sworn enemy of the black marketers in Cordoba. Because he ordered that the goods that were seized should be shared among the beggars and the needy and because he did not beat the female black marketers or dose them with castor oil, these women began to think of him as a saint.

The same corruption and outrages repeated themselves in town after town. Still, the prosecution of these offenses must not have been considered all that important as the press only reported so-called 'crimes of hoarding' that were punishable by death and where the sentence could not be reduced by hard labour, as was the case of two individuals who were arrested in Montemayor for 'contraband in olive oil and chick peas'.[lxxxvii] At the end of 1941, Nicolás Ramírez was arrested in Peñarroya after he was found to have an illegal warehouse containing 250 Litres of olive oil that he was selling at 9 pesetas per Litre, when the legal rate was 3.70 pesetas.[lxxxviii]

The most widespread black-market activities in Southern Spain, however, involved the major landowners who produced wheat and olive oil. Although the entire harvest had to be declared to the National Wheat Board, they only declared part and redirected the remainder to the black market, where it was sold at much higher prices. It is calculated that in all of Spain, from 1940 to 1955, 28.45% of all the wheat grown and at least 15% of all the olive oil produced were sold on the black market.[lxxxix] In Cordoba, the local press published a scandalous report for all of Spain, whereby 405,268 prosecutions for black marketing, mostly wheat and olive oil, were filed up to October 1943; 50% of all individuals charged were found guilty and the remainder absolved. Moreno Gómez has fond memories of going as a child with his mother to buy some olive oil at a black marketer's house, where they were shown to a back room where their two small pitchers were filled.

Black marketing may not have been a popular ploy for alleviating the misery, but nothing in any way as severe as the great corruption of the Regime, which few have studied. As Ramón Garriga said, jail and the firing squad are reserved in Francoist Spain as punishments for all crimes, with a single exception: financial scandals involving the State. Franco, whose conscience is burdened with the execution of professors and physicians for the sole reason that in their youth they were Freemasons, tolerated his associates', including some of his closest collaborators, trafficking in the well-being of the population.[xc]

In conclusion, and returning to the subject of misery in general, the exploitation of poverty under Francoism is nothing more than one of the defining elements of fascism: 'total misappropriation', both material and moral, to quote Ricard Vinyes.[xci] If we speak of Francoist prisons, total misappropriation, targeting fundamental needs, was practiced to the full. 'Misappropriation', or in other words, the moral and material confiscation of food, hygiene and health, was practiced by Francoism in the prisons and on the streets, always with a view to destroying the identity of the defeated, their personality and their dignity. Nazism added a last, terrible symbolic feature, nudity. Skeletal and naked, the human being retains a single possession: the breath of life. Franco contented himself with shaving prisoners' heads, but he made them die of hunger in droves, especially in 1941. A defeated person was either transformed or eliminated. This was one of the several features that linked Francoism to European Fascism.

The Law of Political Responsibilities (LPR).

The instigators of the coup and their accomplices especially, were chiefly motivated by two events: the proclamation of the Second Spanish Republic on April 14, 1931, and the triumph of the Left in the February 16 1936 legislative elections. Francoism was dedicated to erasing those events from the Annals of History and wiping all reference to them from the face of the Earth.

When Franco signed the Law of Political Responsibilities (LPR) of February 9, 1939, he sent a clear sign to the Republicans of what was coming. Above all, this law confirmed that there was not even the remote possibility of an armistice, a fact that would prove to be such a disappointment to Colonel Casado and all those who were blinded by the Caudillo's supposed 'generosity of spirit'. From that moment onwards, there could be no other prospect than a future of unrestrained reprisals and revenge.

Presiding Judge Don Rafael Añino Ilzarbe
Members of the Courts
Don Francisco Díaz Plá, Don Tomás Martin de Barbadillo
SENTENCE

In the city of Seville, the twenty-fourth of January, nineteen hundred and forty-two. Case number 3,672, Regional Court of Political Responsibility, before the Provincial Court of Cordoba, against EUGENIO JURADO POZUELO, deceased, aged 32 years, married, carpenter and resident in neighbouring Villanueva de Córdoba.

PROVEN: That the Permanent Council of War for the region, meeting in Villanueva de Córdoba, in an indictment under the terms of the National Movement, pronounced a sentence on 13 March 1940, approved 21 April next, by which Eugenio Jurado Pozuelo, author of a crime of joining the rebellion, was condemned to death, and executed. The accused owned no property, he left a four-year-old daughter, and his widow, who also does not own any property, lives from her employment.

OUTCOME: The legal formalities were observed.

CONSIDERING THAT in the case of Eugenio Jurado Pozuelo, the proven facts represent a serious breach of political responsibilities, as set forth in Item a), Article 4, Law of 9 February 1939.

CONSIDERING THAT there are no attenuating circumstances.

Examined: Articles 8, 10, 12 and 13 of the aforementioned Law, and other applicable legislation.

IT IS THEREFORE DECIDED THAT we must and do condemn Eugenio Jurado Pozuelo to pay the sum of one hundred pesetas. The Provincial Instructor of Cordoba is hereby ordered to notify his widow and her daughter of this ruling.

THE ABOVE BEING OUR RULING, we so order it and affix our signatures in witness thereof.

Rafael Añino, Francisco Díaz Plá, Tomás Martín Barbadillo)

(signatures)

Secretary

(Signature illegible)

Source: Regional Court of Political Responsibilities, Seville, January 14, 1942. Document given to Moreno Gómez by Eugenio Jurado Pozuelo's family.

The LPR was another turning of the screw in the punishment and reprisals against Franco's political opponents. Secondly, it provided an impetus to his determination that no one should escape the strictures of the New State. Thirdly, it was a reminder and an example, in case anyone had not yet taken due note of the wartime imprisonments, tortures and firing squads. Fourth, it provided a doubly effective mechanism for obtaining funds for building the New Totalitarian State at the expense of 'all those who contributed to create or aggravate subversion…'. Effective retroactively to October 1934 (Article 1), the LPR was an unacceptable judicial aberration in that actions that were legal until then, were now penalized, posthumously if necessary. Lastly, Article 3 of the LPR ratified all the prosecutions and expropriations under the Confiscation of Property decrees.

Under the new law, everything was arbitrary and extra-judicial, not to say illegal. The LPR was designed to demolish, punitive and economically, the very bowels of the Republican half of Spain, by ascribing a kind of collective guilt to 'all those who contributed to the triumph of the Popular Front in the 1936 elections; those who later supported and defended it, and lastly, who once the Nation had taken up arms in July 1936, tried and continue to try to create obstacles to our National Movement (….).' Clearly, there was nobody like the Spanish Right able to use our language quite so cynically as a weapon of mass destruction.[xcii]

Up to 17 supposed political responsibilities were identified in the LPR, some in specific detail, others so totally indiscriminately that they could be applied to almost everybody.[xciii] Those who had fled Spain were given a deadline of only two months to return and turn themselves in. Three types of penalties were set forth: a) economic sanctions; b) possible exile or banishments; and c) possible disentitlement. In other words, extreme sanctions could entail the confiscation of all a victim's property and he could even be stripped of his citizenship.

At first, prosecution for political responsibilities was based on sentences handed down by courts martial. A summary of the sentence was sent first to the Regional Political Responsibility Courts and then to the Provincial Courts (Special Hearings or Courts). Prosecution could also be initiated on the basis of denunciations from private individuals or Francoist authorities. The procedure was grounded on three principal reports: from the Mayor (usually the local head of the Falange Española), the head of the local Guardia Civil post, and the parish priest). That the latter report is included is particularly significant as it demonstrates the active role of the Church in the repressive machine, a truly serious matter regarding an institution that preached a mission of fraternity and charity to all people, without exceptions. In reality, the Inquisition was riding again, this time on the back of Francoism. Together with other features, this was National Catholicism.

The LPR was another of the New State's wide-reaching phenomena against the masses, one that was predominantly dedicated to repression. Francoism's sole expertise was in the field of military disciplinary mechanisms: mass imprisonments, mass court-martials and mass Political Responsibility judiciary procedures. This explains the unmitigated chaos in processing the three phenomena. There had to be continuous reprieves in the jails. The courts martial were overwhelmed (in the 1st Military Region, Madrid, alone, 300,000 individuals were tried, even though many charges were collective. On top of this, there then was the processing of indictments under the LPR, another typical log jam of the time. This is why the government had to call an end to the indictments, which it did April 13, 1945.

Needless to say, the military were the leaders of all these repressive initiatives. The government created the National Court of Political Responsibilities (presided by Enrique Suñer), 18 Regional and their respective Provincial LPR Courts. Each court and likewise, the Court for the Repression of Freemasonry and Communism, always consisted of three Presiding Judges, a military commander, a civilian magistrate and a high-ranking member of the Falange. These courts, contrary to the practice in Germany and Italy, were always subordinate to the military, and understandably continuously subject to the antipathy of the judiciary.

An important legal date was the publication of <u>Order of September 23, 1939</u>, by which all the cheap housing belonging to leftist cooperatives became the property of the National Housing Institute. Equally, if not more surprising, was the handing over to the Falange, of all buildings belonging to trade unions (more than 33 buildings in the province of Cordoba alone).[xciv]

Because of the chaos resulting from the back-up in LPR prosecutions, that law was amended <u>19 February 19, 1942</u>, so that only financially solvent individuals would be arraigned, leaving out all those whose property was worth less than 25,000 pesetas [$57,550]. The effect of this was immediate: 75% of all indictments were withdrawn between 1944 and 1945.[xcv] The reform of this law also lightened the load of the Seville Regional Provincial PR Court, which handled all the LPR prosecutions, as these cases were transferred to the Provincial Courts. In January 193, the Cordoba Provincial PR Court dealt with no less than 10,500 indictments (distributed among the 17 Magistrate Courts of the judicial districts). Furthermore, the repressive madness was so schizophrenic that the same person could be indicted by three different courts: a military court, the Court of Political Responsibilities and the Court for the Repression of Freemasonry and Communism.

The fund-raising reality of the LPR did not, in any way, meet expectations because almost all of the labourers and peasants (many of whom had also been executed by firing squads) who were sentenced to pay a 100 peseta [$230] fine were either dead, insolvent or impoverished. Even such a small sum could prove ruinous to a working-class family as on the whole this was equivalent to one month's wages (6 or 7 pesetas was the average daily wage in 1940). During those onerous years, the Official Bulletin for the Province of Cordoba published extremely long lists of persons fined under the LPR, usually for 100 pesetas [$230] each.

The bulk of the economic repression, however, was borne by the Republican political, intellectual and academic elite, most of whom had had brilliant careers and who, in the case of Cordoba, represented an entire generation, wiped out by Francoism and the Church. They decapitated the thinking heads and destroyed a century's worth of culture.

According to Barragán Moriana's extensive research, 98 persons were sentenced in Cordoba to pay more than 1,000 pesetas [$2,300] each, while 567 were each fined [$2,300] or less. There was only one exorbitant fine (125,000 pesetas [$376,822], the one levied against Dr. Vicente Martín Romera, Socialist Member of Parliament for the Frente Popular, who was executed soon after the military coup. His widow had to pay the fine in order to obtain free use of her property. It is worth noting that these economic sanctions did not just punish the individuals who were prosecuted as they had frequently been executed by firing squad, but their heirs as well. The family of Blas Infante, executed on the orders of Queijo de Llano in Seville, was ordered to pay his 2,000 pesetas [$4,600] fine. Masons, in particular, were usually sentenced to pay extremely large amounts.

In addition to the astronomic fine that Dr. Martín Romera was sentenced to pay, there were others. The family of Antonio Fernández Carretero, schoolteacher of Villaviciosa, executed, had to pay 10,000 pesetas [$23,000]. The family of José Alcalde Machuca, an attorney of Espiel, executed, had to pay 15,000 pesetas [$34,500]. Alfredo Herrara Siles, physician of Posadas, shot when the Francoist troops entered that town, was fined 5,000 pesetas [$11,500] which his family paid. Antonio España Ocaña, a Republican icon and a Mason in Palma del Rio who fled into exile, was fined 1,500 pesetas [$3,450], although we do not know whether this was paid so that his family could have access to his estate, or not. A picturesque individual, Máximo Muñoz López, expert quantity surveyor of Conquista, married in Hinojosa del Duque and who played quite a leading role in Los Pedroches during the war, fled to exile in Mexico owing 15,000 pesetas [$34,500] fine.

Antonio Hidalgo Flores, businessman of La Rambla, a Mason, executed in 1936 (it is said that someone had eyes on his business), was fined 2,000 pesetas [$4,600] which his family paid. The Masons from Lucena who were a breath of fresh air in the middle of the overpowering stench of clerical incense that imbued everything, who followed brilliant careers and who were all executed in 1936, were also fined: Javier Tubio Aranda (250 pesetas [$575]), Domingo Cuenca Navajas (150 pesetas [$345]), José López Jiménez (200 pesetas [$460]), and Anselmo Jiménez Alba (100 pesetas [$230]). One of the Masons in Cordoba capital, Vicenete Lombardia Pérez, executed in 1936, was fined 500 pesetas [$1,150].

Indictments and prosecutions of all sorts rained upon the distinguished Professor of Cordoba and Republican politician Antonio Jaén Morente, also a Mason, who escaped into exile in Ecuador. He was first prosecuted in Cordoba under the Confiscation of Property Order in 1936; then in Madrid, both under the Law of Political Responsibilities and by the Court for the Repression of Freemasonry and Communism in which he was sentenced to 20 years and 5 days in prison.

In the case of his fellow politician, a member of the Republican Left, Ramón Rubio Vicente, vice-president of the Red Cross in

Madrid during the war, the Francoists dismissed the charges against him because he 'corrected his mistakes' by handing over the Red Cross funds when the Francoists entered Madrid and by having saved many right-wngers. Another co-called 'Red Angel'.

Jesús Hernández Tomás, Communist Member of Parliament for Cordoba, suffered the wrath of the Francoists despite his being a very junior member of Parliament, as he was condemned by the Court of Political Responsibilities to pay a million pesetas in fines and banishment for 15 years, all which was in vain because he was already living in exile. Pedro Rico López, another junior Member of Parliament for Cordoba and Republican Mayor of Madrid, was fines the highest amount we know: 10 million pesetas [$22,920 million].[xcvi]

Ramon Carreras Pons was one of those individuals who people now enthusiastically describe as belonging to the 'third Spain'. Professor at the Public School for Teachers, supporter of Leroux and ex-Member of the Constitutional Assembly, he found himself in Madrid in the middle of the war and decided to go to France with his family. He was discovered in La Junquera (Gerona), arrested and tried in Barcelona for attempting to flee and sentenced to pay a 2,000 peseta [$4,600] fine. Another ex-Member of the Constitutional Assembly, socialist Juan Morán Bayo, was prosecuted under the Confiscation of Property Edict in 1936 and the Court of Political Responsibilities in 1940. He left politics in 1936, but the authorities did not consider him sufficiently repentant and fined him 1,000 pesetas [$2,300].

Nationwide, according to information from the Courts, more than 114,000 LPR indictments were dismissed in mid-1941 (only 38% of the total were sentenced). The authorities had calculated that the LPR could affect a quarter of million Spaniards. There is another summary report, cited by Manuel Álvaro Dueñas, that states that 'in the two years during which the Courts were operating, the judges ruled on 38,035 cases. 87,231 remained pending, and 101,440 were dismissed, which represents a total of 188,671 indictments, which if dealt with at an average rate of 19,027 per year, would take nine years and ten months to process'.[xcvii]

As regards Cordoba, based on data published in the Official Bulletin of the Province, Barragán Moriana[xcviii] calculated a total of 6,454 prosecutions under the Law of Political Responsibilities in the province, the following being the most affected judicial districts:

Fuente Obejuna (1,150), Montoro (881), Pozoblanco (808), Posadas (614), Hinojosa del Duque (607) and Cordoba capital (544).

At first, the Regional Court of Political Responsibilities in Seville had its headquarters at Calle Amor de Dios, number 18. The Court's services were reorganized after the 1942 reform and many procedures were decentralized to the Provincial and the Magistrate Courts. In 1943, 17 judicial districts were given added jurisdiction in these matters. A consultation in the judicial district of Pozoblanco in 1980, showed more than 600 such indictments that were later filed in the Province of Cordoba Historical Archives. The following case is an example of an indictment that was brought before the Regional Court and where the sentence under the Law of Political Responsibilities was handed down two years after the victim had been executed.[xcix]

Clearly, the dimension of this brutal economic repression, added to the other repressive mechanisms and measures masterminded by the military, is insane. This was not a productive Spain, but a bureaucratic Spain, in which the military, the Falange, parish priests and a legion of public servants spent day after day preparing and filing indictments with which to prosecute defeated Spaniards. One wonders whether there ever could have been such an extensive repressive and bureaucratic lunacy in old Europe during the other post-war periods.

AUTHOR'S ENDNOTES FOR CHAPTER I.

i   Juan Siméon Vidarte. Todos fuimos culpables. (We were all guilty.) Barcelona, Grijalbo, 978. Vol. 1. p. 923.

ii   Emiliano Mascarasque Castillo. 1985. Interviewed several times by Moreno Gómez in Pozoblanco.

iii   Casimiro Jabonero. Diário del soldadao Republicano Casimiro Jabonero Campo de prisonieros de Lavacolla. Prisión de Santiago de Compostela. 1939-1940. (Diary of Casimiro Jabonero, Republican soldier. Lavacolla concenttration camp. Santiago de Compostela Prison.) Ed. Victor Manuel Santidrián Arias. Ayuntamiento, Santiago de Compostela, 2004. Pp. 93 et al.

iv   Historical Archives of the Army of the Air (AHFA), Ministry of Defence, Madrid. Archive A2035.

v   Laura Contreras, September 1980. Interviewed by Moreno Gómez in Villaviciosa, and to whom she later also wrote several letters.

vi   Members of Catalina Mestre's family report to Moreno Gómez.

vii   Miguel Ángel PeñaMuñoz. Letter to Moreno Gómez, in which he states that the victim was his grandmother's brother-in-law and that he died in her arms.

viii   Luna Gómez Rordriguez. Interviewed by Moreno Gómez in Villanova de Córdoba.

ix   Diario de Operaciones del Ejército del Sur (Daily Operations Report of the Army of the South). Serviço Histórico Militar, Madrid; Partes oficiales de Guerra, 1936-1939, Vol. I. Ejército Nacional (Official War Reports 1936-1939, Vol. I, Spanish Army). Serviço Histórico Militar, Madrid, 1978, 2 Vols.; General José Cuesta Moreno, Deputy to General Queipo de Llano in Seville. Documents. Hechos ocurridos en los puebles de la provincia de Córdoba (Events in towns and villages in Cordoba province). Serviço Histórico Militar, Madrid.

x   Herbert R. Southworth. Myth of Franco's Crusade. Translated into Spanish and French. Paris, France, Ruedo Ibérico, 1963.

xi   Carlos Menéndez Viñuela. Interviewed 1982 by Moreno Gómez in Dos Torres.

xii   Corporal José Navarrete's eyewitness account appears in his unpublished Diário de la guerra (War Diary) dated 30 November 1939, Pozoblanco, which was shown to the author by his son, José, December 2013.

xiii   Gabriel Jackson. Quoted in Una inmensa prisión. Imágines contra el olvido (An immense prison. Images against oblivion). Documentary produced and directed by Carlos Caecero and Guillermo Carnero, 2005. Impulse Records, 2006.

xiv   Antonio D. López Rodriguez. Cruz, bandera y Caudillo. El campo de concentraión de Castuera (Cross, flag and Caudillo. Castuera concentration camp). Badajoz, Ceder-La Serena, 2006, pp. 98 and 169.

xv   Albino Garrido's oral testimony can be heard on YouTube, as part of a 2011 documentary on Spanish TVE2 by Juan Sella and Rafael Robledo, El pesadillo de Castuera Badajoz (The nightmare of Castuera Badajoz), which includes multiple visual recordings of oral testimonies of survivors and relatives of disappeared prisoners at this concentration camp. Uploaded to https://youtu.be/MAatWuuQbbQM 25 January 2014 by CGT Barcelona. Garrido was also interviewed by Luis Sanchez & Maria Ángeles Hernandez on Cadena Sur, Avila Television, following the publication of his book Une longue marche. De la répression franquiste aux champs français. (A long march. From the Francoist repression to the French camps). Translated into French by his son Luís Garrido. France, Privat, 2011, later published in the original Spanish. Lleida, Milenio, 2013.

xvi   Bartolomé Cabrera Peralbo. Interviewed 1983.1985. by Moreno Gómez in Villanueva de Córdoba.

xvii   Manuel Bustos Badia. Interviewed 15 July 2002 by Moreno Gómez in Villanueva de Córdoba.

xviii   ABC, Cordoba, 29 March 1939. Eduarado Valera Valverde had previously served as a Republican governor in 1931. He was later implicated in the Sanjurjada uprising in Seville.

xix   Hilari Raguer, a monk in Monserrat (Barcelona) is one of today's most astute researchers into the Church's role under Franco. La pólvora y el incienso. La Iglesia y la guerra civil española (1936-1939). (Gunpowder and Incense. The Church and the Spanish Civil War (1936-1939). Barcelona, Peninsula, 2001.

xx   Marino Ayerra Redin, a Basque parish priest who rejected the military and the role of the Church in the slaughter that followed, fled to exile in Uruguay in 1939 and in 1940 moved to Argentina. In 1958 he published a book on his experiences in Alsasua, Pamplona 1936-1939. Maldito seais! No me avergoncé del evangelçio. (Be damned! I was not ashamed of the Gospel). Buenos Aires, Periplo, 1958.

xxi   José Manuel Marencio, 2010. Eyewitness account reported in the Minutes of the I Jornada para la Recuperación de la Memoria Histórica (I Conference on Recovering the Historic Memory). Posadas, City Council.

xxii   Minutes of the July 14, 1939 Meeting of the Villanueva de Córdoba City Council.

xxiii   Oscar Rodríguez Barreira. 2013. El franquismo desde los márgenes (Francoism from the edges). University of Lleida.

xxiv   Felix Ruben Dário (1867-1016). Nicaraguan poet, father of the Spanish-American literary movement known as modernismo.

xxv   Paul Preston. El gran manipulador. La mentira cotidiana de Franco (The great manipulator. Franco's daily lie.) Barcelona, Ediciones B., 2008.

xxvi   Antonio D. López Rodriguez. Op. Cit.

xxvii   José Manuel Matencio, Op. Cit.

xxviii   Elisa Carillo, p. 85

xxix   Francisca Adame. In: Mirta Núñez Diaz-Balart al... La grán repressión. Los años de plomo del franquismo (The great repression. The sombre years of Francoism). Madrid, Flor del Viento, 2009, pp. 74-75.

xxx   Rafael García Contreras, Sussurros de libertad. Memorias. (Whispers of freedom. Memories). Cordoba, Puntoreklamo, 2008, pp. 21-23.

xxxi   Jesús Maria Romero Ruíz. Recuperación de la memoria histórica de La Rambla (Recovering the historic memory of La Rambla). La Rambla, City Council, 2010, p. 20.

xxxii   Arcángel Bedmar Gonzálves, Op. Cit., p. 71.

xxxiii   Ibid. Los puños y las pistolas. La repressión en Montilla (1936-1939) (Fists and revolvers. The repression in Montilla (1936-1939). Cordoba, Ed. Lucena, 2009, p. 102.

xxxiv   Ibid. Repùblica, guerra y repressiòn. Lucena 1931-1939. (Republic, war and repression. Lucena 1931-1939). Lucena City Council, 2010, pp. 218 et al. Revised edition.

xxxv   Idem. p. 224.

xxxvi   José Francisco Luque Moreno, Montemayor, 1900-1945. Cuestión social, República, guerra y repressión (Montemayor, 1900-1945. Social issue, Republic, war and repression). Cordoba, Provincial Council, 2011, pp, 23 et al.

xxxvii   Arcángel Bedmar Gonzálves, Desaparecidos. La repressión franquista en Rute (1936-1950) (The disappeared. The Francoist repression Rute (1936-1950). Lucena Cordoba, Rute City Council, 2004, Vol. 2, pp. 84 et al.

xxxviii   Ibid. Baena, roja y negra. Guerra civil y repressión (1936-1943) (Baenam red and black, Civil war and repression). Lucena, Cordoba, Juan de Mairena, 2008, pp. 113 et al. Revised edition 2013.

xxxix   Idem. p. 138.

xl   Television Documentary, 2007. Desafectos. Esclavos de Franco en el Pirineo (The disaffected. Slaves to Franco in the Pyrenees). Regarding a Workers' Battalion in El Roncal-Salazar, employed in the building of a road between these two towns, 1939-1941. Un-named survivors' oral testimonies.

[xli] José Espejo Ruz. Memoria fértil (Fertile memory) & Jesús Maria Romero Ruíz, Recuperación de la memoria histórica de La Rambla (Recovering the historic memory of La Rambla). La Rambla, Cordoba, City Council, 2010.

[xlii] José Maria Romero Ruíz. Op. Cit. Some examples of the La Rambla interrogations, transcribed and published by that author.

[xliii] José Torralbo. Interviewed by Moreno Gómez in Villanueva de Córdoba.

[xliv] Alberto Reig Tapia. Ideologia e historia. Sobre la represión franquista y la guerra civil (Ideology and history. The Francoist repression and the civil war). Madrid, Akal, 1984. Excellent compilation of Francoist proclamations.

[xlv] Pedro Donaire Leal. November 17 1986, Seville. Written eyewitness account sent to Moreno Gómez.

[xlvi] Pozoblanco Municipal Archives, document provided courtesy of Fernando López.

[xlvii] Fernando Sánchez Marroyo, Professor of Contemporary History, University of Extremadura.

[xlviii] Dolores Romero. Interviewed by Moreno Gómez during the 1980s in Espejowhere the family moved permanently because after being released com jail, they were banished from Villanueva. It would be impossible to gather all these reports today. The transition period forced historians to lose no less than 30 years of information, to History's misfortune.

[xlix] Rafael Bedmar Guerrerro, November 1983. Unpublished memoires. He later self-published a second version of these: 1936- Memorias de una guerra (Memories of a War). Cordoba, 2007, p. 45.

[l] Rafael Lemkin. Special Report presented to the 5th International Conference for the Unification of Penal Jurisprudence. Madrid, 14-20 October 1933.

[li] Idem. Axis Rule in Occupied Europe: Laws of Occupation – Analysis of Government Proposals for Redress. Carnegie Endowment for International Peace, Washington D.C. 1944, Part II, Chapter IX.

[lii] Gregory H. Stanton. 1998. Stop Genocide. (Yale University Cent). Translated by Diana Wang and presented as Ocho estadas de genocídio (Eight stages of genocide) at the US Department of State, Washington D.C., 1996. Gregory. Stanton is the founder (1999) and Chair of the International Campaign to End Genocide

[liii] José Francisco Luque Moreno. Op. Cit.

[liv] José Brizuela. In: Mirta Núñez Díaz-Balart et al. Op Cit., p. 179.

[lv] Francisco Poyatos López. Recuerdos de un hombre de toga (Memoires of a man wearing a jurist's robes). Cordoba, 1979, p. 151. Self-published.

[lvi] José Murillo Murillo. Numerous intrviews with Moreno Gómez in Madrid.

[lvii] Joaquim Sama Naharro. Summer 1983. Two morning-long interviews by Moreno Gómez in Cordoba. He was exceptionally clear-minded and did not mince his words.

[lviii] Bartolomé Cabrera Peralbo, November 21, 1985. Interviewed by Moreno Gómez in Pozoblanco, and subsequent intreviews.

[lix] José Torralba Rico. Vidas secretas. Memórias d'un militant clandestí (Memories of a clandestine militant). Mansesa, Barcelona, Centre d'Estudis del Bagaes, 2009, p 32.

[lx] Antonio Bautista Romero, December 27 1985. Interviewed by Moreno Gómez.

[lxi] Beatriz Blancas Pino, May 10 2002. Interviewed by Moreno Gómez in Valencia.

[lxii] Antonio Reseco Rivera. Letter from the victim's grandson, sent August 26 1989, from Barcelona.

[lxiii] Rafael Lemkin. Axis Rule, Op. Cit.

[lxiv] Antonio Barragán Moriana. Control social y responsabilidades polítics. Cordoba (1936-1945). (Social control and political responsibilities. Cordoba (1936-1945). Cordoba, El Páramo, 2009, pp. 119 et al. This study includes a list of the 'economic' edicts pp. 315-316.

[lxv] Francisco Moreno Gómez. 1936: El genocidio franquista en Cordoba (1936: The Francoist genocide in Cordoba), Barcelona Crítica, 2008.

lxvi Antonio Barragán Moriana, Op. Cit., p. 125.

lxvii José Francisco Luque Moreno, Op. Cit., p. 261.

lxviii Antonio Miguel Bernal, José Luís Gutiérrez Molina, Fernando Romero and Cecilio Gordillo et al. Proyecto Rapina (Robbery Project), Seville, January 23 2011. Investigative project created in Seville by the Recovery of the Historic Memory of Andalusia Group, with a view to studying the thefts, confiscations and seizures by Francoists beginning July 18 1936.

lxix Natividad Rodrigo. Testimony collected in Villanueva de Odra by the Recovery of the Historic Memory of Andalucia Group under the Robbery Project. Published by Patricia Campelo in 'Sin muertos, tampoco hay culpables' (If there are no dead, there also are no guilty persons). In: Memoria Pública, March 27, 2012.

lxx Manuel Álvaro Dueñas. 'Control político y represión económica' (Political control and economic repression). In: Mirta Núñez Diaz-Balart et al. Op. Cit., p. 260.

lxxi Antonio Barragán Moriana. Op. Cit. pp. 204-206.

lxxii Dueñas, Op. Cit., p. 261.

lxxiii Antonio Barragán Moriana. Op. Cit. p. 164.

lxxiv Idem, p. 145.

lxxv Rafael Lemkin. October 1933. Vth Internaional Conference for the Unification of Penal Jurisprudence. Madrid, 14-20 October 1933.

lxxvi Idem. Axis Rule in Occupied Europe, Columbia University Press, New York, 1944.

lxxvii Idem. "Genocide", American Scholar, April 1946.

lxxviii Miguel Hernández. Obra completa. III Prosas. Correspondencia (Xomplete work. III Prose. Correspondence). Madrid, Espasa-Calpe, 1992.

lxxix Ana Tudela. "Hambre, cartilla y estraperlo: España no come escrúpulos" (Hunger, ration cards and black market: Spain does not eat scruples). Published on the Internet at http://www.Publico.es April 2 2009.

lxxx Petra Robero Hiuertas. Eyewitness report recorded by her son Bartolomé Pozuela, Summer 2011, in Villanueva de Córdoba.

lxxxi Ana Tudela, Op. Cit.

lxxxii Maria Victoria Fernández Luceño and José Maria García Márquez. Fallecidos en el campo de concentración de Las Arenas (La Algaba, Sevilla), (Deaths in the Las Arenas, La Algaba, Seville, concentration camp), El Mundo, Seville, 2013.

lxxxiii Mirta Núñez Díaz-Balart et al., Op. Cit.

lxxxiv Francisco Moreno Gómez. Córdoba en la posguerra. (La represión y la guerrilla, 1939-1950) (Post-war Cordoba. Repression and the guerilla, 1939-1950). Cordoba, F. Baena, 1987, pp. 296 et al.

lxxxv Córdoba newspaper, May 10 1042.

lxxxvi Interviews collected in Cordoba with the assistance of Rafael González Roldán and numerous correspondence with Moreno Gómez.

lxxxvii Córdoba newspaper, 28 September 1940.

lxxxviii Idem. 4 December 1941.

lxxxix Nueva Enciclopedia Larousse. Barcelona, Plante, 1980, Vol. 8.

xc Daniel Sueiro and Bernardo Días Nosty. Historia del franquismo (The history of Francoism). Madrid, Sedmay Ediciones, 1977, p. 239.

xci Ricard Vinyes. 'El universo penitenciario durante el franquismo' (The prison universe during Francoism). In: Una immensa prisión (An immense prison). Barcelona, Crítica, 2003, pp. 170 et al.

xcii Antonio Barragán Moriana, Op. Cit. p. 186.

xciii César Laiana Ilundáin. Oral testimony. Basque Documentary, Op. Cit.

xciv Antonio Barragán Moriana, Op. Cit. pp. 314 and 204-206

xcv Idem. p. 237.

xcvi Idem. p. 304.

xcvii   Manuel Álvaro Dueñas. 'Por derecho e fundación: la legitimación de la represión fran-
         quista' (Rulings of the courts: the legitimation of the Francoist repression). In: Mira
         Núñez Díaz-Balart et al., Op. Cit., p. 124.

xcviii   Antonio Barragán Moriana, Op. Cit., p. 254.

xcix   Regional Court of Political Responsibilities. Seville, January 14 1942. Document giv-
         en to the author by Eugenio Jurado's family

Campo de Concentración
de
MIRANDA de EBRO

MIRANDA DE EBRO CONCENTRATION CAMP
Built June 1937, the first of almost 200 concentration camps
created throughout Spain under Franco. Last operational
concentration camp, it was closed, January 1947.
Originally built to house some 600-700 members of the International
Brigades, the only camp used to incarcerate non-Spanish nationals
trying to escape from Nazi occupied Europe. During its lifetime,
housed 60,000 prisoners of war of various nationalities.[i]

II

CRUSHING THE VANQUISHED
(A) A CRIME AGAINST HUMANITY.
THE CONCENTRATION CAMPS – ANTI-
ROOMS OF THE MULTI REPRESSION:
ARRESTS, SUICIDE, THE CHURCH
AND NATIONAL CATHOLICISM

*"Those whom you now see exhausted, battered, angry, downcast,
unshaven, filthy dirty, destroyed, are nevertheless – don't forget it son,
never forget it, no matter what happens – Spain's finest, the only
ones who truthfully rose without anything but their bare hands,
against fascism, against the military, against the all-powerful...
each in his own way, to the best of his ability..."*

Max Aub
Campo de los almendros, 1968

The Concentration Camps, First
Circle of Franquista Hell

The penal universe was the hub of the fascist repression and in the descent to the Francoist hell, the concentration camps formed the first great circle around this hub. The defeated suffered a first traumatic shock when they were confronted with the strictures of the New State and a second one when they learned what their role was to be under the new regime. First, as captives in the concentration camps and later, as prisoners in the jails, where they were subjected to indescribable suffering, a tragic ordeal of the kind that is universally recognized as a crime against humanity. The Rome Statute of the International Criminal Court defines a crime against humanity as 'as act committed as part of a widespread or systematic attack against any civilian population, with knowledge of the attack,' namely 'imprisonment or other severe deprivations of physical liberty in violation of fundamental rules of international law',[ii] and other acts.

History tells us that mass concentration camps were invented by an associated fascist regime in Germany following Hitler's rise to power in 1933, when Dachau and Buchenwald became operational. June 1935, the number of political prisoners in Germany rose to 23,000 but dropped to 11,265 at the end of 1938. Following a process of admissions and releases in which they were involved, before and after this date, the number of Socialist or Communists totalled 150,000.[iii] These numbers pale in comparison with the half a million individuals 'concentrated' by Franco in 1939 alone.

Franco is most often ranked as the leader in the fraternity of 20th century European tyrants. The repressive philosophy was the same: the terrifying texts of Generals Mola and Queipo and other high-ranking Nationalist military officers were fully comparable to Hitler's tirades against Jews and Communists.[iv]

Especially interesting is what Eric Hobsbawm had to say regarding the purely political nature of the Rome-Berlin-Madrid Axis repression, one that had nothing to do either with 'common crimes' or 'political crimes of rebellion', in that Socialists and Communists knew that the principal objectives of a fascist regime was to exterminate them. This was why, when the wolves of the fascist regime were set against them, the first concentration camps in the Third Reich would be designed as a place where the Communists could be safely kept under lock and key.[v]

Although the fascist repression in Spain was a solely political extermination, Franco took this one step further. To the eradication of his political opponents, he added the punishment and humiliation of an entire defeated army.

The dark phenomenon of mass imprisonment, with the attending grief and suffering, facilitated several Francoist objectives following the defeat of the Republican army. To begin with, it enabled the extensive classification procedure by which Franco would be able to obtain a census of his political enemies. Next, it served as an instrument that provided various means for attaining other objectives directed at promoting the moral defeat and 'cleansing' of everyone who did not support the movement, those whom it designated as disaffected:

physical destruction: punishment, torture, starvation and, in many cases, death by firing squad;

political destruction: eradication of the entire political or trade union education dating from the beginning of the 20th century and that a great part of the social base of the Republic had acquired by means of many struggles and many doctrines, in other words the annihilation of all the disaffected in the country.

moral destruction: through terror, by which a defeated individual is pushed to the limits in situations intended to destroy a person's coherence, his dignity and self-respect; to force the prisoner or the incarcerated to negate his inner self and abjure his principles. In quite a few cases, Francoism managed to do this. In many others, when faced by truly indomitable spirits, it did not; and

repudiation and reconversion: the nature of which was clearly ecclesiastic and inquisitorial. The practice of repudiation was instituted by the Church during the Inquisition with the *autos de fe*: if the prisoner abjured his 'heresies', he was put to death before he was burnt at the stake; if he did not, he was burnt alive. The Church had considerable experience in repudiations, torture, reconversions, atonements and crucifixions, an entire range of evil practices that were resurrected in 1939. In effect, the Church and its dozens of chaplains working in all the centres of confinement, were totally involved in fulfilling this fourth objective. Their goal was to force the prisoners, not just to express their love for the Caudillo and Totalitarian Spain by using the Fascist salute, singing Fascist hymns and revering the Francoist flag, but also to embrace the principles of National Catholicism by way of the sacraments, compulsory attendance at Mass, confession, taking communion and participating in other religious ceremonies.

The extent of this *physical, moral, ideological and religious assault* was terrible for those against whom it was directed, some hundreds of thousands of men who came from another, totally antagonistic world. Never, during the expelling of the Jews and the Moors, and all others whom fanatic Spain banished from within its borders throughout History, never were so many persecuted by the Church and Francoism in such a terrible manner as were the Republicans.

As stated earlier, the concentration camps were the first encounter the defeated had with Fascist Spain, camps like those that the Nazis had employed since 1933 to segregate and terrorize their opponents, the starting point for the chain of repression. Furthermore, not only was this a form of administrative criminality that was an integral part of a Government project, but the proceedings themselves were unlawful. Hundreds of thousands of men were confined without having committed any crime, without having been lawfully indicted, without having been sentenced by any legitimate court of law.[1] The entire population of the camps was not a population of prisoners serving sentences for crimes; they were there simply to be purged of their 'disaffection' with the regime.

The number of concentration camps in Franco's Spain rose rapidly throughout the entire war. The first camps were officially created by a 5 July 1937 Order from the Secretary of War of the self-appointed Burgos Government. By the end of that year, Franco had already taken 106,822 captives.[vi] The Concentration Camp Inspectorate was created in 1937 and placed under the command of Colonel Luís de Martín Pinillos. If we add the number of camps created March, April and August 1938 (especially in Northern Spain, which had already succumbed to Franco) to the existing camps, we arrive at a total of 72.[vii] In July 1938, the number of incarcerated totalled more than 166,000.[viii] February-March 1939, the number of prisoners in these camps rose to 367,000, to which we must add the 140,000 men taken captive at the end of the war, after the so-called Victory Walk. All these numbers add to a final count of *507,000 individuals incarcerated in camps in Aril 1939*, but the actual final figure was undoubtedly greater. Joan Llarch reports a total of 700,000.[ix] We can therefore estimate that more than half a million men were held captive during the month of Franco's victory; a shocking number. Furthermore, if to this total we add the more than 270,000 (underestimated) individuals incarcerated in Franco's prisons in January 1940, plus those who came and went afterwards, *Franco's Spain incarcerated a million men and women*, not on any given date, but in successive periods during his reign. Spain had become an immense jail. It was as if the order of the day was: 'Everybody, off to jail'.

Javier Rodrigo computes the final number of permanent concentration camps in Spain at 104, plus numerous temporary camps, for a total 188.[x] This number, however, is not complete as

---

[1] The decrees and laws enacted by Franco's regime to justify its actions, were a mockery of the true meaning of due process of law and thus should be considered illegal and illegitimate.

some camps were left out and must be added. For example, he failed to include a temporary camp in Paseo de la Estación, Villanueva de Cordoba, with some 1,000 prisoners, and two more in Cuenca: one at the site of the Seminary containing more than 2,000 inmates, and yet another one in La Serrería. There may be more.

During the war in Cordoba province, additional concentration camps were created in Fuente Obejuna, Cerro Muriano (listed by Joan Llarch as containing 15,000 ex-combatants, although a Jesuit priest whose work he quotes, Rev. Delgado Iribarren S.J., says that this was the maximum capacity of the camp, not that this was the actual number of inmates), Montilla (600), Aguilar de la Frontera (300), Cabra (300), Lucena (300) and Cordoba capital (600) inmates.[xi]

At the end of the war, the larger concentration camps in Cordoba province were located in Valsequillo, La Granjuela and Los Blázquez. When they began to overflow, the great majority of Cordoban prisoners were sent to the notorious Castuera camp. In an earlier book on post-war Cordoba,[xii] Moreno Gómez referred to a total 5,000 prisoners in the Valsequillo concentration camp and 21,000 for all three camps: Valsequillo, La Granjuela and Los Blázquez, a number taken from the work of Rev. Delgado Iribarren S.J., also cited by Joan Llarch in 1978[xiii], but still an estimate.

More recently, Antonio D. López Rodríguez published documented data in his magnificent book on Castuera.[xiv] He states that in the immediate post-war period, 17,000 men were imprisoned in the three Cordoba province camps mentioned above. They came mainly from Cordoba capital but also from Badajoz and from elsewhre, such as the comedian Miguel Gila from Valsequillo who ended up in La Granjela. In his memoires, he described how here there were 12,000 inmates who were taken from the camp every day to where they had to go and dig. Only after they had finished their work, were they given their only meal of the day: two sardines, an ounce of chocolate and two figs.[xv]

Again, according to López Rodríguez,[xvi] the Nationalist Army Corps of Extremadura and Cordoba, stirred by the 'sweet trumpet sounds of victory', captured 62,900 prisoners (37,000 in Extremadura and 25,000 in Cordoba province). The 60th Division took charge of the prisoners in the North of Cordoba: 7,500 in Valsequillo, 8,153 in La Granjuela and 1,342 in Los Blázquez. This shows just how enormous these concentration camps were. The number of inmates varied greatly as many were vetted and others shifted out elsewhere several times a day. For example, 10 May 1939, 2,500 prisoners were transferred from Los Blázquez to Valsequillo. At times, La Granjela held as many as 20,000 inmates.

This was just one aspect of the great humanitarian catastrophe created by Francoism and the regime's avowed intention that no one should be allowed to escape its net. Nobody could ever have imagined just how brutal the victorious armed forces would be. Today, although the academic world avoids discussing the great repression, the historical facts are undeniable, even when they are manipulatred. That a country should tolerate this is totally inexcusable. Today, as the victors continue to demand respect for their fallen, it does not occur to anybody to ask for forgiveness for yesterday's victims.

López Rodríguez also tells of the huge numbers of prisoners captured by the Nationalist Tagus-Guadiana Group of Divisions. A tragically spectacular scenario: half of Spain imprisoning the other half, the so-called 'magnanimous' doing of the Caudillo. April 5, 1939, these Divisions had more than 32,000 rank-and-file soldiers in custody (45,000 twenty days later), plus an additional 18 commanding officers, 700 general officers, 93 administravive officers and sundry non-commissioned officers.

In Badajoz, not including Castuera, April 22, 1939 there were 12,000 prisoners in the so-called Extremaduran Siberia: 3,874 in Caserio Zaldivar and Casas de Don Pedro; 4,290 in Siruela; 651 in Fuenlabrada de los Montes; 502 in Castilblanco; and 2,543 in Palacio del Cijara. In the case of Siruela, in accordance with the segregation and exclusion orders of the moment, the prisoners were removed from within the town and packed into the neighbouring farms of Las Lanchas and La Pachona. The purpose of this major 'disinfection' was to 'quarantine' them to forty years, not forty days, seclusion.

Let us return to Cordoba province. The La Granjela camp was swamped by wave after wave of incoming prisners. Survivors of this period who have been interviewed, all agree that they suffered considerably more from hunger during those days, than from the cold. The victors sent no supplies of any kind to the camps for several days. The first week was one of total fasting. Several days later only, the prisoners received a bit of bread and the famous can of sardines. As Miguel Molinero told Moreno Gómez,[xvii] he and three companions

100

were captured in Regalón, where they remained for three days, without any food. Afterwards, marching under a heavy guard, they arrived in Pueblonuevo del Terrible. They spent a night in an old damp house and the next morning, still without any food, they were marched to La Granjuela where they were given a bread roll and a can of sardines, to be shared among four of them, and then ordered to dig a two and a half metre-wide ditch around the town, surrounded on both sides by barbed wire. They began eating wild grass and carob bean flour. The victors often did not send any food to the camps for weeks on end. Juan Pulido Cantador also told how when he was taken to La Granjuela at the end of the war, they were kept without food for the first six days. When the first food arrived, there was such chaos that the guards sprayed the inmates with their machine guns, and some died.[xviii]

In addition to the lack of food and forced labour, the prisoners were subjected to visits from local Falangistas who would come to the camp with orders permitting them to remove individuals who, after they left the camp, were tortured and shot.[xix] In another letter to Moreno Gómez, Miguel Regalón added that in Paraleda, Los Blázquez and Valsequillo, where he was also interned, Falangistas from Pueblonuevo, la Añora, Dos Torres and El Viso, would present themselves to the head of the camp with authorizations to remove inmates whom they would take to the neighbouring hills and shoot. Whenever they could leave the camp to gather firewood for the kitchens, they saw the deaad and the dying, and once, three hanging from an oak tree.[xx]

One of the most descriptive testimonials of the prisoners' lives in the concentration camps is that provided by José María Carnicer who was taken prisoner at the end of the war and ended up in Castuera, a humiliating hell like so many other camps. He was interned there for three months until he was released on a good behaviour document his father obtained for him, June 1939.[xxi]

"Castuera consisted of a large yard with, in the centre, some flagpoles where the flags of Spain, the Carlists and the Falange waved. The inmates' huts were arranged around the yard. On one side, the kitchen and on the other, the latrines, which were no more than trenches that the inmates had to dig beforehand. If we needed to go to the toilet at night, we had to call the guard and a soldier with a fixed bayonet would accompany us. The camp was surrounded by a double row of barbed wire, a trench 3 metres deep and 6 metres wide, then another double row of barbed wire. And finally, all around the camp, machine gun posts.

At sunrise and reveille, the inmates had to fall in immediately in front of their hut; if we did not, the sergeants would enter through the windows and lash us with whips. Once we had fallen in, we were made to sing the hymn of the Legion, of the Falange, and the *Oriamendi*.[2] We were then marched out of camp, two by two. On one side of the gate there was a mountain of pickaxes and on the other, a mountain of shovels. There we remained doing forced labour until noon when we returned to camp to eat.

Depending on the day, we got either cold or hot food. Cold food was a small can of tuna. When it was hot food, we went to two by two and were given a ladle of water and some chickpeas. In the evening we got either cold or hot food, the opposite of what we had been given at lunch. As far as bread was concerned, we were given less than half a roll. Every hut was allotted 30 or 35 rolls but as there were 80 men in each, we each got less than half a roll.

As far as personal hygiene was concerned, we were never given any water in which to wash and as most of us had no razors, we just let our beards grow. As we could not wash our clothes, they were a mess. As far as bugs were concerned, in addition to being abundant, there were all kinds and colours. One of our pastimes was racing body lice.

When we first arrived at the camp, the guards were Falangistas. When we went out to work, there was a guard for every 8 prisoners. In addition to making sure that we did not escape, another one of their duties was making sure that we did not sleep on the job. These guards always carried rifles. A month or a month and a half later, the Falangistas were replaced by soldiers, and we all benefited from that."

These, and more personal testimonials and letters regarding the horror of the concentration camps, especially the largest, Castuera, Valsequillo, La Granjela and Los Blásquez, are reproduced in their entirety in Appendix I. In every one, the authors speak of the agony of their capture and transfer to the camps and the humiliation and

---

2   The March of Oriamendi. The hymn of the Carlist movement.

mistreatment they suffered once they got there, especially the filth and the pursuit.

Other Cordobans were distributed amongst concentration camps elsewhere in Francoist Spain. In Albatera, Alicante, the following began their ordeal with no hope of salvation: Miguel Ranchal, Socialist, Mayor of Villanueva del Duque (later executed in Barcelona); José Cantador Herros, another Socialist from Villanueva who had served as secretary to the Carlota town council, executed in Paterna; Francisco Jiménez García, comunist, Mayor of Espejo, broken by torture, committed suicide when imprisoned in his town jail. So, it was throughout Spain. Delegations of Falangistas, Guardia Civil and extremists hunted down Republican leaders from their towns. They cast their nets far and wide, just to punish them. The Republicans who remained would fall little by little.

A Falangist delegation went to the Alicante concentration camps in search of some 'heavy weights' from Montilla. There, they captured two Militia commanders – Juan Cordoba Zafra and Manuel Alcaide Aguilar, as well as Manuel Sánchez Ruíz, the Mayor, and Francisco Merino, a town councillor, among others. They took Francisco Hidalgo from Santa Barbara castle.[xxii]

One of the many disappeared in Castuera was the Mayor of Zafra, José González Barbero. No sooner was he imprisoned in the concentration camp that Falangistas from Zafra discovered his whereabouts and sent a party to the camp to get him. They removed him from the camp and executed him in a neighbouring field around April 29 1939. He remained 'disappeared' as his family was never informed of what had occurred, until his death was recorded in the Zafra Registey of Deaths only some ten years later.[xxiii]

The lunatic persecution and frenetic hunt were so brutal and schizophrenic that 500 defeated from a single Cordoba village, La Rambla, were released from no less than 50 different concentration camps all over the country, most of them during the Black Spring of 1939. The abridged list of camps from which almost one hundred of the released were removed is an indication of the enormity of the repressive neurosis.

The notorious concentration camp in the South that most interests us here, is Castuera (Badajoz). The victors began building this camp March 1939 as soon as they began to get an idea of the huge number of persons they would be imprisoning. So high was this early number, so great the unexpected need for a large camp, that the victory bells were still pealing yet the camp was only half built and not particularly crowded as camps go, as on April 22 1939 it housed 6,000 inmates, although more than 20,000 prisoners passed through it between April and September. López Rodríguez says it contained 84 sheds, but some eyewitnesses said there were as many as 92. It was built out of sight of the city of Castuera, in accordance with the rules for segregating and separating the townspeople from the camps so that punishments could be carried out of sight of any possible witnesses, just like the Nazi camps. Two images were posted all over the camp: the crucifix and the flag.

Overcrowding, torture, starvation, infrahuman conditions, physical and verbal violence, bad weather, terror, removals and firing squads were the norm and attest to the severity of the repression. The degree of punishment was much greater than anyone could have imagined. It terrified the inmates who made desperate attempts to escape.

It is important to remember that these inmates were not common criminals, they had not been charged with any crime, nor had they been sentenced by any court of law. They were quite simply 'the enemy', the 'disaffected', whose disaffection began here as they came into contact with the terror, with the Francoist intent to destroy their self-esteem once they had been defeated by the insurgent military, Franco's desire to imprison the entire Republican Army.

Because these inmates were irregular and unlawful prisoners, they were considered exempt from all formal rules and guarantees under law. In other words, they were subject without recourse, to the unconditional and arbitrary whims of those who held them captive. The most heartless hardened criminals were placed at the head of the camps. This was nothing new as since 1936, this was the usual Francoist practice when appointing local military commanders and others responsible for keeping the peace.

In Castuera, in keeping with the dictum of 'power to the most criminal', the newly-appointed head of the camp was Ernesto Navarrete, Guardia Civil commander in Fuente de Cantos who, in 1936, had already ordered the extra-legal execution of 307 individuals.[xxiv] That this kind of criminality was approved under Franco and encouraged by the Falange, is confirmed by the numerous promotions and rewards that Navarrete obtained throughout his career, including his promotion to Brigadier General in 1957 and his being decorated by Franco with the Grand Cross of Saint Hermenegildo[3]. Navarrete's policy in Castuera was to let Falangistas do whatever they wanted, although he was never loath to roll up his sleeves and, given half a chance, beat inmates half to death. He himself programmed the removals and the disappearances.

Under Francoism, every missing person was considered a 'disappearance'. In the beginning, Falangistas were the first to go to the camp looking for fresh meat. When the Guardia Civil took over the job, they became the actual executors of the *paseos*[4] and of the disappearance of inmates whose bodies 'fell' into the wells of some of the surrounding Somoza and the Gamonita mines. Lópes Rodríguez states that all the camp's files have disappeared. The Civil Registers are also totally mute as to the extra-legal executions by firing squad that were frequently carried out in the municipal cemetery, yet another example of the Francoist 'disappearance' strategy. Today, those who persist in denying that Nationalist Spain did shoot the defeated, who insist that individuals whose names are missing from the records just 'disappeared', demand proof in the form of lists or records when they know full well that none exist. As if the people of Badajoz, who have already suffered so much, needed any more aggravation, they still have to contend with the leader of this pack, a fervent supporter of National Catholicism, author of countless articles and a Church historian, Rev Martín Rubio who was born in Castuera.[xxv]

The camp's very first task was to make a general classification of the prisoners, the first step in the general census of the repression. Prisoners were divided into four categories: A, B, C and D. The first two groups were considered 'recoverable'. Group C consisted of Republican leaders and officials, administrative officers and political and trade union leaders. Prisoners in Group D were those also accused of presumed criminal acts. Groups C and D were sequestered in a separate shack (number 70 say some; number 80 say others) and subject

---

[3] The Royal and Military Order of Saint Hermenegild is a military distinction designated to serve as a maximum reward for soldiers who exceeded their military obligations and would serve as examples of their bravery. https://en.wikipedia.org/wiki/Royal_and_Military_Order_of_Saint_Hermenegild

[4] Executions under the unlawful *Law of Escapes*.

to 'removals' and to 'disappearing'. On occasion, the latter inmates might be removed by the Guardia Civil and taken to their hometowns, where they were court-martialled[5] and, in most cases, executed by a firing squad. Neither was life a bed of roses for Group B inmates as they provided the cannon fodder for the beatings, punishments, starvation and sufferings hat often also resulted in their death.

Antonia González and Pablo Ortiz, Instituto de Zafra teachers, presented the extensive and extremely informative testimonial of a survivor, Rafael Caraballo Cumplido, at the 2003 Barcelona Congress on Concentration Camps and the World of the Penitentiary in Spain during the Civil War and under Francoism.[xxvi]

Rafael Caraballo Cumplido was imprisoned in Siruela church and then sent to Castuera April 13, 1939. The first things he reported were the beatings, such as the ones that broke the Mayor of Puebla de Alcocer's spine, and random shootings such as that of a young man from his hut, whose mother had come to see him. He jumped through one of the windows at the back of the shed to giver her a letter, but he was seen by a Legionnaire who shot hi in the back. As they watched. He died where he fell.

Rafael bitterly commented on the Church's role in the middle of this inferno whereby Mass was said in the camp, and everyone had to attend and sing at least five or six hymns. When one of his companions complained that he didn't give a damn whether he lived or not, one of the Siruela priests commented that they should all envy those who were killed.

As to the behaviour of the guards… there was a winch next to the command post where the inmates went to make depositions, and it was used as a gallows where many died. The guards would get drunk and enter the huts, pick some of the men out and take them away, beat them with rods and bring them back in pieces. Some were so unmercifully beaten; it would have been better if they had been killed. Navarrete, the Commander, turned a blind eye. When the camp was dismantled around February 20, 1940, the prisoners were sent either to Castuera or Herrera el Duque, then Puebla de Alcocer, one after another. Rafael was only nineteen years old at the time.

---

[5] Regardless of whether the accused were military or civilian, these courts were always military, not lawfully constituted civil Courts of Law.

Moreno Gómez mentioned how the only way a prisoner could be released from a camp was by providing good behaviour letters of reference from persons of recognized standing in his hometown. The format of these later became more technical than the following early example, and required confirmation by the local triumvirate who controlled the lives and the property of the people: the commander of the Guardia Civil, the Falangista Mayor, and Head of the FET-JONS, and the parish priest. If a prisoner was unable to obtain the reference and could not work in the fields, he was sent to one of the forced labour battalions where he faced years and years of hard slave labour with a pick and shovel.[xxvii]

> *"The undersigned hereby declare that Rafael Cabrera*
> *Caballero, resident in the neighbourhood of this city,*
> *is a person of guaranteed probity regarding the noble*
> *cause of the Nacional Sindicalista, law-abiding and*
> *possessed of an exemplary conduct. The trustworthiness*
> *of all this information is confirmed by the fact that*
> *his relatives have suffered great losses and that they*
> *decry the atrocities committed by the Rojos.*
> *In witness thereof, we sign this declaration in Pozoblanco,*
> *April 2, 1939 III Year of the Triumph*
> *Signed: Antonio Moreno Muñoz and Antonio García.*

When writing this, Moreno Gómez wondered whether today's readers might think he was exaggerating the Church's involvement, but the fact is that young as he was at the time, he remembers having to go to the parish priest to get a Certificate of Good Behaviour for some primary school activity. He also remembers his mother having to pay the priest to get permission to eat meat during Lent (streaky bacon and blood sausage, the only meat they had at the time). These and many other facts are true, no matter how long ago they occurred and how much negationists try to get us to disbelieve them. Because memories are often clouded by the fog of time, before writing of such events authors should perhaps begin with a warning: These descriptions of the past are so strikingly painful that they may hurt some people's feelings.

April and May 1939 were the most terrible months in Castuera camp, as everywhere in Franco's Spain. Included with his descriptions of the beatings, removals of individuals from hut 80 and the disappearances, Antonio D. López Rodríguez describes an execution at the camp in May 1939 that Benjamín Gallardo has never forgotten.[xxviii] At 11h in the morning, some persons who were on their way to visit prisoners crossed the path of the 'truck of death' that was taking some prisoners from the camp. The truck stopped next to the Somoza mine a few kilometres from the camp. The prisoners got off the truck, the guards shot them and threw their bodies into the mine shaft. The unfortunate accidental witnesses who had hidden themselves, got away as soon as they could and fled in terror to Benquerencia.

There is another thing that must not be forgotten. Under Franco, it was the families who were generally considered responsible for providing food and care for the prisoners, not the State. Many relatives of camp inmates from Cordoba, Extremadura and other towns and villages, made superhuman efforts to travel to the camps to their kin. This explains the numerous humble families who, from 1936 onwards, could be seen travelling from jail to jail with food for their relatives. The number of these travellers rose dramatically during the first post-war period. One of the great tragic spectacles of Spain under Franco was of hundreds of women in mourning, often accompanied by young children, walking up and down roads towards the prisons and the camps. Prisoners who were unable to get support from their family usually did not live to tell the tale, as the weak broth of nettles, turnips or carrots that the victors fed their prisoners provided very little sustenance.

More about this later, but for the moment, suffice to imagine the sight of people of all ages, sizes and shapes, trekking around the Castuera concentration camp trying to see and give something to their relatives. Here, too, punishment and repression were extended to the families as the jailors took advantage of the situation. Wardens and guards promised to hand over food, clothes or money to the prisoners, in exchange for sexual favour from their relatives at night-time. Falangistas and drunken soldiers did what they wanted with the families who spent the night camped around Castuera. López Rodriguez tells of an interview with an individual who had been a Falangista guard and who, when remembering how they treated the women who visited the prison, broke into tears. At last! A Falangista who showed some remorse over what he did. Still, a river of tears is not enough to atone for the great humanitarian catastrophe.

To give one an idea of the widespread ill-treatment, López Rodríguez records Ángel Sánchez Santos' heart-rending personal account. Ángel was only eight years old when he accompanied his mother to the camp one day as she took some cigarettes and oranges to his father. The camp was surrounded by a double row of barbed wire and a trench. About 8 metres across in all. At that distance they communicated by calling out to the prisoners on the other side, but because of the combination of the sound of voices and the distance, they practically could not hear each other. Because he was so little, Ángel scrambled under the first row of barbed wire, crossed the trench and then crawled under the second row with the items they had brought his father. As he was returning and had already gotten past the first row of wire, a field guard appeared with a revolver and a whip, and he began to beat Ángel. The field guards were close to each other on the inside and there were more guards on the outside. The man still could not forget the way that the bastard beat him. He was enormous, he was nicknamed *Mulato* or *Javilla* and he was from Zaamea de la Serena. He came at the boy with his whip and beat him furiously until Ángel almost lost consciousness. His father, stood there on the other side, watching everything, unable to speak out. Ángel was finally able to get back to his mother. He supposes that the things he had managed to give his father were taken away from him later, but what he wanted to tell López Rodríguez was about the beating that Mulato gave him. If that is what they did to an 8-year-old, what were they doing to those whom they held captive?

Numerous oral testimonies and eyewitness accounts and other observations, clearly demonstrate that the concentration camp guards executed and 'disappeared' many more inmates than one believes or have been spoken of. In Castuera and everywhere else, witnesses all agree that prisoners were frequently called out from the huts and most simply disappeared, although we know that many of these were executed, as witnesses outside the camp have attested. How can one compute this slaughter? In towns or villages, where people knew each other, Moreno Gómez and other researchers have been able to compile incomplete lists of names not recorded in the Civil Registry. But in the camps, in chance or unknown locations, where not even the prisoners knew each other, they cannot even begin to make these lists, especially as the Francoists erased all traces of their actions wherever they could and Civil Registries do not mention these deceased, For now, however, we must agree that it has been clearly established that countless inmates were slaughtered in Franco's camps, especially during April and May 1939.

ANTEROOMS OF THE MULTI-REPRESSION

"What is incredible is that this can continue that these assassins of the devil are allowed to remain unpunished... That for almost forty years they have tormented the Spanish people, without any shame or humility. The historical verdict shall be terribly severe."
Olaf Palme, *1975 xxix*

Hunting and Capturing Rojos

After so many years of loss of memory and induced amnesia, of dismemory, very few Spaniards today really know what the military coup was all about, what the war was like, nor what Francoism stood for.

Filipism and Carlism wagered on forgetfulness, the so-called fallacy of progress: look to the future and forget the past. With so much emphasis on forgetfulness, Spaniards are left with very little on the horizon for the future and almost nothing regarding the past. Dismemory explains why today nobody really knows who was and why so many action groups and individuals are striving to restore those memories today.

Francoism made its mark on our history, and it did so in a terrible way, at the cost of the humanitarian catastrophe that Moreno Gómez is at such pains to reconstruct. Francoist terror was no less that which is said about it; it was so much more than that.

"We must kill kill, kill. Do you know that? Our program consists of exterminating one third of the male population of Spain."

Thus spoke the aristocratic Lieutenant-Colonel Gonzalo de Aguilera y Munro, 11th Count of Alba de Yeltes, one of Franco's press officers, a man who knew very well what Franco thought and said.xxx Earlier, we discovered what Francoist concentration camps were like and the first shock with which the defeated were confronted upon their return home at the end of the war.

With victory, Franco and his supporters had only just begun to put his formal program into effect. The phenomenon of the concentration camps the Francoists employed as a means for their repression was, alone, a thousand times worse than the comparatively few Republican arrests in 1936. On its own however, the tragedy of the camps is too great to enable a proper comparison of both sides' actions. Right-wingers who were arrested by the Republican government following the uprising (the role of the *checkas*[6]) never suffered such torments, nor anything that was even remotely close.

The Falangistas had become so involved in implementing the first part of Franco's agenda – eradication of all who opposed Francoism – that they eventually realized that they were neglecting all the plans they had laid for the National Trade Union Revolution, the so-called Glorious Revolution.

The Francoist victory bells were pealing throughout the country and the incitement *Go, and capture Rojos* rang in everybody's ears. The victors, inflamed with a desire to exterminate their opponents, did not even stop to savour their victory; directing all their energy to setting the slaughter in motion.

In 1939 and 1940, the media provided the public platform for this persecution mania. Every day, the Falangist press, most especially *ABC Sevilla* and *Diário Azul* of Cordobaxxxi, published detailed reports of arrests in the towns and villages of Cordoba province. This showcasing acted as an unrelenting advertising campaign for the regime, whose effect was to add fuel to the repressive machine everywhere. Such a spectacle of accusations, denunciations and captures would certainly not have occurred or been permitted in the Republican press during

---

[6] The *checkas* were extra-legal courts set up by some leftist political parties and trade unions and, although responsible for the deaths of many right-wingers during the first months following the coup, were falsely accused by Francoists of having assassinated hundreds of thousands of individuals in areas of the country that supported the 18 July 1936 miltary coup.

the period immediately following the coup that Francoists called the Red Terror, when tempers ran high. Yet another of the many differences between the two regimes.

On the contrary, in 1939 the Francoists campaign of punishment and extermination was carefully expanded and orchestrated by the press that, against the name of every individual arrested, added a list of supposed crimes, none of which had been either corroborated or judged in the Courts. Another Francoist juridical irregularity was the flouting of the rule of the presumption of innocence unless proved otherwise, as individuals were considered guilty from the onset, just because they belonged to a Republican political party.

The *Diário Azul* newspaper could not contain its delight at the continuous arrests in Cordoba, as when it headlined the May 5 1939 news item rejoicing at the capture of Alfredo Caballero Martínez, ex-councillor for the Frente Popular in Cordoba, ex-director of the Metalworkers Trade Union and provincial secretary of the PCE[7] since October 1936. Remarkably, such was the chaos in the implementing of the repression, that Caballero was one of the few who escaped being executed. Two days later, the paper reported on the execution by firing squad of the past commander of the Garcés Volunteer Battalion, Manuel Palos Cosano, another ex-director of the Metalworkers Trade Union.

These news items were not in the least informative; they were entirely accusatory. Rather than just informing its readers, the Francoist press cheered and encouraged the repression. Suffice to see how *Diário Azul* reported actions in Castro del Rio, bad-mouthing, if not outrightly slandering, those who had been arrested:

The Castro del Rio Guardia Civil has arrested the following notable extremists who continue to speak their minds after returning from the recently liberated zone:

- Francisco Carpio Algaba, age 25, bachelor. Used violence to force several landowners to give him great sums of money and later voluntarily joined the 85th Red Brigade.
- Gabriel González García, age 38, member of the CNT union. Enlisted with the Rojos and arrested José Puebla Centella, whom he tied up hands and feet, robbed and imprisoned.

---

[7]   *Partido Comunista de España.*

- José Rinoso Ortega, age 32, secretary and member of the CNT. This individual, under the orders of the Red Council of War, was commander of the fighters at the Bujalance-Baena crossroads barricade He later became a member of the Andalusía-Extremadura Red Militia, with the rank of Lieutenant. October 21, 1936, he and his fellow fighters[8] attacked Castro del Rio, entered the town and got as far as Calle Casas Nuevas.
- Rafael Navajas Rosa age 30. Enlisted with the Red Army and was appointed responsible for the stolen cattle that he later sold for 69,000 pesetas. Enlisted in the 215th Red Brigade that was garrisoned on the Teruel front...[xxxii]

On other occasions, the headlines reflected the newspaper's disrespect for the prisoners and its support of the widespread practice of informing on individuals, a practice fanatically followed by the entire right-wing sector of society.

PINKO DENOUNCED [xxxiii]

José Payán Porcel informed the Commissariat of an individual named Francisco Lanauro García, age 41, who, according to the informer, carried a rifle in the service of the Red cause at the beginning of the Movement in the village of Villaharta, which required his wearing Republican badges on the lapel of his coat. He has been placed at the Civil Governor's disposal.[xxxiv]

As a result, simply because he was wearing Republican badges, Francisco Lanauro began a long ordeal of imprisonment and torture. His ultimate fate is unknown.

Another article appeared in June announcing that Falangista services in Cordoba capital had captured an anarchist from Torres Cabrera. Manuel Lucena Padilla was accused of having entered Torres

---

[8] The official Francoist report described them as guerrillas, although there were no 'guerrillas' at the time, just volunteer fighters who attempted, unsuccessfully, to liberate Castro del Rio from the Francoist troops who had taken the town.

Cabrera with 50 militiamen, pillaging the town and arresting Juan Bautista del Rosal Luna, whom they killed down by the river, near Santa Crucita.[xxxv]

The *ABC Sevilla* daily was, as a rule, like all the Francoist press, every bit as engaged in publishing news reports of these tragedies. The following article was headlined in April 1939:

IMPORTANT ARRESTS

Cordoba, 21. The Information Brigade of the FET-JONS has been doing a praiseworthy job of performing law enforcement services of great importance.

Police officers Crespo and Pinazo arrested an individual named Joaquín Moreno Herencia, born and resident in Castro del Rio, who is charged with having been President of the Trade Union Party before the beginning of the Glorious Movement. In this position, he dedicated himself to all kinds of pillaging and arrests of noted individuals, among them the Inspector of Security and the government delegate in that town, Sebastián Velasco, who as the object of the Red fury, was tormented and assassinated, Joaquin Moreno Herencia was Head Executive of the Grocery Store Committee.

Also worthy of note is the aforementioned brigade's arrest of the female President of the Libertarian Youth of Cordoba, Consuelo Fernández Sánches, the 'Cordoban Passionaria'.

In May, *ABC Sevilla* published the news that the Information Brigade of the FET-JONS had arrested Manuel Bellido Moreno, a high-ranking member of the Republican SIPM[9] active in the Southern zone, between Porcuna and Villa del Río. He, Rafael Galesio, Pedro Sánchez and Ramón Cano Moya, were together accused of having

---

9   SIPM - *Servicio de Información e Policia Military* – Military Police and Intelligence Corps.

murdered Lieutenant Colonel Luís Pastor Coll, commander of the Nationalist 1st Regiment of the 34th Division.[xxxvi]

A few days later, *ABC Sevilla* reported that the Falange Brigade had arrested a common criminal, Manuel Velasco *El Esparter* Munõz, who had killed a neighbour in 1934 and was serving a sentence in Ocaña in 1936. Also arrested was Francisco Jurado *El Recovero* Gutiérrez, accused of participating in the incidents at Belalcázar as a Lieutenant in the Republican Militia. The same newspaper continuously published similar news reports, especially regarding events that occurred in Cordoba province, but goodness know why only here, as identical hunt and capture programs were actively pursued everywhere in Spain.

October 31, 1939, the examining magistrate in Peñarroya determined that Juan Peñas Pérez, from Hinojosa, was guilty of 'belonging to the 193rd Brigade of the Marxist Army', and Pablo Herruzo Nogales, of belonging to the 69th Brigade.

The Hunt and Capture Rojos Syndrome continued to inflame Francoists during the entire Year of Victory, when these kinds of unfounded accusations were consistently made against prisoners. December 1929, *Diário Azul* reported that Lisardo Cano Fernández, age 32, from Pedroche, was arrested in Cordoba capital for having participated in the murder of Antonio Cabrera, a priest from Pedroche.[xxxvii]

In a news report regarding Villaviciosa, the paper reported that Rafael Cuevas Alcaide was arrested for having escaped from Pozoblanco jail where he was being held. Taking advantage of the dark and the rain, he went to his family's house in Villaviciosa, but the Guardia Civil found out and he was captured.[xxxviii] The commander of the post at the time was Sergeant Romualdo Reyes who in 1936 had led the insurrection in Adamuz. When the uprising in the village was quelled, Cuevas was released from jail.

December 19, the paper published, with its usual delight, the capture of Miguel Justo Sánchez Garrido, age 33, who had been a member of the local War Committee and had fled into the countryside. Rural Policeman José Hernández received a reward for capturing him. Miguel Justo ended his days viciously tortured and facing a firing squad in the Pozoblanco cemetery, yet his Civil Registry record states that he died of a brain haemorrhage.

This family was one of those which suffered the most in Pozoblanco. His father, Justo Sánchez, was shot under the Law of Escapes on 6 June. His sister Josefa lived for two years under a sentence of death. His wife Juana Serrano León was condemned to 12 years in jail. His brother-in-law Juan Casijo González fled to France but was captured by the Nazis there and transported to Mauthausen concentration camp where he died. The only members of the family who remained at liberty were Miguel Justo's mother and his baby daughter, although his mother was stripped of all her property and household goods.

Also published that December was the chief of police in the capital, José González Lara's announcement that Bernabé Cano Delgado from Adamuz had been arrested because he had killed the village telephone operator's son aat the beginning of the war.[xxxix]

Municipal police and Falangistas competed with each other in hunting and capturing Rojos., flinging accusations right, left and centre to improve their standing with the Regime and receive medals and rewards. The Falange group in Cordoba capital and the Police Commissariat worked without respite. Policemen Aparício Romero, Heredia Espinosa, Ruíz Molina, López Linares and González Lara, to mention but a few of the more notorious, laboured night and day to ensure that no eminent Republican remained hidden in the city. At the ed of the year, *Diário* Azul published all sorts of libel and insults against José Álvarez Torquemada, a Cordoban mechanic who, after having fled Cordoba in 1936, had helped Antonio Jaén Morente organize the Cordoba Volunteer Battalion in Madrid.[xl]

The climate of persecution continued with the same intensity in 1940, in keeping with the regime's ongoing extermination program. January, *Diário Aul* published the arrest of Rafael Muro López – *Carasucia* (the repressors always enjoyed giving nicknames to those they captured), a Freemason and member of the JJ.LL[10] born and resident in Cordoba, on the grounds that he had been involved in burning the altars in San Agustín Church on the evening of July 18, 1936, a fictional event that is not recorded anywhere. He was also accused of supposedly having killed Dios Rios, a Guardia Civil, on

---

[10] JJ.LL *Federación Ibérica de Juventudes Libertarias* (Iberian Federation of Libertarian Youth).

Calle Empedrada, before the war. He was arrested by a municipal policeman, Rafael del Olmo Garcia.[xli]

October, the paper headlined the news that a high-ranking Marxist in Belalcázar had been captured. This was no less a person than Antonio Vigara Regidor, schoolmaster and ex-president of the *Casa del Pueblo*[11] who had been living in hiding in his house for the past year and a half. Vigara Regidor was one of the great intellectuals who helped create the Republic, a wise and venerable man, whose only crime was his ethos. He was arrested by a municipal policeman, Fernando Ballester, at the height of the plague of informers and accusers, Lieutenant Ortega and Sergeants Escobar and Rodrigues were sent to capture him, on orders from the Spanish Legion which at the time was garrisoned in that town The right-wingers in Belalcázar could not contain their delight.[xlii]

November, more news from Belalcázar- The police in Cordoba arrested Higinio Amaro *El Mocho* Rodriguez for committing blood crimes in Belalzázar. August 1939, after right-wingers in Belalcázar launched a series of attacks and counterattacks in support of Franco, Republican troops were sent down from Madrid to quell the uprising. Although there were many dead on both sides, no fewer than 158 victims were Francoist insurgents. As a result, the Falange accused the entire Republican sector in the town, almost without exception, of blood crimes.[xliii] In 1941, we note yet another example of the hunt and capture of the defeated, with the report of the arrest in Jaén, of the 'the head Rojo in Peñarroya, Eutimio Romero González, President of the Communist Party', who was accused of every killing in the mining zone.[xliv]

While the press launched its propaganda campaign proclaiming Francoist successes in arresting the defeated, the military government published numerous judiciary decrees and edicts threatening to declare that all the defeated were to be treated as rebels. In effect, the BOPC[12] records dozens upon dozens of subpoenas from the military courts that were operating in Cordoba capital and province during 1939 and 1940, as well as numerous search and capture warrants from respective examining magistrates and military judges.

---

[11] Socialist community social club.
[12] Official Bulletin for the Province of Cordoba.

March 13, 1940, the examining magistrate in Fuenteobejuna, Eloy García Pérez, issued a warrant for the search and capture of Manuel Caballero Pizarro from the village of Cuenca, a village in which Francoists had previously killed a many people in 1936. The examining magistrate of Cordoba capital, Fernando Sepúlveda Courtoy, issued a similar warrant for José Martínez Santiago from Posadas. This individual, another of the lay Republican saints, had served as town councillor in 1931 and Mayor in 1936, was a member of the Unión Republicana and highly respected Master of the Abril Masonic Lodge in Posadas. In other words, his entire curriculum made him a perfect candidate for the National Catholicism scaffold. Already October 1936, when Guardia Civil Lieutenant Colonel Bruno Ibáñez was subjecting Cordoba to a bloodbath another crime against humanity, the Catholic newspaper *El Defensor de Cordoba* came forward to add fuel to the extermination fire with this headline: 'Freemasonry, that is the enemy'.[xlv] *Don Bruno*, as Ibáñez was infamously known, took note and liquidated every Mason he could lay his hands on.

In El Viso, José Romero Peñas, the local examining magistrate, cornered the arrest of several individuals who had escaped from Santa Eufemia, including Manuel Fernández – *El Secretario*, Norberto Castillejos – *Veneno*, and Luís Blanco Martín, who had fled to the mountains where they survived until after 1947.[xlvi]

Earlier, August 4 1939, the examining magistrate in Belalcázar, Manuel Márquez Rubio, exhausted himself ferociously issuing warrants and searches all over the place when many prisoners he had condemned to death escaped from the Divina Pastora school in Belalcázar, where they had been imprisoned. What followed could have been a scene from a movie. Fifteen escaped, three died whilst escaping, and another was captured (he was latter shot in November). The remainder fled to the hills, and some became famous for their anti-Francoist activities during the following years.[xlvii]

The judge continued to hunt high and low for Dionisio Castellano *Zabarza* García, Luís Hidalgo *El Huevero* Escribano, Eduardo Bejarano *El Portugués* Medina, Artero Paredes *El Fiscal* de la Cruz and Ángel Torrico *Nene de la Carmela* Garcia. These men did not go like so many sheep to the slaughter; they paid dearly for their lives.

One of these, Ángel Torrico's story could be the subject of another movie. On the evening of September 18, 1947, his group of guerrillas was surrounded by the Guardia Civil in the Las Hoces mountains of El Viso del Marqués (Ciudad Real). During the fight, all except one (*Pedro el Cruel*), were fatally wounded. When Sergeant Ruano rapidly moved closer to his prey to watch them die, Ángel Torrico, although already in his death throes, summoned enough energy to shoot and kill the bloodthirsty Ruano. Both died: executioner and victim. At least this Republican was avenged in death.[xlviii] Not content with this bloodshed, the Belalcázar judge continued with his hunt for Manuel García *Quivicán* Peco and Francisco Paredes *El Bizco* de la Cruz.[xlix]

In Obejo, warrants were issued August 9 1940 against two members of the Barrios family: Bartolomé Barrios Herruzo and Luciano Sánchez Barrios. In Villaviciosa, Julio Nevado Martínez and Gabriel Pozuelo *El Gato* Expósito, were subpoenaed September 11 1940. Meanwhile, many of the hunted who could, fled to the hills.

The brutal pursuit of the defeated by Francoists led many individuals to take extreme measures for survival, such as those who camouflaged their and their family's identities and lived under assumed names during the forty years of the dictatorship, Such was the case of the ex-president of the Villaralto War Committee, Manuel Muñoz *El Preso* Sánchez, an active leader of the local Socialist Youth who had been imprisoned after the October 1934 strikes, and was released from prison following the Frente Popular elections.

Villaralto Falangistas unsuccessfully searched high and low for Manuel Muñoz after 1939, but he had obtained false documents for himself, his wife and son, and they were living in a discreet and nomadic fashion. They first went to Los Blázque where they worked making charcoal pellets, then they traded tobacco on the black market. Finally, they settled as farmers in the province of Malaga until the whole family moved to Salamanca, where Manuel Muñoz died. Shortly before his death, he revealed the family's true identity to his children. His widow and children returned to Villaralto after the death of Franco to join relatives and obtain correct identity documents.

A similar case was that of the only survivor of the *Los Jubiles* voluntary militia unit, José Moreno *El Quincalllero* Salazar, born in Bujalance, an 18-year-old youth who managed to escape from prison in Cordoba capital not long before he was scheduled to be hung, by pretending to be one of the bricklayers who were working there. In 1987 Moreno Gómez found him in Osa de la Vega, Cuenca, where he was living under the pseudonym of Antonio Pérez Sánchez. They met and he recounted his experiences; his oral testimonial is recorded in Moreno Gómez's book on the guerillas.

Suicide, a last resort against Fascism.

The relentless harassment of the defeated also fostered another tragic phenomenon: suicide as a last resort against Fascism. One of the consequences of the defeat and the wave of repressive measures was the considerable rise in the number of suicides, which is estimated at some 30% above the average for previous years. Moreno Gómez remembers that at the beginning of the 1950s there were frequent reports of people committing suicide by hanging or throwing themselves down wells. Today this is almost an unknown fact.

The dictatorship's incarceration of half of Spain, exclusions of all kinds, the social vacuum, beatings and humiliating situations, the lack of any means of survival, the absence of a future and drama of broken families, painted such an unbearably gloomy picture that a good many people were driven to take their own lives; not only during the hard times immediately following the defeat, but for decades afterwards, during the 1940s and 50s, until the explosive phenomenon of mass emigration provided an escape valve.

Many suicides occurred during the first weeks of victory as a last protest against fascism, because of the despair of having been defeated and the impossibility of finding a means to leave the country. Everyone has heard of the shocking scenes in the Port of Alicante where numerous ex-combatants, officers, administrative officers and Republican authorities, preferred to shoot themselves rather than fall into the clutches of the Francoists. Witnesses have told how every so often you would hear a shot on the quay, and someone would fall lifeless.

Later, as defeated soldiers and civilians painfully made their way back to their hometowns during the Black Spring of 1940, many contemplated suicide – as told to Moreno Gómez by the eyewitnesses he interviewed. Other desperate individuals jumped from bridges, threw themselves before trains or shot themselves with the last bullet in their guns. Persons who had suffered reprisals from leftists in times past, never had to endure such terrible moments, a fact that there is no possible comparison of either regime's reprisals.

From the very beginning of the insurrection in 1936, the family dramas that the advent of Francoism created in many homes added to the number of suicides, principally in Andalusia. In his book on the Francoist genocide in Cordoba,[1] Moreno Gómez tells of Manuel Landrove Pouzo, a municipal employee belonging to the UGT who was executed by firing squad in Cordoba, September 3 1936. After returning home from the cemetery where she had gone to recognize his body, his widow hung herself but not before she had hung each of her three children. The widow of Manolo Reyes, an elegant gypsy of Cordoba who was executed because he was recognized in a photo with Fernando de los Rios, committed a similar act. Dr Carlos Zurita, acting as a coroner at the time, told Moreno Gómez that his widow looked at him with such hatred at the cemetery that he will never forget. The day after her husband was shot, he learned she had hung herself after first hanging her seven children, one by one.

Although Moreno Gómez did not wish to make an exhaustive study of this subject in the province, he felt that some stories were necessary to give readers an example of the extreme despair that engulfed the disaffected during the immediate post-war period. It was at this time when the most suicides occurred in the home and in jails in both the province and the capital. This was a direct consequence of the brutal repression and torture meted to the imprisoned, particularly to individuals of some standing in pre-civil war Spain.

Gabino Cabrera Expósito, Communist leader from Villanueva de Cordoba, Captain in the famous Garcés Volunteer Battalion and member of the War Committee in 1936, committed suicide by throwing himself down a well on the evening of April 24 at Number 5, Calle Iglesia (known as La Preturilla, later Bar Balagón, today a supermarket), the headquarters for the Nationalist SIPM and the Guardia Civil, under the command of Corporal Martínez Galiániz. Broken by ten days of savage beatings while waiting for his turn to be beaten yet again, he dragged his bloodied body towards a well in the patio and threw himself down it. His widow, who was still living in 1987, told Moreno Gómez that when she and the family went to the cemetery to recognize his remains, his body was one entire wound. He had been hung from his feet and his screams could be heard outside, in the street. Gabino's sister Francisca Cabrera was arrested because she left her house dressed in mourning for him; they shaved her head and forced her to drink castor oil before releasing her.[13]

Francisco Jiménez García, was Mayor of Espejo in 1936 and a Communist who was highly respected by the workers, before and during the war, when he belonged to the Villanueva de Cordoba Provincial Committee of the PCE. When Franco declared victory, he was arrested in the Port of Alicante and sent to Albatera concentration camp. Falangistas from Montilla and Espejo went there to get him. February 27 1940, he was subjected to such continuous and terrible beatings in Espejo jail that he took his own life, although some witnesses say that he was beaten to death with sticks and then hung to simulate a suicide. One of his torturers was nicknamed *Mata alcaldes*, meaning 'Killer of Mayors'.

In Puebloblanco, Aniceto Villarreal Jurado, ex-councillor for the Frente Popular, was one of the Republicans most hunted by the Falangistas. They were unable to capture him alive because he first committed suicide by tying himself to some rocks and throwing himself into the cemetery water tank. In reprisal they shot his brother Eleuerio and his nephew Pedro Villarreal Moreno.

Another famous case in Pozoblanco was that of Rafael Bueno Roldán, a Republican doctor aged 53, who died in the local jail November 13 1939, after refusing medical assistance. Officially, he died from septicaemia. He had been condemned to death.

In Dos Torres, another much hunted individual was Eulalio *El Preso Medrán,* for his anti-insurrection activities in 1936. He evaded capture July 14 1939 by taking his life in a small hut on the edge of the town. In Puente Genil, Francisco Delgado Morales slashed his wrists in jail, but they stitched him up and then shot him November 6 1939.

Fear of the influence of the Church often heightened the feeling of despair such as when Bartolomé Rey Torres hung himself August 1 1939 because he was terrified that he would suffer reprisals for having

---

[13] When Gabino's sister discovered that a historian was writing on these subjects, she met frequently with Moreno Gómez and spoke freely about the events during that period, no matter how painful the memory of the suffering she and her family had endured. She belonged to that class of defeated whose moral integrity remained strong and could never be discredited, no matter how much the victors tried to erase the historic memory of the times. Moreno Gómez is honoured that she attended the launching of his first books.

been involved in burning the San Sebastian Religious Brotherhood's flag three years earlier.

In Castro del Rio. Where the CNT was dominant in the pre-war government, the number of prisoners tortured and executed created such a climate of terror that Moreno Gómez knows of at least three suicides in the local jail between 1939 and 1940. Felipe Agulera Arroyo, a farm worker, committed suicide May 11 1939 by putting his head into a prison latrine; either that or he collapsed in the latrine, headfirst, after a torture session.

There were several executions in Castro del Rio at daybreak December 5 1939: six prisoners, led by Manuel Castro Merino - *El Abogado de los pobres*,[14] were shot against the cemetery wall. One prisoner, Francisco Torronteras García, who had also been condemned to death, decided that he preferred to die in bed, so he cut his wrists with a razor during the night. His body was found the next morning. Another suicide, José Sánchez, put an end to his life and to the torments he had been suffering by throwing himself down a well in the Las Monjas Convent prison, August 18 1940.

In Villafranca, Francisco Jurado Fernández, one of the great heroes of the Republic and a member of the Villafranca Volunteer Battalion, chose an extreme means to end his life. He had decided that the only way to end the beatings he was regularly receiving was to pretend that he was running away from the guards as they led him to another interrogation. When in the middle of the morning June 9 1939, he was on his way to make yet another statement and receive the subsequent beating, he ran down the street, handcuffed, knowing that there was no possible salvation for him. There was none; they shot him in the back under the Law of Escapes.

Not all took their lives in prison. In Añora, Antonio *El Zumbo* Oviedo, was repeatedly summoned to make statements at the Guardia Civil barracks and then beaten, until one day, after receiving yet another summons, he hung himself in his bedroom at home.

When another member of the Villarelteño family in Villanueva de Cordoba that had suffered so greatly, Hilario Gómez Luna, returned home in 1946 under parole, he took his life unable to bear the loss of so many members of his family: his son in jail, his father, two of

---

14 Affectionately known as 'the people's lawyer'.

his brothers and a brother-in-law executed. It was just too much for a human being to bear.

The Cordoba Provincial Prison was a cemetery of living men. Examination of the list of prisoners who died from starvation and privations (no less than 756 individuals died in this way, 25% of the prison population in the 1940s), indicates that several bear the hallmarks of suicide. This agrees with eyewitness reports of prisoners taking their own lives by hanging themselves or leaping off the top floor of the jail onto the gallery below. Unfortunately, the inability to contact relatives of the deceased makes it impossible for Moreno Gómez to confirm the accuracy of the following 'official' causes of death, adding to the infinite number of unknown tragedies for which there is no closure.

- Francisco Priego Parado: fracture of the top and base of the skull, August 29, 1941
- Bernardino López Morales, from Villanueva del Duque: skull fracture, April 1, 1943
- Joaquín García Lázaro, from Adamuz: rupture of the liver and pulmonary injury, October 17, 1944, possible indicative of torture as the contributing cause.
- Rafael García Gutiérrez, from Villanueva de Cordoba: skull fracture, April 11, 1945
- Francisco Romero Paredes, from Belmez: fracture of the spine, June 7 1946.

Although there were other frustrated suicides in Cordoba prison, such as Juan Sánchez Pozuelo, from Villanueva de Cordoba, who after being condemned to death, threw himself from the top floor, did not die from the fall and was still executed by firing squad. It was not unknown for the prison authorities to issue their own version of an inmate's death, preferring to call it suicide to cover up their own culpability. In Almodóvar del Riom the official version of Ángel Plazuelo Lozano's death in 1939 was that he committed suicide by throwing himself down a well. According to fellow prisoners however, he died from a beating and his body was then thrown down the well to simulate a suicide.

Years later, there was a similar official version for the death of Vicente Mudarra Cañete, from Almodóvar, ex-Socialist councillor in

1931, who was said to have committed suicide in the Dueso, Santander, jail by throwing himself off the fourth-floor gallery. However, whether this was suicide or murder remains as much of a mystery today as it was then, because he was scheduled to leave prison on parole only a couple of days later.[li]

Referring to a case outside Cordoba province, there is the tragedy of José Gómez Osorio, the last Republican Governor of Madrid, who was executed in February 1940. His entire family – wife and children – among these the well-known Socialist Sócrates Gómez, were imprisoned in different jails. No longer able to bear such misfortune, his 19-year-old younger sister took her own life.[lii]

Moreno Gómez hopes that this small, but significant sample, will give readers an idea of the number of lives that were lost in Spain from suicide, so damaged were the minds of those who took their lives because of the military coup, the war, the extermination, the terror and the unbearable living conditions that Franco's military forced upon the defeated and their families. It seems that the Four Horsemen of the Apocalypse had been sent against half of Spain. We have glimpsed only the tip of that monumental iceberg. History will never be able to describe its true magnitude.

Caídos por Dios y por España
¡Presentes! Francoist battle cry

The 1939 orgy of repression was fuelled by another vindictive ingredient, the Francoist battle cry: *All those fallen in the line of duty for God and Spain. Present arms!* There undoubtedly also existed a component in Franco's plans demanding a furious vengeance that was more Judaic than evangelic in nature. Even so, one must be careful and not make assumptions because the Francoist repression was so much more than a matter of revenge.

In the first place, because the persecution had been programmed from the very beginning, before there was anything to avenge. Nothing had occurred in more than half of Spain – Galicia, Navarra, Soria, Zamora, Salamanca, the Canary Islands, etc., that had to be avenged, yet these regions were not exempt from the extermination. The Francoist repression clearly pre-dated any feeling of revenge and was at the margin of any vengeance. If there had been occasional incidents in some places that could be used by Francoists as a pretext for getting even, so much the better for the program.

During the post-war period, part of the Francoist propaganda paraphernalia included banners calling for revenge for those who had fallen in the line of duty for God and for Spain, waved in profusion on every public occasion. There is another aspect that needs to be made clear: the expression *Caídos* represented all the Francoists who had died for the Nationalist cause in battle, as well as in every way as right-wing victims of Republican, or Rojo, actions against Franco and the regime. There was no reason to include Heaven in their cry, but as Francoists started the fire and there was a risk that some of them might be singed, it was wise to include the religious aspect. Just in case. After all, one cannot expect to launch a military coup where everyone is unscathed.

Peter Anderson, at a recent conference, defended a thesis according to which the repression and punishment were a consequence of the existence of Caídos (which he mis-translated as *martyrs* instead of *fallen*, as martyrs was not the standard interpretation of the term in those days).[liii] According to Anderson, the repression was accomplished in revenge for ad fuelled by memories of these 'martyrs'.[15] This conclusion, although applicable within the confines of his study, could lead to historical misinterpretation because Anderson's work did not contemplate, let alone explain, the Francoist extermination on dates and in places where there were no Caídos (Galicia, Castilla la Vieja, Navarra, Canary Islands). Such an oversight could mislead one to conclude that first there was a Republican repression, the cause, followed by the Francoist repression, the effect thereof. In fact, the Francoist repression was planned to cover the entire country, with or without any Caídos, as occurred in numerous towns and villages where the Left never bothered anyone belonging to the Right.

The truth is that it was exactly the other way around. There is no possible hypothesis that can excuse Francoism from its nature as the aggressor or deny that at all times, all supporters of the Republic were the prime target of Francoist aggression True, there were victims on

---

[15] Anderson's study of part of the Pedroches district, Cordoba province, in villages where some Francoists died as a result of the so-called Rojo Terror, to a certain extent influences his conclusions. Despite the misnomer, this is not to say that the work itself does not address some very interesting issues that are worthy of reflection.

both sides, many more on one than on the other, but there is only one reason for that: this was a military coup. Had there been no coup, there would have been no battles, no dead no fallen or Caídos, on either side.

True, in some places the Francoist repression was more ferociously enforced than in others, as it was fuelled by the funerals and memorials to their fallen. There were numerous post-war exhumations and reburials, with the ceremonial transfer of coffins from one place to another, all of which excited the Francoists and added an emotional aspect to the repression, but only in some places. Undoubtedly also, post-war calls for punishment from angry relatives of Caídos served to expand the vindictive social base of Francoism, those who were accomplices in the general repression, The repression was orchestrated from above and from below; it was the widespread phenomenon of social complicity that totalitarian regimes always encourage.

Celebration of the memory of the Caídos became an instrument of unification between the affected base and the supporters of the totalitarian State. It created a network of demagogic loyalties, rooted in the heart of Francoism, ready to attack the demonized defeated with full force from every direction. It was a gigantic culture of punishment, not one of forgiveness, that eventually superseded the official program of extermination that had been the fundamental principle of the Fascist Movement since 1936.

So that the repressive impetus should not flag as time passed, what better than to continuously remind the people of the Caídos. Immortalize the memory to perpetuate the punishment «*So that the death of so many will not have been in vain; always alert and vigilant, in memory of the Caídos*» as Spaniards were repeatedly told. Memory that called for revenge. On the other hand, today we ask for a democratic memory: *truth, justice and reparation* was the call raised by Carlos Jiménez Villarejo and others in Madrid, September 19, 2012, in the presence of ex-Judge Baltasar Garzón, in an appeal for the creation of a Committee of Truth as recommended by the United Nations.

Reactionary Spain, that for decades fuelled the memory of its fallen with a spirit of revenge, today cries out with prejudice as the heirs of defeated Spain attempt to disinter existing and newly discovered mass graves and bury their dead. Closure in the name of a Democratic Historic Memory. The Church adds to the intolerance of any memory of the defeated every time it evokes the memory of the Caídos and calls for them to be beatified and canonized. The Roman Book of Saints is a constant exercise of memory.

Nowadays, society and some general public appear to believe that although memory of some is welcome, remembrance of others is not. Post-war reprisals were verbally fuelled by continuous allusions to 'victims of Rojo brutality'. In Cordoba capital, José Maria Herrero was the single victim of the Right, (as compared to 4,000 loyalists murdered in the city by the insurgents). The Cross to the Caídos on the La Malamuerta tower is a monument to the only Caído in the city. Inside the Mezquita-Cathedral, one can still see a large stone plaque bearing the names of 102 members of the clergy 'murdered by Communism' in the province. Many years had to pass before the first loyalist Walls of Remembrance, engraved with the names of several thousand individuals murdered by Francoism in Cordoba, could be erected in 2011 in both the city's cemeteries.

The post-war Francoist press marked the anniversaries of the 1936 anti-insurgency fighting in Cordoba province by publishing elaborate obituaries with the names of Caídos. These occasions were celebrated everywhere with outdoor memorial masses perfumed by gunpowder and incense – the crucifix and the sword – presided over by local dignitaries: movement leaders, the Mayor, representatives of the Falange, the armed forces and the Guardia Civil, clergy, the local military commander and more... The twelve o'clock Sunday mass in San Marco Church in ultra-Catholic Lucena was a great Nationalist spectacle. First came the military parade (volunteers, Guardia Civil, Falangistas, municipal police and paramilitary militia), followed by a band playing martial tunes, everyone singing *Cara al Sol* and *Oriamendi* and shouting Viva! to the Virgin of Araceli, the awesome pomp and circumstance of the Church. Again, the crucifix and the sword, the scent of gunpowder and incense, bonding with a common purpose. The most surprising was that, as the religious processions left the church, the population greeted them with the raised arm of the Fascist salute: anyone who did not do so was strong-armed by believers or Falangists. Of all the priests in Lucena, all able practitioners of the Fascist salute, the one who stood out was Federico Remoro who served as chaplain in the Fascist militia and, with a revolver tucked under his belt, participated in patrolling the streets.[liv]

Meanwhile, many towns and villages were busy erecting monuments to their Caídos. July 17 1939, the Pozoblanco Town Council voted to 'erect a monument to the martyrs of the Red Horde'. September 1940, construction began on a huge mausoleum in Villanueva de Cordoba, budgeted at 80,000 pesetas, under the direction of a committee presided by Bartolomé Torrico.[lv] November 1941, in Palma del Rio, another wealthy landowner, Félix Moreno Ardunuy, inaugurated a great monument to 42 victims of the Right, paid for out of his own pocket. He probably did so to salve his conscience as it was he who August 27 1936 ordered the machine gunners on his property to mow down more than 200 loyalist townspeople who were imprisoned there. Similar monuments were inaugurated in 1941 in Pedroche, Montoro, Hinojosa, and elsewhere, mostly in the North of Cordoba province, a region that had been a Republican stronghold. In the rest of the province, where Francoist rule had prevailed since the first day of the insurrection, the monuments were erected earlier. In Montilla, a Cross to the Caídos was placed on Calle Obispo Pérez Muñoz in 1937. May 1939, Salesian monks from Montilla inaugurated the Cross of Student Caídos, inscribed 'Don Bosco Alumni – Martyrs to God and to the Fatherland'.

1939, the Year of the Victory, was marked by the constant to and fro of Nationalist exhumations and re-burials. The body of Captain Cortés in La Cabeza Sanctuary was disinterred at the end of May 1939 and re-buried with a solemn mass and military honours in Andújar. October, it was the turn of the remains of General Sanjurjo that were carried in triumph to Burgos. The most elaborate of all these ceremonies was the transfer of the remains of the dictator José Antonio Primo de Rivera, Father of the Falange, from Alicante cemetery to the El Escorial monastery. His coffin was carried on the shoulders of Falangistas, who walked non-stop day and night, in the Nazi manner. As the cortège passed through towns, it was greeted with scenes of extreme fanaticism and raucous cries for vengeance. Republican prisoners were taken from the jails and executed by out-of-control firing squads.

In Belmez, June 2 1939, there was an impressive funeral and re-burial of the remains of 41 Caídos who were disinterred and brought from the neighbouring town of Belalcázar where they had been 'assassinated'. Placed in ten coffins, a prayer vigil was organized in City Hall. The re-burial, a crowded affair, was presided by the local authorities, military and Falangistas. Later, when the regime started executing Republicans from Belmez, they were taken to Belalcázar cemetery where they were shot by firing squads as a ritual of revenge. Similar vindictive acts occurred in other places as well, such as Santa Eufemia, where Republican prisoners were executed in a mine in Agudo (Ciudad Real) that had served as a prison for several right-wing individuals in 1936.

Another packed funeral was celebrated 3 October 1939 in Villanueva de Cordoba for 20 individuals whose bodies were transferred from the cemetery in Jaén where they had been shot in 1936. The prayer vigil was held in Cristo Rey Convent church, as the main parish building had been turned into a warehouse and market during the war and had not yet been restored to its former use. A Jesuit priest from Villanueva, Rev. Bernabé Copado, ex-chaplain to Lieutenant Colonel Luís Redondo's Nationalist paramilitary militia, celebrated the funeral, a vivid demonstration of National Catholicism. Rev. Bernabé led the funeral cortège all the way to the cemetery where he pronounced a heated patriotic-religious speech in the manner of an avenging angel. This fanatic priest, an enthusiastic Francoist activist, closely followed the extermination of Republicans in Villanueva and more than once attended Francoist Council of War meetings at the Casino with landowners, Army and Guardia Civil officers, as together they drafted the blacklists of those who were to be executed. He did not even lift a finger to save a first cousin of his, Francisco Copado. The executions programmed in the Council of War commenced a few days after the mass funeral in Villanueva. According to Catalina Cantador, whose father was executed by firing squad, the day they brought the remains of those who had been previously buried in Jaén was a day of panic for the families of the defeated prisoners whom they tortured all through that night.

In Villa del Rio, the victors forced several of the defeated to dig up the graves of some right-wing Caídos, according to José Luís Torralba who told Moreno Gómez the story of his father, Bartolomé Torralba Pastilla. The war over, Bartolomé handed himself into the local authorities in the belief that although he would be arrested because he had served as an ambulance driver, as his conscience was clear that his hands were not bloodstained, he would be pardoned. The victors, however, treated him mercilessly, beating him daily until his body was one entire sore. He was forced to disinter the dead that anarchists had killed in Villa del Rio, with his bare hands and no tools of any kind, whilst they whipped him, in the full sun, without a drop of water all day. Four months later, a compassionate soldier from his village arranged for him to be transferred to Cordoba Prison where he was executed June 8 1940.[lvi]

Likewise, excited by the spirit of exalting the memory of the dead, the press echoed repeated calls for revenge without quarter and no justice for the defeated. Felciano Delgado, member of a reactionary family from Belalcázar, published the following in the Cordoba press:

> "Anniversary: Three years ago, on a hot August day, the Marxist beast entered this town of Belalcázar, annihilating everything in its path... People of Belalcázar... do not forget the lesson, if you find yourselves possessed of envy, ineptitude or despicable thoughts, ignore them and join your Caídos and shout: 'Long live Spain! Long live Christ the King! Martyrs and heroes, present arms!".[lvii]

Causa General Decree of 26 April 1940 instructed the Ministry of Justice to determine the political responsibilities of the Frente Popular Government and to initiate criminal proceedings against the 'subversive forces that in 1936 openly acted against the existence and the essential values of the Fatherland, fortunately saved *in extremis* by the Liberating Movement'.[lviii] The Causa General archives, consisting of thousands of reports from municipal authorities throughout the country, contain lists of alleged crimes committed by Republicans against people and property, intended to demonstrate the truth of unfounded Francoist propaganda claims that the Rojos had assassinated hundreds of thousands of persons who did not support the July 18 1936 uprising, the frequently proclaimed 'Red Terror'. These same archives contain reports of the multiple exhumations of those Caídos following the victory. A typical record for Almadén (Ciudad Real) states:

- 21 August 1939, were exhumed and re-buried the remains of the following persons from El Viso, Cordoba, who were shot at km 17 on the road from Almadén to Saceruela during the civil war [12 are named].[lix]

The record concludes by declaring the 'situation of persons accused by the Almadén Municipal Authorities on 20 February 1940', listing 37 individuals, 7 of whom are recorded as having been executed, 115 imprisoned, some charged in absentia and some freed. This and other records draw attention to the fact that local institutions, not just private individuals, also informed against the defeated, considerably exaggerating the facts as in this case, where 37 inhabitants of Almadén are jointly accused of having killed 5 persons.

The above is but a brief outline of the events of the Black Spring of 1939. It also illustrates how, in all of Spain, the machinery of the extermination program was prepared and controlled by the military. Everything in that great humanitarian catastrophe was military. One is reminded of the words of Felipe Acedo Colunga, a Francoist military prosecutor: *We first have to disinfect the soil of the Fatherland. This is the task – sorrow and glory – that fate has entrusted to military justice.'*

The Church
At the root of the new Totalitarian State

Franco's so-called 'justice' was at the same time military, Falangista and ecclesiastic. The tripartite grouping of the military coup: barracks, casinos and church vestries, cannot be ignored or forgotten. Some authors have written about "The Church on Franco's side", but this needs to be rectified: the Church was not an adjunct of Francoism, nor was it a complement; it was an integral part of Francoism and one of the pillars of the regime. Logically, the Church acted in other ways, not overtly but instead as this institution did best: slyly, behind the scenes, in the confessional, from the pulpit, issuing bad conduct reports, through its sermons, chaplaincies in the prisons and in the multiple ways by which it influenced, supported and incited its flock.

Franco's principal ally could be found in the Church hierarchy. The Church helped Franco validate his military coup, justifying it by describing it as a crusade, more for foreign consumption than for domestic purposes, so that uncountable numbers of Catholics worldwide supported Franco. The Church hierarchy blessed Franco theoretically and judicially, quoting Canonic Law. They blessed him physically; saturating him in incense and raising canopies over his head as if he were the Holy Grail, saluting him with right arms upraised and finally, voting in favour of his being awarded the Supreme Order of Christ[16] in 1954. The Church prostrated itself before the supreme lawbreaker of the 5th Commandment, the proponent of his own Final Solution, the greatest war criminal in the History of Spain.

At the end of his life, the controversial Miguel de Unamuno who, whatever his faults, was possessed of a great intellect, commented on the blood pact between Franco and the Clergy, describing it as a stupid regime of terror in an African, pagan and imperialistic military manner. Referring to the execution of individuals without due process of law as unjustified murder, Unamuno stated that there was nothing worse than the combination of the mentality of the barracks and of the church vestry, because traditional Spanish Catholicism was barely Christian. He predicted that the coming dictatorship would be the death of liberty and the dignity of Man. None of those who had emigrated, he said, would ever return to Spain. They would not be able to return, unless it was to live banished from their homes and reviled. "Poor Spain, poor Spain", he lamented.[lx]

The Catholic Church, well entrenched in the eye of the hurricane, spoke more of Franco than it did Jesus Christ. It was euphoric, as Leopoldo Eijo y Garay, Archbishop of Madrid, one day told Franco in the Las Salesas Church: "Your Excellency, I have never burned so much incense with as much satisfaction as I did yesterday".[lxi] The April 16 1939 front page of the *ABC* Seville newspaper was covered with articles and photos of Franco's great victory parade, including one in which the dictator's silhouette stands out, showing him under a religious canopy, surrounded by priests.

The Church did the military establishment a great favour as it provided the Francoist regime with the doctrine that it lacked. The military launched this great adventure without any set project in mind unless it was the typical barracks rhetoric of order, discipline and harsh punishments. The military establishment arrived on the scene with an empty discourse that wanted for direction. Little by little, with

---

[16] The Supreme Order of Christ, formerly the Order of the Knights Templar, was the highest order of chivalry awarded by the Pope.

the assistance of the Church, the Regime fashioned its discourse on a lot of Tridentine Catholic doctrine, a little traditional Calderonesque thought, some imperialistic notions and conservative beliefs, and a great deal of totalitarian dogma emanating from Rome and Berlin. The Church contributed with its congregations and impassioned masses, acting as if it were the great Spanish Fascist party (much more so than the Falange). This was how National Catholicism came to be founded.

Never was Spain so asphyxiated by incense and smothered by cassocks; not even during the darkest years of the Inquisition. By and large, the Church provided Francoism with the ideological leadership that it lacked, at the same time that the military took command of the practicalities of the repression.

The Church lay the table of Francoism with a wide range of retrograde material against the modernization of Spain, a modernization that had been counting on the first third of the 20th century to bring the country out of its state of underdevelopment. The Church weighed in with a Manichaean vision of the not unexpected social and political confrontation, seeing demons everywhere with every strike or slightest demand on the part of the workers. Anyone who expressed the least disagreement was branded by the Church as Godless. This was its eternal paranoia against heresy that it countered with the threat of the rack and the pyre in the background. The bishops brandished the flag of reactionary Spanish thought with a bigotry that abhorred liberalism, constitutionalism, modernity, laicism, freethinking and every new social theory introduced during the 'damned' 19th and 20th centuries.[lxii]

As Moreno Gómez stated earlier, the Church continuously prostrated itself before the Dictator as much as Franco turned to the providers of holy water to legalize his military coup The best example of this was the pomp and circumstance when the Church blessed the Caudillo in Santa Barbara Church, Madrid, May 20 1939, at the same time that the Law of Escapes was being liberally applied in the towns and villages of the Centre and South of the country.

Not in modern times had there been such pageantry! Nineteen bishops, led by Cardinal Gomá, received Franco on the stairs of the church. Leopoldo Eijo y Garay, Bishop of Madrid-Alcalá, handed Franco the silver aspergillum so that the master of lives and properties could dip his fingers in the holy water and cross himself. Next, Franco and his wife entered the church under a majestic canopy, flanked on all sides by bishops, whilst the organ blared the monarchist anthem. The highpoint of the celebration was when the Dictator placed his 'bloody' sword on the alter to the Christ of Lepanto, whilst monks from Silos Monastery who were also taking part, intoned arcane antiphonies from the Middle Ages. Franco, who could only manage to pronounce half a dozen clear lines in a row (public speaking was not his strongpoint), declared that this sword had "heroically vanquished the enemy of Truth in this century". The ceremony ended with Cardinal Gomá blessing Franco before God and History.

The following day, the newspaper *Arriba* interpreted this ceremony as the simultaneous reincarnation of Julius Caesar, Charlemagne and Emperor Carlos V, in the person of the Caudillo. The day before, another reporter, Ernesto La Orden, had dedicated a series of rhymes to that day's parade, in the style of the *Song of the Cid*. Even poor Don Rodrigo found himself enmeshed in the Francoist delirium tremens. This is what Hilari Raguer, a monk from Montserrat, called the "debauchery of National Catholicism".[lxiii] The Church, prostrate before its new idol, the new golden calf.

In Cordoba, the Church hierarchy, lacking a local Franco-like personage, had for some time bent over backwards to honour Colonel Bruno Cascajo, the Military Governor of the province, a leader of the coup and the perpetrator of the genocide in the city. The Church's worship of the regime is illustrated in an October 1 1936 photograph published by the *Diário Azul*, showing Bishop Adolfo Pérez Muñoz, surrounded by canons of the church, giving the Fascist salute as they leave an official reception in honour of a Francoist celebration.[lxiv] During these dark days of October, there was no doubt as to the Church's approval of Colonel Cascajo, as Bishop Pérez Muñoz publicly congratulated the brave saviour, a General by popular acclaim, who with a certain hand at decisive moments was entrusted with a heaven-sent mission that he accepted without reservation, defeating, as if he were another courageous and heroic David, the Red Goliath in Cordoba (....).[lxv]

It was not only the bishops who praised the winner of the civil war in those days of victory, when church bells rang out everywhere. The supreme leaer of he Church himself, Pope Pius XII, sent Franco his own messages of congratulations.

- "... Our fraternal congratulations for the peace and victory with which God has deigned to crown the Christian heroism of your faith (....). This nation, chosen by God, has just shown this century's followers of materialistic atheism, outstanding proof that the eternal values are above everything else."

This was the first of two messages Pope Pius XII sent to Franco in 1939, no greater proof of the Church's blessing of a war criminal, leader of a Fascist regime, instigator of a river of blood and a vale of tears. Not a single word of forgiveness or reconciliation. On the contrary, a call for 'justice for the crime'. Yet, when the other European Fascist regimes were defeated in 1945, Pope Pius XII had no qualms in declaring that there would be no peace until there was charity and forgiveness, sentiments that he did not extend to the Spanish defeated in 1939.

All the clergy, from the highest member of the Church hierarchy down to the lowest parish priest, with very rare exceptions, were at the service of the Dictator. The parish priest became yet another local authority: when new Francoist municipal officers were invested, this was in the presence of the parish priest as well as the local military commander and the head of the Guardia Civil. The ubiquitous good conduct references (especially 'bad' reports) had to be certified by the parish priest. Rather than being engaged in their roles as managers of Divine Affairs, priests became executives and supervisors of the repressive machine. The summary records of the military courts also contained certificates issued jointly by the victorious tripartite: the head of the local Guardia Civil, the Mayor and the parish priest. Again, the barracks, the casino and the church vestry ruled hand-in-hand.[17]

The Church made remarkable statements regarding good conduct certificates. The Archbishop of Santiago de Compostela, Tomás Muñiz Pablos, issue a circular September 14 1936 according to which

---

[17] Moreno Gómez consulted many of these summary reports in the Pozoblanco Judiciary Archives regarding indictments for Political Responsibility, that today are filed with the Cordoba Provincial Historic Archives. The immediate post-war summary reports contained a high percentage of death sentences. He consulted them when they were kept in Pozoblanco, despite some opposition from the archivist who told him that from then on, he had to forget these matters. To which Moreno Gómez replied that doubtless, he meant that that one must forget them now and re-write history later. Moreno Gómez' research was becoming a never-ending obstacle course. Francoism, which wrote its own fallacious History, embellished with bells and whistles, has always opposed to the defeated writing theirs. Power against memory.

it was scandalous that a priest might save a parishioner's life by issuing a favourable report, not that he could virtually sentence him to death with a negative one. He further instructed his priests to abstain from giving certificates of good religious conduct to anyone who belonged to a Marxist organization and in all cases, to certify behaviour 'diligently without any further deliberation, without contemplating any kind of human sensitivities'.[lxvi] It was well known that a negative report (does not go to mass, for example) implied the execution of the individual, and so it frequently happened.

Military and ecclesiastic authorities did not conceal their collaboration and greeted each other warmly when participating in public events. The clergy attended political acts as honoured authorities and the military occupied front row seats during religious acts and ceremonies. Funerals, numerous and performed with much pomp and circumstance in the days following the victory, were held exclusively for the victors' Caídos. The Catholic Church never again prayed for the defeated, yet it has continued praying for Francoists for more than half a century.

The matter of crucifixes in public places, a subject that was fiercely debated during the Republic, a democratic regime that tried to anticipate the separation of the Church and State and that which is the norm today, became a burning issue. Crucifixes were hung all over the place. Not only in school classrooms, but in the most unexpected places, including City Hall. In the latter, religious paintings of the Sacred Heart of Jesus, a typical Jesuit memorial, were commonplace. One city hall unanimously deliberated that the image of the Sacred Heart of Jesus should be exhibited with all solemnity at City Hall to make amends for the sacrilege committed in that house by the Marxist Horde and make it clear that the New State desired to inspire the Spanish people with the sacrosanct rules of the Christian Religion and to shelter all its subjects under the protection of the Sacred Heart of Jesus.[lxvii]

The above text brings many surprising reflections to mind: one, that sacrilege may be committed in a church but not in City Hall; two, that Francoism talks of subjects, not citizens; and three, that it is not up to the Municipality to preach the Christian Religion. All of this is politico-religious bunkum, something that was never preached in Spain, not even under the Inquisition.

In Montilla, the crucifix returned to preside over the Municipal Meeting Room October 1936. On the same day, priests blessed crucifixes for all the schools, as were the ones that the women who worked in the Soup Kitchen were required to wear. December 1936, blessed crucifixes were given to every soldier. November 1936, when a teacher requested that an illustration of the Sacred Heart of Jesus be solemnly hung in his house, the entire National Catholic retinue went to celebrate the event. The famous archpriest Luís Fernández Caso, in an inflamed oratory, took advantage of the occasion to exhort everyone to do the same in their homes as it was time that the Sacred Heart of Jesus reigned over everyone, everywhere. It was no longer only schools that were turned into monasteries, but entire municipalities as well.

In most Cordoban municipalities, the religious image that had been enthroned still hangs on walls. The process of dismantling Francoism has been extremely slow, in accordance with the parsimonious spirit of the transition. Even today, this image still hangs above the councillor's seats in Pozoblanco City Hall.

These images also hung on the walls of premises belonging to the Falange, such as Post Offices and factories like the Electro Mecánica. Accordingly, May 7 1942 the local press printed an article reporting that the Electro Mecánica celebrated its anniversary with a mass and the solemn hanging of the image of the Sacred Heart of Jesus, a formal act attended by the municipal authorities led by the recently retired genocide Cascajo and Féliz Romero Menjíbar, secretary to Bishop Adolfo Pérez Muñoz.[lxviii]

Clerical harassment targeted three groups in particular: all children in general, the prisoners who packed the jails, and humble people in the countryside. In other words, against the social base that had supported the Republic. As far as poor children and children of the imprisoned went, one and the same where poverty was concerned, they were placed under the auspices of the so-called infamous Social Welfare program, where they were fed but primarily, brain washed. Each roll of bread was accompanied by a class in religious doctrine in preparation for their First Communion (actually, mass christening services for children of 'Rojo atheists'. They were taught numerous New State martial hymns: *Cara al Sol, Falange, Requetés*, etc., according to a perfected plan for indoctrinating (i.e., 'disinfecting') the children of the Rojos, upon whom it was impressed that their parents were 'lost sheep', sinners who bore all the hallmarks of Hell. Democrats were fodder for the devil, Fascists were not.

The Church forced it way into and controlled the private and public lives of the entire population. All civil registry marriages were declared null and void. The Church issued fire and brimstone ultimatums for church weddings, stating that it considered civil registry marriages as no more than sinful ordinary cohabitation. At the same time, it organized a wave of mass christenings as under the New State, if a child was to have any kind of right, he had to have been baptised.

The press was a good reporter of National Catholicism harassment. One headline in the *Diário Azul* stated that 23 marriages of persons remaining in the Rojo zone were celebrated and 40 children received the sacrament of baptism in the town of Espiel.[lxix] This was followed by another headline reporting 63 marriages and 150 baptisms in a single day in Villa del Río.[lxx]

After each christening, parish records erased the 'pagan' or proletarian given names and replaced them with names from the Church-approved list. Many names under which children had been christened according to leftist tradition or European laicism, such as Libertad, Aída, Lina, Germinal, Jaurel, Bebel, Floreal, and so forth, were viciously eliminated. For example, the last four had been used to christen the sons of Miguel Ranchal, the Socialist Mayor of Villanueva del Duque. The children were renamed Miguel, Juan, Antonio and José, respectively.

A notorious case of the religious coercion and manipulation of children was recorded in Montilla, December 1936, when Ágnela Zafra, a ten-year-old girl, was forcibly christened with her teacher as godmother, followed by a celebration in her house. Ágnela was the daughter of the great Socialist leader and ex-Member of Parliament, Francisco Zafra Contreras, who had been taken in handcuffs to Baena where he perished in a mass execution of more than one hundred defeated. One of his other children, her brother Francisco, was shot in Cordoba. Ágnela's christening and First Communion, a few days later, was considered quite a coup for Montilla's National Catholics.

It is presumed that similar, long forgotten, cases occurred all over Spain. Let us continue in Montilla. Socialist Rafael Baena Cruz, imprisoned in Montilla jail, was promised that he would be saved from imminent execution if he agreed to marry in the Church. The poor man accepted and archpriest Rev. Luís Fernández Casado celebrated the wedding. As soon as the ceremony was over, Rafael was taken away and executed. A few days later, the archpriest callously christened Rafael's three children.

Rev. Casado did not rest in his mission of redeeming children of Rojos. He christened the three children of another Socialist leader, José Gama Rodriguez (executed in July 1936). He warned Antonio Luque and his wife Rosa Gómez who had married in the civil registry, that they should remarry in the Church if they wanted their daughter to be admitted to school. Another, no less rabid parish priest, Rev. Rafael Castaño, forced Manuel Aguilar to baptize his daughter so that she could attend school. Elsewhere, Araceli González, who had spent the war in Jaén and whose brothers Juan and Manuel were imprisoned in Mauthausen, where they died, had no sooner returned to Montilla after he was released than he was visited by several 'virtuous' young ladies (Sunday school teachers, daughters of famous landowners and winemakers in the region), who warned him that he had to baptize his children if they wanted to attend school and receive food cards and vouchers for municipal benefits.[lxxi] Religious blackmail, purely and simply.

Holy Sunday (Sunday after Corpus Dei) May 26 1940, 18 Republicans were executed in Villanueva de Cordoba just before the ceremonial religious procession. One of these was Eugenio Jurado, a high-ranking member of the PCE, who left a daughter named Aida. She was soon included in the group of children who were to be forcefully christened. The Sunday School teachers who prepared them acted as godmothers to the Rojo children as the Church did not allow family members who were 'contaminated' by Marxism to do so. Eulia Santos was godmother to little Aida who was rechristened Isabel and her original name erased from the records.

In practice, those who were destined to be killed and their families were subjected to al kinds of religious coercion. Antonio Cordoba Gálvez, 28 years old, was executed November 1939. He was already on his way to the firing squad when he was forced to marry in the church and rechristen his son Lenin Cordoba Polonio to Antonio forthwith.

Even the few letters the condemned to death were allowed to write, were manipulated by the prison chaplains. Carlos García Herrador, who was executed in Cordoba June 18 1941, was forced to dictate his letter of farewell to his wife and daughter to the prison chaplain, Rev. Alfredo, a Carmelite. As the chaplain wrote what he felt like, it appeared more like a letter from a nun than from a Republican facing imminent death. The priest wrote that Carlos insisted that he forgave [his executioners] with all his heart, that he wanted his wife to do the same, and that he had a message for his daughter to the effect that she should be educated in the Christian faith. The executioners felt that it was important for the family of their victims to forgive them, otherwise one day they might be attacked in a fury of revenge.

The Church's involvement in the conflict of christenings and church weddings pre-dated the civil war when, during the Republic, many employers would only employ persons who had been married in the church. With the coup in 1936, the dispute went much further and in towns that came under Fascist control, several people were shot solely because they had only gotten married in the registry office. As the Church has, as have other religions, always been obsessed with matters involving sex, priests in Cordoba capital and elsewhere in Spain drew up blacklists of individuals who had not been baptised or married in the church, or who may have had a clandestine love affair. Added to these, any perceived breaches of the 5th and 6th Commandments provided fodder for the lists that another priest, Rev. Ildefonso Hidalgo, presented to Don Bruno Casajo. This Nero-like Lieutenant Colonel of the Guardia Civil was responsible for the genocide in Cordoba, the city that he drowned in a relentless blood bath.

> **Government of the Nation**
> **Ministry of Governance**
> **DECREE of January 12, 1940, confirming the**
> **absolute prohibition of Carnival celebrations.**
>
> As the so-called Carnival celebrations have been banned for several years and there being no reason for changing the said decision, the Ministry resolves that the said ban shall be maintained and that all Authorities under this Ministry shall be reminded that such celebrations are absolutely prohibited.
> Madrid, 12 January 1940.
>
> Signed: SERRA SÚÑER

Another anathema of the Church during Francoism was the observance of pagan celebrations, such as Carnival.[lxxii]

Post-war, all traces of laicism were eliminated without second thoughts. The church wedding became compulsory; no excuse or pretext was accepted. Church regulations were published in the Official Government Bulletin, as were long lists of forbidden behaviour as in the matter of dances, the length of a woman's skirt, the wearing of short sleeves in church (girls of all ages, no matter how young, had to cover their arms when attending mass), movie theatres (lights were placed in the back rows to prevent temptation) and prostitution during Lent.[18]

No matter how tight the Church's control on the everyday lives of the Spanish people, its power over Education was even more intransigent, to the extent that the schools were run in an almost monastic manner. Catholic doctrine was a major course subject, as was Sacred History, and teachers were required to organize all kinds of devout activities. Classes began and ended with prayers. Pedro Cantero Cuadrado, Archbishop of Zaragoza, praised the 'Christian and Christianizing' work of the Ministry of Education by declaring that coeducation was banned in Secondary and Primary schools. Crucifixes reappeared, private libraries were purged, assistance was

---

[18] Also, during the Republic, there was less tolerance of extra-marital behaviour; conservative civil authorities organized 'anti-prostitution crusades'.

given to ecclesiastic universities and chapels were built in Catholic universities and schools of higher education, proof that the Ministry was no longer the bastion of laicism and now served Catholic Spain.[lxxiii]

Although children and students were favourite targets for the Church's religious brainwashing, its activities in the Francoist jails were of considerably greater concern. The prison universe was the basic theatre of operations for the Francoist repression, the focus of actions against the physical integrity of the defeated and against their moral fortitude. If the physical integrity was handled by the military and the Falange (punishments, torture, and firing squads), the Church took care of the ideological repression (brainwashing, re-Christianization and 'disinfection' of 'dissolutionary', i.e., democratic ideals) through coercion and the forced observance of religious practices.

Some ten years after the re-establishment of Democracy in Spain, the position of the Benedictine Abbot of the Valle de los Caídos sanctuary (and likewise, of almost all the clergy and doubtless also the military) continued to be clearly pro-Francoist and opposed to all the humble homages to victims of Francoism. Not long ago the Laín Entralgo family gave Moreno Gómez a letter dated August 23, 1986, that the Benedictine Abbot Manuel Garrido Boñano sent to Rev. Pedro Laín Entralgo, reproaching him for participating in the inauguration of a monument to the victims of Francoism in Dos Hermanas (Seville) August 1986. He reprimanded Rev. Pedro for extolling the so-called Rojo sovereignty, falsely claiming that they burned churches, imprisoned a multitude of persons and assassinated many thousands of innocents solely because they were priests, members of the clergy (monks and nuns) and practicing Catholics.[lxxiv]

After thirty years of Democracy, a young priest serving in the Valle de los Caídos, briefly replied to some opinions asked of him in a 2009 television documentary:[lxxv]

- What does Franco mean for Spain today?
  "He is an example for those who govern us and an example for the Spanish and for Catholics. And this must be taught and well-taught, without changing things. When history is changed, it is no longer history; it simply becomes a story."

- Who is changing history?
  "Today it is politicians who are doing so, and the media, and the schools. Many people tell me 'Look, father, I had the best years of my life during the days of the Caudillo'. And others tell me: 'I only discovered that we had been living under a dictatorship two years after Franco died, when they began to say so. I lived under a dictatorship for seven years and unfortunately have not been able to live under it for longer."

These opinions are not only those of this priest. The tentacles of National Catholicism still flourish today. Suffice to mention the ongoing integrationist activities of another priest, Rev. Ángel David Martín Rubio, the so-called 'historian' of Badajoz who continuously promotes his unacceptable ideas in his personal Blog on the Internet.

To conclude this brief description of the Church's role during the dark and very complex world of Francoism, we need to take a look at how this was seen in some Catholic sectors throughout Europe at the beginning of the civil war. Of special note is the exemplary case of an academic who had served as a canon of Cordoba cathedral during the 1920s. Rev. José M. Gallegos Rocafull was a brilliant scholar, Professor of the San Pelagio Seminary and later Professor at the Universidad Central de Madrid, where he was when war broke out. He did not delay in expressing his censure of the military coup and his support of the Republican cause. October 13 1936 in Madrid, Rev. Gallegos Rocafull and a fellow academic, Rev. Leocadio Logo, published a manifesto entitled *Palabras cristianas* [Christian teachings], where these learned scholars cite Papal and church texts in their condemnation of the uprising:[lxxvi]

- "The Church will never stop teaching the respect and obedience due to the established power..." Collective declaration 20 December 1931 by Spanish bishops.
- "...the Church has always condemned the doctrines and the men who rebel against the legitimate authorities." Pope Leon XIII, Au milieu.
- "The truth is that a few men have burdened the shoulders of the innumerable multitude of proletarians, a yolk that is

very little different to the one of the slaves." Pope Leon XIII, Rerum Novarum.

- "The economic organization violates the true order when capital enslaves the workers." Pope Pius XI., Quadragesimo Anno.

Bolstered by this baggage of authoritative ecclesiastical pronouncements, the Cordoban canon stood up to public opinion, speaking strongly against the coup and in favour of the Government, the only clergyman in Spain who dared to do so. Gallegos Rocafull travelled across Europe during the war, noting the for and against swinging of clerical opinion regarding the Spanish case and that this was by no means unanimous.[lxxvii] He was especially scandalized by the fact that priests were going to the front bearing arms, with a revolver tucked in their belt. Once again, the Spanish Church followed the Pope's lead in supporting his policy in all ecclesiastical matters, whereby infantrymen, in the Medieval manner, could be compared to Archbishop Turpin in *La Chanson de Roland*, a swordsman who with a few blows of his sabre charged against more than four hundred Saracens. During the days of the Francoist victory and the National Catholic paraphernalia, Gallegos wrote that he was pained by this abandon, this hard-heartedness, this loathing of the defeated, so alien to Christian charity.[lxxviii]

From Paris, Gallegos Rocafull was informed that Irish Jesuits were beginning to object to the actions of the Spanish clergy and had published statements such as the following, in their magazine *The Messenger of the Sacred Heart,* to the effect that the real cause, albeit an occult one, is that the war was undoubtedly the social injustice condemned by the Popes.[lxxix] Jesuit General Rev. Leodokowski's response was rapid and furious as he rushed to the pulpit to tell the Irish to shut up. As always, 'the reasons of high politics' that are always 'dishonourable' when they are not rotten and pestilential.[lxxx] March 1937, Gallegos Rocafull met with Jacques Maritain, one of the great representatives of French Catholic existentialism who greatly impressed him with his modesty, his sincerity and his sweetness, that conveyed a profound and solid spiritual life.[lxxxi] In turn, Maritain, in his prologue to Mendizábel's work on the Origins of the Spanish Tragedy, refuted the assumption that what happened in Spain was a

Holy War; that it was by excellence temporal and that all war implied political and economic interests, greed of the flesh and the blood.[lxxxii]

Regardless, the Spanish Church sank into this can of garbage up to its thighs. In his 1938 book *Les grands cimitières sous la lune* [The great cemeteries under the moon], Georges Bernanos, another notable French Catholic existentialist, gave the most tremendous tongue lashing ever to a group of clergymen, a literary expression to the shocking crimes that Francoists were committing in the Canary Islands.[lxxxiii].

Whilst travelling in Europe September 1938, Gallegos Rocafull attended a homage to François Mauriac who had strongly expressed, in the French newspaper *Le Figaro,* his disgust at the violence wielded by the insurgent rebels that June. In that article, Mauriac said that the assassinations committed by the Moroccan troops who had pinned a Sacred Heart to their robes, the systematic purges, the cadavers of women and children left behind by German and Italian aviators at the service of a Catholic leader who claimed that he was a soldier of Christ, all of that was a horror at another scale.[lxxxiv]

Gallegos Rocafull was also impressed by the words of another famous French Catholic philosopher, José Bergamin, who also spoke at the homage to Mauriac, when he stated that Christ was not at the orders of any general nor of any dictator, and that furthermore, the priests and monks who were assassinated, as well as the victims of the summary judgements, the secret executions and the innocents machine-gunned on their doorsteps, were equally entitled to the love and mercy of Christ.[lxxxv]

Clearly, opinions in Europe were far from unanimous regarding Franco. The Patriarch of Lisbon, Cardinal Cerejeira, criticized the excesses of the political Catholics in Spain, remarking that they did not understand the meaning of Christ. They were working consciously or unconsciously to de-Christianize Catholicism.[lxxxvi]

None of the above drew the fury that followed the Archbishop of Zaragoza, Rigoberto Doménech's medieval message blessing the violence in which he said that violence was not carried out in the service of the municipality, but legally, to the benefit of order, the Fatherland and Religion.[lxxxvii] As Hilary Raguer pointed out in his book on the Church and the Spanish Civil War, quoting Southworth: "It was in Spain, during the civil war, that the union between the Catholic Church and the fascist movements was sealed with blood."[lxxxviii]

Gallegos Rocafull's opinions did not go unnoticed by the 'hierarchy of the crusade', who considered that he was Cordovan although born in Cádiz. As early as 1937, the Bishop of Cordoba, Adolfo Pérez Muñoz, suspended him *a divinis*, but it was not until the 'Triumphant Year' of 1939, by which time Gallego Rocafull had fled to Mexico, that the Bishop of Cordoba's wrath exploded. Bishop Adolfo, such a keen displayer of the Fascist salute when standing next to Cascajo, the genocide, accused Rocafull of promoting the Rojo Marxist revolution, when what Gallegos Rocafull was actually supporting was the constitutional Government of the Republic. The Bishop sent the following message to Mexico, formally expelling Rocafull from the Church:

> "You are hereby barred, with a perpetual and unlimited suspension of every kind; that is, not only *a divinis* but also *ab officio et beneficio*, including all the distributions and administration of assets you have benefitted from under Canon Law 2222, for your enormously serious and highly scandalous transgression of paragraph I of Canon Law 141 through your spoken and written activities in favour and in defence of the Rojo Marxist revolution that the Pope and the Spanish bishopric have condemned." [lxxxix]

This infernal labyrinth was just the tip of the iceberg of reprehensible activities with which the Spanish clergy cloaked itself in its synergy with the Dictator, Moreno Gómez addresses other aspects of this when he discusses the number of executions, the microhistory of the prisons, the pestilent world of denunciations, the manipulation of children and the unholy maelstrom of the purges.

The religious manipulation of children

To begin with, some background data is essential. Many poor children, sons and daughters of the imprisoned, were scattered across a multitude of reception centres and for whom there is no exact data. These centres belonged to or were controlled by the Social Welfare authorities, religious organizations in towns and villages, charity soup kitchens for children of all kinds, and so forth. The few available statistics refer to children of the imprisoned, but only of those prisoners who came under the Redemption of Sentences Through Work regime, of which there were few, very, very few. The official report to the government on these children, the 1943 *Memoria*, indicates that 9,050 children of prisoners were taken in during 1942 and 12,042 in 1943.[xc]

Furthermore, the Regime enacted another key measure (Law of March 30 1940), that decreed that all children over the age of three years had to be removed from the prisons. These children were taken away by truck (in Saturrarán, Amorebiera, and elsewhere) and in many cases, their mothers never heard of them again. Finding them, if at all possible, of course, was made more difficult by Law of December 4 1941, which legalized the Church's changing their names.

The ensuing reality was that these children and many, many more, filled religious schools, convents hospices and orphanages, all of which were under the tutelage of the Catholic Church, thus providing the favourite grounds for the Church and National Catholicism's work in the ideological repression of the defeated through their children. The children of the defeated who were placed in religious schools and welfare institutions provided them with perfect 'Dianas' with which to shoot arrows of re-education and propagation of the victorious ideology. Mónica Ordoño's research into the Catholization work of the Welfare Authorities[xci] demonstrates that in addition to data regarding baptisms, First Communions and church weddings, there was a separate section devoted to the number of religious vocations sent to the Seminary. It is also interesting to note that children of Rojos are recorded as more frequently enrolled in religious institutions, convents and seminaries than in public schools.

Juan Caunedo Domingues' excellent documentary for the Madrid Forum for History, relates two cases in those days when parents were imprisoned, and their children frequently placed in a seminary or convent.[xcii] The first case is the mother's story:

"A mother was allowed to see her son who had come to visit her. He was about twelve years old, which means that he had been removed from the jail (and his parents) because he was more than three years old.

He was wearing a cassock, the uniform for seminary students. During the entire visit, the woman held her tongue and held it well, but when she returned to the cells, she exploded in anger – they had killed her husband and turned their son into a priest. Children were dressed as priests and nuns when they were taken to visit their mothers in prison, and they were taught that they had 'to pray for all the sins of their parents'. Although the intent was to break down a family and deconstruct that family's offensive ideals, all they achieved was to deconstruct a political ideology."

In the next case, another witness from Gijón reports on similar religious manipulations she was subjected to as a child:

"I was taken to a religious school in Oviedo. After eating, we said an *Our Father* in thanks for the food we had received and another for 'His Excellency the Head of State, Lord of the Armies'. When we prayed for our father and mother we had to add 'kill them Franco'. When I refused to pray, the nuns would force me to pray alone if I wanted to go to recreation. When we went into the town, we stood out and people jeered at us and called us 'the hospice girls'.

The nuns would punish us by making us kneel on dried chickpeas and hold out our arms in a cross, and they would take our morning and afternoon snack away from us. They wanted me to become a nun in Pelayos, as another girl there had done; on Sundays we went to Mass there, so we could see her cloistered behind the grills, praying. They were determined that I should become a nun but all I wanted was to go home to my grandmother.

The above witness' brother was sent to a Jesuit school in Gijón, as they did in those days when they tried to regain control of the orphans of the civil war through religious brainwashing. (Both were charming people from Asturias whose names escaped Moreno Gómez – if there was a fault with the documentary, it was that it did not sub-title the names of those who were interviewed; it only briefly listed them at the end).

Montse Armengou and Richard Belis produced a magnificent documentary on Francoism's lost children, for Barcelona television[xciii], which provides additional details of the forced religious vocation of prisoner's children, together with personal memories similar to those reported in Caunedo's documentary.

In the documentary, Teresa Morán, speaking with the great moral strength typical of Republican women, remembers that Santis' husband and two guerrillas who were with them were executed and she was taken to Ventas prison and condemned to death, but her sentence was later commuted. Occsionally, some relatives would bring her news of her children who had been sent to different schools. One day she was called in to the visitors' room. What a surprise! What a rare event! When she got there, there was her eldest son, dressed as a priest, accompanied by one of his teachers, a priest, as they had just come from the Seminary. She entered the room and saw her son wearing a cassock and... what can I say? "Traitors!" She shouted. "How could you, they killed your father!" It was the worst thing she could ever imagine. Seeing one of her sons dressed up like a priest, accompanied by one of those who had murdered his father and had now taught him to hate him.

Around 1942, four girls from El Viso de los Pedroche were 'netted' and taken to a Barcelona convent, among them, the sister of the young guerrilla José *Rios* Murillo. The latter and his father had fled to the mountains in mid-1941 but there is no information regarding her mother. Moreno Gómez has no idea why the girls were taken. Perhaps to punish their parents.

As National Catholicism was determined to increase the number of its followers, the Church resorted to a massive campaign for religious vocations and by the end of the 1950s, seminaries were bursting at the seams. Most vocations came from humble families and from those whose leftist parents had, or had not, been converted by the Church's constant pressure on every aspect of their lives. Juan A, Bustos Casado, one of many members of the Communist Party but of no known political activism, was executed June 12 1940 in Villanueva de Cordoba. He confessed to Rev. Marcial and asked him to look after his children. One of his sons, Gaspar, was sent to the seminary and he is still a practising priest in Cordoba.

Not all Rojo children entered the church, however. Manuel Rubio, whose father Antonio Rubio Cobos from Pedroche was executed March 26 1940 in Villanueva de Cordoba, was taken to the Salesian seminary, where he remained for several years, acquiring a very high level of academic education. With no vocation for the priesthood, he left the seminary to teach in a private school in Villanueva de Cordoba, where Moreno Gómez was one of his students. A history of the religious vocation of Rojos children whose fathers were either imprisoned or executed, would be very extensive and interesting. Sadly, it is too late to undertake the necessary field work; too much time has passed to discover what needs to be known. Spain's everything except exemplary, revolutionary 'Transition' has destroyed our memories and our history.

The Church's first great channel for preaching religion to children (and also, partly to their relatives), was the Falangista welfare institutions in which many priests and nuns served. The Francoist media contributed with stories of the food rations the soup kitchens provided, but most especially, by publicising the welfare institutions' support of the Church's campaigns for *en masse* christenings, communions, confirmations and church weddings. From 1939 to 1940, the Madrid Social Welfare Institute was the venue for 9,872 baptisms, 6,642 First Communions and 1,116 church weddings. There was a total of 24,513 christenings in all of Spain in 1940.[xciv]

There were no bounds to National Catholicism's harassment of children, the forcible imposition of a new, doctrine on the minds of children who came from Republican, worker and lay backgrounds, and who had never heard of such things. The theories of God's punishment, the fire and brimstone terror of Hell and similar philosophies had a major impact on the minds of children in Welfare Institutions, as we see in Ernesto Cabellero Castillo's *Memoires*[xcv].

During those years, the Church employed all kinds of coercive ways to impose its authority and customs. As described earlier, if workers wanted to be hired by the Señoritos, they had to declare that they were Christians, to attend Sunday Mass regularly, and if married, to have been wed in the Church.

The Church's second great channel for converting Spain was the missions. After National Catholicism declared that Spain was a Mission country, the principal religious orders went to all the main towns throughout the country organizing missions. The towns and villages of Cordoba province were given a good dose of this proselytizing. The Francoists' favourite religious order was the Society of Jesus – the Jesuits, founded by a Spanish soldier turned priest in the 16th century and martial in nature. Next came the Order of Preachers – the Dominicans, created in the 13th century.

Ernesto Caballero describes the first of three missions which appeared in Villanueva de Cordoba in 1943 or 1944.[xcvi] These were young Jesuit priests intent on Christianizing every leftist individual who was still alive and not yet imprisoned, as well as his relatives and the relatives of those who had already been executed. The imprisoned were Christianized in the jails.

A second mission in Villanueva de Cordoba arrived at the end of March 1954. In addition to novenas, confessions and mass communions, schoolchildren were sent down the streets in processions, carrying small statues of saints and waving Vatican flags. This was the work of the Dominicans who turned the entire town's routines upside down so that everyone could attend dawn rosary prayers and other religious observances.[19]

The last mission in Villanueva was in Spring 1961, under the Jesuits, and it was more elaborate. Having brainwashed the children, the Church knew that it had its best guarantee for the future, therefore as a rule, children were given leading roles in almost every religious celebration. Schoolchildren put on plays in the town square, illustrating the seven sacraments. Everything was directed at the masses, and everyone attended *en masse*, which illustrates the Church's enormous capacity for mobilizing entire communities.

National Catholicism at war with Education.

Right from the early days of the military coup, the Church actively collaborated with the terrible crusade to purge civil servants and employees of the most varied kinds of businesses and organizations,

---

[19] In 1951, a Dominican priest came to town to lead a novena and attracted a young man to the Order, Pedro León Moreno, said to be the son of a Rojo, but quite contrary, his father was a Francoist municipal policeman. The friar had earlier served as a missionary (Honduras, Venezuela...) and had brought needy people from Latin America to Seville where the Order financed their studies. Moreno Gómez recently spoke with him, and he got the impression that he was a very high-ranking member of the Order.

of non-believers in the Francoist regime. In Cordoba, for example, Banesto employees had to go to the military authorities and ask them to endorse their positions. Teachers were included in this kind of general purgatory directed at eliminating anyone who might have expressed lukewarm feelings towards the Glorious National Movement. It was a search for unwavering support. When the Nationalists captured a town, Republican municipal policemen and employees were forced out of their jobs en bloc and many were sent on the road to the cemetery. Considered spoils of war, the jobs would be filled by insurgents and supporters.

Law of February 10 1939 introduced the purging of civil servants. Later, Law of August 25 1939 decreed restrictions to all *oposiciones*[20] and applications for public employment, while reserving 80% of all public jobs for select supporters of the Regime, namely war wounded, ex-prisoners of war, orphans of Caídos, and more. The remaining jobs were set aside for supporters of the Cause. The defeated were totally excluded. Purging of the press was declared May 24 1939: any journalist who wished to retain his job had to sign a statement detailing his reaction to the events of July 18, obtain sworn references of good behaviour and comply with other requirements. The Order for the purging of doctors was dated October 6 1939. Lastly, came the Orders purging of members of other professions such as lawyers and football referees.

Despite the general purging of the entire country, it was the teaching profession that would suffer the most. In 1936, the Church and Franco battered Education with the greatest purge in Spanish history, one that had no equivalent with that which was going on in other professions. This was reactionary Spain and the Church's great crusade against teachers in general, and it had much to do with the ensuing cataclysm. There is much to be found in this respect in recent studies that must be consulted, especially the work of two Morentes: Morente Valero (Granada) and Morente Díaz (Cordoba).[xcvii]

That teaching was the first professional class that the supporters of the coup called to order in Cordoba city, is highly significant. July 3 1936, by order from Colonel Cascajo, several teachers in the

---

[20] Competitive public examinations required for employment in the civil service, state teaching positions, and more.

capital were urgently summoned to report to the new Mayor, Salvador Muñoz Pérez. In his office, the Mayor reproached them for being responsible for the dissemination of Marxist propaganda in the schools, he ordered them to hang a crucifix in each classroom and to teach religion, as reported July 24 1936 in the *Guión*, a newspaper whose Director, Antonio de la Rosa, was a passionate member of the extreme right wing. As these teachers left the office, they were arrested and soon afterwards executed by a firing squad: Agapito de la Cruz, Enrique Fuentes Astillero and Juan García Lara. Many more were later sacrificed as the barracks, the casino and the church vestry joined forces in their persecution.

This was a monstrous and terrible inquisitorial process that one finds impossible to believe ever happened. This was not a persecution in the manner of the ancient sackings of the 19th century; it was more than just a purge. This was the case of a much more profound crusade: of the Church's persecution of heretics, the clergy's violent retaliation against laicism, and the settling of political scores. A persecution along the lines of European fascist anti-culture.

The burning of books that marked the beginning of the IIIrd Reich, also occurred in Cordoba. Goebbel's declaration "When I hear the word culture, I immediately put my finger on the trigger of my gun," became the byword of Nazism, of Francoism and of Fascism in general. In Seville, when an elderly 72-year-old teacher José Sánchez Rosa was arrested by Falangistas, he was placed, because he was ill, on a mattress on the back of the truck that was taking him to be executed, surrounded by his books. When the military assassinated Federico García Lorca, Franco destroyed an icon of Republican genius and culture. Likewise, in 1936, the multicultural literary circle of Cordoba was destroyed when Francoists murdered its most distinguished members: the poet José Maria Alvariño, poet and bookseller Rogelio Luque, sculptor Enrique Moreno, music professor Aurelio Pérez Cantero, and many more. Clearly, this crusade against the teaching profession was no more than another chapter in the eradication of culture in Fascist Europe. 'Darkest Spain' was back with its destruction of culture, modernity and freethinking. More than a purge, which it also was, it also possessed the persecutory nature of European totalitarianism, a feature that cannot be ignored.

Francisco Morente Valero also commented on the striking similarities between the Italian Fascist and the Francoist schools of thought. In both cases, Education had to be converted into a propaganda instrument for thr Regime.

In eliminating or punishing teachers, the Francoists were not initiating a class war, but a war of ideologies, a religious war, and a war against culture. There was a common denominator to the persecution of culture, teachers and books: an ideological and a religious war, whose features are still apparent today.

One such example is seen in Amenába"'s movie *Ágora* (2009), depicting Christians in the IVth and Vth centuries, who in a flood of fanaticism led by Saint Cyril of Alexandria, razed libraries to the ground destroying all of Alexndria's ancient knowledge, immolated and excoriated Hypathia, the astronomer-philosopher. Clearly, the 1936 religious war was not the first time that the Church would move like a tsunami against books and teachers, as it also did during the Inquisition in the 15th and 16th centuries.

Manuel Morente Díaz described the purging of books and libraries in Cordoba in detail.[xcviii] In Moreno Gómez' earlier book on the Francoist genocide (2008) he mentions some of Don Bruno's neurotic proclamations against books. The Nazis did the same, May 10 1033, when they built a huge pyre of more than 25,000 books, under Goebels' approving eye. In Cordoba capital, the birthplace of sages such as Seneca and Maimonides, there were several pyres of burning books during the early days of the military coup. At first, it was Falangistas who burned the books, the kiosks and bookstores. Soon afterwards, the Carlistas also built a pyre. Privately-owned libraries, such as the one belonging to Antonio Jaén Morente on Calle Juan de Mena, were seized and burnt on Las Tendillas square, as they did those belonging to Dr. Vicente Martín Romera, Francisco Azorín, Eloy Vaquero, and so many other Republican scholars.

In 1940, a fanatical Jesuit priest who preached at several missions in Cordoba capital, wished to celebrate his inquisitional passion with another burning of books in Las Tendillas square, books taken from piles in the La Corredora square. Earlier, at the end of 1939, several other Jesuit missions in Puente Genil organized another burning of books. José López Cavilás, who was a child of the time, describes one of those Nazi-like pyres as a great fire of an impressive pile of books, in front of the church on Compañia square. The parish priest had rallied all the neighbours to bring books for the pyre and fear did the rest. It was intended as a public show of hatred of culture that the Church approved of. Never since the days of Almazor had so many books burned in Cordoba as when Bem-Abi-Amir destroyed Alhaquem II's 400,000 book library, burning everything that did not have a religious content. The Francoist scene was repeated several times throughout Spain.[xcix]

The anti-library persecution almost led to the execution of Carmen Guerra, the Director of the Cordoba Provincial Library. The Catholic newspaper *El Defensor de Córdoba* heartily fuelled the Francoist demand for the purifying destruction of books so as to eradicate the bad seed, the poison, these tools for disseminating dissolutionary ideas and the vehicles of modernity and laity in Spain.

The Seville purging Committee, with branches in Cordoba, was created under the tutelage of four types of authorities: military leaders, Falange, Church and the Catholic Association of Fathers of Families. In Seville according to Morente Díaz, one director was Antonio Domínguez Ortiz, for the Falange, and Manuel Gómez Rodríguez, a teacher at the seminary, for the Church. In Cordoba, Canon Féliz Romero Menjíbar represented the Church.

As Francoists destroyed books and culture, so they did mountains of teachers. In 1936, the Francoist so-called National Defence Junta began by decreeing the removal of all teachers who had leftist ideas, for which purpose it ordered military commanders and the Guardia Civil to make the appropriate inquiries. The Purging of Teachers Committee came into action at the beginning of 1937 with the implementation of State Technical Junta Decree 66 of 8 November 8 1936 (BOE November 11, 1936). In addition to the persecution and the removal of teachers, the initial visual symbol of the process was the reintroduction of crucifixes in schools.

Whenever a school was left without a teacher, the parish priest took his place. Later, army corporals or officers or members of the Catholic Action, would be instructed to do so. The open, modern, lay or liberal Republican school was erased from the face of Spain, as the country became militarized, or beatified (as what today we would call a Koranic school). Freethinking disappeared under Franco and the bishops.

Although purging teaching was part of the overall program for the repression of the working class, it had an added feature, the ideological repression of schools because of pressure from the Church. New Totalitarian Spain demonized the Republican school and proposed to strip it immediately of all modern, lay or European influences (such as Voltaire), of Krause philosophy and the Free Teaching Institution principle, in other words, of all democratic, workers' and equalitarian ideals. Franco himself, who lacked even a minimum idea of what education was, proclaimed in 1937 that the new school would be based on three principles: patriotism, the absence of all foreign influences, and Catholic values. He was particularly obsessed with vilifying the modernity that a free teaching institution represented, declaring that it furthered the 'pedantic and pseudo-intellectual promotion of every anti-Spanish action in matters of culture and teaching'.[c]

Spokespersons for the Confessional State such as the Archbishop of Zaragoza, did not mince their words when making antediluvian declarations regarding matters of teaching and education, as did Minister of Education José Ibáñez Martín (1939-1951), in 1943 when he asked how could a teacher who does not know how to pray, shape the soul of a child. That is, he said, the fundamental problem of Spanish education.[ci]

As José Casanova so ably described them, priests were nothing else than talking heads who also bequeathed a rosary of anti-modernistic declarations, of heated praise of Saint Joseph of Calasanz pedagogy and sovereign idiocies regarding the training of teachers and basic problems of education.[cii]

The greatest tragedy was the mass murder of teachers and pprofessors during the first months of the military coup and of some during the post-war period, and a few more who died of starvation in prison. During the first days of the insurrection, the instigators of the coup suspended the salaries and jobs of 54 Cordovan teachers, September 15, 1936. Incited by the Catholic newspaper *El Defensor de Córdoba*, the military government fired every teacher who belonged to the FETE[21] union at the same time that the firing squads were busy in the cemetery. Moreno Gómez' research and Manuel Morente Díaz's

---

[21] *Federación Española de Trabajadores de la Enseñanza* – Spanish Federation of Education Employees.

extraordinary work, point to the fact that many more than thirty public school teachers (and some private teachers) were executed then. All crimes are repugnant, but murder of teachers is the nastiest of all.

Moreno Gómez is reminded of the son of Claín, the writer, Leopoldo Alas Argüelles, Rector of the University of Oviedo, who was taken hostage by Nationalist General Aranda soon after the military coup. Despite international petitions for his release, Claín was executed at 6 p.m., February 20 1937, in the Oviedo prison patio. Again, military savagery such as seen in Rome and Berlin. Also, the appalling murder of Federico García Lorca, icon of Spanish Letters, at the hands of the most rotten right-wing element in Andalusia. Like these, so many, many more...

True, some teachers were also fired from their jobs in the Republican zone at the time, but more for social class reasons regarding what party they belonged to, not because of their profession. On the other hand, Francoists all over Spain shot teachers, solely on the excuse that they were poisoning their students by sowing 'dissolutionary', i.e., democratic, ideas. Table 2[22] provides a descriptive list of some of the teachers slaughtered in fascist Cordoba, based on Morente Diaz's lists and on Moreno Gómez' own research.

Manuel Torralbo, aged 29, a private school teacher, and five other unfortunate men certainly did not imagine the tragedy when they were arrested on the afternoon of June 7 1948. Nobody in the village has ever been able to explain the reason for that particular Francoist crime, except to suggest that some of these were leftist sympathizers and had perhaps even belonged to the PCE. Manuel Torralbo was teaching class in his school on Calle Navas when several Guardia Civil appeared and ordered him to send his students, young and old, home. Somewhat naïvely surprised, he asked: "Will it be for long?" "Yes, for some time," they replied.

They and another six individuals (including Catallina Coleto, aged 52, mother of seven young children, a humble hard-working woman married to *El Ratón*, a member of the guerrilla), were taken to the mobile headquarters at the Fuente Vieja schools where they were interrogated whilst the telephone rang non-stop with calls to and from Seville and Cordoba. The Francoists were about to teach the defeated

---

[22] TABLE 2. Names of teachers executed in Cordoba or died later of other causes 1936-1948.

another hard lesson. Ángel Fernández Montes de Oca, Lieutenant Colonel in the Cordoba Guardia Civil, gave the order. In the early hours of dawn June 8 1948, the unfortunate six were taken along the Villanueva to Adamuz road and just two kilometres later, at the first curve – the Los Almagreras bus stop – their cries and tears echoed in the silence of the night as those innocents were robbed of their lives.

Nowadays, every time that Moreno Gómez drives past that curve on the road to his family's country cottage in the village, he thinks of them. Others might not remember them, but he does. He writes these words in memory of them, not as a simple note of historical research.

So far, this has been an elementary synopsis of the crimes against Cordoban teachers. No matter how painful this overview of what occurred in Andalusia, it pales against the impact that the far-reaching Francoist crusade for purging the entire bureaucratic process would have on education throughout Spain. Decree 66 of 8 November 1936 (published in the BOE 11-11-1936), created four committees or Courts of Inquisition, charged with implementing the purge: A, B, C and D. The last one, D, was charged with purging the Primary Education sector.

In Cordoba province, Committee D first consisted of Irmina Álvarez, Assistant Director of the Teacher Training College, José Priego López, Head of the Primary Education Inspectorate, and in addition to another member, Lieutenant Colonel Juan de la Cuesta, an ignorant lunatic who incessantly declared that the teachers were very bad.[ciii] Irmina Álvarez, however was soon replaced by Ángel Cruz Rueda, Director of the Instituto de Cabra and a Francoist to the bone, who set up his private purgatory in the Instituto Provincial Góngora in Cordoba city. Another member of this provincial Court of Inquisition was Joaquín Velasco, Vice-President of Catholic Action, whose father was Mussolini's official representative in Cordoba.

The lawful Director of the Instituto Provincial Góngora, Antonio Jaén Morente, was in Madrid on July 18, thus fortunately escaping certain execution had he been in Cordoba when the insurgents marched in. In his absence, the Francoist faculty of the Instituto convened a shameful Court of Honour October 17 to strip him of his academic qualifications and expel him from the Instituto. Their deliberation was later officially confirmed by Decree from the Burgos

Government barring Antonio Jaén Morente from holding any kind of academic or teaching position in Spain.[23]

As far as all of Spain was concerned, Francisco Morente Valero's research shows that 20,435 purging indictments were filed in 14 Spanish provinces, that is against more than 80% of all teachers. Of those indictments, 75% were resolved in favour of the teacher and 25% were sanctioned in different ways: dismissal, absolute disqualification and/or loss of rights (2,021); suspension of job and salary or temporary disqualification (1,044); transfer to other locations (1,983); disqualification for administrative positions (608); other sanctions (235). In one town, Aguilar de la Frontera, as many as 20 teachers of both sexes were indicted; 12 of those sanctioned were female.[civ]

Those percentages are in agreement with Manuel Morente's study of Cordoba, where of the 814 teachers in the 1936 census, 205 were indicted by the Liquidation Committee, that is, 25% of the total.[cv] Morente cites an interesting case in Villanueva de Cordoba, where Vicente Pascual Soler, a famous teacher, was indicted in La Rambla where he had taught for some years, on the grounds that he was a member of the Socialist party. His case was later resolved in his favour. What is not generally known is that May 1946, the lunatic Guardia Civil commander in Villanueva de Cordoba had included Vicente Pascual in an absurd raid where he and another 15 persons were handcuffed to each other, marched through the village and then imprisoned in Cordoba where they remained for a year until their cases were tried. After his release, Vicente swore that he would never return to that Villanueva gone mad. He remained in Cordoba, where he became Director of the well-known Academia Espinar.

To better understand the manner by which Francoists purged Education, it is worth recalling the individuals who denounced those who were indicted, in order of importance: 1) the military and the Guardia Civil; 2) members of the Church; 3) sundry local residents who toadied to the Regime; 4) the municipalities, that actively collaborated in the repression on the side of the Francoists; 5) and

---

[23] 24 February 2016, the Instituto Séneca, heir to the Instituto Provincial Góngora, convened an Extraordinary Meeting of the Faculty to revoke the 17 October 1936 decision and formally reinstate Antonio Jaén, Full Professor and ex-Director of the Instituto Provincial, as a member of the faculty.

curiously, in some out-of-the-way places, the Falange. One enthusiastic accuser of teachers was Rev. Paulino Seco de Herrera, San Nicolás parish priest in Corodoba city, who was very well known in the literary circles organized by teacher and poet Rafael Olivares Figueroa. The latter, who wrote the prologue to the well-known book *Canciones morenas* about martyr José Maria Alvariño, was able to escape into exile but was not forgotten by the parish priest who, in a damning report, described the poet as 'notably extravagant'.

In the specific case of 144 secondary school teachers and similar professionals, denunciations were filed against 82, that is 57% of the total. Finally sanctioned, 44, or 31%.

As to the catalogue of accusations against the teachers and professors, aside from accusations of a political nature (followers of Marxism, Socialism, Communism, supporters of the Frente Popular, or leftist sympathies in general), the most curious were accusations of a religious nature, which bordered on the grotesque: middling believer, indifferent, not seen in church, has no religious ideals, indifferent to religious practices, distanced himself from the Church during the Republic, anti-Catholic ideas, atheist, doubts the existence of God and traditions of the Fatherland, criticizes the Society of Jesus in a book, does not allow his students to kneel during the viaticum (last rites), does not kneel during processions, May 1936 refused to kneel when a procession was going past, etc., etc. As a result, idiocies like these which lacked in consistency or any semblance of truth, brutally cut short the professional and family lives of many teachers, honourable men and women possessing indispensable culture. The destructive whirlwind of culture in the hands of European Fascists converted Spain into a cultural wasteland from which, possibly, it has not recovered nor will every recover

The number of calamities that fell upon Education as the authors of the coup did their best to demolish an entire generation of dedicated, irreplaceable teachers, is so vast that it is difficult to grasp. Likewise, the countless number of teachers who, notwithstanding pain, misfortune and heartbreak, refused to be broken, as illustrated by the following stories.

Ángel Carmona Jiménez, whom the coup caught in his school in Cordoba, went to Montemayor, his hometown, where he heard that Modoaldo had been executed. From there, he walked to the

Republican zone where the FETE charged him with organizing schools for children who had been evacuated to the north of Cordoba province. He then enlisted in the fight for the defence of Madrid until a hand grenade amputated his right hand. Thus mutilated, he went to France where he organized children's camps, In the middle of 1940, he decided to return to Spain where Franco received him in the Miranda del Ebro concentration camp from which he was released and permitted to return to his village under house arrest. He was eventually able to rebuild his life in Cordoba and in Barcelona, but he was never allowed to return to teaching.[cvi]

Another famous teacher from Montemayor who lost his right arm, Alejandro Cabello Sánchez returned on a stretcher to France where he had trained, on the point of death. Once recovered, he went to Hungary where he was graduated in Agronomy in Budapest, then to Cuba where he obtained his PhD in Biological Sciences from La Habana University. He taught at both these institutions.

This section concludes with some deeply moving texts by Antonio Jaén Morente, taken from a four-page leaflet he wrote and that was published in Valencia by the Izquierda Republicana National Council, *Estampas da Guerra*[cvii]. These are immortal writings in a Ciceronian style worthy of those times and a living homage to the teachers who were martyred by Spanish fascism during the cataclysm that decimated Cordoban Education. One of the texts, *Marruecos una escuela sin maestro* (Morocco, a school without a teacher), is an elegy to Antonio de Ontavilla, a teacher in Alcánzarquivir, Morocco:

> "Ontavilla was in the bloom of youth, 25 years old. He had a big heart. Republican ideals, left of the left. He spoke easily and his pen was fair and informative. A fighter of the kind who, like so many others, hearing treason germinate in the mouths of the fascist crowd, sent warnings to Madrid calling for help. Therefore, no sooner did we win the election in Februry, he openly asked for the removal of certain members of the military whose names he indicated. He had to flee to Tetuán, as his life had been threatened."

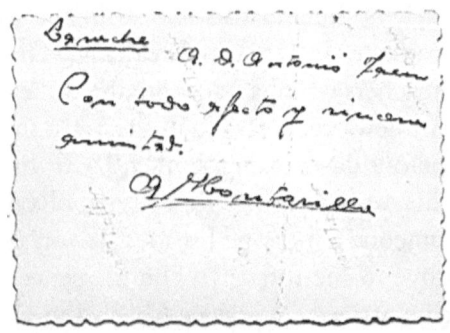

**Handwritten note on the back of a postcard that
Antonio Ontavilla signed and sent to
Antonio Jaén Morente from Larache, Morocco, late in 1936:**
*"With great affection and sincere friendship
Antonio de Ontavilla"*
Courtesy of Gorrell Jaén Family Archives

War broke out whilst Ontavilla was on home leave in Galicia, visiting his ailing mother. Denounced by local Francoists, he was captured and returned under guard to Tetuán. Antonio Jaen's description of the prisons in Alcázarquivir and the schoolmaster's ordeal is horrific:

> "The makeshift prisons in the regular army barracks were packed with men who supported the left, that is, anyone who had had any contact with the Frente Popular, who were tied to the troughs. Ontavilla and his companions were savagely beaten by the very Arabs that he had taught and were now fighting on the side of Fascist Spain. (...) He was marched along the streets of Alcázar, hands tied and guarded on either side by two Nationalists, followed by the crowd yelling and shouting insults... During a halt in this terrible procession, he was stopped in front of the door to his own school, where his guards called out to the children: 'Come and look at your teacher!'
> Then came sham justice: the court-martial. The children of his school placed some posters on the street corners on which they had written 'Reprieve and Pardon' and it is said that some mothers and

fellow teachers shouted this appeal in the streets. It was September. Dawn was breaking. The female Spanish teachers who had accompanied the children during this new ordeal, could not restrain their tears. A little before dawn 4 September 1936, in the last hours of his captivity, Antonio de Ontavilla managed to give someone a message for me: "If you are saved, say goodbye to me forever". The messenger arrived late, but he arrived. I received both the messenger and the message. Teachers of Spain, record this in the chronicles of Spanish Education's great martyrdom. Take note of 'The passion and death of Antonio Ontavilla, teacher'. Ontavilla's grave is unmarked; no headstone bears his name. On a day that October, the students without a teacher, timidly walked through the gates of the cemetery. In the children's hands, flowers from African fields."

Lastly, Moreno Gómez attempts, yet again, to impress upon the reader, the tragedy that drove Spanish education to the edge of the cliff, by quoting another of Antonio Jaén's extraordinary texts, entitled *Galicia mártyr. Estampas de Castelao.*[24]

---

[24] Galicia martyred. Illustrated with Castelao sketches.

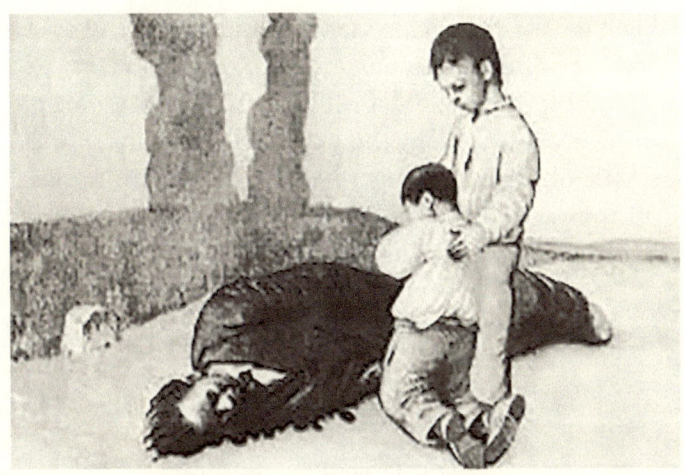

"THE LAST LESSON"
Pen and ink drawing by Alfonso Daniel Rodríguez Castelao[25]
Illustration for *Estampas da Guerra* by Antonio Jaén Morente
Courtesy of Gorrell Jaén Family Archives

The above pen and ink sketch of a fallen teacher and two children looking down on him, by this Galician artist and eminent politician, is one of many by which Castelao used his art to denounce the cruelties of the Francoist regime. Entitled 'The last lesson', Moreno Gómez concludes this section with a last, heartfelt text by Antonio Jaén Morente.

> "Fascist Spain has become a great necropolis of teachers. In every region they have been the favourite target, because they were masters and seed of the Republic...
>
> How I weep for you, Juanito García Lara, teacher, President of the FETE, assassinated in Cordoba... If you could only see 'The Last Lesson', you would recognize yourself. It is your last lesson also. Thus you gave it, executed by a firing squad against the white walls of the cemetery in Cordoba, and like you, Antonio Reina, and Agapito de la Cruz, and Enrique

---

[25] Galician politician, painter, author and physician. A founder of the Galician national party, Member of Parliament for Galicia under the II Republic.

Fuentes, and Modoaldo Garrido, and Augusto Moya, and... I cannot continue.

The list from Cordoba, like these *estampas*, is torn from my heart, from the depths of my pain... The scythe of death no longer harvests the fields; it digs up the earth, opening holes. Death sows the seeds. Buried today, they will germinate tomorrow."

Antonio Jaén Morente, Valencia, 193

Table II.
Teachers executed in Cordoba Province the first months of the coup or died later of other causes. 1936-1949

SOURCE: Moreno Gómez research; + Brothers; FETE – Union of Teachers

Name	Age	School	Date died	Cause	Place
Agapito de la Cruz López de Robles	42	President of FETE Cordoba	17 Nov 1936	Executed	Cordoba
Antonio Baena Moreno	34	Pozoblanco	17 Nov 1941	Exe-cuted	Cordoba
Antonio Martínez Gutieérrez	?	Public School No. 3	1 Aug 1936	Executed	Cordoba city
Antonio Mendez Gómez	34	Industrial Elementary	28 Sept 1936	Executed	Cordoba city
+ Antonio Molina Fuentes	25	City Councilor	30 July 1936	Executed	Cordoba city
Aurelio Pérez Cantero	27	Dir. Symphony Orchestra	8 Aug 1936	Executed	Cordoba city
Blas Gajero López	26	Montoro	9 June 1941	Executed	Cordoba
Eduardo Ruiz Yepes	26	La Rambla	16 Aug 1936	Executed	Cordoba
Enrique Fuentes Astillero	46	Instituto Seneca	20 Aug 1936	Executed	Cordoba city
Fernando Ferdández de Haro	55	Private. Villanue-va de Cordoba	26 May 1949	Executed	Villanueva de Cordoba
Fernando Mata Povedano	?	Montemayor	26 Sept 1936	Executed	Cordoba
Francisco Dueñas Llergo	25	Pozoblanco	4 Sept 1936	Executed	Cordoba
Francisco Duque Iñiguez	41	Nueva Carteya	9 Apr 1937	Executed	Seville
+ Francisco Molina Fuentes	35	Puente Gentil	30 July 1936	Executed	Cordoba
José Gómez Cárdenas	41	Nueva Carteya	9 Apr 1937	Executed	Cordoba
José González Cantillo	35	Nueva Carteya	9 Apr 1937	Executed	Cordoba
José Pérez Arenas	33	Nueva Carteya	27 Dec 1936	Executed	Cordoba

Juan de Miguel Budia	23	Palma del Río	? 1936	Executed	Cordoba
Juan García Lara	29	Instituto Seneca	17 Aug 1936	Executed	Cordoba city
Juan Robles Relaño	28	Aguilar de la Frontera	Summer 1936	Disappeared	Aguilar or Cordoba
Manuel Camacho Parejo	61	Escuela Normal	2 Aug 1936	Executed	Cordoba city
Modoaldo Garrido Diez	41	Vistahermosa, Director FETE	10 Aug 1936	Executed	Cuesta del Espino
Pedro Aljama Siles	29	Fuenteobejuna	7 Oct 1936	Executed	Fuente-obe-juna
Santiago Dionisio Gil Diaz	33	Fuenteobejuna	16 Oct 1936	Executed	Fuente-obe-juna
Tomás Cortés Rodrígues	29	Priego	13 Aug 1936	Executed	El Tarajal
Adalberto Serrano Rodas	54	Peñarroya	?	Encephalitis	?
Antonio Fernández Carreiro	56	Villaviciosa	? 1939	Tortured	Linarea jail
Benito Cordobés Herencia	52	Espejo	6 Aug 1936	Presumed executed	Montilla
Galo Adamuz Montilla	45	Belmés, active member FETE	7 May 1942	Pneumonia	Cordoba
Juan Manuel Nacarino	63	Private teacher, Palm del Rio	16 Oct 1941	Uremia	Cordoba
Manuel Torralbo Cantador	?	Private teacher, Villanueva de Cordoba	? 1948	?	Cordoba

AUTHOR'S ENDNOTES FOR CHAPTER II

i   https://www.colectalia.com/en/info/INFO/articles/franco-s-concentration-camps
ii International Criminal Court, Legal Texts and Tools. Article 7(e). Consulted on the Internet at:https://wwwice-cpi.int/NR/rdonlyres/ADD16852-AEE9-457-ABE79CD C7CF02886/ 283503/RomeStatutEng1.pdf.
iii Ángel Viñas. En el combate por la historia (In the fight for History). Barcelona, Pasado & Presente, 2012, p. 20.
iv Antonio Elorza. 'Genocides'. In Hispania Nova (Online Contemporary History Magazine in Spanish) at http://erevistas.uc3m.es/indLex.php/HISPNOV/index. Number 10, 2012.
v   Eric Hobsbawn. Interesting Times: A Twentieth-Century life. Spanish translation. Barcelona. Critica, 2002, p. 72.
vi Pedro Pascual. "Campos de concentración en España" (Concentration camps in Spain), Historia 16, Year XXV, number 310, February 2002.
vii Idem. "Campos de concentración en España y Batallones de Trabajadores" (Concentration camps in Spain and Forced Labour Battalions). Minutes of the Congress on Concentration Camps and the Penitentiary World in Spain during the Civil War and Francoism. Barcelona, Crítica, 2003, pp. 359 et al.
viii Javier Rodrigo. Los campos de concentración franquistas, entre la historia y la memoria (Francoist concentration camps, history and memories). Madrid, Siete Mares, 2003, p. 221. This number is surely greater if one takes into account the great many who were taken prisoner after the battle of Teruel and the disaster in the Lower Aragon province.
ix Joan Llarch. Campos de concentración en la España de Franco (Concentration camps in Franco's Spain). Barcelona, Producciones Editoriales, 1978, p. 80.
x   Javier Rodrigo, Op. Cit., p. 366.
xi Pedro Pascual. Minutes of the Congress Op. Cit., p. 366.
xii Francisco Moreno Gómez, Cordoba en la posguerra. (La repressión y la guerilla, 1939-1950), Op. Cit., p. 41.
xiii José Ángel Delgado Iribarren, Jesuitas en España (Jesuits in Spain). Madrid, Studium, 1956, p. 235. Cited by Joan Larch, Op. Cit., p. 41.
xiv Antonio D. López Rodríguez, Op. Cit., pp. 167 et al.
xv Miguel Gila. Y entonces nací yo. Memorias para desmemoriados. (And then I was born. Memories for those who have forgotten). Madrid, Temas de Hoy, 1995. & Isális Lafuente. Esclavos por la patria. La explotación de los presos bajo el franquismo (Slaves for the Fatherland. Exploitation of prisoners under Francoism). Marid, Temas de Hoy, 2002, p. 142.
xvi Antonio López Rodríguez, Ibid.
xvii Miguel Regalón Molinero. Personal testimonial sent to Moreno Gómez from Valencia, 31 August 1984. These men are all since long gone. The historian is left with the satisfaction of having saved this account of their sufferings. Not everyone forgot. Many managed to talk about them.
xviii Juan Pulido Cantador. Several interviews by Moreno Gómez in Villanueva de Cordoba, 1983.
xix Miguel Regalón Molinero, Ibid.
xx José Maria Carnícer Casas, written testimonial sent to Moreno Gómez from Eus, Tarragona, October 19, 1987.
xxi Idem.
xxii Arcángel Bedmar, Los puños y las pistolas (Fists and revolvers), Op. Cit., p. 102.
xxiii Idem.
xxiv Antonio D. López Rodríguez, Ibid., p. 201.

xxv   Ángel David Martín Rubio. http://desdemicampanario.es/autor/angel-david-mar-
      tin-rubio/
xxvi Antonia González & Pablo Ortíz Romero. Memoria y testimonia del campo de con-
      centración de Castuera (Memory and witness account of Castuera concentration
      camp). Minutes of the Congress on concentration camps and the world of the peni-
      tentiary in Spain, Op. Cit., pp. 240 et al.
xxvii Pozoblanco Municipal Archives.
xxviii Antonio D. López Rodríguez, Op. Cit., p. 236.
xxix Olaf Palme. 1975. Quoted in El final del silencio (The End of Silence). Documentary
      by Maria Carmen España, 2011.
xxx   Julián de Casanova. La Iglesia de Franco (Franco's Church). Annotated edition, Barce-
      lona, Crítica, 2005, p. 277.
xxxi Founded in Cordoba in 1936 as the official organ of the FET-JONS and published
      until 1941. Predecessor of today's Diário de Córdoba newspaper.
xxxii Ibid., 11 June 1939, p. 7.
xxxiii Ibid., 15 June 1939, p. 12.
xxxiv Ibid., 23 June 1939, p. 3.
xxxv ABC Seville, April 23, 1939.
xxxvi Ibid., 12 May 1939.
xxxvii Diário Azul, Cordoba, 10 December 1939.
xxxviii Ibid., 21 December 1939.
xxxix Ibid., 21 December 1939.
xl   Ibid., 27 December 1939.
xli   Ibid., 5 January 1940.
xlii Ibid., 16 October 1940.
xliii Ibid., 7 November 1940.
xliv Ibid., 19 September 1941.
xlv   Francisco Moreno Gómez & Juan Ortiz Villalba. La masonería en Cordoba (Freema-
      sonry in Cordoba), Cordoba, F. Baena, 1985, p. 254.
xlvi Francisco Moreno Gómez. La resistencia armada contra Franco (Armed resistance
      against Franco). Barcelona, Crítica, 2001. This book deals with the entire phenome-
      non of the guerrilas in Cordoba and other provinces in the Centre-South of Spain.
xlvii Ibid, pp. 45 et al.
xlviii Ibid., pp. 583 et al.
xlix Boletim Oficial Provincial. Cordoba, January 1 and August 1 1940.
l   Francisco Moreno Gómez. El genocidio franquista en Córdoba (The Francoist geno-
      cide in Cordoba), Op. Cit., p. 235.
li   Antonio Ramos Palomares. Eyewitness testimonial to Moreno Gómez, Almodóvar del
      Rio, 1982.
lii   Eduardo de Guzmán. Sócrates Gómez, de la derrota a la represión (Sócrates Gómez,
      from the defeat to the repression). Tiempo de Historia, number 62, January 1980, p.
      16.
liii Peter Anderson. Universidad Complutense of Madrid 11 November 2011 Confer-
      ence. Author of The Francoist Military Trials. Terror and Complicity, 1939-1945.
      Routledge, London 2009.
liv   Archángel Bedmar Gonzalez. La Luz Sepultada (The Buried Light). Published by the
      1st Historic Memory Congress, Aguilar de la Frontera, 27 September-October 7,
      2006, p. 65.
lv   Córdoba, 4 November 1941.
lvi   Juan Luís Torralbo. Email to Moreno Gómez 4 November 2013.
lvii Feliciano Delgado. Diário Azul. Cordoba, 13 Ausut 1939.
lviii Francoist Decree of 26 April 1940. Reproduced in Fuentes para la Historia de la II
      República, la Guerra Civil y el Franquismo (Sources for the History of the II Republic,
      the Civil War and Francoism): at http://fuentesguerracivil.blogspot.com

lix   L. M. Mones Oviedo. "70 años después. Ley de Memoria Histórica" (70 years later. The Historic Memory Law). Feria y Fiestas, Almadén, 2009. Magazine kindly sent to Moreno Gómez by Ángel Hernández Sobrino.

lx   Julio González Gil. "El último testimonio de Unamuno" (Unamuno's Last Will), uploaded to YouTube 2 November, 2007.

lxi   Ramón Serrano Súñez. Entre el silencio y la propaganda, la historia como fue. (Between silence and propaganda; The story of how it was). Memorias, Planeta, Barcelona, 1997, p. 272. Quoted by Julián Casanova, Op. Cit., p. 19.

lxii   Manuel Álvaro Dueñas. "Por derecho de fundación: la legitimación de la represión franquista" (By rightful foundation: the validation of the Francoist repression). In Mirta Núñez-Diaz-Balart et al., Op, Cit., p. 69.

lxiii   Hilar Rager, Op. Cit., p. 399.

lxiv   Diário Azul. Cordoba, 3 October 1936. Quoted by Franciscco Moreno Gómez in Cordoba en la posguerra. La represión y la guerrilla, 1939-1950) (Post-war Cordoba. Repression and Guerrilla, 1939-1950). Cordoba, F. Baena, 1987, p 35.

lxv   Francisco Moreno Gómez. La Guerra Civil en Cordoba 1936-1939 (The Civil War in Cordoba 1936-1939), Cordoba, 1985.

lxvi   Hilari Raguer, Op. Cit., p. 212.

lxvii   Minutes of the Villanueva de Cordoba Municipality, 19 December 1939.

lxviii   Cordoba, 7 May 1942.

lxix   Diário Azul. Cordoba, 16 October 1939.

lxx   Ibid., 29 March 1940.

lxxi   Arcángel Bedmar. 'El nacionlcatolicismo en Montilla y Lucena durante la guerra civil' (National-catholicism in Montilla and Lucena during the civil war). In: La Luz Sepultada, Aguilar de la Frontera, Sept-Oct, 2006, pp, 77-78.

lxxii   BOPC Boletí Oficial del Parlament de Catalunya (Official Bulletin of the Parliament of Catalonia), 23 January 1940, reproduction of the Boletim Oficial del Estado of 13 January, Year V, number 13.

lxxiii   Julián Casanova, La Iglesia de Franco (Fran's Church), Op. Cit., p. 318.

lxxiv   Abbot Manuel Garrido. Valle de los Caídos, 23 August 1986. Letter to Rev. Pedro Laín Entralgo.

lxxv   Marisa Paredes and José Luís Peñafuerte. Los caminos de la memoria. (The paths of memory). Filmed documentary, 2009, in Spanish. Can be downloaded free from: http://cinepeliculasflv.com/21524-los-caminos-de-la-memoria-online-peliculas-gratis-hd-espanol.html.

lxxvi   José M. Gallegos Rocafull. La pequeña grey. Testimonios religiosos sobre la guerra, war). Barcelona, Península, 2007, pp. 211 et al.

lxxvii   Idem, p. 203.

lxxviii   Idem, p. 199.

lxxix   Idem, p. 85.

lxxx   José Luís Casas Sánchez. "La memoria histórica del exilio republicano. El caso del canónigo Gallegos Rocafull" (The historic memory of Canon Gallegos Rocafull), In: La Luz Sepultada, Op. Cit., pp. 47 et al.

lxxxi   José M. Gallegos Rocafull, Idem, p. 88.

lxxxii   Idem., p. 91.

lxxxiii   Idem., p. 169.

lxxxiv   Idem., p. 183.

lxxxv   Idem., p. 181.

lxxxvi   Idem., p. 198.

lxxxvii   Heraldo de Aragón newspaper, 11 August 1936. Cited by Julián Casanova, Op, Cit..

lxxxviii   Cited by Hilari Raguer, La Pólvora, Op., Cit. p. 28.

lxxxix   José M. Gallegos Rocafull, Op, Cit., p. 206.

[xc]   Memoria que eleva al Caudllo de España y a su Gobierno el Patronato de Redención Sentences Through Work Group for 1943, to the Caudillo of Spain and his Government), Madrid, 1944.

[xci]   Mónica Orduño Prada. Ell Auxilio Social (1936-1940) La etapa fundacional.(Social Welfare (1936-1940). Foundation.) Madrid, Librería Libre Editorial, 1996, p. 263.

[xcii] Juan Caunedo Domínguez. Sombra, niebla y tiempo (Shadow, fog and time). Documentary by this free-lance producer and director for the Madrid Forum for Memory. Moreno Gómez was unable to obtain any additional information.

[xciii] Montse Armengou & Richard Belis. Los niños perdidos del Francoism (Francoism's lost children). Award-winning television documentary, Televisó de Catalunya, Barcelona, 2002.

[xciv] Ángela Cenarro Lagunas. "Historia y memoria del Auxilio Social de la Falange" (History and minutes of the Falangista Welfare Institution). In: Pliegos de Yuste, numbers 11-12, 2010.

[xcv]   Ernesto Caballero Castillo. Vivir con memoria. (Living with memory). Cordoba, El Páramo, 2001, pp. 66 et al. Son of Julián Caballero, ex-Communist Mayor of Villanueva de Cordoba who, at the time of these events, was a guerrilla comander in the north of Cordoba province.

[xcvi] Idem., p. 67.

[xcvii] Francisco Morente Valero. La Escuela y el Esado Novo. La depuración del Magisterio Nacional (1936-1943) (The School and the New State. The purging of the national teaching professions (1936-1943). Valladolid, Ámbito, 1997. & Manuel Morente Díaz. La depuración de la enseänza pública cordobesa a raíz de la Guerra Civil (The purging of public education in Cordoba as a result of the Civil War). Cordoba, El Páramo, 2011.

[xcviii] Manuel Morente Díea. La mala semilla. Depuración de libros y bibliotecas en Córdoba (The bad seed. Purging of books and libraries in Coordoba). Cordoba, ECO magazine, number 8, 22 June 2011.

[xcix] José López Gavilán. Aquellos duros tiempos. Anecdotario (Tales from those hard times). Cordoba, 2004, p. 135.

[c]   Manuel Álvaro Dueñas, Op. Cit.

[cI]   Idem., p. 319.

[cII]   Ibid.

[cIII] Morente Díaz, Op., Cit., p. 276.

[cIV]   Rafael Espino Navarro. Asociación para la Recuperación de la Memoria Histórica de Aguilar de la Frontera. La tiza roja (Red chalk), published on the association's social media site.

[cV]   Idem.

[cVI]   Luque, Op., Cit., p. 234.

[cVII] Antonio Jaén Morente. Estampas da Guerra (Postcards of the War). Valencia, Izquierda Republicana, 1938. Moreno Gómez has three sets of Estampas, kindly given to him by Manuel Toríbio García, Cordoba historian and Antonio Jaén Morente's biographer.

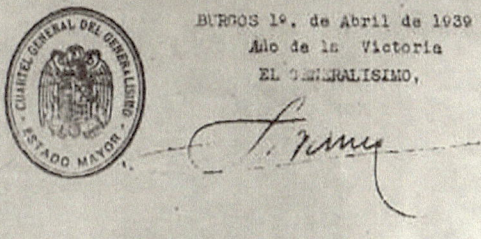

CUARTEL GENERAL DEL GENERALISIMO   SECCION DE OPERACIONES.

ESTADO MAYOR

PARTE OFICIAL DE GUERRA

correspondiente al día 1º. de Abril de 1939.- III Año Triunfal

En el día de hoy, cautivo y desarmado el Ejército rojo, han alcanzado las tropas Nacionales sus últimos objetivos militares.

LA GUERRA HA TERMINADO.

BURGOS 1º. de Abril de 1939
Año de la Victoria
EL GENERALISIMO,

HEADQUARTERS OF THE GENERALISIMO
CHIEF OF STAFF.   OPERATIONS SECTION

OFFICIAL WARTIME DECLARATION
1 April 1939 - Year of Victory III

On this day, having defeated and disarmed the Rojo Army, the National troops achieved their final military objectives.
THE WAR IS OVER.

Burgos, 1 April 1939
Year of Victory
THE GENERALISIMO
/s/ *Franco*

III

CRUSHING THE VANQUISHED (B) COURTS-MARTIAL, TORTURE AND EXECUTIONS

MILITARY TRIBUNALS AND COURTS-MARTIAL

> *"By exterminating one third of the male population in Spain, we shall cleanse the country and get rid of the proletariat."*
> **Lieutenant-Colonel Gonzalo de Aguilera**
> **to John Whitaker**, War Correspondent
> Chicago *Daily News*, 1936.

The unrecorded slaughter in Andalusia: Enforcement of the *Lei de Fugas* (Law of Escapes) April/May 1939

When the expression *extermination* appears in this study, it refers to the Francoist interpretation of what Franco believed was an 'essential extermination', that is, the annihilation of everything and everyone that Franco's Regime considered most evidently represented the Republic's social base and its elite.

In all the studies of the Francoist repression, one has yet to find an author who has paid due attention to this subject. Academics today still pay little or no heed to this unusual phenomenon of post-war Spain: the wave of extra-legal summary executions that the victors perpetrated April and May 1939, as well as the multiple summary executions in the concentration camps, another matter regarding which still little or nothing is known.

The Center-South of Spain, the last region to fall into the hands of the insurgents, was further subjected to an additional extra-legal form of indiscriminate slaughter, what the people called *paseos* [*sic*. walks] and reacted to with a terror similar to that which they felt during the early days of the insurgency. It was as if the unlawful Nationalist wartime proclamations, supportd by Francoist public opinion, were again being applied as the law of the land. Undoubtedly these schemes were not apparent to everyone, although we hear of frequent references to an earlier euphemistic Law of Escapes, a traditional tool of repression, whereby a person was shot without further ado, usually in the back, an execution justified on the grounds that 'the prisoner was attempting to escape'. This stratagem was activated with great enthusiasm at the end of the war by the euphoric victors and it would again emerge in an out-of-control manner, during the so-called Triennium of Terror (1947-1949).

The application of the _Lei de Fugas_, as soon as victory was declared, was in line with the victors' first punitive objective, the implementation of actions to be served as a bloody aperitif paving the way for the bureaucratic dealings of so-called Military Justice. The victors were loath to wait for due process of law; they wanted to teach the defeated a quick 'cleansing' lesson. This would indicate that although due process began to be legislated in February 1937, the swift enactment of legislation to substitute the military's wartime proclamations was not of great concern to the victors. Besides, everyone knew that this legalization was a farce because whenever the Regime was in a hurry, it resorted to extra-judiciary shortcuts and direct elimination, as it did between July 18 1936 and February 1937. Observance of due process of law was ignored April and May 1939 and again later, during the 1947-1949 Triennium of Terror.

The fact that the 1939 Law of Escapes was applied in almost every town and village that was finally occupied by the insurgents throughout the Centre-South region, demonstrates that this was not a spur-of-the-moment, improvised series of actions, as some authors have said, but rather that it was an intentional, clearly programmed, planned and carried out objective.

The records tell us, for example, that in Villarroledo, Albacete, these crimes were perpetrated without due process of law in the Los Barreros region, although there appears to be no detailed study of this. There is, however, more detailed information for some villages in outlying regions of Extremadura, near Badajoz.[i]

In Villarta de los Montes, for example, the new Falangista Mayor, Carlos de Rivas Molina, who had already held this position during the two previous dark years of the insurgency, ordered an immediate cleansing of the town. Speaking from the Town Hall balcony, he told the townspeople that since the Nationalists had the balls to win the war, they now had the balls to clear up the town. He really meant it.

16 May 1939, without waiting for a military tribunal and not beating about the bush, he ordered a raid against 23 individuals from Hoya de Fernando, who were to be 'taken for a ride', the infamous *paseo*.[1] The next day at dawn, the prisoners were removed in two groups from the garage-cuum-prison in which they had been held. The first victim was Republican ex-Mayor, Julián Molina, an ordinary breeder of goats, who had earlier strived to prevent any kind of violence in the village. No matter. The rightwing supporters of Franco were determined to wipe the slate clean of all Republican office holders, authorities and other civic leaders, just because of who they were. They then turned to relatives, friends and neighbours, as we see in this case where the raid included two of the ex-mayor's brothers: Aurelio and Lisardo. Julián Molina's son Honorio was also scheduled to be shot but he managed to escape and flee to the mountains. Julian Molina's wife Marciana Merino died in Mérida jail and one of their daughters, Eleonor, died at home after being subjected to grievous torture. That is how Francoism decimated entire families and why, for this and so many more reasons, we speak of 'extermination' in this book.

In Tallarubias, still in Badajoz province, in one of these raids May 17 1939, the prisoners were taken for a paseo to Cuesta de la Escalera on the banks of the River Guadiana. Killed were a Republican Guardia Civil, Vicente Montalbán Prieto, a Justice of the Peace, Ángel Fajardo, several ex-town councillors, two or three ex-municipal policemen, and several others. Their bodies were left where they fell, unburied, at the mercy of the wild animals. Ignacio Cendrero was killed later, his body tied to a donkey and paraded around the town.[ii]

There were similar cases in every village in the region: Navalvillar de Pela, Casas de Don Pedro, Fuenlabrada de los Montes, etc. There were so many that Ángel David Martín Rubio, pro-Franco Church historian, could not avoid talking about these paseos in his work[iii], curiously referring to data from the Causa General investigation archives.[2] He mentions that regarding Esparragosa de Lares, the

---

[1]   At noon, late afternoon or at night, a knock on the door would announce the arrival of a truck with soldiers to take someone 'out for a ride', or *paseo*. Execution usually followed afterwards by the roadside or at dawn the next day, in the jail.

[2]   The *Causa General*, or General Cause, is the name of an extensive investigation by the fascist Ministry of Justice in 1940, regarding the 'criminal acts committed throughout the country during the Rojo rule'. These archives can be consulted on the Spanish Archives website at www.pares.mcu.org .

archives record two burials May 1939 in some very deep trenches on La Horca farm, and in Peloche, two burials in the field of individuals whose names he was unable to determine but who were definitely disaffected with the National Cause, had enlisted in brigades of Rojo volunteer militia and were executed when this village was 'liberated'.

He also states that in Fuenlabrada de los Montes, there is an entry referring to 17 initial executions by firing squad when that village was also 'liberated'. These are said to have died of 'acute hammorhages caused by a traumatic agent unleashed by soldiers belonging to the military police'. These individuals had been arrested by the Francoist military and, generally speaking, before and during the Glorious Uprising, all were extremely prominent individuals who were disaffected from the Holy Cause and had also been directly or indirectly involved in the murder, imprisonment and seizure of persons of authority in this village.

Jacinta Gallardo's research in Badajoz province villages[iv] confirms the harash reality of the consequences of the victory suffered in Cordoba capital and province. When Villanueva de la Serena was occupied by Nationalist troops in the summer of 1938, the first bloody tribute to the Cause was the summary execution of 59 individuals. 50 villagaers were victims of the Law of Escapes in 1939: 30 in April and 20 in May. A ceremony honouring the 300 persons assassinated by Francoists in this village was celebrated October 20, 2012. Of these, the deaths of only 125 are recorded in the municipal Civil Registry. This gives the reader an idea of the failings of Civil Registry offices in all of Spain, a feature that has been more than demonstrated in numerous published essays and articles.

A recent article by Agustina Merino Tena[v] provides additional significant data regarding Villanueva de la Serena. Athough earlier, Jacinta Gallardo had managed to calculate that 110 persons had been executed: 93 men and 17 women, at first, Agustina Merino was able to document as many as 282 executed: 259 men and 23 women. May 2012, further research produced a new total: 291 executed, plus an additional 27 missing whose actual fate was unknown. Later again, 318 victims in all, plus 18 also sentenced to death and executed on unknown dates, for a sum total of 336. Considering that this was a tiny village with a population of only 15,000, the total number of executions is especially appalling.

One cannot overstate the extreme imçportance of this research and the overwhelming weight of the conclusions, as ongoing research into the Francoist repression continues to bring new data to light by which the numbers are constantly corrected upwards. This is a dificult historiographic undertaking and an increasigly important one for which there will be no end as long as there are attempts to determine all the resting places of the Republican fallen and, with their identification, reduce the number of *disappeared.*

In Navalvillar de Pela, at least 68 are known to have been taken for a paseo, force-marched to the tune of the victory. There were many more 'irregular' victims on various dates, some because of reprisals during the persecution of disaffected fugitives, *manu militari,* and only 38 by a court-martial. One individual from Navalvillar taken for a paseo May 14 1939 was the Republican ex-Mayor, Lorenzo Gallardo. An additional 33 victims fell under the Law of Escapes in April and 35 in May 1939.

In Orellana la Vieja, 63 were taken for a paseo in 1939: 9 in April and 54 in May. Such a mass raid was repeated in Peña del Mentiro May 13 (48 men) and May 14 (6 women). Under the excuse that the arrested were being transferred to a prison in Puebla de Alcocer, three trucks were filled with prisoners. The third truck was stopped on the orders of the parish priest, Rev. Ramón Cordero, and the prisoners were executed on the side of the road. That night, Falangistas ran all round town hunting down the relatives of the victims and the imprisoned and beating them with sticks, thus adding to the terror by which they ruled the townspeople. In 1941, at the Las Gargáligas farm, Juan Cerro and his wife Brígida Ruiz Sierra were arrested because they were suspected of being in contact with freedom fighters. She was shot under the Law of Escapes, and he was 'made to disappear' in the neighbouring hills.

In Don Benito, which had already suffered so much in 1938 (191 victims, summarily executed without due process, including several dozen women), Jacinta Gallardo lists 61 victims under the Law of Escapes in 1939: 46 in April and 15 in May. Their cause of death is officially given as an 'act of war', although this occurred April 1939, long after the end of the civil war. In all, Francoists executed 309 persons in Don Benito, of which 49 were women.

Juan Casado Morcillo, who had served as Civil Governor of Badajoz province, with headquarters in Castuera since March

1937, is another who 'disappeared' under the Law of Escapes. His grandchildren recently contacted Moreno Gómez for assistance as their investigations into the circumstances surrounding his execution and place of death had led to naught. They only knew that he had been arrested by Nationalist troops in Calzada de Calatrava, from where he was taken to Don Benito prison. May 10, 1939, their grandmother was given her husband's clothes but no details. He was presumably executed in Castillo de la Encomienda.[vi]

Historians and academia must stop ignoring the events of the April and May 1939 Black Spring. This period needs to be studied in greater detail. As shown, the entire Centre-South region of Spain that was the last to fall to Franco, was forced to make a first blood tribute to the victors, in the form of extra-legal post-war summary executions. In the small village of Chillón, Ciudad Real, local Falangistas and military declared that although a first lesson was being taught to all the towns and villages and that 'they were not finished'. Some months after the end of the war, June 2, 1939, 9 prisoners from the jail were chosen for a paseo, mostly farmworkers and people of no political importance, but including Manuel Puebla, a schoolmaster (another victim of the crusade against teachers) and Bernardino Gallego, a 17-year-old youth. The prisoners were taken from jail at dawn in several vehicles and told that they were being transferred to Almadén. When they reached the El Contadero farm (outside Almadén), they were executed, 3 June 1939. News of this crime against humanity spread rapidly. When the relatives of the deceased reached the locale, they found their bodies partly buried and eaten by wild animals.

Proof that the pack of hounds of victory thirsted for blood and that this occurred throughout the Centre-South region, is the case described earlier in Chapter I regarding the slaughter of 50 sailors of the Republican fleet, executed on board their ship 14 April 1939 as the Francoist Admiral Moreno was bringing the Navy back from Bizerte, Tunisia, to Cádiz.

The peculiar hypothesis that all post-war executions were the outcome of lawful summary procedures is totally incorrect. During the post-war period, orders were also given in an irregular and arbitrary manner for extra-legal executions. The applcation of the Law of Escapes is well documented in the territories that were the last to fall to Franco, most specifically in the north of Cordoba province, as recorded in the orders and communications that the military authorities sent to the Civil Registry. In all these documents, there is a profusion of the following stated cause of death: 'shot whilst attempting to escape from the guards who were accompanying him/her'. Again, we note the Francoist Regime's older and the newer, even more forcible, use of terror.

There was nothing improvised in this latest burst of terror; it had been carefully planned for the period following the victory, as was the terror that was unleashed at the beginning of the coup in 1936. The lists of those who were to be taken were prepared at Military Police Headquarters and the executions were carried out by a unit of soldiers led by an officer under the orders of the local Military Commander. These units worked in close collaboration not just with the Military Police, but also with the support of the civil authorities and the Falange, who officially represented the local 'Information' or 'Purification' Committee.

In Los Pedroches country in Cordoba province, the first implimentation of the repressive plan, the so-calleld 'bloody aperitif', was directed from Pozoblanco, especially in April 1939. Some of those executed from the townships in this country were listed twice: once in Pozoblanco and again in their respective hometowns. This was the case of 6 individuals from Villanueva de Cordoba, 6 from Villaralto and 2 from El Viso. It appears that all who were arrested were taken to Pozoblanco, where the Military Pokic and Falangistas took care of eliminating them.

Bartolomé Cabrera Peralbo tells of his personal involvement in the first raids in Pozoblanco when he was taken by a couple of soldiers to the Party prison where he and seven others occupied the next to the last cell on the left, at the entrance. Every evening at 8 p.m., a truck came to the door of the jail and waited with its engine running and they would hear the bolts being drawn from the doors of other cells from whre they removed prisoners whose names were on a list. April, his brother-in-law, Domingo Sánchez Redondo was one of the chosen. To this date, this family have had no further news of what happened to him, other than that he fell under the Law of Escapes, like so many others.[vii]

Although the above is an approximate idea of the extra-legal repressive actions that occurred throughout Spain during the Black

Spring of 1939 as part of the first 'cleansing' operation, before the numerous courts martial began to act, these were more in the nature of isolated, improvised incients. In the Centre-South region of the country, however, finally occupied at the end of the war, these paseos took place during the Victory Walk, in the afternoon, at noon, or at nightfall, and ended either in a roadside ditch or in the local cemetery.

The repressive phenomenon itself was much more widespread and not always under the guise of a paseo and the means employed were more brutal and ruthless. Máximo Castro, who had belonged to the local War Committee in 1936, was one of the first 13 individuials executed in Pozoblanco. All witnesses for Rufino Fernández *El Poleo* Alcaide agree that he died from a beating. Gaspar Jiménez was killed just because he was engaged to Tomasa Díaz, Director of the female branch of the Young Socialists, a twenty-year-old girl who was executed in October after unbelievable humiliation and torture. Sebastián Márquez Romero, a nineteen-year-old from a well-to-do family, was never forgiven by the Pozoblanco Falangistas for having spent the entire war in the Republican zone, was executed April 13.

Francoist disrespect for its victims knew no bounds – Nationalist Corporal Rejas, from Triana in Seville, was frequently heard singing during the executions in Pozoblanco. The first allegations, jokingly known as 'paper kites of death', were often signed by a Salesian priest, Rev. Antonio do Muiño. This was the true face of victorious Francoism, not the aesthetic visions blurred by the fog of the passage of time that those involved still attempt to present today.

One of the first residents of Torrecampo to be imprisoned in Pozoblanco, where he was executed, was José Romero Moyano, President of the local branch of the Spanish Socialist Workers Party – PSOE. Acisclo Romero Luque, another resident from Torrecampo, was taken from the fields where he was watching his goats and taken to Pozoblanco. A humble peasant belonging to the UGT, he was denouced by a right-winger from his village (Antonio Cantador) because when the latter was imprisoned in 1936, Acisclo who was serving as a guard at the jail, did not let him go home to sleep at night. Acisclo's daughter, when going to see him in Pozoblnco jail, arrived when he was being taken to the cemetery in a truck. He was executed in Pozoblanco April 15, together with Miguel Romero Vila who for a time had served as Mayor of Torrecampo during the war.

Acisclo left ten orphaned children with no more of a Last Will and Testament than the following lines that he was able to get to his wife, written on a piece of paper he had hidden in a bit of bread:

> "I am accused of crimes that I did not commit. In addition to stealing the cape that covers the statue of the Virgin, that I stole blankets and bedspreads from houses, and dozens of silver forks. Lies, all lies. The greatest lie of all, was Antonio's complaint. We will never see each other again. Resign yourself and continue to watch over our children. Don't tell anybody about this paper."[viii]

The list of the first victims of the Law of Escapes in Pozoblanco and Los Pedroches April 1939 was obtained from the following sources: Civil Registry for Pozoblanco and neighbouring townships, Pozoblanco Cemetery Registry, research for my 1987 book, Gabriel García de Consuegra's book and numerous unclassified notes Fernando López from Pozoblanco lent me.[ix] In total, 101 individuals were slaughtered under the Law of Escapes and the repression (some due to the savage beatings received), almost all April 1939.

Attention is drawn to the number of 'unknown' entries for this period in 1939, which is how the above burials are identified in the Book of Burials. On the other hand, the deaths recorded in the Civil Registry appear to inbdicate a greater involvement of the military as these individuals are recorded as executed in the cemetery, whereas the references to 'unknowns' state that they died in the gutters and in the countryside, thus indicating that the latter were presumably the work of civilian vigilantes or Falangistas. The Law of Escapes would return to Cordoba during the Trienniun of Terror (1947-1949), whenever Francoists were in a hurry or when a crushing action against the disaffected population was desired and the formalities of military tribunals considered as no more than so many bits of paper.

TABLE III
Victims of the Law of Escapes March/April 1939 in
Pozoblanco, Villanueva de Cordoba & Neighbouring Townships

Pozoblanco (identified 14, unknown 34)	48
El Viso	10
Torrecampo	9
Villanueva de Cordoba	11
Villaralto	3
Villaharta	3
Hijojosa del Duque	2
La Granjuela	2
Adamuz	1
Santa Eufemia	1
Villanueva del Duque	1
Montoro	1
Conquista	2
From outside the province	13
Total	**101**

The first bloodshed in the north of Cordoba province also occurred in the entire recently occupied Centre-South region of the country: Northeast Badajoz, Toledo, Ciudad Real, Albacete, Cuenca, Guadalajara, and so forth. There are, however, very few studies to provide the exact number of those who died from torture and the Law of Escapes. With victory, the lives of the defeated became totally worthless. Mass imprisonments, general humiliation, weeks of starvation, the Law of Escapes, paseos and savage beatings were their fare.

The following is an example of a report sent by local Military Police Headquarters to the Civil Registry[x], giving the names of individuals killed under the Law of Escapes:

"To the Municipal Judge of Pozoblanco
As required for purposes of the Civil Registry records, I hereby transmit the list of individuals from your town who died as a result of the wounds they suffered when attempting to escape from the guards attached to these Headquarters. Their names are written in the margin of this notice.
May God watch over you for many years.
Pozoblnco, 25 April 1939. The Year of Victory
Sergeant (illegible) on behalf of the Senior Captain."

Names written in the margin: Antonio Herruzo
Cejudo, age 26; Máximo Castro García,
age 28; Diosdado García Cruz, age 49

Table IV
April 1939 Victims of the Law of Escapes in Cordoba Province

POZOBLANCO CIVIL REGISTRY AND BOOK OF BURIALS

Name, age, profession, where died/buried

31 March 1939
　　Tomás García García, 75, farm worker, Pozoblanco

6 April 1939 at 8 p.m.
　　Manuel Linares Ruiz, 52, farm worker, El Viso
　　José Ollero Sepúlveda, 47, farm worker, El Viso
　　Cristino Cebrián González, 41, farm worker, Pozoblanco
　　2 Unknown. Common grave.
7 April 1939
　　Esteban Pérez Ruegón, 48, Villaharta
　　Antonio Ortiz Arias, 58, Pozoblanco
8 April 1939
　　3 Unknown. Common grave
10 April 1939 at 8 p.m.
　　José Romero Moyano, 30, peasant, Torrecampo
　　Felipe Madueño Orellana, 25, Villaralto
　　Juan Serrano Tornero, 28, Pozoblanco (at 3 p.m.)
　　Antonio Bociagas Ruíz, Santa Eufemia
　　1 Unknown, in Pozo del Concejo, Pozoblanco (Eulalia Grande's nephew).
11 April 1939, at 6 p.m.
　　Gaspar Jiménez Cebrián, 25, manual worker, Villanueva del Duque
12 April 1939, at 4 p.m.
　　Antonio Blanco Balsera, 51, miner, La Granjuela
　　2 Unknown. Common grave
13 April 1939, at 1 p.m. and 4 p.m.
　　Juan Rísquez Ranchal, 25, farm worker, Torrecampo
　　Francisco Romero Sicilia, 42, farm worker, Torrecampo
　　Sebastián Marquez Romero, 19, shopworker, Torrecampo
　　José Sierra Murillo, 32, farm worker, La Granjuela
　　3 Unknown. Common grave.
14 April 1939
　　Antonio Conde Gutiérrez, 27, farm worker, El Viso
　　Andrés Pérez Alamillo, 33, Torrecampo
15 April 1939
　　Aciclo Romero Luque, Torrecampo
　　Miguel Romero Vila, farm worker, Torrecampo
　　Juan Puerto Zarco, 55, farm worker, Torrecampo
　　2 Unknown. Common grave.
16 Aril 1939
　　Manuel Rodríguez Fernández, 23, Constantina
17 April 1939
　　1 Unknown. Common grave.

18 April 1939
  Manuel Gómez Carrillo, 53, El Viso
  Emiliano Jordán Obejo, Torrecampo
  7 Unknown. Common grave.
19 April 1939
  Justo Casado Escribano, 56, Villaharta
  1 Unknown. Common grave.
20 April 1939, at 3 p.m.
  Alfonso García Torrico, Villaharta
  Modesto Rubio Rubio, 21, Villaralto
  Rafael Luna Gómez, 22, Villaralto
  Antonio Muñoz Gómez, 24, Villaralto
  Paulino Sánchez Muñoz, Villaralto
  Manuel García García, Villaralto
  3 Unknown. Common grave.
21 April 1939
  Blas Risco Salazar, 49, Siruela (Badajoz)
22 April 1939
  8 Unknown. Common grave.
25 April 1939
  Antonio Herruzo Cejudo, 26, Pozoblanco
  Máximo Castro García, 28, Pozoblanco
  Diosdado García Cruz, 49, Pozoblanco
  Pedro Fernández Calleja, 30, Pozoblanco
  Domingo Sánchez Redondo, 52, Pozoblanco
  Manuel Rubio Garrido, 25, Pozoblanco
  Diego Alcaide Porras, 34, Pozoblanco
  Rufino Fernández Alcaide, 49, Pozoblanco (from torture)
28 April 1939
  1 Unknown. Common grave

VILLANUEVA DE CORDOBA CIVIL REGISTRY
Name, age, profession, where died/buried
Author's comment in brackets

27 March 1939
  Miguel Salado, 40, farmer, refugee from Almodóvar del Rio.
  cause of death: gunshot wound in the precordial region
  (Two Moroccan soldiers killed him before they raped the women on the farm)

1 April 1939
  Pedro Capitán Moreno, 45, cattle breeder
  died in the farm worker. (He was later found in a well).
8 April 1939
  Benito Pozuelo Regalón, aged 28.
  died in hospital from pulmonary congestion (Tortured)
20 April 1939 at 4 p.m.
  For trying to escape from their guards (according
to Military Police Headquarters, M.S. 8):
  Juan Huertas Torralbo, 45, manual worker, Villanueva de Córdoba
  Juan Pedraza Garcia, 44, businessman, Villanueva de Córdoba
  Diego Montoro Luna, 66, cattle breeder, Villanueva de Córdoba
  Manuel Cruz Coleto, 25, bricklayer, Villanueva de Córdoba
  Sebastián Molinero García, 45, manual worker, Villanueva de Córdoba
  Bartolomé Calero Sánchez, 46, manual worker, Villanueva de Córdoba
24 April 1939
  Gabrino Cabrera Expósito, 37, blacksmith
  died in La Preturilla jail (Torture and suicide)

23 May 1939, at 5 a.m.
Miguel Huertas Caballero, 43, bricklayer
cause of death: cerebral haemorrhage (Tortured)

MISCELLANEOUS COURT RECORDS - Deaths

CONQUISTA COURT
12 April 1939
Juan Luque Moreno, 24, outside the village

MONTORO COURT
20 April 1939
Miguel Cañuelo Lozano, 30, baker, Marmolejo

VILLARALTO COURT
25 April 1939 in the outskirts of Pozoblanco
The 6 individuals listed for Pozoblanco on the 20th of the month
Also Felipe Madueño Orellana, 24, manual worker, Villaralto; recorded in the
Pozoblanco Book of Burials as 10 April at 4 p.m.)

EL VISO COURT
14 April 1939
José Aranda Murillo
Rafael Díaz Helvio

2 May 1939 – between Pozoblanco and Villanueva, by soldiers who were guarding them.
Manuel Teno González
Juan García Rubio
Marino Ramírez Moyano
Juan Muñoz Hoyo
Also José Ollero Sepúlveda and Manuel Linares
Ruiz, recorded in the Pozoblanco Book
of Burials 6 April

ADAMUZ COURT
15 April 1939. Cause of death: gunshot wounds
José López Ayllón, farmworker
1 Unknown, in the farm worker.

HINOJOSA DEL DUQUE COURT
19 April 1939, in La Gutierra
Sebastián Martínez García, businessman, Hinojosa
Prudencio Garía Gómez, 50, retired Lieutenant, Santander
Pablo Gómez Leal, 18, manual worker, Hinojosa.

Several cases were recorded for Villanueva de Cordoba during
Black Spring 1939 such as that of Benito Pozuelo Regalón, aged 28,
died from torture April 7, officially from pulmonary congestion but
most probably from torture, and that of Captain Gabino Cabrera
Expósito of the Garcés Batallion, who committed suicide because of
the beatings and torture, recorded April 24. The description of Miguel
Hertas Caballero's ordeal in Villanueva de Cordoba, which led to his

death May 23, is of particular interest as it describes an all-too-typical nightmare in Spain.

Miguel Huertas Caballero, having spent a month and a half in Valsequillo concentration camp, obtained the required good behaviour references, was released and returned to his hometown where he was amost immediatley re-arrested. It appears that José *Laurentino* A. Díaz' widow, a woman who spent her time falsely accusing individuals, had denounced him. What had happened was that the day that Republican forces counter-attacked and re-captured the town from the Nationalists in 1936, some members of the Republican volunteer militia had found Laurentino, a wealthy landowner, hiding in his house on Calle Real. They took him and the parish priest to Plaza Laguna del Pino where they shot them both. Although those responsible were miners from elsewhere (Puertollano), Laurentino's widow still accused Miguel Huertas of having been seen on Calle Real that day. That was enough. To make matters worse, he was a Communist and in 1936 had gone to Madrid where he enlisted as a volunteer in the 5th Republican Regiment and spent most of the war fighting. Migel Huertas did not resist his first savage beating. Mortally injured, he was taken from La Preturilla prison to the hospital where, he died some hours later from a 'cerebral haemorrhage'.

There was much talk in Villanueva about the killing of a group of six prisoners April 20. When that morning their relatives went to the prison with their breakfast, Fructuoso, the municipal policeman, told them that they had gone to Pozoblanco to make their statements. But the prisoners were already dead, executed outside town at the beginning of the road to Conquista, next to the second group of water tanks, where they were buried. This is one of the many unmarked mass graves that still exist in Cordoba province. At least three of the victims (Juan Huertas, Diego Montoro and Juan Pedraza) were killed because of the widow Diaz' accusations that on the day that her husband was killed, somebody saw them on Calle Real when he was taken from his house. (This was not surprising, as that day, all the working men in town were out on the streets.)

Of the above, the victim that the people still talk about today was Manuel Cruz *El Picachos* Coleto, aged 25. He was accused of being a deserter, because although the war caught him doing his military service in the Francoist forces, in Cádiz, when the troops passed through the El Muriano region, he took advantge of this to cross over to the Republican zone. Furthermore, apparently some years earlier he had had an argument with a powerful local Falangista, Diego *El Chunga*, the person who would be responsible for Manuel's ordeal. At the end of the war, he was arrested in Torrenueva, Ciudad Real, and sent to a concentration camp where he remained for a few weeks. Released April 18, he returned to Villanueva with his wife, Isabel González Jurado. As she told it,[xi] they arrived home at 10 p.m. The next day, the Guardia Civil (led by El Chunga) came to arrest him and take him to the infamous a Pretuilla prison. There, they beat him viciously, cut off his testicles and applied all sorts of torments. They kept him alive during the night just to make him sufffer. When he was taken to be executed at dawn the next day, he had almost bled out from the amputation of his genitals.

Now for all those who today continuously speak with fervour of 'psychological support' to relatives of victims, do stop and remember how it was in those days. Whilst Juan Pedraza, one of the above vicims, was still in the realm of the living, a municipal policeman, Emilio *El del Lunar,* went to his house, grabbed Juan's daughters and put them to work as servants for local Falangistas.

There are at least three cases of the Law of Escapes in Hinojosa del Duque. In the case of one of those, Pablo Gómez Leal, executed at the age of 18, the Civil Registry indicates that he died in La Cutierra. An email Moreno Gómez recently received from one of Pablo's nephews, following a notice he had posted online on the *Todos os Nombres* project website, states that Falangistas shot him at the door of his home, just because of his kinship with two local leftist leaders, his brothers Antonio and José, *Los Vidal*, both of whom were also executed.[xii]

Names that will be forever unknown are those of the 'disappeared' prisoners who were taken and summarily executed *in situ*, in the Valsequillo, Los Blásquz, La Granjuela, Cerro Muriano and other provincial concentration camps, and in Castuera. There are numerous testimonials that in all these camps, prisoners were plucked from their huts and shot.

In Castuera, there was a continuous procession of trucks leaving the camp, packed with prisoners on their way to their execution. Castuera eyewitnesses have told us that a hut that also functioned as chapel was an anteroom to death as those who entered it were never heard of again All of this reinforces our belief that the true dimension of the Francoist genocide shall never be totally quantified. The Francoist authorities did an excellent job of getting rid of all the camp records and other documentation. A totalitarian system, we know, not only engages in the physical elimination of persons as it takes pains to erase all evidence of their crime.

The Farce that was Military "Justice"

> *"There were 100 of us in our cell; there are only 73 now.*
> *Dear foreign comrade, we three are also sentenced to death and*
> *We shall be executed either tonight or tomorrow.*
> *You, however, may survive; if one day you are free,*
> *you must tell the whole world of how we were killed here,*
> *just because we wanted freedon, not Hitler."*
> *Signed: Three Republican Guerrillas*

Letter written on a little ball of paper that the condemned men managed to send to Arthur Koestler in the cell next to them. March 11, 1937, Seville prison.[xiii]

Proliferation of Military Courts

The Franoist Regime's first great purge of the social base of the Republic (genoide, purely and simply) that began July 17 1936, was enforced with no more rule of law than the fledging insurgent government's wartime decrees, the so-called *Bandos de Guerra*.[3] One hundred thousand persons were exterminated by Francoism on the basis of those decrees alone.[xiv] As far as Andalusia is concerned, it was only February 28 1937 that ex-Republican General Queipo de Llano, presumably on orders from Burgos, decreed the abolition of his wartime decrees and introduced the subject of preliminary pretrial hearings and the intervention of military tribunals. (Of course, 'military justice' is no Justice, just as what military music is to Classical

---

[3] The crucial *Bandos de Guerra* were:
National Defense Junta decree of July 28 1936, approved as Decree number 79.
Decree of August 31, 1936.
Decree from General Franco number 55 of Novembr 1, 1936.

Music; in, other words, nothing). All the military governors of the provinces under Queipo's command were sent the following telegram:

"All authorities within my jurisdiction are hereby ordered to abstain from directing the application of my wartime decrees in which the ultimate sentence is imposed; they must henceforth follow the legal procedures indicated by the military Legal Advisor and gather the greatest possible amout of evidence against all the detainees so that the urgently re-established courts-martial may rapidly proceed with the due sentencing. Receipt requested."

Theoretically at least, Queipo's wartime decrees ceased to exist March 1937 when the military courts became fully operational. We must remember, however, that the Francoist courts were already functioning in 1936 in a few exceptional cases such a those involving loyal members of the militiary, Republican authorities and other influential individuals, Still when it suited Francoism, it resorted to shortcuts to eliminate persons, without any fear of reprisal or due process of law, by means of unlawful arrests and raids, and the extra-legal application of the Law of Escapes.

This was the scene for the events of the Black Spring of 1939, and the pre-trial hearings of hundreds of thousands of individuals indicted on military grounds. In Madrid alone, at the end of 1944, 128,000 hearings or indictments are recorded, almost all collective. More than 300,000 persons were investigated, according to data on file at the First Military District Archives in Madrid. If to this sum, one adds the numbers for all the 9 Military Districts, the total is quite simply, horrifying and difficult to accept. In Cordoba province alone, more than 26,000 individuals were indicted in 1939, 37,000 in 1940 and so on afterwards. The closer we come to compting the multitude of Spaniards who were tried by Francoist military courts, the more astronomical the numbers.

The first thing that strikes historians is the proliferation of courts martial in 1939. All the victorious military considered themselves as being on duty and following orders when they sent people to the firing squads. Today, we are baffled at the thought of where did so many high-ranking officers come from to enforce the victory and as mgistrates, apply Franco's justice against so many defeated compatriots. The whole of Spain was filled with military judges, military magistrates, military tribunals and courts-martial. Franco mobilized the army and the Guardia Civil for his *final solution*, the definitive, physical and moral extermination of all Spanish Republicanism. The armed forces became the Dictator's Pretorian Guards. Individuals were executed without any fear of reprisal, they were tortured without exception, there were mass imprisonments, prisoners were starved to death, all the disaffected (including many who were neutral) were made to bite the dust as the wholespread wave of terror came into being. This is no exaggeration; that is exactly how it was. It was the final solution for the program that had been started in 1936. If anyone still doubts this, Moreno Gómez reommends that he read, *Testimonios de mujeres en las cálceles franquistas,* [xv] and after studying this Bible of Francoist repression, he might speak with some knowledge of the facts. When a biased historian does not listen to the victims (the so frequently despised oral sources), the rest is no more than cacophony in his ears.

The entire repressive machinery was in the hands of the military legal system, beginning with Queipo's wartime decrees in 1936, and following the victory, with the courts martial. Furthermore, as military justice is not very good at making distintions, everyone was accused of the crime of military rebellion, in accorance with the Code of Military Justice (articles 237 et al., namely articles 86, 287, 288 and 289). This code was complemented by a maelstrom of repressive legislation, which indicates that not only did the wave of repression not diminish with Franco's victory, but it grew and grew until it was a sweeping, all-inclusive attack against defeated Spain.

The general post-war repressive program was structured according to a wide range of major legislation:[xvi] the Law of Political Responsibilities (February 9 1939); Law for the Repression of Masonry and Communism (March 1 1940); Laws for the Security of the State (July 12 1940 and March 29 1941); Law amending the Crime of Military Rebellion (March 2 1943); Decree-Law on Military Rebellion, Banditry and Terrorism (March 18 1947); Law of Public Order (July 30 1959); Decree on Banditry and Terrorism (Sepember 21 1960) and Law 15/63 creating the Court of Public Order (TOP). As Captain Díaz Criado at Queipo de Llano's General Headquarters stated, the purpose of all this legislation was: "So that thirty years from now, not a soul will remain alive". How distant these barbaric laws were from the 'fair and wise laws' of the 1812 Constitution.

As the military coup itself was illegal, its repressive legislation was tainted from the onset. Furthermore, it fell into the ridiculous aberration of so-called 'reverse justice', that is, the insurgents who led the coup condemned as rebels all those who had remained loyal to the constitutional government and were opposed to the military coup – the Glorious Movement as they called ait. International jurisprudence has not taken this type of reverse justice seriously, nor given due consideration to the tragic implications of this way of thinking. Likewise, if Francoist legislation was corrupt from the beginning, much worse was the jurisprudence that regulated the courts martial whose farcical nature has been pointed out by everyone who has studied the matter. Franco's so-called military justice was a farce in terms of the pre-trial investigations, a travesty of due process of law in the way it accepted denunciations and unsubsantiated accusations, a disgrace regarding the crimes allegedly commited by the defendants, and a shambles as to how the courts martial handled the proceedings.

The Francoist legislation was only repealed by the Spanish government thirty-two years after the death of Franco, under <u>Law 52/2007</u>. At the same time, Article 3 of Law 52 reversed the rulings of the Francoist courts and the sentences of the courts martial, declaring that they were illegal because they were 'fraudulent in form and substance'.

As everything and everyone came under military Francoist justice, one could say that civilian jurisprudence went on holiday for almost forty years. Even the theft of a bag of acorns could be considered an act of military rebellion if, to this crime, the prosecution added the charge of having given a handful of acorns to a guerrilla. It was both a barracks *and* a convent because the Church worked hand-in-hand with the Regime as it also stoked its own fire.

With victory in their hands, Francoists were faced with the massive task of organising the trials for more than 280,000 prisoners that they had arrested during the first year of the National-Syndicalist Revolution. Ten provisional Inspectorates, in addition to the existing ones for the military regions, were created, as well as a multitude of military courts that found themselves totally overwhelmed during the first three post-war years.

An idea of the intensity of the repressive furore can be seen by the fact that in Cordoba, no less than 20 special military courts were set up in the capital and another 15 throughout the province, where they worked feverishly, especially in the surrounding hillsides, the last area to be occupied. Added to these, were the numerous Political Reponsibility Courts and the Courts for the Persecution of Fugitives.

The military courts for important towns also covered smaller neighbouring towns and villages: Luque and Valenzuela in Baena; Nueva Carteya in Montilla; Fuente Palmera and Hornachuelos in Posadas; Espejo at Castro del Rio; several neighbouring towns and villages in Pozoblanco; Pedroche, Torrecampo; Adamuz in Villanueva de Cordoba, and so forth. There were temporary military courts in Baena, Bujalance, Hinojosa, Montilla, Montoro, Obejo, Puente Genil, Villaviciosa and some other towns, but these are difficult to identify.

In Barcelona city, 10 military courts were created at the beginning, and another 15 in the province. In Malaga, no less than 67 military courts were opened in the city and province. In Cartagena, the Francoist Naval authorities created an unspeakable number of military courts of their own: 57. Alicante was served by 22 military courts: 8 in the capital and 14 in the province, just to mention a few examples elsewhere in Spain. The operating costs of the military courts were borne by the town and city councils in which they were located, as were the meals and other expenditure for the courts martial.

An Examining Maagistrate responsible for updating the individual case files, presided over these courts. Pre-trial hearings were held *in camera* because lawyers were not involved, nor was there any reason for them as the accused were only permitted to say a few words in their defence during the court-martial, let alone when they were being interrogated. The Examining Magisttrate was assisted by a secretary and a clerk. Also at the Judge's orders were civil, Falangista, military and Guardia Civil staff who would 'facilitate' the accused's declarations, whips in hands. This does not mean that all the so-called evidence was obtained by beatings, but there were a great many vicious beatings as we shall see. Torture was such an essential part of the committal proceedings that it was the rare pre-trial hearing that was not accompanied by some savage torture of the accused.

The Examining Magistrate, almost always a member of the military, or a loyal Francoist magistrate or attorney, was often appointed from amongst persons whose families had suffered or been victims in the days of the fight against the coup, and who sought revenge. Those who also thirsted for the blood of the defeated and who lacked all scruples had a whole future before them. The Examining Magistrate, who himself usually had personal victims to revenge, was both judge and jury. These judges were contaminated by their own interests in the cases before them and when called to do so, rarely deprecated the charges they were examining. Besides, few accused or their families would ever dare question or appeal a magistrate's ruling.

Committal proceedings were rushed summary proceedings, without any formalities or due process. In a same committal dossier, there might be a document stating that the accused was a militant Communist; another document stating that he was a militant of the Izquierda Republicana; and yet another stating that he was a militant Socialist. The information was usually unsubstantiated as it was usually based on uproven declarations or accusations. In other cases, presumption was the rule: everyone had been involved in strikes during the Republic, therefore everyone was guilty of 'extremely bad behaviour' at the very least.

An Examining Magistrate's kinship with a leftist victim was apparent in quite a few cases in Cordoba province, but this did not always work in an accused's favour. Such was the case of Manuel Fernández Contreras, a barber, who was arrested during the post-war repression. An attorney from Pozoblanco, Juan Calero Rubio, had been appointed Judge of Military Court number 11 in Villanueva de Cordoba. Although the Judge had lost a son-in-law during the revolt against the uprising in his town, he was able to escape to the Francoist zone with the help of some leftists, including the barber. When the Judge was called to try Fernández, although he acknowledged that Fernández had helped him escape to the 'other' side, he never forgave him for not having saved his son-in-law and did nothing to prevent Fernández' execution.

If Francoism placed these magistrates at the vanguard of the killing machine, it was because each was known to lack the minimum sense of clemency or humanity. Juan Calero, mentioned above, was one of the most despised, amoral and cruel of these. Arriving in Villanueva de Cordoba as a Military Judge in the first days after the victory, he sowed terror in a jail that was packed with men and women prisoners from Villanueva, Pedroche, Torrecampo and Adamuz. Not only did he organize numerous pre-trial summary hearings that dictated several death sentences, but he also initiated a program of beatings and torture that still haunts the townspeople, beatings that he himself joined in. In 1940 he fell into disgrace with the Regime for beating a priest, but the exact reason for his final discredit is unknown. Some say that it could be because he ordered the execution of the Head of the Post Office in Villanueva, Misael López Díaz, whose family included a Francoist soldier who had managed to get his sentence reduced. Whether it was because of this or some other misdeed, the fact was that Calero had done something terribly wrong and decided to end his life with a fatal dose of poison August 28 1940, aged 53. His evil doings were many indeed, such as asking the daughters and wives of the condemnd for sexual favours in exchange for saving the lives of their loved ones. According to a relative's testimonial it was said that he asked Juan Escoriza Segura's daughter for sex but still executed her father. To Moreno Gómez' knowledge, the judge never saved anybody, and he never heard whether any woman ever agreed to the judge's demands.

Demetrio Carvajal Arrieta, a Clerk of the Court appointed Mayor and local head of the Pozoblanco Falange as soon as the town was occupied by Francoist troops, was responsible for his own reign of terror in the Cordoba militry courts.

In Rute, Bernabé Andrés Jiménez, Captain in the Military Corps of Justice, was a member of several of the courts martial most responsible for sentencing the accused to death. For example, in the first of his involvements in Baena, May 20 1939, he sent the famous workers' union leader José *El Transio* Joaquín to the wall In another court-martial in Puente Genil in which he intervened, Antonio Romero, a distinguished sexagenarian, recognized icon of Republicanism, past Mayor of Puente Genil on two occasions during the Republic, totally unaffiliated with any anarchical event or political movement, was sentenced to death by strangulation. After several tragic interventions involving this Francoist magistrate, Andrés Jiménez was appointed Judge in Aguilar de la Frontera and from there, to the Cordoba capital courts. His bloody curriculum elevated him to the benches of the Spanish Supreme Court and eventually, to his being honoured by the Regime with the Great Cross of Saint Raimundo of Peñafort. His portrait continued to hang in the Rute City Council assembly hall until 2005 when it was removed after an eloquent citizens' protest organized by Pascual Rovira. This image is now stored in Moreno Gómez Black History Archives.

In Baena, we also note the family relationship of the military judge and right-wing victims of 1936, when the War Inspectorate of the Army of Operations in the South appointed right-wing Manuel Cubillo Jiménez to the position of Examining Magistrate. A man possessed of a fanatical and fervent desire for vengeance after losing his wife and three sons in the anarchist killings at the San Francisco Asylum, he ensured that practically nobody connected with the left survived the wartime and the post-war repression. In this, Cubillo was assisted by Luis Cordoba García, Secretary of Baena city council.

In Castro del Rio, there also was a kinship between military judge Manuel Criado Valenzuela, who also served as Secretary of the city council, and was a right-wing victim of the fighting in 1936 when both his father-in-law and a brother-in-law were killed in battle. This judge was further assisted in keeping the people of the village in line by his Falangista brother-in-law Pedro Luque.

Cavalry Colonel Carlos Palance y Martínez-Fortún from Lucena presided over several courts that sentenced the accused to death.[xvii] Born in the village of Jauja (Lucena), he was the prototype of the great landowner and a rabid anti-Republican after he had numerous run-ins with the Lucena city council September 1931, because of his treatment of his farmworkers. In 1936, he was one of the leading conspirators of the military coup in Cordoba. Not only did he represent the perfect union between the agrarian oligarchy, the judiciary and the military, but his hatred of the Rojos was also fuelled when one of his brothers, Fernando, was shot in Guadalajara, a victim of the so-called Red Terror. Such individual characteristics were always of particular interest to Franco when it came to filling judiciary positions, as they were the perfect guarantee of implacable and vindictive repressive actions, devoid of any scruples.

A remarkable case of a Francoist Examining Magistrate was that of José Aparício de Arcos, from Aguilar de la Frontnera, who was posted in Montilla where he sentenced quite a few Republicans to death. What is peculiar in this case is that his brother Rafael, an eminent lawyer and a leading figure of Cordoban Socialism who held several top positions during the Republic, was executed in Cordoba capital by Fascists uring the summer of 1936. The idelogical abysm that separated the brothers was tragically typical of Spain in those times.

One of the military judges who specialized in trials against people from Posadas, was lawyer and wealthy landowner Fernando Sepúlveda Courtoy, a ruthless individual at whose farm Nationalist General Varela stayed whenevr he visited Cordoba. A fanatical Falangist, he was an intimate friend of the appalling Don Bruno and the bloodthirsty prosecutor Rafael de La Lastra y Hoces, whose rabid harangues (the highly controversial 'rantings of Don José') are described ahead.

Amid the general furore, however, some court officials kept a degree of common sense. José Espina Almansa, an attorney who had been a PCE candidate to city council in the 1931 municipal elections (how the world turns....), retained some embers of his ideals and was influential in lessening some of the sentences given to important Communist leaders, such as Adriano Romero Cachinero and Alfredo Caballero Martínez. Curiously, considering that José Espina had also been a member of the Abril Masonic Lodge in Posadas during the 1920s, it is a wonder that he escaped being purged by the Francoists and that he was ale to serve as a military court secretary.

The substance of the committal proceedings was based on three sources: the declaration or interrogation by the Guardia Civil, information from the Falange, and the number of real or bureaucratic, almost always false, accusations or allegations. In the case of the Guardia Civil declaration, thousands of lives depended on the mood of the local commander. For example, in the case of one of the most disitnguished residents of Pozoblanco, Bartolomé Fernández, ex-Major of a Volunteer Militia Regiment and as circumspect, respected and professional an individual as any, the Guardia Civil concluded that, because of his actions and bad behaviour, he was considered a disaffected danger to the Regime and society. This report was signed by the morose commander of the post, Andrés Arévalo García. Still, by some happy quirk of fate, he was not executed, as in the judicial circles of the time prisoners were classified as belonging to one of three groups: unimpressed, disaffected and dangerous. The latter inevitably doomed.

The second source of information was the report from the Falange Department of Information and Investigation (SHF). The reports from the local branches, the *Junta Local de Información*, also played a major role in thousands of lives. As an example, the case of another accused from Pozoblanco, Ventura Redondo Fernández, which reads:

"This is a person with leftist ideas, although we cannot specify the party to which she belongs. At the beginning of the Glorious Movement, she joined the Red Corps of Carabineers as a volunteer and therefore she is considered disaffected with our Cause.

Signature illegible.
For God, Spain and the National-Syndicalist Revolution.
Viva Franco! Viva Spain!"

The number of contradictions between reports regarding a single individual show how frequently there was a total disregard for the truth, as the slightest indication of accuracy shone by its absence. An example of this are the evidentiary reports presented at the 1945 pre-trial hearings of three volunteer militia from Villanueva de Cordoba. The Guardia Civil report regarding one, José *El Lobito* A. Cepas, states: "Was a Communist and participated in the arrests of individuals who were later shot." The Falange report states: "Belonged to the Izquierda Republicana" and adds the more serious crime: "Participated in assassinations".[xviii] Noticeably, the Falange reports went straight to the jugular. In effect, the Falange killed many more people with its reports than it did with its revolvers. The same can be said of the Francoist Guardia Civil, whose reports were more lethal than their rifles.

The third source was civilian denunciations and, frequently, reports from the parish priests. The latter are almost always present in the files of those indicted under the Law of Political Responsibilities, but less often in more conventional trials. Even so, in all cases the reports from the clergy were of great import and were also responsible for thousands of lives. Sometimes, the priest simply wrote the damning words: "I do not know the accused", implying that this person never went to Church and as such, there was sufficient reason for his arrest. What a disaster: condemned to death because he was a leftist and to add insult to injury, because he was an atheist as well!

Many priests became so deeply involved in this maelstrom of extermination that they informed on matters that were none of their business and well beyond their religious obligations. Rev. José Armario, parish priest in Morón de la Frontera, turned to gossip to write the following report on an accused: "Having been duly informed, I can say that Pedro *El Carabinero* García Flores, a metalworker by trade, was an eminent member and acctivist of the CNT in this city, that he made bombs and threw them against the Guardia Civil barracks and participated in the attack on the same, August 16, 1939. The Year of the Victory."[xix]

These accusations were totally falso. García Flores had left for Ronda when the war broke out amd that was all. So much for priests and the 8th Commandment. (Thou shall not bear false witness.) As far as the 5th Commandment is concerned, Moreno Gómez could only make cynical remarks in the light of the orgy of bloodshed.

Widespread flood of denunciations and false witness statements

The Francoist Regime decreed a *General State of Denunciation* as if it were an emergency measure and promoted it through the media with mantras suas as 'It is everybody's duty to report every indication against the Marxist Horde'. Every Falangista, just because he was one, was required to present denunciations, so many denunciations per person. Every right-winger was also expected to follow suit. During the first months of 1939, the Falangista daily paper in Mataró, as did the press all over the country, repeatedly published articles encouraging denunciations with headlines such as: 'Franco's Justice needs and asks all Spaniards to coooperate'. The same newspaper later insisted 'We remind and encourge all people of their obligation to cooperate with the implementation of Justice'.[xx]

Denunciation was the spark that triggered the motor of the extermination machine. For an individual to be considered 'worthy' by the New State, nothing better than to volunteer as an informer. Right-wingers who did not present denunciations were considered 'neutrals', as if this were a sin by omission. The state of generalized denunciation involved a far-reaching objective extending this complicity and collaboration to the Regime's entire social base. This collaboration, more rural than urban, was an important factor in ensuring cohesion within the Regime, and this cohesion was rewarded with high marks for loyalty and unwavering allegiance. This way, the victorious base became full members of the Movement as they contributed to the great nation-wide cleansing and extermination operation. This is in keeping with repressive reigmes the world over who have always looked to integrate their followers in their task of general repression by means of denunciations, rumours, gossip and whispers, with falsehood as the main ingredient.

In the Spanish case, some civilian accusers attempted to withdraw their allegations when they saw the very serious and unjust results, but they were so threatened by the authorities that they gave up trying. What's more, allegations from authorities, almost always very vague and clearly false, were often changed by the Examining Magistrates, with terrible consequences for the accused.

In reality, Falangistas and unconditional supporters of Francoism cooperated with the Regime in its arbitrary use of denunciations as a tool with which to ensure the great extermination or macro cleansing of dissenters, staging farcical pre-trial hearings and proceedings whose result was decided upon beforehand. A remarkable case of the capricious use of allegations and total disregard for the truth, was Juan Cantador Zamora's court-martial May 14, 1940. The entire case against him was based on Luísa Doctor, a 1936 widow's, presumed allegation. When she was asked in Court whether she recognized the accused, Juan Cantador to everybody's surprise she said no. The judge then asked the widow why she had accused him of having killed her husband, to which she replied: "No sir. Judge Juan Calero must have misunderstood me. What I said was that he was known as one of Juan Elías' kindly tenants". The ensuring situation was ridiculous. The Examining Magistrate repeatedly threatened the widow in an attempt to force her to retract her statement, to no avail. Manipulation of the allegation was obvious, but setting the record straight also was of no use, as Juan Cantaor was still executed in Cordoba September 12, 1941. Clearly, his elimination had been decided beforehand and the faked allegation was just for show.

Moreno Gómez wrote the distinguished professor Vicente Pascual Soler, Director of the Academia Espino in Cordoba city, regarding this particular and other cases and this is what the professor replied: "A career Examining Magistrate, Juan Calero Rubio from Pozoblanco and well-known in Villanueva, substantiated the charges against the accused. This representative of so-called Francoist justice took his job seriously with the collaboration of a typist and a couple of henchmen who would torture the accused. With such persuasion, the magistrate obatained the accuseds' signed confessions that were considered evidence enough to justify the sentence that had previously been decided upon for each case".[xxi]

The dark world of informants and accusations encouraged by Francoism is surely the most foetid and nauseating illustration of the character of those responsible for so-called Francoist justice. Their route down the moral sewer is the most stinking, sometimes ridiculous, always tragic, path ever imagined.

As previously mentioned, the first allegations that were examined usually were the reports from the Guardia Civil and the Falange. In addition to the bizarre examples described, there were some equally peculiar situations influenced by National-Synicalist thought in Montilla. Here, as in every town and village, local authorities received a multitude of requests for information for releases from concentration camps, forced labour battalions, and so forth. A sample of the answers from Montilla informers is a true anthology of the nonsensical. As everywhere, prisoners were accused of 'bad or extremely bad moral behaviour'. As Membership of, or association with, a leftist political party, trade union, or service in a civil or military position, was classified as 'undersirable socio-political behaviour' or 'without shame'. One accused, Aurelio Casas Cordoba, was also charged with being an 'atheist from birth'.[xxii]

Faced with this flood of denunciations of supposed crimes and outrages, it is useful to look at some data to determine what exact percentage of prisoners could actually be guilty of the supposed 'blood crimes' with which everyone was also accused. There is some data that helps illustrate this. According to the 1937 *Memoria de la Inspección de Campos de Concentracion y Prisioneros,*[4] the number of Franco's prisoners that year was 106,822, classified A, B, C, or D (D being those accused of civil crimes). Of the total number of prisoners, only 2,282 were accused of specific crimes such as seizure of property and registers, burning of religious images or sacrilege, and of course, blood crimes. Of those executed, only 2.13% were accused of common law

---

[4] Minutes of the Concentration Camps and Prisons Inspectorate.

crimes and not all of these were blood crimes.[xxiii] The accusation 'was involved in assassinations' was the joker in the pack that was used in most of the cases that received the death penalty. During their trial, the only thing those unfortunate individuals could say before they were executed, was to claim that the accusations were false. To no avail. Everything had been decided, and well decided, beforehand.

One lethal accusation by Francoist justice was that the prisoner had served on a local War Committee. When accused of this, nobody was safe. *Causa General* records clearly illustrate the purpose of this stratagem as they include detailed lists of all the leaders of working people in a town or village. Each list identifies twenty to thirty individuals who are accused of having belonged to a town's War Committee, when in actual case, it never comprised of more than five or six people, including the town Mayor.[xxiv] In Bujalance, for example, there is a list of more than twenty individuals. There are a dozen names on the Santa Eufemia War Committee list; El Viso, more than 15; Dos Torres, more than 13; Pedroche, almost 30; Villaralto, 20; Fuenteobejuna, 12; Villaviciosa, 13; Villa del Rio 17; Montoro, almost 20; Hornachuelos, 33; and so on. The fake Committee lists reported in the Cause General records were noting more than blacklists of individuals slated for execution.

If in a town some crime had allegedly been committed against one or more right-wing individuals, every Republican in the town was accused of having taken part in the crime. There are the collective denunciations we find in both official reports and in allegations from private individuals. Here again, the specialist data is the Causa General records. It is well uderstood that the real purpose of the Causa General investigation was, in addition to searching for foundations of a Francoist martyrdom, to collect the reports and denunciations that were submitted to it by Mayors, Falange Information Service, Guardia Civil Municipal Police, Courts and other sourcs. The Causa General files represent a repertoire of allegations in a town, based on which the investigators drew up a list of the most notable leftist individuals in the town and on the basis of which those individuals who were on the respective blacklist were accused of being collectively guilty of the alleged crimes(s).

That is what occurred in Posadas, for which a list of 87 right-wing victims in 1936 concludes with the accusation and the statement that 'all the accused have been imprisoned or executed'. In Cardeña, with 4 right-wing victims, the Causa General accused no less than 22 farmworkers 'all who belonged to the left wing in the town'. The Causa General also collected accusations from private individuals, such as the one filed by Elisa Gallardo Velarde, from Pedroche, against 22 persons for the murder of her husband in July 1936. One of the persons on her list was Alfonsa de la Remonta, whom she also accused of taking religious pictures home to burn to cook the stew.

Attempting to analyse the repertoire of allegations and accusations found in case files is tricky. To dare to draw the profile of an accused based on what appears in his file is dodgy because although one can read what the Francoist machinery states, one cannot know what an accused might have said, because he was never allowed to speak in his defence. A recent publication in Madrid, *Los fusilamientos de la Almudena* [The Almudena executions] by Manuel Garcia Muñoz, is totally misleading. The author gives a list of individuals executed, accompanied with an extract of the accusations against each one, many of which are clearly invented. The result is astounding. Instead of providing a list of judcial victims, it looks like a list of devils, terrorists and thieves. To make matters worse, the author attached the derogatory nicknames that Francoists liked to give everybody. This is intolerable. In addition to being a historical aberration, the entire work is offensive, purely and simply because the majority of the denunciations and accurations were false, and especially because we have no idea of what version of the events the victims could have given had they been permitted to speak.

An example of how the denouncers' omnipotent powers enabled them to become virtual owners of lives and property, occurred in Madrid. Agutina Sánchez Sariñena, whose husband was killed in the war, lived with her mother-in-law, Josefa Perpiñán. A neighbour of theirs who coveted their apartment, made a double denunciation in two separate courts, falsely accusing both women of having been leaders of the *Checa*[5] on Calle Fomento, when neither had ever heard of that street. They were tried by two courts-martial January 12 and 17 1940. Agustina managed to be released. Not so her mother-in-law

---

[5] Communist police headquarters in Madrid at the beginning of the insurrection.

who was executed July 24 1940, despite the fact that it was proven that she had spent the entire war far from Madrid, in Denia, Alicante.

Another example from Madrid shows how when Francoists could not locate an individual they were hunting, they would content themselves with a relative and make him pay. Hilario Collado, a farmer from Santa Olalla, Toledo, where his son Eugenio Corruco appears to have killed a baker in 1936, was imprisoned in Madrid. As Eugenio had fled to the mountains in 1939 and they could not find him, the Guardia Civil arrested his father instead and sent him to Madrid to be tried. The Mayor from his hometown sent a nasty report in which he stated that Hilario encouraged his son Eugenio to murder Juan Sánchez, the baker. The Mayor had invented the bit about the father encouraging his son and it was that which led Hilario to the grave. In the court-martial, the prosecutor blew the facts out of proportion as usual, stating that: "The accused encouraged his son Juan Sánchez to murder the baker from that town, and in effect he died because of this encouragement".[xxv] In other words the baker did not die because Eugenio was supposed to have shot him, but because his father supposedly encouraged his son to do so, wen the poor man was probably more involved in tending to his goats at the time. Yet another example to show how anyone who still doubts that Francoist justice lacked even a modicum of credibility, is incredibly naïve.

On the other hand, there were many cases of fanciful accusations that carried the death penalty where the accused escaped execution for one reason or another. Such was the case of Miguel Hernández who was sentenced to death for being a writer and poet of the revolution. He was not executed simply because Vicente Alexandre and other literary notables appealed for his release *in extemis*. Sadly, the result was the same because he died in prison of starvation. Antonio BueroVallejo, the playright, whose death sentence was commuted despite his having been arbitrarily accused of being an active Rojo propagandist, escaped falling off the edge of the judicial cliff by remaining in prison until 1946 when he was released.

A case in point from Reus, Tarragona, illustrates the frivolity of Francoist denunciations and acusations, for which there was only one reality: the trials were a total farce and case files were only created as cover-ups. On this occasion, Falangistas wanted to get rid of Ferrán Fontana Grau, an outspoken author and playright. Failing a specific accusation, he was indicted on the grounds that he was 'the author of several plays of a revolutionary nature and tendency, that he directed several of these in Reus theatres; that he also wrote verses of the same nature and tendency, such as the one entitled *Canción de la retaguarda* (Song from the rearguard), a poem that sums up the charges he is accused of.' The prosecutor asked for 20 years in prison, but the court-martial sentenced him to death. He was executed August 8 1939 in Pilatos prison, Tarragona.[xxvi]

On the other hand, many allegations were attributed to a case arbitrarily because there was such an avalanche of dossiers, such huge numbers of prisoners and accused, that it was humanly inpossible to put any order in the allegations or judiciary inquiries received or asked for. The entire Francoist bureaucratic sector was in total chaos and the only clear objective was that peopke had to be eliminated in one way or another, and the more, the better.

Take the case of Francisco Copado Sánchez of Villanueva de Cordoba, who was sent under arrest to be tried in Valencia. A chauffeur by profession and a Communist of little importance, he was nevertheless a candidate for the firing squad because in his pre-trial hearing, it was alleged that he was known to handle dynamite and that in this capacity he had been involved in the leftist revolt in the town. All of this was false, and Copado's wife did everything she could to save him. Maria discovered that Juan Fernández had denouncd him, but when she spoke with him, he assured her that he had nothing against Francisco. He just happened to be at the Headquarters when he was told to sign the accusation because the Court needed two signatures. He agreed to accompany the Copado family to the Examining Magistrate Juan Calero and withdraw his accusation, The judge rose to his feet angrily and furiously accused the family of dithering. Terrified at the outburst, Juan Fernandez kept quiet. Actually, someone else was responsible for Copado's demise. His accuser, Manuel Rodriguez Moreno, stated without proof that he had heard rumours tha Copado was seen on the streets the day of the revolt. Nothing specific, just insidious allegations.

It is worth looking a bit closer at this case, to see just how the dark world of the informants worked. It also so happened that Francisco Copado was first cousin to Rev. Bernabé Copado SJ, the famous Jesuit chaplain to the Nationalist Redondo Unit conscripts. Cpado's wife decided that she should go to this cousin and enlist his help in obtaining her husband's release. Francisco respinded to Maria's plea in a letter to his wife from Modelo jail in Valencia, in which he praises her attempts at getting his release, although he is resigned to his fate as he is aware that his demise was pre-ordained.[xxvii]

As it turned out, Maria had no time to travel to Málaga or do anything else, because two days later, January 11 1939, Francisco was executed in Paterna, Valencia, cemetery. There had been no commutation or review of his sentence. The pre-trial hearing that he mentions in his letter had nothing to do with his Jesuit cousin, who did go to Villanueva. The hearing was convened based on nothing more than a petition for Francosco's release, signed by one hundred right-wing and apolitical persons. The document was sent to Valencia, but it was accompanied by an insidious notation from Mayor Gregorio Pedraza who stated that the petition was a fraud and that many of the signatories were not right-wingers, but leftists. With a stroke of his pen, the Falangista Mayor destroyed all of Maria Copado's efforts and anything his Jesuit cousin might have been able to do had he so wished.

Francoist justice was not just a farce, a staged performance, or a display of vindictive rancour. Even worse, if this was possible, the allegations and accusations on which may of the sentences were based clearly indicate the extreme retroactivity of the emotions involved, as many accusations centered on events that dated back not just to 1934 but much further back, as far as 1931. The purpose was clear: the Republican system was not just to be demolished, it had to be purged and with it, the entire social base that sustained the Republican Regime. Demolish the system, demolish the base. This was "the work – the nightmare and the glory – that destiny required of military justice" in the words of Felipe Acedo Colunga, a military prosecutor and one of the Spanish war criminals who should have been allocated a seat at the Nuremberg trials.

Contrary to some authors, Moreno Gómez prefers never to use the word 'revolutionary' to describe the actions of those who fought against the military coup. The people who fought against the coup, Republicans, did not initiate a revolution; they acted in defence of the Republic, in defence of democratic Spain.[6] If there was a true revolution, it was the

---

6   When Moreno Gómez refers to Republicans, he is almost always speaking of the

National Syndicalist and National-Catholicism's involvement in the military uprising. These were the true revolutionaries, those who lay Spain flat on its back. The idea of a revolution was constantly promoted by Falangistas and the Regime's propaganda that bandied the idea of a leftist revolution to scare the population, promote hatred of the disaffected and justify the military insurrection. European Fascists of the day were so terrified of the presumed leftist 'revolution' that they went wild with their attempts to destroy the existing democracies, and Spain suffered the cconsequences of the insanity. One is reminded of Estaban Ibarra's famous words: *A ghost is running around Europe.*[7]

The menu of allegations and accusations, extracted from more than half a thousad proceedings Moreno Gómez consulted at the Pozoblanco Dirstrict Court, 1987, is extremely revealing:

- The <u>most lethal accusation</u>, punishable by death, was having 'belonged to a War Committee'.

In the early days of the uprising, trade unions and other civilian groups raised Regiments of volunteer militia to supplement the regular Republican Army. Although untrained fighters, they proved to be excellent soldiers and fighting men. These units were each commanded by a War Committee, usually civilians, who organized the units, appointed officers, obtained weapons and supplies, and so forth, were considered dangerouis 'revolutionaries' by Franco's Regime and its supporters.

- The <u>second most serious accusation</u> was 'exhibits extremely bad conduct', in other words, was a leftist or had been a militant

---

population of southern Spain, because he is an expert on the *Hispania ulterior*, south of the River Ebro, not on *Hispania citerior*, the region north of the River Ebro

[7] Ibarra, Esteban. "A ghost is running around Europe, the spectre of xenophobic populism who dangerously feels the totalitarian tsunami that wants to destroy historic democratic achievements, especially those toward universal human rights. The new extreme right continues its long march against the institutions in all Europen countries encouraging intolerance and hate, contaminating parties and democratic institutions across Europe." Ibarra's statement regarding 21st century xenoophobia, is nonetheless reminiscent of what was occurring in Spain during and after the civil war. *Xenophobia in time of crisis.* Madrid, 24 March, 2011. Posted by cristobalgomez and downloaded from https://movementagainstintolerance.wordpress.com

in a workers' party or a trade union (Izquierda Republicana and Unión Republicana were included here).

This extremely bad behaviour included having participa-ted in strikes or demonstrations as long ago as 1918. Just what did the Regime want to achieve with all these retroactive declarations of terrible conduct? No more nor less than to penalize all those who were involved in the entire trajectory of the trade union movements since their inception. The destruction of the unions was at the heart of the military coup's objectives.

- Likewise, there was a <u>third class of allegations and accusations</u> that had grave consequences for those whom they described as having 'stood out as a 'propagandist' for the left' during the war or, more seriously, before the war. Along the same lines but much, much worse, was 'has spoken at meetings or was a spokesperson for the Frente Popular'. Having served in any public position during the Republic and before, was a serious crime, even if it was only having been elected town councillor or a member of the Country Council. Woe to anyone who had ever been elected a Member of Parliament.

If the above were prime targets of the Francoist judiciary, high officials – mayors, governors, officers in the Republican Army, political commissars and suchlike – were particularly hated. In the concentration camps, the classifying Juntas graded all these leaders and local authorities as Class C criminals whose lives were cannon fodder for the firing squads.

- The <u>fourth class of allegations and annotations</u> in the case files of those who took up arms and fought for the Republican cause. 'Enlisted as a volunteer in the militia' was intolerable and counted as an immobale aggravating factor. Having 'belonged to the Corps of Guerrillas of the Republic' and been 'involved in sabotage' were firmly proclaimed serious offences. As almost nobody accused of these allegations escaped death, many of these 'children of the night' fled to the mountains.

- Lastly, a series of <u>less specific allegations</u> added to the above list of crimes, allegations and accusations with more specific military and political connotations, there was a series of less specific allegations:
  - was a friend of Marxist leaders
  - participated in some kind of anti-clerical action': burning of churches and religious statues, insults to the clergy, that which they called profanations, and similar acts. There are numerous photographs of protestors in Malaga wearing chasubles and surplics, which led many to the firing squad.

Other peculiar accusations were levied against honourable Republicans who had absolutely nothing to do with any kind of violence but whom the Regime demanded should be removed from the scene. Examples of these are: 'is vain', 'insulted the Generalissimo', or simply 'is a danger to the Glorious Movement', or a plethora of cases bearing the surreal accusation: 'left town and did not avoid bloodshed'.

Equally bizarre, an apparently positive declaration in favour of an accused, Internalethal, such as when one prosecutor argued that: "If the accused saved one person, two things must bconsidered. Was he a very influential person? If so, if he saved one, why didn't he save them all?" This is what happened to Joaquín Pérez Salas, who protected many right-wing individuals and used this argument in his defence at his court-martial. It did him no good. He had forgotten that there was no excuse for his having been a member of the Republican military. He was executed in Murcia.

When it came to women, the choice of denunciaations and accusations was equally peculiar:

- is influenced by her husband's ideas or is married to a leftist
- supported the *Socorro Inernacional Rojo* [8]
- incited men to take up arms
- carried the flag in demonstrations or has embroidered a Republican banner

---

[8] International Red Aid, an international service organization created by the Communist Internationa in 1922 as a kind of political Red Cross.

- organized plays to raise funds for the *Casa del Pueblo* or Peoples' Army
- was a militant meber of a leftist political party or organization
- participated in creating registers, in seizures or testified in a lawsuit against somebody from the Right
- spoke against the National Cause, and so forth...

The entries in women's case files go far beyond the known boundaries of stupidity, such as in the case of Eustaquia Encinas Olmo[xxviii] who was accused of 'talking with Marxists', 'participated in demonstrations agaist the imprisonments', 'bore false witness against Lucas Díaz Fernández who was murdered during a bombing'[9] and lastly, 'the parish priest says he does not know her', in other words, she never went to mass. Very bad. She was sentenced to 12 years in prison for 'gossiping'.

Moreno Gómez studied numerous Causa General (CG) and other case files containing examples of the above allegations and the unacceptable manner by which these have been processed. A few of the more notable, from Pozoblanco, Villanueva de Cordoba and other Cordoba province towns, as recorded in the respective CG files, are desribed in some detail.

Some of the CG files bear the notation *educated*, which was especially grevious. As far as Francoists wree concerned, such an accusation was fundamental to its repressive orgy as the Regime believed that the uneducted masses were influenced by their leaders and by distinguished individuals whom they charged with swindling, twisting and poisoning the minds of the workers. Involvement in any cultural activity, especially teaching, was considered by the Francoist courts as a major aggravating factor. Franco, in his daily confirmation of death sentences, almost never commuted the sentence of anyone whose file bore the indication 'educated'. Schoolteachers and members of the *Centros Obreros*[10] were regularly executed. Cordoba courts were no exception to the rule. "Death to the intelligentsia!" shouted the generals and the Falangistas who were against all forms of education

---

[9] Actually, he was a Francoist who was killed by the Nationalist Air Force when it bombed Pozoblanco.

[10] Workers' Community Centres.

for the lower classes. *"We don't want men who study, just bullocks who work." xxix*

If the tragic story of Antonino Varo, whose story is told in Appendix II, were not enough, his son-in-law, Enríque Ramírez Dópido, Head Postmaster of Pozoblanco, was less fortunate. Despite the fact that he was imprisoned by the insurgents from July 23 1936 to August 15 in the afternoon, after town surrendered to the Nationalists, according to what he himself wrote in an appeal, his accusers fabricated another lot of false allegations for which he was arrested:

- was one of the crowd who took members of the right-wing on *paseos*
- inspired and morally responsible for the excesses committed by the Red committees
- ordered the arrest of Moisés Moreno Castro and was responsible for his execution
- acted as the go-between for the besiegers of the town, and more...

Just when Enríque hoped he was going to be released, the train for Valencia on which Moisés Moreno travelled had already left. Enríque Ramírez was unable to clear himself of the lies. When April 3 1940 his appeal reached Franco, he simply worte *enterado*[11] on the file. Enríque Ramírez was exeuted in Pozoblanco April 12.

To this additional tragedy, the victors added the econnomic ruin of Antonio Varo's family and business associates, as the Falangistas were not content to be left empty-handed when there were assets to be seized. Enríque Ramírez and his father-in-law Antonio Varo owned a flour factory in Fuentes de Andalusia, in partnership with José Madueño, whose story aso appears in Appendix II. The latter's house, the finest in Pozoblanco, was seized by Antonio Calero, the leader of the Falange. Other Falangistas confiscated the flour factory. Antonio Calero also bought and paid Enríque Ramírez' widow peanuts for an orchard. Another example of how the elimination of persons was not solely fired by the Fascist cleansing furore, but that it was also led by greedy locals.

---

[11] Noted.

During the Black Spring of 1938, there was an avalanche of reprisal allegations and accusations in Pozoblanco from relatives of right-wing victims, not just of the fifty who had been killed there August 1936, but also of the more than one hundred who were sentenced by the Republican People's Court in Valencia, most of whom were executed in Paterna September 1936. Causa General records state that 25 Republicans were accused by relatives of 61 right-wing victims, namely:

- Socialist Antonio Baena (executed) was accused by relatives of at least eight right-wing victims, all from outside Pozoblanco and absolute strangers to him. Yet there was not a single accusation against him from the relatives of some 55 right-wing victims in that town.
- Communist Bautista Herruzo de la Cruz (executed) was accused by relatives of another eight victims, especially of the Muñoz Cabrera brothers and the afore-mentioned Moisés Moreno.
- Another Socialist Mayor of the town, Rafael Rodriguez Tres Cuartas Redondo (in exile in France), was accused by relatives of four victims.
- Elías de la Cruz Gutiérrez (also in France) was accused by relatives of six victims.
- Assorted individuals from Pozoblanco whom we know were unconnected to any kind of violence, such as Antonio Varo, Mayors Rafael Rodriguez and Antonio Márquez, among others, were nevertheless accused by many relatives of right-wing victims.

In Belalcázar, as in many other townships, the municipal archives contain copies of the multitude of daily reports that the Francoist Mayors signed, as requested by the military courts. During the first months of the Regime, these courts continuously sent the Mayors urgent requests for information regarding everyone they considered a Rojo and the Mayors responded with totally negative information and allegations, truthful or not. Following is an example of one such report for Alfonso *Sincolor* Paredes Medina:

"I, the undersigned Mayor, by virtue of my office, am honoured to make public the following information:

ALFONSO PAREDES MEDINA is an individual with leftist ideas and an active propagandist of the same. When the Glorious National Movement began, this individual fled to a farm named Malagón, where there was a concentration of Marxists from that town. He joined the Rojo column that assaulted and occupied the town August 13 and 14 1936, committing all kinds of murders and arrests and was seen in the patio of Federico García-Arévalo's house during the morning of August 13 1936. He engaged in plundering properties, and he went off to join the Rojo Army, in which he remained until the end of the war, when he was arrested. He later escaped from the Hinojosa del Duque jail in which he had been imprisoned.

In witness of the truth of these accusantions against this inividual

(signed)
Manuel Escribano Medina, Mayor
Belalcázar, March 12 1940." xxx

Alfonso Medina, the great freedom fighter from Belalcázar, had already been sentenced to deaath but had managed to escape from prison and flee to the mountains where he led a large group of Republicans who had also fled. His end is a mystery; he disappered and nothing more is known of him.

The municipal reports, an 'official' variation of the allegations and accusations, were added to other reports from the local heaads of the Falange and the Guardia Civil, the parish priest and private citizens. All these were filed together with the accused's statements (always obtained after a good beating). In the town of Belalcázar, among those who appear to have signed the greatest number of private denunciations were members of the Delgado Gallego right-wing family and the head of the municipal police, Fernando Ballester Tobajas.

It is difficult to get to the truth of each event because to do so, one must listen to both sides, which is all but impossible as so many years hav passed since then.[12] Furthermore, we are faced with the well-oiled extra-judicial mechanics of a repressive program that caused the death of many innocent men, without any due process of law or even the minimal legal proceedings that would make it possible for historians to obtain a clear image of the facts.

Undoubtedly, the great majority of post-war victims can be attributed to reasons of an exclusively political nature and not to a desire to purge all those who were alleged to have comitted presumed crimes of blood. Nobody should be surprised at this conclusion, given that the thousands who were exterminated by Francoism behind the lines of battle during the three years of the war, were also killed for political reasons and not on the basis of any lego-juriciial criteria of any kind. The carnage in Galicia, Leon, Zamora, Valladolid, Burgos, Logroño, Navarra, Zaragoza capital, Canary Islands, Ceuta, Melilla, Corodoba capital, etc., etc., did not obey any judicial criteria. It was spurred purely by the political motivation of an extermination programmed by the authors of the military coup.

Many membaers of the Frente Popular who did their very utmost to prevent all kinds of violence, were killed in an arbitrary and cruel manner for political reasons alone. As far as Cordoba province is concerned, a few namrs immediately come to mind: Miguel Rachel (Mayor of Villanueva del Duque), Pedro Torralbo Gómez (Provincial Deputy from Villanueva de Cordoba), Francisco Díos (Capitan Paco, from Villafranca), Eduardo Bujalance (Member of Parliament, from Hornachuelos), Antonio Baena (from Pozoblanco) and many, many more.

Obviously, the Francoist repression was not looking for the simple cleansing of individuals allegedly responsible for blood crimes. Likewise, the military courts never showed any interest in the truth, simply because the principal objective was the ultimate dissolution, not to say extermination, of the Republican social base and the Republic's

---

[12] The entire dark world of denunciations in Pozoblanco can be consulted in the Causa General document boxes for Cordoba, National Historical Archives, Madrid. The Causa Examining Judge, or his collaborators, did a considerable job of obtaining exhaustive details regarding the multitude of allegations. This document appears to begin in Pozoblanco September 18 1941.

civil and military leaders, at the same time that the rising pillars of the New Regime were cemented with violence and terror.

During the post-war period, in towns where there had earlier been attacks on right-wingers, mostly by individuals who were inflamed with indignation against the military coup, those events, the work of a minority, were used as a pretext for the execution of hundreds of Republicans, including hundreds of innocent people. This occurred in Baena where the tragic killing of right-wingers in the San Francisco refuge, the work of only a handful of people in reprisal for the slaughter, by Framoiststs, of hundreds of workers in that town the same day, was used as a pretext for executing hundreds of working men, both during and after the war.

In the same way that in Spain today, when there is talk of the Francoist killings, latter-day Francoists wave the flag of the Paracuellos massacre by Communists November 1936, to shut people up. The same is true regarding the case of the San Francisco Refuge in Baena, where Francoists continue to refuse to admit, let alone accept, their cold-blooded execution of 700 leftist townspeople in reprisal for 80 right-wing victims. Sadly, shut-your-mouth exhortations continue to work efficiently against those who are searching for the true historic memory.

Systematic torture

Torture is one of the most heinous crimes aga:nt the dignity of Man. Raphael Lemkin who, in his studies on genocide, first called this a crime against humanity as he included torture in the same group of *Acts of brutality by which an individual is wounded in his dignity.* [xxxi] The crime of torture as a crime against humanity is defined in Article 7 of the <u>Statutes of the International Criminal Court</u>, Item 4. This humiliating crime was included in <u>The Nuremberg Trial Proceedings under Crimes Against Humanity</u> (August 1945): namely murder, extermination, enslavement, deportation and *other inhumane acts* committed against any civilian population, before or during the war, or persecutions on political, racial or religious grounds (Article 6.A.c).

To the list of Crimes Against Humanity, the Charter of the International Court adds: 'Leaders, organizers, instigators and accomplices participating in the formulation or execution of a common plan or consiracy to commit any of the foregoing crimes, are responsible for all acts performed by any persons in the execution of such a plan.' In Spain, it was the authors of the insurrection (military, Falangistas and Church) who were responsible for those crimes, yet to date, none of these three groups has asked for forgiveness, as did the perpetrators of the crimes of apartheid in South Africa. As for the Church, its proclivity for violence dates back through the ages, not just to the days of the Inquisition, but even further back, to the 5th century A.D. when St Cyril drove the Jews out of Alexandria.

So-called Justice under Franco provided the same personal guarantees as did the Inquisition. Likewise, that ancient form of Church justice was based on torture, as were the Francoist judiciary military proceedings, even though trial by torture was abolished in Spain in the 1812 Constitution. If the first punishment against the defeated was a general imprisoning in jails and concentration camps, torture was the first rung to Franco's hell and justice. Every case file tht Moreno Gómez was able to consult, without exception, included information obtained by torture in the anteroom of all the suffering that was to follow.

Francoism triggered that which could be described as the horrendous storm of a general state of torture. The brutal methods of torture practiced by the Inquisition pale in comparison to those applied under Franco, particularly after his victory in 1939. There was less torture during the war, as the Regime appeared to content itself with mass executions by firing squad – the macro cleansingg of the first stage. In the post-war period, however, there was an explosion of a vindictive neurosis, the malicious desire to punish the defeated before execution, the application of absolute cruelty, the total contempt for the individual.

This was not, as Moreno Gómez stated earlier, simply revenge for behind-the-lines and battlefront right-wing vicims. It was so much more: vengeance against those who had opposed the immediate triumph of the insurgents, who had resisted the coup and who fought in defence of the democratic government. For the victors, opposition to the Glorious Movement was an unpardonable affront, an insult to the Church that, together with the Berlin and Rome Regimes, had blessed the insurgency. There could be no pardon for the defeated fighters who had taken up arms in defence of some democratic freedoms that Francoist looked upon as being of no more interest than a pile of sodden papers on the altar of totalitarianism.

Francoism justified torture on two beliefs: the impunity of the victors and the necessary demonization and dehumanization of the defeated. The great conditioning factor was the generalized belief that the victors enjoyed absolute impunity for their ations. This what Judge Baltasar Garzón wrote in his book *El alma de los verdugos* [The soul of the executioners][xxxii] and so eloquently explained in his interview on Radio Argentina in 2008.[xxxiii]

> "Who were these men who, after kissing their children as they left their homes, went to work as exemplary employees, torturing and murdering political prisoners?... People have become convinced that they were ordinary people, who at a given moment in time went beyond the barrier restraining perversion that Man also carries within him, and if once he has passed this hurdle, cannot stop... (for the simple reason of political disagreement) ... That, associated to the climate of impunity... When this situation disappears, when he no longer holds the reins of power in his hands... is when cowardice appears, there is a hiding, a non-acceptance/recognition of the facts."

As regards the belief in the need to dehumanize the defeated, this assumption could be attributed to the Catholic Church. Throughout History, this holy institution has been a master in degrading the image of the heretics, the heterodox or, in our cae, the dissenting Republicans. The Church villified them, demonized them, declared them to be abominations and dehumanized them. From then on, all barbarities were possible.

For the repression and torture to reach a maximum level of efficienecy, during 1939 when systematic torture was most brutal, the Regime began by ensuring that the cruel and savage punishments were first enforced in every imprisoned person's hometown, where hatred and calls for revenge would stoke the general fires of political vengeance, before those who had been arrested were transferred to prisons in the capital cities. It was not unpremeditated. Well into the 1940s, the idea was that the euphoria of victory would feed a sadism embellished with the most perverse accusations so that the defeated individual would suffer the full weight of the purge where people knew him. Later, when prisoners were transferred to provincial prisons, they stood out less as individuals, although the threat of brutal beatings and unbridled attacks by those who wielded the rods always hung over their heads.

Systematic torture, especially widespread during 1939-1940, was another of the great post-war differences compared to the repression of Republicans in 1936, when extra-legal raids were carried out with some celerity and only rarely involved the added sadism of torture. (There was, however, a precedent for the military's use of torture when, in 1934, right-wing government troops savagely repressed the miners' uprising in Asturias. Several military were later tried for 'the Asturias excesses'.)

There were three types of torture. The first, shock or vindictive torture, seen in the towns in 1939, a violent reaction of Nationalists triggered by anger repressed during the war and personal hatreds, as well as revenge for the sufferings of right-wingers. The most savage beatings were given by guards with whips and rods in the jails or when Falangistas went into the municipal holding cells and kicked and slapped the prisoners or took some outside and beat them senseless. One way by which the victors, who were often drunk, celebrated their victory.

Then there was the judiciary torture, applied during the instruction of a summary procedure, when the accused was beaten in the presence of the Examining Magistrate, on his orders. Again, nothing was improvised or uncontrolled. The declaring defendants often ended up unconscious from the blows. If an accused fell to the ground, he was roused with buckets of water and again beaten, until he made his 'voluntary' statement. The imprisoned, when they were called to 'make statements', that is, to be interrogated, trembled with fear. When they returned to their cells, they were frequently so battered as to be almost unrecognizable. When they could, their cellmates tried to cure their wounds with water mixed with salt and vinegar. More often than not, the unfortunate prisoner never made it back to his cell.

The purpose of judiciary torture varied: to compel the accused to sign the record which many refused because of the number of lies that were included, although after a good dose of 'syrup of the lash', they ended up signing whatever was put in front of them; to get an accused to confirm that he was guilty of the crime for which he was charged; to force him to give the names of friends or neighbours who might also be involved; and to get a prisoner to speak and provide some kind of information. Falangistas, local right-wingers, municipal police, Guardia Civil and all kinds of 'volunteers' took part in this judiciary torture. There was no doubt that this torture was institutionalized as an integral part of the judiciary mechanism. Seven years later, in 2011, ex-prosecutor Carlos Jiménez Villarejo correctly described certain eighty-year-old active members of the Supreme Court with shady pasts, as torturers' acocmplices.

The accused were led from this world under a mountain of denunciations and false accusations, that they signed as true under duress. Although in 1939 no one believed that a condemned man could ever protest the false accusations, this was occasionally possible in 1940, when we find case files for some condmned in which they express their unhappiness at the false allegations and the beatings they received.

Arcángel Bedmar describes the case of José Tarifa Galvez who sent a letter to the Cordoba Wartime Court of Appeal July 20 1940, from the Castro del Rio jail, requesting a review of his case on the grounds that: "All this is totally false, because I was forced to admit to this in the Guardia Civil barracks after I was beaten with a lash".[xxxiv] It did not do him any good because the Francoist avenging angels paid no attention to such trifling details. August 6 1945, another prisoner declared to the Baena Examining Magistrate that he 'neither confirmed nor amended the declaration he had made to the Guardia Civil in Baena on the 31st of the previous month, because although he recognizes his signature on the document he is shown, he signed this under duress. That the statement he wrote was dictated by the magistrate as his, but he was forced to sign it and four blank pages of paper as well'.[xxxv]

Confirmation that torture was an approved and widespread practice is present in the instructions that the Regime sent to the prisons in the mid-1940s, telling them to lay off the mistreatment, not because they were concerned with the prisoners, but because they were afraid that details of their actions would be reported outside Spain. The Director Gemeral of prisons sent a first 'most private and confidential' circular August 4 1944, stressing the need to avoid mistreatment. As that was generally ignored, he sent another circular November 22, which also had little effect.

In third place, there was the <u>police torture</u>, the kind that the National Corps of Police practiced at police headquarters in capital cities and major towns. Police torture was much more refined than other kinds, in that it involved the use of electric current and all kinds of torture contraptions that also resulted in a number of deaths. More techniques inherited from the Church's Holy Inquisition. Little by little, police torture became the dominant instrument of terror, especially when in 1942 Blas Pérez, Minister for Government, enthusiastically dedicated himself to fine-tuning the Regime's terrible police policies that were aimed less at the events since 1936 and much more at any clandestine outbreaks of political unrest.

There is something, however, that must be made clear: although police torture prevailed in the capitals, in the towns and countryside it was the Guardia Civil under the orders of Camilo Alonso Vega who, during the Triennium of Terror (1947-1949), went much, much further, launching a program for the renewed application of the Law of Fugitives with such force that in all Spain, more than one thousand persons (160 in Cordoba alone) were executed without due process of law, by means of the extra-legal paseo.

Lastly, a fourth kind of terror that we could call *maintenance torture*, was routinely employed, throughout the post-war period, by the Guardia Civil in villages where they had their rural headquarters. This was not a very refined form of torture, mainly beatings with rods, lashes, slapping and punching, as well as some cases of hanging the prisoners by their wrists and other practices to make them talk. There were two very specific objectives to this torture. One, the persecution of escapees and the more or less severe punishments inflicted on relatives of the escaped or of guerrillas who were presumed to be in contact with their kin, and country folk in general. Sometimes, a person was beaten to death, and some were hanged. The second objective was the *paso de lista*, or the posting of a list at the door of the jail in the evenings. All those whose names were on the list, described as Rojos, included the sons (even children) and relatives of guerrillas, amd especially *libertos*, prisoners who had been released from jail or were out on parole. Every afternoon, some of those were chosen for a session of beatings, just to remind them who were the victors and who were the defeated.

Parallel to the 'official' mistreatment of prisoners, the phenomenon of drunken Falangistas visiting the jails at night-time to beat the inmates was widely known throughout Spain. Similar behaviour by Falangistas and young supporters of the Regime is documented in the prison of Aldeanueva de San Bartoloomé.[xxxvi] We also know that in Quintanar de la Orden (Toledo), they crushed the inmates' feet to prevent them from escaping.

As Moreno Gómez already observed when he spoke of the concentration camps, the reality of torture was a new descent into the hell of the Francoist repression. As he proceeded with his investigation of the tragedy of Francoist mass torture, he was eventually driven to the deepest levels of this hell, into the cesspits of society. In Cordoba province, many townships suffered terribly: Puente Genil, Palma del Río, Posadas, Castro del Rio, Bujalance, Villaviciosa, Fuenteobejuna, Hinojosa, Pozoblnco and Villanueva de Cordoba. If he centred his research and comments on Villanueva de Cordoba and its outlying townships, it was becausea it was here that he was able to obtain the greatest number of eyewitness accounts in the town considered by Franco to be the most important nucleus of Communism in Andalusia.

Painful as it was, his investigation brought with it an extraordinary retrieval of the facts, numerous personal testimonials and the occasional finding of a notarized act of Francoist cruelty and extermination that would have otherwise remained in oblivion. He was able to come close to the truth of what had happened both during and after the end of the civil war by interviewing survivors and relatives and people who knew the victims personally, although he was unable to confirm much of what he learnt from official documents or Civil Registry records. Recording the testimonials of survivors was essential as it would have been naïve to think that the oppressive Regime would keep an official record of the torture. After all, nothing new can come to light if all that we do is simply affirm that in defeated Spain, the defeated were massively beaten by the victors.

The following monograph that Moreno Gómez grouped around the villages, towns and cities of Cordoba province where these events occurred, are a sample of the personal testimonials collected and reported in greater detail in the Appendixes, together with the Causa General case number when available. For decades, the victims who managed to survive and their families have lived in enforced silence, and it is time that History is allowed to speak for them. The individual stories of these unsung heroes must be told to substantiate this proof of the Fancoist reality and, by helping recreate the Historic Memory, offer some form of closure to Spain. Some accounts are paraphrased.

Those who today attempt to sweeten the pill of the past and who persist in denying the facts must be prevented from supporting the current allegations that what went on under Francoism was little more than some friendly rough and tumble between individuals of different opinions. Does anybody still doubt that the purpose of the purge was to exterminate everyone who had formed the social base for the Republic?

During the early post-war period, the preferred method for interrogating prisoners was beatings and whippings. The more aggressive, or refined, methods of the political police wre not applied until later years.

Marcos Ana, the poet arrested in 1939, reported that in those days, the methods were brutal, not refined: they just beat you ferociously. Many who fell into their hands died. In the end, the third time round, you just lost consciousness and the torture ended. The second time around, in 1943, it was much worse.[xxxvii]

Puente Genil

In Puente Genil, where almost a thousand Republicans had already been murdered by troops from Africa under General Antonio *Castejón* Espinosa in the first days of August 1936, those who escaped that attack suffered yet another flood of repression after the fall of Malaga February 1937. Many were captured and taken to Cordoba, to a new wave of executions. Clearly, despite the wartime genocide in Puente Genil, the Francoists were not satisfied with the results, so they continued with the bloodshed in the immediate post-war period, with mass arrests of all those who returned from the Republican zone (the system of vigilance in the Station was continuous). The most savage torture was applied to the returnees and there was a new slaughter of the defeated. Rafael Bedmar Guerrero, who managed to survive the mistreatment he was subjected to, provided a heart-rending description of his ordeal and the sufferings of his fellow prisoners.[xxxviii]

Few went quietly when they were arrested during the victors' maelstrom of arbitrary violence. José Mora Valencia, a Socialist railway worker who had been the Chairman of the local War Committee, was beaten time after time until he was finally executed April 24 1940, as was another distinguished Socialist, Justo Deza Montero, who had been a Frente Popular councillor and also a member of the War Committee.

In Puente Genil we note 39-year-old local blacksmith Francisco Palos Gálvéz' desperate resistance as he was ready to make those who came for him pay heavily for his life. November 16, 1939, two municipal policemen (José Palos Ramírez and Juan Mendoza Calvillo) went to Francisco's house, intent on arresting him. Warned that they were coming, he was waiting for them in his living room where he disarmed them easily as they entered, then killed them with an iron bar. They were soon followed by a group of Falangistas led by Mayor Jesús Aguilar, revolver in hand, who shot the defiant Republican.

Baena

The survivors of the July 28 1936 brutal massacre in Baena at the hands of the Nationalist troops from Africa under Eduardo Sáenz d Buruaga, suffered the usual reprisals levied at the prisoners of war, and in the immediate post-war period, at the defeated who returned home. These reprisals were aggravated by the Baense bourgeoisie's unabated eagerness to obtain revenge as they launched themselves like hyenas agaist the defeated. As occurred elswhere, the bourgeoisie remained behind the scenes, furthering their ends by hiring civilian vigilantes, thugs who openly arrested people and beat them liberally with sticks.

There was no escaping the beatings and torture. All the prisoners were systematically beaten in the Guardia Civil barracks during their interrogations. One survivor testified as to how they beat him on his back whilst he knelt on a stick. José Padillo Marín, a member of the Zarabanda family and another survivor, was tortured as he hung from his arms until, broken, they allowed him to die in hospital, His father Agustín and his uncle Manuel were also executed.[xxxix]

The townspeople of Baena still remember the sadism of José *El Moraíto* Rabadán, a Falangista prison guard who delighted in throwing the food that relatives brought the prisoners down the toilet. Other notorious persecutors, enflamed by the *vox populi*, were Manuel *El Conde* Rojano, Antonio *Faroles* Morales, the guard *Papafritas*, the municipal policeman Amador de los Ríos (who before the war had been arrested for murder), Pascualito *El Sacristán*, Rafael *Cordelillo* Santiago, and Cristóbal *El del Bacalao*, to name but a few. The last two of these had defected from the Republican CNT and in their savagery, excelled the Falangistas with whom they sought to gain brownie points. The bourgeoisie-hired thug arrangement was not a trivial relationship; it presumed the delegation of power and an extension of complicities. This could be found all over Spain, where the upper class delegated their power to this group of henchmen, sleuths or killers for hire, fully confident that they would obtain optimal results. To this day in Andalusia, there is an oral tradition that a Señorito's foreman is always much more cruel than his boss.

Castro del Río

Francoism had a special mission in Castro del Rio, a region where Cordoban trade unions and the CNT had considerable influence. The townspeople sadly remember that prisoners held in the town jail or in the Convent of the Nuns were brutally beaten when they refused to answer searching questions such as: "Who were those in the square on the day that the Guardia Civil barracks surrendered?" or "Who were those who went to the El Garabato farm?" There had been right-wing victims in 1936 in both locations where local gangs of hoodlums such as the *Gallito*, the *Blanca*, the *Potrilla* and others acted as executioners.

In his research of the events in Castro del Río, historian Francisco Merino writes about how sometimes a prisoner's shattered body could not survive the beatings, which is how Juan Rojano died Julay 26 1939. Others would return to their cellmates, dragged by their tormentors, bleeding from their mouth, nose and ears. Felipe Aguilera Arroya was thrown into a latrine, where he died an inhuman death. The beatings went on for months. The unfortunate recipients, semi-conscious from the blows they had received, would utter the first names that came to mind when they were interrogated, even if the accused were innocent.

In this way, every day new working men were arrested, and they swelled the list of the condemned.[xl]

Also, according to Merino, an inmate only known as *El Moño* died because of torture and another, Miguel Márquez, was tormented with nails before he was shot by firing squad. The appalling treatment of the defeated often led to suicide or to desperate attempts to eacape, sometimes successfully, sometimes not so, with tragic results. One extremely unlucky individual, Andrés *El Colorín* López, was discovered as he had everything ready for his escape from the Castro del Río jail. He was shot in the act and all the prisoners were forced to march past his body. Another, nicknamed *Sobraguisos*, esaped from the jail through the roof but he was later caputred on a neighbouring farm and shot.

Bujalance

No description of the repression in this small town would be complete without mention of José Moreno Salazar, a freedom fighter who fled to the hills after the war to join the Los Jubiles guerrillas. In addition to the book of memoires he published in 1985, he also wrote Moreno Gómez with details of how prisoners were treated in Bujalance.

> "Prisoners were packed in the Party jail on Calle Zarcos, and behind it, where they set up the Military headquarters where they were tried and tortured and taken to be shot at dawn by a firing squad, against the cemetery walls. Many of these executions were attended by a prostitute, so-called *La Pepilla*, whose lover had been executed in 1936 by the people's volunteer militia.
>
> The worst tortures however, were carried out in Calle de Las Cadenas at the Guardia Civil headquarters. The most sadistic of all the Guardias Civil is the one whom I frequently mentioned in my book of memories, a man by the name of Requena. Civilians also tortured many prisoners, namely Falangistas such as *Praíto*, Marcelino *El Carcelor*, someone calleld Ríos, especially the last one who

became famous as a blackleg during the years of the Republic, during strikes. Killed by beatings – I remember someone called Trigueros who left the Guardia Civil headquarters dead; and in my memoires. Never forgotten, Francisco Milla's mother and sister who were beaten to death by Requena and who died in my arms, when I was imprisoned in the Hospital."[xli]

Fuenteobejuna

In Fuenteobejuna, Claro González returned home April 18 1939 and the next day was already imprisoned in the City Hall building's improvised jail. The beatings were terrible, and they had no other medicine than salt and vinegar to heal their wounds. They lived between beatings and hunger. Felisario Cidoncha, another inmate who was cruelly tortured, was shot by a firing squad August 4 1939.[xlii]

Villanueva de Córdoba (also Valle de Pedroche, Adamuz and other outlying townships).

Enforcing Francoist law in Villanueva de Cordoba were: Corporal José Martínez Galiániz (head of the local Guardia Civil post), Lieutenant Leopoldo Mena (SIPM Lieutenant), Captain Ignacio Pizarro (local military commander) and the infamous Judge Juan Calero Rubio. Added to this cluster of Francoist judiciary and para-legal officials, was a clutch of henchmen and followers eager to thrash the imprisoned defeated: low-class rabble with guns tucked in their waistbands, hired thugs and staunch supporters of the landowners who hired them. Relatives of right-wing victims of 1936, such as Pepe Delgado, Pedro Serrano and Vicente *Salado* Muñoz, also participated in the torture, seeking revenge. Although the tormenters included several landowners who were also knwn as staunch Francoists, some such as Matías *Malaleche* Pedraza had no relationship whatsoever with any right-winger who had suffered in 1936, quite the contrary. The latter's father, Dionisio Pedraza was, in fact, protected by the loyal militia because of his excellent relationship with working men during the Republic.

A selection of the eyewitness accounts from survivors of this climate of violence and torture, especially in the outlying towns, is included in Appendix II. They describe the two types of mistreatment already discussed: vindictive beatings and sadism by Falangistas, and the judiciary beatings during the pre-trial hearing interrogations, especially meted out to those who had been released from a concentration camp then re-arrested as soon as they returned to their hometowns.

Moreno Gómez has certain knowledge of at least five prisoners who died from torture between 1939 and 1940 in Villanueva de Cordoba. These, in particular, died from torture in 1940, whilst the Spanish Legion African troups were garrisoned in the town and the surrounding countryside. The first of the three victims, Juan Fernández Moreno, aged 54 years, was beaten savagely for no apparent reason at Leigon Military Headquarters and left to die on a bit of waste ground August 24, 1940. Likewise, two days later, Juan *Horozco* Cantador Cachinero, Octobr 16. Next, the Leigonnaires killed José Huertas Valverde by forcing water down his throat through a tube, until his throat was torn by the aggression.

The number of individuals responsible for the brutality in Villanueva de Cordoba is never-ending. Already mentioned is Matías Pedraza, the landowner who led the night-time visits by drunken Falangistas to the local jail, where he amused himself by stepping on the inmates, lashing at them with his riding crop or with willow canes, and according to a witness, throwing their lunchboxes containing food at their heads, always with bursts of laughter, insults and threats. When those who were most persecuted – Grancisco *Biatas*, Eugénio *Palmera*, Lopez *El Dinamatero* Igañez, one *Cucharas*, and others became unconscious, they were taken out to the patio where they were left on the floor to recover, until the next night.

An unforgettable event in Villanueva that was often mentioned by eyewitnesses, was the night that the Falangistas entered Fuente Vieja jail for some entertainment and, to please the Señoritos, prison guard Frascarro *El Tiraor* ripped Francisco Illescas' ear off. This unfortunate inmate had served in 1936 as a guard over right-wingers who had been arrested, among them *El Tiraor* who was now getting his revenge. Several surviving fellow inmates remember that night as one of the most dramatic in the jail.[xliii]

Far too often, the desire to arrest and punish the greatest number of Republicans meant that many individuals were arrested for either unknown or totally futile reasons. Pedro Molinero, ex-activist of the Unión Republicana, a thoughtful and careful individual whom Moreno Gómez interviewed several times at the beginning of the 1980s, was arrested in 1939 for unknown causes. On other occasions, allegations of childhood disagreements as in Antonio Pedraza García's case, or simply having stood as a prison guard during the Republic, were sufficient cause for imprisonment.

Blas Arévalo Carbonero, brother-in-law of Nemesio *El Floro* Pozuelo, the Communist leader, was denouncd by *El Tiraor* and by Roque *Castilla,* because the latter's son had been arrested by the Civil Government, apparently when Blas was standing guard over right-wing prisoners. Apparently, the young man was handed over to a guerrilla group and he was killed near Cardeña. In 1939, the victors tried every means to determine exactly what had happened, without success. Blas knew nothing about this, so could confess to nothing despite brutal torture. According to his family: "He was beaten so badly he didn't even let his own children hug him". This poor man, who had no notable political role, was executed without any specific charge and without any proof of his having committed a so-called blood crime. This was not unusual, because as Moreno Gómez said earlier, of the 106,000 imprisoned in 1937, only 2.13% had actually committd any kind of civil crime.

It is presumed that matters were similar in 1939. Cecilio Ruiz, a Socialist, was aged 55 when he was arrested by Diego *El Chunga,* accused by the Falangista Lara, and executed. He was not a leading politician, but he had also stood guard over right-wing prisoners in 1936.

Not infrequently, many years after the events, when relatives of executed disaffected heard of Moreno Gómez' research, they made special trips to speak to him, sometimes just in an attempt to ensure that what happened to their father, brother or uncle should not go unrecorded nor their memory forgotten. Often, these surviving relatives were sons and daughters of a deceased who, as children, suffered untold horrors they could not fully comprehend at the time. This was the case of José Torralbo Rico, Antonia Sánchez Cerezo and Catalina Cantador Romero.

As they proceeded to cast their net as far and wide as possible and sow terror in the towns and villages, Falangistas cooked up a variety of schemes to serve as a a pretext for raids among people of little or no political significance. In summer 1939, there appeared some posters in Villanueva proclaiming 'Viva! Negrín[13] is returning to Spain!' a fanciful creation of the Commander of the municipal police, Bartolomé Baerenguer. This bad taste 'joke' cost José Luna Mata no less tthan 30 years in jail and numerous beatings:

> "They took me to the barracks and began the interrogation. First, they took my jacket off. They sat me on a stool and several Falangistas stood behind me. Before they asked me anything, the beatings rained down on me. 'This is the welcome', they said. That is what it was. They then interrogated me regarding the posters, and I denied havig printed them or even seeing one, which was true. The beatings returned and continued for a couple of hours. Several times I fell to the floor from the blows, where they kicked and stomped on me. When they saw that they could get nothing from me, they hung me from my wrists and continued whipping me until I lost consciousness. When I regained consciousnss, I was soaking wet from the bucket of water they threw on me.
>
> The interrogation ended and they took me to Fuente Vieja prison. There were more than 500 men there. The women's jail was also packed. My comrades came to tend to my injuries, and they said: 'You were lucky. When we removed your shirt, your skin did not stick to it. They didn't beat you all that much'. They massaged me with vinegar, which is all that they had. We repeated this operation every day for the comrades who came from the barracks, from headquarters or after suffering the nighttime attacks of the Falangistas. I can prove that the bodies of many comrades were live flesh after their skin peeled off from sticking to their shirts."[xliv]

---

[13] Juan Negrín Lopez, Republican Prime Minister 1937-1939.

Everyone Moreno Gómez interviewed in Villanova remembered that the week of October 12 to 19 1939 was particularly horrific in terms of the victors' increased brutality, especially when Juan Calero was the Examining Magistrate at Military Court 11. The ex-prisoners call this the 'tragic week' because almost all of them were re-interrogated at La Preturilla and at Military headquarters. Sebastián Gómez' testimonial[xlv] is significant in that it confirms that military judges took part in the beatings.

Judge Calero, taking it upon himself to act as both judge and jury, often wielded the lash. In doing so, he ushered in a post-war administrative reign of terror under a totally militarized justice that was wholly dependent on and subject to Franco's totalitarian Regime. Antonio Rubio Cobos, for example, was tortured until they broke an arm and several ribs because he refused to sign the statement written by Judge Calero. He was executed March 1940. His young son Manuel Rubio was taken in by Francoist 'benefactors' and pressed to commit to a life in the Church at the Salesian Seminary. Instead, he abandoned his studies when he was well advancd in his training and became a celebrated teacher at the Villanueva Public School where he was Moreno Gómez' first teacher.

A similar story of a Francoist military judge participating in the torture and immersing himself in a whirlwind of hateful actions, was recounted by Francisco Poyatos, an eminent Cordoban attorney who witnessed a court martial in Cordoba against 19 prisoners, as part of Examining Magristrate Gregorio Prados' Court. He was the only trained career attorney present there because the others were all active military officers. The pre-trial hearing was interrupted for a break, after which the magistrate turned to one of the accused, from Belalcázar, and scolding him shouting: "So you are what's-his-name!", physically launched himself against the accused, slapping him in the face.[xlvi]

Prisoners from Pedroche were also taken to Villanueva de Cordoba jail and it was there that Falangistas from their town would go to beat them up. Francisco Romero Cachinero told Moreno Gómez of a man from Pedroche (or Torrecampo), whoe name he could not recall and who was beaten to death in Villanueva jail.[xlvii] Miguel Regalón later identified him as a fellow from Pedroche, whose surname was Regalón, like his, who was caught with a razor. They beat him so savagely that he died the next day.[xlviii]

Falangistas from Torrecampo were accustomed to travelling to Villanueva to torture fellow townsmen who were imprisoned there, although it is true that the Villanueva tormentors, quite content to torture everybody they could lay their hands on, did not require any help from outside. For example, they savagely beat people from Torrecampo such as Sebastián Luque and his son Manuel, as well as Ricardo Ranchal, whose nails they pulled out, among other ordeals. One of Manuel Luque's worst tormentors was a right-winger called *El Colodro*. All these unfortunate men from Torrecampo ended their lives in front of a firing squad.

As an example of those who managed to survive the appalling abuse, Moreno Gómez earlier told of José Garcia *El Perica* Coleto, who managed to escape from a line of prisoners 1939 and fled to the hills swearing that one day he would kill his tormentor, Fructuoso *El de los Dientes* Reyes. The oppressors in Villanueva were so enraged when they heard of this, that in revenge they carried out a major raid in the town to punish as many individuals as they could, taking their prisoners to the Guardia Civil barracks where they beat them mercillessly. According to statements from his family, one of these, Acisclo Cruz Villarreal, was beaten unconscious then dumped outside on the street at the door to the barracks, from where his family picked him up.

Cordoba capital

The events in Cordoba capital prison were no less heinous, as we hear from Rafael González Roldán regarding his uncle Manuel González de la Fuente. As Franco had proclaimed that 'anyone whose hands were not stained with blood could return in peace', his uncle returned to Spain where he was arrested and imprisoned. Not only was he frequently beaten in the jail, but they also took him from the jail to the Guardia Civil barracks, hung him from his ankles, and left him unconscious. Rafael, who was also imprisoned in the same jail, remained by his side all that last night, wrapping him in a shroud and praying for him, until the next day they took his uncle out amd buried him, ignoring his relatives who were waiting to receive his body at another door.

Moreno Gómez was unable to find any official or Civil Registry records for the above person nor regarding Juan Sánchez Cabrera from Villa del Río who, according to José Merino Campo's testimonial, died from being beaten with rods in the Cordoba Provincial Court. As the whereabouts of the remains are still unknown, they are considered as 'disappeared'.

Pozoblanco

The following cases have all the appearance of death from torture, although the usual euphemisms for the cause of death were entretd in the records. Nonetheless, Moreno Gómez' experience over the years regarding the customary treatment of prisoners, the lack of specialized medical care for their wounds, as well as numerous individual testimonials, indicates that such cases may not always escape the 'clinical eye' of the historian.

- In Peñarroya-Pueblonuevo. Dionisia Alcántara Calvo, 'due to a collapse', according to a report from the Judge Advocate, December 16 1939.
- In Pozoblanco. Gervasio Martínez Hidalgo, from 'epileptic attacks', according to a report from the Military Police, September 3 1939.
- In Pozoblanco. Juan Álvarez Pozo, from Hinojosa, from 'cardiac asystole', according to a report from the Military Court, October 14 1939.
- In Pozoblanco. David Cuello Amadeo, from Alcaracejos, from 'acute endocaraditis', according to a report from the Military Court, December 17 1939.
- In Pozoblanco. Rufino Fernández *El Poleo* Alcaide, 'died attempting to escape from the guards who were leading him', although actually he died from torture April 25 1939, as confirmed by witness statements.
- In Bujalance. Pedro Alcaraz Mira, from 'cardiac arrest' May 22 1940, according to the Civil Registry.

During the first days of the terror, a leader of the reprisals in Pozoblanco, Corporal Rejas, commander of the Military Police Headquarters in Costanilla del Risquillo, was noted for his cruelty. Interrogations and the consequent beatings took place in the so-called Palace and at Las Monjas school. Captain Bautista Herruzo de la

Cruz of the Republican Pedroches Battalion was one who suffered the most. After each session, several men had to carry him out, dripping with blood and unconscious. The Falangisatas had decided that he had informed against the right-wingers from Pozoblnco who were imprisoned in Valencia, many of whom were shot September 1936. The intellectual Antonio Porras' son Rafael was also sorely beaten in Pozoblanco prison after he was brought under arrest from Madrid. Although he was tried in Pozoblanco and found not guilty of the charges against him, his enemies denounced him again and again until he was finally returned to Madrid where he was executed May 19 1943. Pedro Villarreal's sister testified that he was carried from the sessions 'bloodied like Christ.' The same punishment was meted to Francisco *El Endeble* Díaz Pastor, according to statements from his widow. Although Moreno Gómez was reliably informed that Severo García Gonzalvo was brutally tortured until he died, the records state that he was executed November 29 1939.

Dos Torres

Ordinary citizens who were caught committing small thefts in the countryside (mainly acorns and olives) and were continuously and vigorously persecuted by the agrarian bourgeoisie and the Guardia Civil at their orders, were often objects of beatings, an endemic evil in the rural world. During the post-war, the number of such thefts went through the roof because of the dreadful widespread poverty. The victims of these severe punishments were most often women (also some young men) whose husbands were imprisoned and who had nothing to eat in their homes. The masters of the public order not only punished these women with the lash, but they also stripped them naked, shaved their heads, forced them to drink castor oil and put them to work sweeping streets, cleaning military headquarters and private Falangista homes. July 1941 in Dos Torres, women were frequently subjected to a usual form of fascist entartainment: stripped naked, they were marched down the streets and forced to sing *Cara al Sol* until they reached the Dos Torres Casino, where Señoritos laughed at the show.

In Dos Torres, the Panzurrie brothers, both Communists, together with Pedro *El de la Filomena* and *El Trapero*, were tied to a window of their mother's house so that she could be sure to hear everything, and there they were savagely beaten. Afterwards, a 2nd Lieutenant in Dos Torres ordered their execution on the pretext that they were in cahoots with others who had escaped to the hills.

Belmez

Vicente Blanco, a leading Communist was one of the brutally mistreated in Belmez. A curious fact was that during the sessions of torture, it was not only the usual local tormenters who participated in the beatings but also several right-wing women who hammered him with nails and other stabbing objects. He was later executed in Cordoba.

Almodóvar del Río

Antonio Ramos Palomares', an anarchist leader from Almodóvar del Río, story of what happened to him after he was release from the army, when he returned to his hometown after the Republican defeat, is reproduced herein.[xlix]

Montilla

Arcángel Bedmar, historian of the Cordoba Campiña region, refers to cases of torture in Montilla as a widespread phenomenon common to all Francoist prisons.[l] Francisco Cordoba Gálvez was beaten, hanging from his arms, until he lost consciousness. Emilio Montoro Delgado was tortured by a Guardia Civil corporal in the Military Court building. Rafael García Espejo was continuously beaten and during one session they tore off his moustache. Manuel Ruz Aguilar was tied to a fig tree and beaten non-stop. Antonio Pérez Lao arrived in prison with a broken leg and despite that, they still beat him. He died from the punishment, shortly after he was released in 1943. His father had ben executed six years earlier. The same author reports the oral testimonial of Francisco Carmona Priego, from Montilla, when in the Worker Soldiers Disciplinary Battallion of Cerro Muriano, Cordoba, he witnessed a sergeant breaking an inmate's back by beating him with a pickaxe handle. That pickaxe handles were often used for beatings was not unusual, especially when prisoners were working in the countryside outside the camps, as Antonio D, López stated regarding a case in Castuera when the guards 'broke the handles of the pickaxes as they smashed a prisoner's ribs'.[li]

Elsewhere in Spain

The above examples are from Cordoba province, but what of the events in the light of the general panorama of Francoist torture in all of Spain? To Moreno Gómez' knowledge, such a monographic compendium has not yet been published in Spain. It is, of course, essential that one obtains the testimonials of the victims themselves and not base one's account solely on Francoist written documentation as it would be naïve to think that the oppressive Regime would keep an official record of its actions. Nothing new can come to light if all that we do is simply affirm that in all defeated Spain, the defeated were massively beaten by the victors. Here are a few such accounts for the region outside Cordoba province.

Navalvillar de Pela, Badajoz

In Extremadura, as in Cordoba province, many defeated who were freed from Castuera concentration camp were re-arrested as soon as they returned to their hometowns. This happened to Valentín Jiménez Gallardo and to several residents of Navalvillara de Pela, Badajoz. They were re-arrestedm beaten and imprisoned in Puebla de Alcocer jail from where they managed to escape and flee to the hills.

Santiago de Compostela

Casimiro Jambonero's memoires of his experience in Santiago de Compostela prison, in his book *Diary of a Republican Soldier*, are similar accounts of the mistreatment of prisoners elsewhere. After dinner, some guards from Labacola came to remove one of the inmates who had been arrested with him and he and his fellow inmates did not know why or where they were taking him. At the end of the day, he returned from Labacolla, and they saw that everything had imagined did occur: his face was totally misshapen from the blows he had received, his head was swollen, and he had been badly beaten all over his body.[lii]

Puerto de Santa Maria, Cadiz

Cordoban Rafael Sánchez Guerra, imprisoned in Puerto de Santa María, Cádiz, met a fellow prisoner in the infirmary who had served as a Guarda de Asalto in Valladolid. When he commented on his fellow inmate's pitiful physical condition, the latter replied:

> "You see me like this now, because when these bastards' movement broke out, I was arrested by the Falange and taken to one of their unofficial jails where they broke my back with their beatings. I lay between life and death for several days and I have had the incredible bad luck of surviving. My brother, poor soul, was more fortunate because when they were beating his back with sticks, they smashed the bottom of his skull and he died instantly."[liii]

Málaga

In his book *Diálogos con la muerte*,[liv] Arthur Koestler, Hungarian-born British journalist and war correspondent, wrote several essays illustrating the Francoist treatment of prisoners that he witnessed when he was arrested in Malaga February 8 1937, the day that the city fell to the Nationalists. During the week that Koestler was imprisoned in Malaga, before he was transferred to another prison in Seville mid-February 1937, he again had a chance to be dumbfounded by Francoist brutality.

Moreno Gómez mentioned several cases when prisoners went crazy and totally lost their minds because of their sufferings and he will mention a few more later on. For the moment, however, a personal experience of his father's in Villanueva de Cordoba around 1940 and that he, Alfonso Moreno Zamora, told him by chance in 1995. One day, the Guardia Civil ordered his father and Miguel Gutiérrez Marín, both of whom were not much more than 20 years old at the time, to take some cattle from the Los Pobos fields to the Guardia Civil barracks in Montes de Adamuz. Halfway there, for no apparent reason, Miguel began to sob, and he cried non-stop for several hours but there was nothing Alfonso could do to console him. The Guardia

Civil's order had been a last straw and it was only much later that his father discovered what was behind it all.

In those days, the young Miguel Gutiérrez was a close friend of Juan *Hebrero* Caballlero Coleto, who escaped to the hills at the end of the war. They had agreed that Miguel would take him food to a hidden spot and the guerrilla would collect it come nighttime. Unfortunately, the Guardia Civil got wind of this. They called Miguel to the barracks and beat him within an inch of his life. The young man went mad and never recovered. Never, during all those years, never anybody, not even Moreno Gómez' family, had ever told his father why it was that Miguel Gutiérrez had lost his mind, even though they lived next door.

Undoubtedly, the legal framework can be studied at any time. Not so the recovery of the facts and the memories of the victims and their relatives. Far too much time has passed. This is what academia and historians should have done during the years of transition, instead of dedicatig themselves to developig the art of 'disrememberring'. Ángela Cenarro, Professor of Contemporary History at the University of Zaragoza, has commented on how historians should proceed with their research into the legal framework. She stresses that if they limit themselves to enumerating the subhuman conditions to which thousands of anti-Francoist prisoners were subjected during those days, they run the risk of getting lost among so much misery and desolation, without adding anything to the knowledge of the legal and institutional framework.[lv]

Moreno Gómez did descend into the misery of prisoner torture, but he did not get lost. On the contrary, as he expanded his field work, as he sank even deeper into the abyss of pain and suffering, he became increasingly aware of the importance of the researcher's duty to study the legal and institutional framework in order to discern the true historical facts. More so in the case of the Francoist legislation of the 1940s, when the demagogic rule of redemption, its altruistic decrees and regulations, hypocritical social welfare programs, represented nothing more than worthless pieces of paper for a propaganda that had nothing to do with the reality; worse still, that concealed it. What Ángela Cenarro said is all very true, but at the end of the day, she concluded that Francoism said one thing openly, to the public gallery, whilst it acted in a totally different manner when it came to the punishment and extermination of the defeated.

Moreno Gómez cannot protest too much at the sovereign stupidity of a recent newspaper article by the much admired, highly commended Antonio Muñoz Molina, to the effect that the people of Spain have always spoken openly about what happened during the war and that the idea that there was some kind of enforced silence about the events was a total myth created by foreigners. Molina declared that all that was said about there being 'two Spains' was absolute garbage and that the significant factor of the war was not the actions of the people who acted in all consciousness, aware of what was at stake. Instead, the activities of the thoughtless, irresponsible and amorphous people, that stupid thing called 'third Spain', those who were forced to go and fight and who did not give a damn for whatever it was that they were supposed to be fighting for.[lvi] So much for Muñoz Molina's inspired analysis of the platitude that *"nobody likes to go to war"*.

Today, we are again far too often faced with the twisted tendency of certain intellectuals from the 'progressive' thinkers who believe that they are possessed of the gift of happily rising above all extremes, disseminators of equidistance and authors of false revisions of the facts. Those whose work is coated with the varnish of a post-modern, banal and superficial – to the point of nausea – aestheticism and whose view of the History of the civil war and its aftermath apparently lies halfway along the road between Gila's War[lvii] and Berllanga's La Vaquilla.[lviii]

Military Tribunals and Courts Martial

"The 40,000 executed by the firing squads during the Commune guaranteed sixty years of social peace."
José Calvo Sotelo
Leader of the *Renovació Española* and the *Bloque Nacional*, speaking in Parliamente, referring to the example of the Paris repression in 1871.[lix]

Linguistically, in English as opposed to Spanish – and *de facto*, it is difficult to distinguish between the expressions 'Military Tribunal' and 'Court Martial',[14] particularly when it comes to the composition and proceedings of both under the farcical Francoist rule of law and

---

[14] In Spanish>English<Spanish, Military Tribunal is the reciprocal translation of Court Martial and vice-versa.

the organized repression that governed the country. Before the Regime passed the <u>Decree on Banditry and Terrorism</u> (September 21, 1960) and <u>Law 15/63</u> creating the <u>Court of Public Order</u> (TOP), all offenders were considered bound by the Francoist military rule of law, regardless of the nature of the offense of which they were accused.

If one can be permitted a linguistic *nuance*, one might ay that the so-called Military Tribunals were regarded as Higher Courts, whose Judges administered, not usual military justice with due process, but a repressive and vengeful, considerably more vicious and speedy justice. They ruled as if from 'on high' as both judge and jury, on the basis of summarized reports of interrogations by Examining Magistrates who also, in countless courts martial, handed down sentences. In actual fact, it was these courts martial, often presided by the very same judges and with the same composition but technically on the next rung down the administrative ladder, that provided the public face of Francoist justice. Magistrates were dedicated to speedily prosecuting and sentencing groups of prisoners, rather than individuals, non-stop and on the basis of summary proceedings and interrogations, in a desperate attempt to keep pace with the astronomical number of accused whose trials overwhelmed the entire court system.

Itinerant and Standing Military Tribunals

To the multitude of military tribunals created especially for the great purge and cleansing of Republicans during the post-war period as part of the Regime's programmed extermination (35 courts in Cordoba province: 20 in the capital and 15 in townships; Malaga 67; Alicante 22; Cartagena 57, etc.), we must add the creation of other military courts. It would appear that the principal occupation of the victorious military in 1939 and following years, was the appalling mission of handing down death sentences (the only accepted sentence) right, left and centre, sending thousands of men and quite a few women to the wall, and serving the Dictator in the foulest manner possible, the annihilation of their own countrymen.

That is how victory was administered, with a great campaign involving the extermination, punishment, harassment and exclusion of the half Spain that had been defeated, a campaign that exhibited a typical Fascist characteristic of invoking of the root of the revolution,

Fascism's *raison d'être*. Not only was the Regime's program typical Fascist purging of possible revolutionaries, it also was the repression of an entire class of activists for workers' rights of trade unions and leftist political parties, all of whom suffered the most brutal punishment in all their history. This was the genocide of an entire sector of the population for social and political reasons. A genocide that was triggered by the military coup in 1936 and that spread throughout the country in 1939 with the end of the civil war.

The sequence of events was as follows. Once there was a sufficient number of indictments or pre-trial hearings in the standing military magistrate courts, a date was set for prosecuting the accused, usually collectively. Itinerant military courts of law were created, and they were sent to large towns in the province of Cordoba, to examine and rule on the summarized files of those who were scheduled to be tried by the notably farcical courts martial. The courts that went to the towns often travelled from one to another, whilst those in the capital were more permanent in nature and they usually limited themselves to receiving the lines of prisoners pre-destined for 'the only sentence'.

The itinerant courts worked intermittently, that is, whenever they had enough cases to rule on. Two or three times a month they went from town to town, beginning with those where the most cases were waiting to be prosecuted: Castro del Río, Bujalance, Puente Genil, Baena, Peñarroya, Hinojosa, Pozoblanco, Villanueva de Cordoba, etc. The cost of accommodating and feeding the courts was paid for out of the municipal budgets (hotels, meals, telephone calls, and so forth, for which there are numerous invoices in the Minutes of the sessions books.)

A Military Tribunal was composed of a Presiding Judge (the highest-ranking officer), three (sometimes, two) Members, a Rapporteur, a Prosecutor and a Public Defender (the lowest-ranking officer). Seven in all, all serving military, except when there was an exception, and the role of the Presiding Judge was given to some local magistrate in recompense for his being an 'undisputed supporter' of the Movement. Evidently, there was ' military justice' and 'justice' that depended on the victor's extermination program.

On the date of this writing (2019), there still are some Francoist magistrates wearing judges' togas who were promoted to high judiciary positions, especially during the days of the TOP (Court of Public

Order). However, as the transition period proved to be the cover-up that was expected, none of these magistrates was, or has been, asked to explain his muddy past.

Moreno Gómez obtained details of the composition and activities of quite a few Military Tribunals in Cordoba province by consulting several summary reports of cases in the newspapers, even though the press very rarely reported details as the Cordoba *Azul* did the following composition of the court that sat in Baena May 20 1939.[lx]

Sometimes, as in this case, the Judge Advocate for the Army of the South, Lt. Colonel Ignacio Cuervo, was also present. The scenario was typical of every trial in Baena. Manifestations of solemnity, before and afterwards, always in response to the official desire to conceal the details of the shocking July 28 1936 Francoist massacre.

10 May 1939	Baena
PRESIDING JUDGE **Colonel Evaristo Peñalver, Guardia Civil**	
Members	Captain Enrique Vilches Captain Rafael Mariscal Captain Baltazar García Valdecasas
Rapporteur	Bernabé Andrés Perez Jiménez
Prosecutor	José Ramón de la Lastra y de Hoces
Public Defender	Lt. Fernando Moreno González Anleo
Also present: the Judge Advocate for the Army of the South	Lt. Colonel Ignacio Cuervo

On that occasion, several hundred unarmed men lying on the ground in the town square, were individually executed with a single shot to the head, one by one, by Lieutenant Pascual Sánchez Ramírez of the Sáenz de Buruaga Guardia Civil. One of many crimes of genocide that should have been tried before a Spanish Nuremburg. As they escaped the massacre in the town, a group of angry anarchists killed 81 right-wing prisoners held in the Asilo de San Francisco.

The *Azul* described the frenetic activity of the above Tribunal as "Having acted with extraordinary diligence in administering Justice and after recently judging crimes committed in Bujalance, Castro del Río and Posadas, the Court returned to Cordoba Saturday 20th in the early evening."[lxi]

Waving metaphorical banners recalling the events in the Asilo, the great landowners of Baena and their entourage still today continue to express their desire to suppress all mention of the Francoist massacre.

In Baena, everyone is involved in the cover-up and if mention of that fateful day comes up, they quickly change the subject to what happened in the Asilo. Not a word regarding the Francoist massacre in the square. This attitude has become so pervasive that when a few years ago, a German film team went to Baena to make a movie of the Francoist massacre (for whom Moreno Gómez was engaged as a specialist advisor/consultant), the local powers that be, so-called 'democratic' authorities included, made their work so difficult, their lives so very impossible, that the Germans abandoned their project and left. To this day, there still is no other kind of law in Baena than the law of the great landowners and olive-oil producers.[15]

The Baena Tribunal, presided by Evaristo Peñalver and embellished by the spectacular oratory of Prosecutor José Ramón de la Lastra (a latter-day version of *Fray Gerundio de Campazas*[16]) sat September 1939 in Puente Genil, where the eminent Antonio Romero, ex-Republican Mayor, was condemned to death by garrotte. Another frequent member of this court was the tireless Torquemada from Rute, the aforementioned Bernabé Andrés Pérez Jiménez. Captain of the Judge Advocate General's Corps, his 'blood' achievements were so appreciated that he was promoted to the position of Supreme Court Judge, He also intervened in numerous courts martial in Baena, Puente Genil, Cordoba and elsewhere.

The Tribunals were put to work very early on in post-war Pozoblanco, beginning May 18 1939. Lt. Colonel Rafael Carbonell Morand, Lt. Colonel Rafael Mora Sánchez and Commander Ramón Navarro de Cáceres were the first presiding judges. Judge Mora also presided over Tribunals in Villanueva de Cordoba, with some variations in the composition of the Court. It is curious that a same military officer acted as a Public Defender for the accused in some

---

[15] When Moreno Gómez' book *La guerra civil in Cordoba* was published in 1985, in which he related the Francoist horrors that were committed in Baena, this implacable bourgeoisie set out to find an author to write a counter-book. They still have been unable to find one. Arcángel Bedmar, in 2008, put a full stop to the subject with a new book along the same lines as Moreno Gómez', *Baena, roja y negra* (Baena, red and black), *Op. Cit.* The Socialist Mayor timidly attended the launching of Bedmar's book; the local neonazis boycotted it. Democracy is still not widespread in Spain.

[16] Fictional hero of a mid-18th century novel by José Francisco de la Isla y Rojo (*Historia del famoso predicador fray Gerundio de Campazas, alias Zores*) depicting a typical preacher who used high-faluting, often offensive, illogical and ludicrous language when addressing his congregation.

Tribunals and as a Prosecutor in others. At the same time, those who acted as Public Defenders were usually the lowest-ranking officers.

Except for the Presiding Judge, the rest of the members of the Court appeared mostly interchangeably from one court to another. The rulings or sentences decreed by the Tribunals were of the maximum severity, almost always condemning the accused to the 'ultimate' or 'only' penalty. A lower sentence of Life Without Parole, or thirty years, was considered a special favour.

November 1939, Commander Francisco Ferrán held Court in the mountains of Cordoba, where he presided over the Pozoblanco Tribunal. In 1940, the most active Tribunal in the Sierra mountain region (the Campiña was subject to less 'avenging justice' pressure because much of the population had already been massacred in 1936), was presided over by Colonel Luengo Benítez, a leading Cordoban Fascist and founder in the capital of the Nationalist Battalion of Cordoba Volunteers in 1936, whose fighters later entered many towns in the valley of the Guadalquivir River, spilling blood and spitting fire.

This Court arrived in Villanueva de Cordoba 1 April 1940, intending to hit hard on the famous Communist stronghold of Villanueva, the so-called Red Town, where it proved to be especially vindictive and ruthless against eminent Communist leaders. Severe sentences were handed down from April 23 to 27 and from April 30 to May 4. May 1, Militia Captain Pedro Torralbo Gómez, past Communist city councillor and Provincial Deputy was sentenced to death. A little more than a year later, in June, he and several others died in a mass execution in Cordoba cemetery. In June, Judge Luengo Benítez continued his repressive judiciary activities in Pozoblanco and again in Villanueva. March 14 1940, Judge Ricardo Rivas Vilaro presided over the Villanueva de Cordoba Tribunal.

Lastly, of the many Tribunals that sat in Cordoba capital, the February 5 1941 one was presided over by the recently promoted Judge Aguilar Galindo, known for being a notable hardliner hawk during the bloody hours of July 1936. He was also responsible for the genocide in Fernán Núñez, where row upon row of dead Republicans lined the gutters on the road to Cordoba.

5 February 1941	Cordoba capital
PRESIDING JUDGE **Colonel Manuel Aguilar Galindo**	
Members	Lt. Diego González Rodríguez Lt. Francisco González Cáceres Lt. Filiberto Agregano Gasco
Rapporteur	Captain Luís Mendieta
Public Defender	Lt. Ignacio Alfaro Guzmán
Examining Magistrate	Lt. Antonio Corredor de la Cruz

As described earlier, the composition of the Tribunals was almost exclusively military and they were always presided over by a high-ranking officer, with only the very occasional presence of civilian magistrates. Court officials could be attached to one or another court, or change about, as they went about their repressive pilgrimage over the entire Sierra in 1939 and 1940, after which the headquarters for the purging were centred in Cordoba capital where all those who were arrested September and October 1940 were taken.

Courts martial

The courts martial were the staged public face of Francoist justice. The first Francoist courts martial all over Spain began in May 1939. All historians have qualified these staged events as 'judiciary farces' hostile to those who were loyal to the constitutional Government. If the summary prosecutions already represented a multitude of fundamental defects and judicial abnormalities, including the application of mass torture, the courts martial added to the number of legal irregularities. Nonetheless, despite the present Democratic Government's repeal of the Francoist judicial legislation, it has been unable, or has not dared or had the mettle, to rescind the courts martial sentences and these are still considered valid under law, after thirty years of democratic rule.[17]

An almost daily post-war spectacle were the chains of prisoners tied to each other two by two, as they were led each morning from the jails to the place where the courts martial were held. In Cordoba capital, trials and courts martial were held in the Military Tribunal Hearings

---

[17] Finally, this changed 19 October 2022 when the Senate approved a revised, definitive version of the Democratic Memory Law, which declares the 1936 military coup illegal, and all sentences handed down by the Francoists courts null and void.

Building, Avenida del Gran Capitán number 4.[18] In Madrid, the accused were taken in automobiles to Las Salesas. In other towns, they usually went on foot. In Cordoba province, the most picturesque places were chosen for these staged events. In Baena, it was the Theatre on Calle Alta, the venue for ostentatious legal farces, worthy of Valle-Inclán's pieces of theatrical nonsense such as *Los cuernos de don Friolera*[19]. It was here, to this theatre, that the non-violent and Messianic anarchist leader *El Transio*, was brought and exhibited as an *Ecce Homo* before a packed audience, to hear that he had been given three death sentences, to the raucous delight of the Baenense bourgeoisie present.

In Castro del Río, courts martial were held in the Jesús Nazareno Hospital, except for the first one which was accompanied by special propagandist pomp at the Teatro Cervantes. In Pozoblanco, as in Montilla, Puente Genil, Peñarroya-Pueblonuevo and elsewhere, the venue for the farce was the City Council General Assembly Hall. In Hinojosa del Duque, trials were held in the main meeting room of the El Gato Casino, a bar on the main square. In Villanueva de Córdoba, chains of prisoners were marched along a considerably tortuous route through the city, from Fuente Vieja prison to the Torres events hall on Calle Concejo. All along this route, obviously extremely humiliating for the accused who were displayed as an example to the population, relatives would come out at the crossroads (mothers, wives with their children, and other kin) in an attempt to catch a glimpse of their loved ones for far too many, for the last time, all of which added to the prisoners' suffering.

The proceedings were open to the public. Whenever the accused had some standing as a public figure, the venue would be packed with a typical public: Falangistas, Señoritos, ladies from the bourgeoisie and especially, relatives of right-wingers who had died earlier, dressed in full mourning and demanding justice. Also present were the denouncers who reported the crime and the witnesses to the accusations. Witnesses for the defence, who very few called, did not dare to appear in the midst of such an extremely hostile environment. Occasionally, relatives of the accused would creep in timidly and stand at the back of the hall where they would shed silent tears at the fate that was about to be meted to their kin. In 1980 Moreno Gómez was able to obtain

---

[18] Later, *Delegaciòn de Hacienda*, a branch of the Government Tax Office.
[19] "The cuckholding of Don Frilera". Today the theatre is a venue for wedding receptions.

some valuable eyewitness testimonial as to the nonsensical opulence of those trials and how, on the rare occasion when the crimes for which a prisoner was accused were proven to be false, this was usually ignored, and the prisoner would still be executed. Such was the case (and there were thousands of similar cases) of Juan Cantador Zamora from Villanueva de Cordoba, whose daughter Catalina Cantador Romero attended his trial.[lxii]

> "24 May 1940, [my father] was tried by the military court in this town and sentenced to death. At the court martial, the prosecutor asked Luísa Doctor, the accuser, to point out which of the accused men sitting on the bench had killed her husband. The good lady said she did not recognize any of them and that she had heard nothing said about him. The Judge then asked the accuser if she had every spoken to the accused, to which she answered that she might have spoken to him as she might have spoken to anyone.
>
> Judge: Then, why did you say that he had killed your husband?
>
> Accuser: No sir. Judge Juan Calero must not have heard me correctly. What I said was that it was someone from outside, one of Juan Elías' agents."

The accusation was clearly disproved at the trial, but as these stagings were farces and the outcome had been already decided upon, the accused was still executed. When the prosecutor asked Catalina's father why he had signed a confession, he answered just like thousands of other accused: "I was beaten until I signed". The prosecutor then asked him, heaven knows why:

> "Prosecutor: Were you ever beaten?
>
> Accused: Yes sir. It would be impossible for me to name all those who beat me as I often lost consciousness. The examining magistrate himself beat me."

Catalina continued:

> "When the court martial was over, my father was taken to Fuente Vieja prison. No sooner did he arrive there, that a 2nd Lieutenant, the Judge and José Higuera entered the patio and, in the presence of the other prisoners, beat my father until he lost consciousness. I could hear his screams from outside the building, on the street. From the day on which he was condemned to death until September 16 1940 when he was transferred to Cordoba prison, we never saw him again."

That was how Francoism 'tried' the accused and how many ordinary people died during the post-war period. This could be the subject for an entire treatise on the reality of Francoist justice, not just because of the tragedies, but also the ridiculous scenarios that Francoists staged to give an appearance of 'legality' to the extermination. Farce, nonsense, burlesque and spoofs, were some of the means the military employed. The above case clearly shows (a) how it was by pure chance that an accused's innocence, not his guile, was proven, but never to any avail; (b) that the accused were tortured or beaten until they confessed; and (c) that relatives or friends of the prisoners were said to have made accusations that they denied or whose meaning was twisted by the Court to its advantage.

A distinctive feature of Francoist trials was their collective approach. It was a rare occasion that a court martial was convened for a single accused. Quite the contrary. The bench of accused was always occupied by several individuals, in greater or lesser numbers, all whose fate was decided in less than an hour. This administration of group justice was another affront to the individual defendants. The Court centred its attention on the group of accused and not on the individual characteristics of any given prisoner. By grouping several cases, the court martial was politicized and promoted the desired depersonalization of the prisoners. The Francoist apparatus and the justice that depended on it only saw the accused as yet another clutch of Rojos, on whose extermination depended the security of the Regime. Examples of single individual trials were rare in Cordoba.

As soon as the accused had been seated in the courtroom, tied to each other two by two and guarded by Guardias Civil and Falangistas, the Court took its seat on the podium and began the formal proceedings as expected. The Rapporteur would read out a so-called summary for each accused and not the complete summary accusation. The summary condensed and simplified the accusations and the charges, disregarding the fact that these were summary accusations obtained by the examining magistrates after torture was applied, based on manifestly false data obtained without the minimum due process of law and signed by an accused after he had been beaten and tortured until he confessed to what they had written in his stead.

Once the summary was read out and each accused's name was briefly mentioned, almost as if it were just another name on a list of prisoners in a jail, the Prosecutor would ask the accused a few superfluous and redundant questions that never met the burden of proof, expanding or simply paraphrasing that which had already been written in the summary accusation. Few witnesses for the defence were brave enough to appear.

Nothing was proven and there were no amendments to the charges, little or no variation in the accusations or allegations, The courts martial decisions were not based on proof; the indictment was considered enough on its own and the only persons from outside who were allowed to speak were the accusers and witnesses to the charges.

The Rapporteur and the Prosecutor limited their interventions to inflating certain aspects of the summary reports and to emphasizing the working-class connections of an accused and his 'extremely bad behaviour' – all the way back to the days of the Republic, and even earlier to the days of the so-called Bolshevik triennium (taking part in strikes, demonstrations, embroidering flags, etc.) - with special emphasis on any rank a defendant may have held in the Republican Army. There was an evident insistence on these political and ideological features, rather than on determining the truth of the supposed commission of blood crimes. The courts martial took little notice of anyone who might have committed an ordinary blood crime, preferring to focus their attention on purging the country of everyone who may have committed what was considered a political crime. In other words, the courts martial were no more than instruments for the political purging of the country of anyone who might be disaffected.

The highpoint of the trials was the intervention of the Prosecutor, who spoke aggressively, vehemently giving full rein to every available Falangist or National Catholic argument. In post-war Cordoba, none of the prosecutors could reach the heights of neurotic rhetoric as much as or better than José Ramón de la Lastra y Hoces, attorney, Marquis of Ugena, grandson of the Duke of Hornachelos, wealthy landowner and ex-President of the Agricultural Association – the personification of the landed gentry and the decadent aristocracy, with all the vices of Spanish caciquism. Don José's interventions in trials were notorious.

A partial transcript of one of these, a May 21 1939 article in the Cordoba *Azul,* gives details of the most infamous court martial in Baena, where the principal 'anarchist' leaders of the town were eliminated. Judge Evaristo Peñalva, a Colonel in the Guardia Civil, presided over the court martial.

After pondering the merits of Baena's 'heroic fight against criminal Marxists' (Don José appeared to be referring to the townspeople's violent reaction to the Francoist army's slaughter of several hundred unarmed men lying in the main square), he went on and on in this vein without this having anything to do with the accused on trial.

> "The Prosecutor continued with his brilliant allocution. He decried the arrival of the Republic, considering it the door that opened the way for Russian cruelty and criminality in Spain. The Republic served only the Masons and the Jews, and created the concentrated hate of the Army, alluding to Pemán's statement that 'The Levites were the damnation of Spain and the Army its salvation'.
>
> He affirmed that everyone and everything were conspiring against Religion and the Army, the most fundamental and serious pillars of the Fatherland, the military ad the religious, so perfectly defined by José Antonio Primo de Rivera.
>
> The Frente Popular, with its Masonic and Jewish coercing agents, had as its sole mission handing Spain over to Russia. The Caudillo's, to whom all of Spain owes so much, mission was heaven-sent as he was entrusted by God with saving the Fatherland.

It now happens that the Marxists claim they are not Nationalists. Can there be anything more beautiful than this description, so that even the bad patriots are unable to apply it to themselves?

The Red beasts murdered 90 people in the San Francisco de Baena Convent. Corners, cloisters and rooms, filled with the pain of the tragedy. Killing for the sake of killing. They knew not what they were defending. He who said that the Rojos were defending themselves, did not know against what. The fact of the matter is that criminal instinct found the open dark door and murder was committed with satisfaction.

Spain is saved. The Sacred Heart of Jesus has fulfilled its promise. God watches over her. We now see those who obeyed like wild beasts, and those who beyond the frontiers enjoy a life of wealth and well-being. These did not hesitate to become criminals when they received their orders to do so.

The Prosecutor referred to the 1934 government banners, the ones that led to the events of 1936. This 1936 must not again happen. He asked for justice with serenity, his hand on his heart in obeisance of his duty.

He declared a welcome and praise to heroic Lieutenant Pascual Sánchez Ramírez, Military Commander of the town of Baena and its stalwart defender, who has already been honoured by the Generalissimo with the Military Medal. Men like these – he said – honour the Fatherland that gave them birth.

We, he added, represent Spain, against criminal Marxism, the hammer and the sickle, the indignity and unconfessable designs of the ominous triangle as opposed to our Cross of Redemption. 96 murders have left their mark on Baena, a city deserving of better luck, filled with bitterness and pain. He connected the deeds and monstrosities committed, signs of unheard of and horrible cruelty.

This is the time for exact, considered, inflexible justice. The deeds are covered under Article 235 of the Code of Military Justice and are punishable under Article 538.

At this point, the Prosecutor and the Court stood up. So did the public. It was a very emotional moment. In the name of the Caudillo, he asked for the death sentence for all the defendants."[20]

The newspaper headlined that ostentatious and grotesque court martial: "36 death sentences for 20 accused". Several accused were sentenced to death more than once. Once would have been enough. After this theatrical description, the press rarely printed such reports regarding other courts martial. It is presumed that the authorities issued orders for prosecutors and the press to be more discreet.

When faced with examples of forensic oratory, the court martials always supported the Prosecutor's conclusions and felt no need to change them based on the defence's succinct remarks. The possibility that a meek intervention from the military defender might influence the court's decision was extremely unlikely, particularly with such meek requests for clemency "considering that we are living in times of strict and inflexible justice, but also of generosity and Christian forgiveness."[lxiii]

The individual nature of each of the accused was totally buried amid the prosecutor's political ranting. As far as he and the Court were concerned, the accused were not tangible persons, just elements of anti-Spain and of a perverse system (Republican democracy), vermin to be demonized and then exterminated without remorse.

The figure of the Public Defender in the court martial was relegated to a marginal position, He was either a 1st or 2nd Lieutenant, rarely a Captain, who spoke timidly, briefly and even as if by rote. What could a military 'defender' say in favour of all those individuals described as

---

[20] The journalist misconstrued or falsely reported many of the facts the prisoners were anarchists (the usual Francoist description for anyone against the Regime, the disaffected), not Marxists, who were being tried. Also, it was Francoists who described themselves as Nationalists, not the other way around. The number killed by the angry townspeople was 81, not 96, and not a single individual was proven guilty of this; 500 was the number of unarmed men executed by the Francoist Military Commander in Baena town square. The journalist appeared unaware that it was home-grown disaffected, not Marxists from Russia, who were being tried on this occasion, unless he chose to spout the official Francoist line rather than speak the truth.

monsters, over whose heads the gallows already hanged? The Public Defender's words were addressed to a collective, he made no reference to any individual prisoner. He limited himself to simply asking for a reduction of the sentence to a less drastic punishment, but frequently, not even that, just for the indulgence, clemency or Christian charity of the Court. His were purely pro-forma interventions that totally ignored individual accusations, without even the slightest suggestion that any of these charges had even been examined.

This latter observation regarding the lack of examination applied to prosecutors and timid public defenders alike. None ever showed the least desire to clarify the charges, either during the interrogations or during the trials. Nor was there any desire for clarification nor real wish to defend the accused. Such was the case in the Baena court martial that was adjourned for a break because another court martial was scheduled for 1 p.m. The records for this second court martial describe the actions of the defence in the following terms: "The Public Defender, Lieutenant Bernal, after a few brief words, asked the Court's indulgence for the accused.". Just like Cicero in his defence of Milo or the poet Arquias.

Three days after the end of the second court martial against seven accused from Valenzuela, *Azul* newspaper reported yet another example of the aristocratic prosecutor José Ramón de la Lastra's extravagant rhetoric.[lxiv] During this trial as in the previous one, the aristocratic Prosecutor alluded to the Freemasons, a subject that was totally irrelevant to the cases on trial and totally unknown to the ordinary workers who were sitting on the bench of the accused. In his earlier intervention, the Prosecutor alluded to the 'hammer and sickle and the ominous triangle'. During his second intervention, he continued to spout the Regime's neurotic, obsessive ranting against the Masonic motto of 'liberty, equality and fraternity'. Both references that meant absolutely nothing in the rural world of a handful of anarchist working men, most of whom were farmworkers and labourers.

The anti-Masonic obsession was a scarecrow that the Caudillo and the Church (especially the latter) constantly brandished, then and during the next forty years. In the Dictator's last, extremely brief declaration in 1975 at the Plaza del Oriente, he returned to the scarecrow when he referred to 'the Jewish-Masonic-Marxist conspiracy'. It was an obsession that would darken his deathbed, as all the ghosts of the past prepared to 'welcome' him from beyond the grave.

An especially dramatic moment during the court martial was when the Public Defender for the accused stood up to speak on their behalf in this totally hostile environment, as a court martial rarely allowed an accused the right to speak for himself. At the very most, the Court gave the floor to only one accused to speak on behalf of all. What could one man say on his behalf or on behalf of them all? Here too, another form of depersonalization under Francoist 'justice'.

Whenever an accused managed the courage to say a few words, he almost always denied the charges against hi, drawing attention to the false nature of the accusations. If the Presiding Judge replied that the accused had signed a confession, the prisoner replied that it was because he was beaten until he did so. Occasionally, from amongst those who sat on the accused's bench, one heard the more eloquent voice of someone accustomed to speaking in meetings and in public. Inevitably, the Presiding Judge would immediately tell him to shut up.

In his book of *Memoires*, reproduced in part herein, Rafael Bedmar Guerreiro reconstructs some of Prosecutor La Lastra's allegations at his trial in Cordoba Court October 25 1939, when Rafael spoke in his own defence. The Presiding Judge was the omnipresent Evaristo Peñalver.

In an undated document of memoires given to Moreno Gómez, Adriano Romero refers to his trials, one of the rare individual courts martial perhaps because the accused had been a Member of Parliament and he was to be made an example.

> "The trial lasted three hours, something that was completely unusual, and at the end, when I was asked if I had anything to say, I stood up ready to speak some truths, but I was only able to say a couple of words before I was forced to sit down, and the session was adjourned."[lxv]

In his memoires, he added that Juan Escribano, a friend of his who attended the court martial in the Cordoba Audiencia in 1941, reminded him that he also declared that he had been a communist since he had reached the age of reason and that the Court could only condemn him for that and not for any blood crime of any kind.

Everything that occurred during the trials was a grotesque travesty of traditional judicial procedures. Franco's typical military justice was no more than a pro-forma vehicle for eliminating the Regime's opponents, procedures where no one troubled to do any prior work to prove the charges, without any desire to clarify the events, that included the absurd intervention of the military Public Defender. In José Subirats' magnificent book of memoires, [lxvi] he cites the bizarre case during a court martial in Tarragona,[lxvii] of a defender who asked for the death sentence for the person he was defending contrary to the Prosecutor's recommendation. The accused was taken to the firing squad. Subirats also recalls his own court martial (August 10 1939) which lasted little more than thirty minutes and was over so quickly that the Rapporteur barely had time to read out the charges against the 15 accused.

The Judge Advocate's Court in Tarragona convened daily collective courts martial, one at 11 a.m. and another at twelve noon. Another farce, a sham, grotesque. After glancing over the indictments, the Court would convene a mini council that would meet briefly and dictate the sentences on the go. The accused were swiftly despatched back to their prison with their death sentences under their arms, and then directly to death row where they were held incommunicado from the rest of the inmates.

Lastly, the remarkable relationship between the local oligarchy in each town and the Regime's judiciary apparatus was particularly noteworthy. The most unyielding local members of the right-wing and the Falange collaborated closely with the military judges both during the instruction of the summary proceedings and later, with the special Military Court that sat in their town for two or three days. In a great many cases, the Examining Magistrate and other officers of the Court were members of the local oligarchy, such as Judge Manuel Cubilli in Baena, Prosecutor de La Lastra in the capital, and so forth. In any case, those responsible for justice frequently asked for and always received support, advice and instructions from the local Francoists.

In Villanueva de Cordoba, for example, it is well known that the members of the military Magistrate's Court and the major landowners of the town met regularly in the Sepúlvedas' home and in the Casino, meetings that were frequently attended by the Jesuit priest Bernabé Copado, S.J. On these and other occasions, those present fine-tuned the details of the repression, drafted the blacklists, agreed upon the death sentences and decided on any possible combination of or changes to the accusations that were levied against the defendants. When we examine the Francoist repression, at no time can we lose sight of the close collaboration between the military courts and the landowner bourgeoisie in southern Spain, along the lines of that which has been described as Rural Catholic Fascism.

Examination of sentences and appeals for clemency. False rhetoric and an arbitrary approach.

The possibility of a condemned person's appealing his sentence was not contemplated by the Regime until the beginning of 1940. Besides, in 1939 such a possibility did not exist for the defendants who were usually executed a few weeks after they had been condemned to death. Even when an accused had a very slight opportunity to defend his case during the court martial, he frequently found that all doors for any kind of an appeal for an examination of the procedure or a reassessment of his sentence had been closed. Furthermore, the accused's generally low level of formal education and the even lower level of literacy of any relatives who might help them outside the jail, proved to be a major barrier to any bureaucratic management of an appeal.

Of special note was the dedication of Communist leader Matilde Landa, who in Ventas, Madrid, and Palma de Mallorca prisons, devoted herself to drafting appeals for sentence revisions for her companions in prison. She was one of the great women of the Republic, with an impressive trajectory, who tragically took her own life in prison when she could no longer bear the Church's insufferable attacks against her.

January 25 1940, the Presidency of the Government issued an executive decree creating Provincial Committees for Examining Sentences (CPEP)[lxviii] specifically to 'abolish, as humanly possible, any inequalities that might have occurred and did occur in many cases for various reasons, given the confusion resulting from the lack of uniformity of criteria to judge and punish similar crimes of the same gravity.' These were not beneficial measures from the totalitarian New State (as some historians have suggested), just simple measures to unify the procedures involving the geographic distribution of the numerous military courts throughout the country, as well as the two great classes of individuals sent to trial: military and civilians. [lxix]

Civilians received the harshest sentences. This was never a question of mitigating the outcomes of the judicial procedures. Suffice to remember the variety of civilians who from the onset were excluded from any possibility of requesting an examination or review of their sentences.

The CPEP, invented for fine-tuning the repression with the excuse of the need to uniformize procedures, began operating February 1940 and lasted five years until the committees were disbanded February 25 1945. They were followed by the Central Bureau for Assessing Sentences (CCEP) attached to the Advocacy of the Ministry of the Army. As these Committees were created by Executive Decree from Franco's personal office they were, consequently, autonomous bodies that functioned independently of the judiciary summary proceedings and the military tribunals. Despite that, they strongly influenced the sentencing, which itself was a juridical irregularity under the very laws of the Regime that apparently did not even respect its own legislation.

According to Judge Juan José del Águila, the CPEP was a hybrid body with an administrative structure and a so-called jurisdictional soul. In case of a disagreement between the Judiciary Authorities and the Judges Advocates, the cases were sent to the Supreme Court of Military Justice which resolved the matter behind closed doors in accordance with a Decree of June 3 1942.

In the wake of the decree creating the CPEP committees January 1940, the Minister of the Army, General Varela, dedicated himself to issuing communiqués regarding the Assessment of Sentences, which were not published in the B.O.E. (Official Bulletin of the Army) and only added considerable turmoil to the judicial tangle, which became immersed in the most absolute subjectivity. To complicate matters relating to the CPEP and communiqués, the decrees were in violation of the Regime's own Military Code of Justice (1870) that in Article 176 states:

> 'No military officer may be punished by a penalty that was not previously set by law for such an offense before the date on which the offense was committed. The only sentences that will be examined are those imposed by the Courts as a result of a judicial proceeding.'

Juan José del Águila then writes:

> "It is clear that the new penalties (the ones established by the CPEP and the many communiqués) are determined after the fact, after the supposed crime had been tried and punished by the sentences from the Courts Martial, and that many of the standing penalties could not be applied by the Courts, as they did not meet the specific terms of the recently created Committees for the Examination of Sentences."

The Regime's judiciary shambles aside, just who could appear for a review of their sentences and who were affected by the wide array of exclusions? The January 25 1940 Decree contained nine instructions, some of which were truly amazing, such as the Second Instruction that stipulates that "a Committee may not discuss facts that have been declared as proven and must limit itself to drafting a proposal, whether it agrees with the judiciary decision or whether it deems it convenient to propose that a sentence be reduced or commuted" or Instruction Nine that further clarifies the role of the Committees as being "to draw attention to the politico-social and moral background of the accused or his personal behaviour 'before the Movement', or the impact of his actions for or against the National Cause".

Clearly, the mission of the CPEP, far from introducing compassionate measures, was nothing more than a fine-tuned bureaucratic regulation to unify and typify every detail of the colossal machinery of Francoist justice, whilst retaining the severity of the sentences handed to 'qualified' disaffected and to a certain extent, to separate them from 'neutral' disaffected and run-of-the-mill criminals. It is also important to remember that the mission of the CPEP only applied to cases directly related to the war, so-called military war crimes, and not to all post-war prosecutions for a variety of other reasons, namely, the 'ordinary conviction' of non-military individuals who as 'disaffected' civilians, nevertheless continued to attract the Regime's anger and for which they were equally severely punished.

By 1944, the CPEP committees impacted the sentences in 107,983 cases of ordinary convictions, of which 2,269 involved career soldiers and 104,702 disaffected civilians or members of the militia. In other words, a quarter of all those arrested by Franco. For some unknown reason, only 2,000 dossiers of the more than 30,000 convictions appealed in Cordoba, were examined. The CPEP was dissolved February 24 1945.

Despite the constraints on the condemned and their apparently hopeless situation, numerous prisoners attempted a new means of salvation – written appeals for some sort of clemency. Since 1939, those condemned to the 'ultimate penalty' searched within and outside the jails for literate individuals, preferably who also knew how to type, to help them submit their appeal, although as we have seen, those appeals very rarely had a positive result. In the famous case of the 'Thirteen Roses', the night before the women were executed in Madrid, numerous written appeals were still being written and submitted but these were not even looked at or processed. Likewise, thousands of cases all over Spain. In all the examples of such appeals that Moreno Gómez found in Cordoba, when the condemned had a high political profile, the death sentences were carried out. Two examples of such appeals in 1940 are addressed to the Judge Advocate for the Army in the province of Cordoba. (Those filed in 1939 were addressed to Lt. Colonel Ignacio Cuervo, Judge Advocate for the Army of Occupation, and in Peñarroya, Captain Francisco Casas Ochoa, Delegate Judge Advocate).

Eugenio Jurado Pozuelo, a prisoner from Villanueva de Cordoba who had been condemned to death, submitted a last-minute appeal which he wrote with the assistance of his brother José, ex-Mayor for the Lerroux Party.[lxx] The Delgado family and Juan Lucio alleged that Eugenio had testified to the charges against several right-wingers who were tried in Jaén, one of whom was Juan A. Delgado Fernández, whose brother Pepe Delgado, was Eugenio's principle accuser and who participated in the terrible beatings that he received.

If not actually ignored, Eugenio's appeal was probably never heard, because just a few days later, Holy Sunday Mary 26, Eugenio Jurado and 18 others were executed at dawn. A few hours later, those responsible for their deaths participated in an ostentatious religious parade up and down the streets of the town.

Another such appeal, Ex-Militia Captain Pedro *Cuadrado* Torralbo's, was also of little avail although there was more time for it to be heard, as he was executed a year later in Cordoba cemetery. His family kindly gave Moreno Gómez a copy. (The texts are of both his and Eduardo Pozuelo's appeals are reproduced herein.)

Everything that Pedro Torralbo had written was true. Apparently, however, he was unaware of the Regime's plan for the extermination. Pedro Torralba was wasting his time claiming that his was a pacifying role, which it was, and that he was opposed to the crimes and the mistreatment of the prisoners. His problem was that he was a political figurehead and Francoism was liquidating every leading leftist politician, every trade union leader, all Republican authorities and all commissioned career officers in the Republican Army. The charges against him were false – which probably did not surprise the Regime. Another factor against Pedro was that he, the victim, had the audacity to ask the oppressors to declare his sentence null and void (they never did so). The very most that Pedro might have hoped for was a commutation of his sentence. Nothing more.

Lastly, it was futile to hope that the totalitarian Regime would interview the witnesses for the defence and accordingly, revise the charges or make inquiries. The Regime almost never called witnesses for the defence, nor could it be bothered to do so. Furthermore, if those witnesses were not called during a court martial, it was even less likely that they would be called for a sentence review. Pedro, like other victims, was apparently unaware that the trials were a sham and that there was no such thing as justice under the self-styled Francoist Rule of Law.

Pedro Torralbo's life had been earmarked for liquidation during the Casino meetings and the outcome was decided beforehand. With or without denunciations and accusations, the simple fact that he was classified as a community leader (not just as an active member of a political party, but that he had also served as a Town Councillor in 1931, Provincial Deputy to Parliament in 1936 and Captain in the Militia, all qualifications that classified him as belonging to Group C of outlaws), indicates that there was nothing that could be done. That was that. Pedro Torralbo fell under a hail of bullets June 3 1941 in Cordoba capital, after having first been sent to prison in Burgos where he suffered unmentionable punishments, as was the exterminators' usual practice.

Some leaders who were saved.

Despite the Regime's persecution mania, some leading Republicans still managed to survive after they were embroiled in the oppressive labyrinth, as Moreno Gómez was able to corroborate in some cases in

Cordoba province. Luck, especially, made it possible for a few leaders amongst the mass of thousands upon thousands of prosecutions, to escape the ultimate sentence. Good fortune and the fact that their trials were postponed to a later date. There was no possible escape from the trials held in 1939, but as time went by, some holes began appearing in the net, as in the case of the Socialist Member of Parliament, Eduardo Blanco Fernández from the mining region of Peñarroya, whose court martial was delayed to June 11 1943 Had he been tried earlier, he would not have escaped the death penalty.

According to Eduardo Blanco himself, his salvation was also due to the efforts of some right-wingers (not all were bloodthirsty), whom he had assisted during the war, namely Carlos Calatayud, Professor at the Instituto de Peñarroya, a very active traditionalist since before the war who owed his life to Eduardo Blanco when in 1936, he prevented the execution of the right-wing prisoners from Peñarroya-Pueblonuevo. Furthermore, on several occasions, Eduardo formally opposed the local War Committee's intentions, as he prevented attempts by outside militia, especially Jaén Militia, from attacking the jail. When Peñarroya was evacuated October 1936, these prisoners were transferred to Ciudad Real, and the majority survived. Nonetheless, meritorious acts were of little importance to the Regime when it was the case of a prominent individual. To have been a do-gooder was not enough. It was also necessary that nobody in the barracks, the casino or the vestry had decided on his liquidation and that there was no allegation or accusation of import against him. A whole set of miraculous events had to occur for a prisoner to escape death.

Eduardo Blanco was arrested in Ciudad Real at the end of the war. He remained there for a year and was later transferred to a jail in Peñarroya that had been set up in the old Miners Union building when the furore of the immediate post-war period had begun to die down a bit. From there, he was sent to Cordoba prison where he was tried by court martial, as he reports:

> "I believe that the CEDA Deputy, Laureano Fernández Martos, intervened in my favour, as I was tried in Cordoba instead of Seville, where it would have all been much more difficult for the Court. The Prosecutor also was clearly in my favour, and I was

sentenced to thirty years. When I was asked which books I had read as a student, I replied Victor Hugo's, whose book *Les Miserables* I read when I was 16 years old, when I worked down the mines.

In the jail I was immediately put to work in the canteen, where I helped with the accounts. However, when I discovered some incorrect entries and tried to get to the bottom of the misbehaviour, I was transferred to Burgos at the instigation of some nuns who were stealing wine and watering down what was left."[lxxi]

Undoubtedly, Eduardo Blanco had to be blessed with a powerful 'helping hand' because the result of his trial was by no means usual or normal. His sentence reads:

"The National Movement found him in Madrid, from where he went to Pueblonuevo during the first days of the war, where he later became a member of the Board of the Rojo Committee. In all the positions that he held, despite his extremely leftist ideas, as is repeatedly apparent throughout, the accused always appears to have acted in favour of right-wing individuals, opposing all criminal acts against them, such as those that a Committee of Rojos from Peñarroya attempted to commit against some right-wingers. IT IS THEREFORE DECIDED: THAT the accused Eduardo Blanco Fernández shall be condemned to the sentence of LIFE IMPRISONMENT..."[lxxii]

Such a result was unheard of in courts martial during the immediate post-war period, especially because he also was a Frente Popular Member of Parliament, belonged to the local War Committee, was a founder of the Terrible Battalion and later, Civil Governor for Republican Cordoba. One cannot understand how it was possible that his case was not tried until mid 1943.

The fate of the majority of the Republican leaders was much, worse.[21]

---

[21]   Moreno Gómez was able to locate Eduardo Blanco in 1981, in his house in Madrid.

More challenging was the grim situation of Bartolomé Fernández Sánchez, founder and Commander of the Pedroches Battalion, Socialist from Pozoblanco, Militia Major (September 14 1938) then Senior Officer of the Cartagena Base during the dark March 1939 days of the Casadistas,[22] when he was arrested and imprisoned by the Nationalists. He, too, was tried by court martial very late in the day, March 16 1943, and sentenced to death. The principal obstacle to his defence was a denunciation against him by a local ruffian, José Plazuelo, who falsely accused him of commanding the squad that executed 18 right-wingers in Pozoblanco September 20 1936. On that date, Bartolomé Fernández Sánchez was actually in Madrid where he was negotiating the purchase of weapons for the Pedroches Battalion. Besides, firing squads were always commanded by lower-ranking officers, never Commanders, and Bartolomé Fernández had been the Battalion Commander since August 31 1936. He dedicated himself with determination to preparing his appeal.

The day after sentencing, March 17, Bartolomé sent an urgent request to a right-wing friend, José Elías Cabrera Caballero, who had considerable influence in Pozoblanco, asking him to convince the local Pozoblanco authorities to lessen the accusations against him. He then approached a series of influential contacts in Seville and Madrid, in order to get a reduction of his sentence. Lastly, he was able to put together a portfolio of signatures of right-wing individuals who had received favours and special attention from him during the Republic.

This was the usual route taken by numerous condemned throughout Spain, with varying results. In addition to these measures, it was usual for a condemned's family to go into action right, left and centre, to collect signatures, recommendations, documents and other material in support of the appeal. During the post-war period, thousands of wives travelled from office to office, often subjected to

---

He could not have more kindly given him access to al the material in which he was interested. Eduardo said that he was honoured by Moreno Gómez' research as this was the first time that anybody had interviewed him regarding his past history. He spoke of his existence and of his presence in a Socialist fora of faithful followers of the Transition policy of 'forget the past', he one of the last survivors of the Frente Popular Parliament. Eduardo left them quietly, without a single gesture of acknowledgement from his companions.

[22] Supporters of Republican General Segismundo Casado who unsuccessfully tried to negotiate a surrender with Franco.

humiliations, as never before. Every effort, every request was directed at finding evidence to weaken the accusations and denunciations and the bitter milk of human kindness that permeated the reports from the Falange, Guardia Civil, Mayor and other local authorities, as well as the parish priests who in some townships, added as much fuel to the fire as they could. Bartolomé Fernández' wife also duly travelled to the Military Headquarters in Seville as part of the so-called 'bureaucratic pilgrimage'.

Bartolomé Fernández' case was also favoured by a most unusual reference: one from the famous Republican Colonel Joaquín Pérez Salas[23], who took full responsibility and exempted Fernández of all blame for the crime he was accused of. A copy of the letter Pérez Salas wrote Bartolomé Fernández June 16 1939 from San Julián castle, and Jaime I barracks in Murcia where he was imprisoned, is reproduced herein.

30 June Pérez Salas signed a declaration in favour of Bartolomé Fernández, in which he stated that Bartolomé always followed his orders as is military practice, and that as his commanding officer, he assumed full responsibility for all Bartolomé's actions. This document, written when Pérez Salas had only a few days left to live – he was executed August 4 1939 in Murcia cemetery, is of particular historic interest although it was unable to produce the desired effect. When the document arrived at Military Court Number 16, unknown interested parties made it disappear. Yet another insight into the convenient disappearance of documents that could be favourable to an appellant whose fate had been decided beforehand. Despite everything, Bartolomé Fernández managed to save himself, because it took so long for his case to be tried and thanks to the help of some right-wingers who owed him a favour.

Chance and good fortune also enabled other important labour leaders in Cordoba to be saved. This was the case of Adriano Romero Cachinero from the large Communist stronghold of Villanueva de Cordoba, who belonged to the Central Committee and the Political Bureau of the PCE during the Republic, travelled to the USSR in 1933 and in 1936 was elected Frente Popular Member of Parliament for Pontevedra (Galicia). The outbreak of the war found him in Almeria

---

[23] Celebrated Republican Artillery officer who stood out for his actions during the Battle of Cordoba and his leadership during the victorious Battle of Pozoblanco in 1937 against Queipo de Llano's Nationalist troops.

July 18 when, after offering his services to the Civil Government, he contributed to maintaining law and order in that province. He commanded the 55th Brigade and held the rank of Militia Major on the Malaga front. At the end of the war, he was arrested March 11 in Ciudad Real by Casadistas who handed him over to the Nationalists on the 28th of that month.

Adriano Romero remained in prison in Ciudad Real for a year. He was only called to make a declaration once, March 1940, when his case was claimed by a special judge from Cordoba, and he was transferred to the old prison in the city. Again, all these delays were always beneficial to the accused. In this case, a new factor appeared in Adriano's favour. The Judge's secretary, José Espina Almansa, an attorney and influential professor was an old friend of his who, in 1931, had belonged to part of a little-known Communist candidacy to the municipal elections in Cordoba capital. Not only had he had a connection with the PCE and the CNT before then, but it is also possible that he was a member of the Abril Masonic lodge in Posadas. José Espina had been on the 1936 blacklists of those to be shot in the capital, but he was saved by the Presiding Judge. How and why José Espina became a supporter of the New State remains a mystery. The fact is, that José Espina, secretary to a Francoist Cordoba judge, saved not only Adriano Romero but also Alfredo Caballero Martínez who had been Provincial Secretary of the PCE during the war. The respective indictments were 'fixed' and the charges against them were toned down. According to Adriano's own testimonial.[lxxiii]

The first conclusion from Adriano Romero's court martial is that even if the accused, as in his case, could sometimes refute the false accusations against him, this practically was never possible in trials from 1939 to 1941 and also unlikely later. In his case, even after the false accusations were withdrawn, he was still condemned to death August 1941. Adriano Romero would surely have been executed had it not been that two months after the trial, he was extended another lifeline in the form of a military attorney who pushed for a revision of his case on the grounds of the lack of evidence and failure of due diligence. This was most unusual. A new judge was appointed who took advantage of the requests from a Granada judge to hand over the case file, which took the prosecution out of the Seville jurisdiction where Antonio was in serious danger. He was transferred to Granada July 1942.

In Granada, the military judiciary process was, as expected, identical. The Examining Magistrate did not even ask Antonio to make the customary declaration, he closed the case file without comments and handed the case over to the Judge for the court martial. A false denunciation by someone named Maldonado Castillo, from the Vélez de Benaudalla Catholic Trade Union, implicated him in blood crimes in this town. Although Antonio was able to refute this false accusation, this second court martial September 1943 sentenced him to death for the second time. Six months later, thanks to help from the Consul for Venezuela in Granada and the defence counsel's excellent work, his sentence was reduced to thirty years in jail. Adriano Romero was finally released from jail on parole March 1946. Behind him, a long string of obstacles that he was able to overcome, almost by good fortune alone. The numerous delays in the prosecution, his transfer to Granada in 1942 and the assistance of several influential individuals, made his survival possible.

Other eminent leaders must be added to the list of those who fortunately survived. Miguel Caballero Vacas, one of the founders of the PCE in Villanueva de Cordoba, several townships, Cordoba capital and Seville. His activities during the war were centred in Jaén. At first, he was sent to Puerto Real, Cádiz, concentration camp where he survived under a pseudonym among four thousand other prisoners. Later, at the end of 1939, he was sent to Rota Cádiz, concentration camp. In 1940 he was transferred to the Barrios, Cádiz, Workers Battalion where he was recognized and denounced by an individual from his hometown. He was sent to jail in Algeciras, accused of promoting the death of Villanueva de Cordoba right-wing prisoners in Jaén (1936) and of being an agent at the service of a foreign power.[lxxiv] The Prosecutor was asking for two death sentences by hanging, when the claims of a well-known judge from Jaén resulted in his transfer to Santa Clara Convent prison July 1940. A new prosecution and back to the beginning. The claims from military judges from another judicial jurisdiction, with the consequent delays in scheduling the trial, were again of great benefit to the accused. Miguel Caballero did not appear before a court martial until the late date of November 3 1943, when he was sentenced to thirty years in jail.

Artillery Captain Francisco Blanco Pedraza, Colonel Pérez Salas' right-hand officer in the defence of the Republican zone of North of Cordoba province, was another one of those who was remarkably saved, unlike his chief Pérez Salas. Seconded since 1936 to the Cordoba front, he was noted for the accuracy of his canon fire. He commanded the 88th Mixed Brigade in 1937 and later, the 38th Division of the VIII Army Corps. June of the same year, he was promoted to Artillery Major, then Artillery Commander for the entire Army of Extremadura. In the end, all these senior military officers supported Casado's coup. The end of the war found him in Puertollano, and it was a tragedy that fate made him lead the surrender of Puertollano to General Yagüe's troops 29 March 1939.

When he was arrested, Blanco Pedraza was sent to Cordoba prison. His case (number 25.419/39) was heard 28 September 1939 in the Artillery Barracks, and he was sentenced to death. There he suffered the unbridled attacks of all his triumphant ex-brothers at arms, who demanded his head at all costs, as they did with Pérez Salas in Murcia. The speed of Blanco Pedraza's court martial made everyone fear the worst, but in his favour was the fact that his brother Juan Rafael, a Falangista, had been a victim of the Republican repression in Madrid in 1936. Armed with this information, his mother urgently travelled to Burgos where she presented a single argument: 'The Rojos killed one of my sons. Are the Nationalists now going to kill my other son?' It appears that General Varela was moved because he recommended the commutation of his sentence effective April 1940. After several years in jail, the dedicated officer who had served so brilliantly in the artillery on the Cordoba front, Captain Blanco Pedraza, rebuilt his life in Madrid teaching mathematics.[lxxv]

The following is the remarkable story of a Republican who, for some years in hiding, managed to escape the extreme cruelty and arbitrary actions of the Francoist Regime

Ex-Member of Parliament Martín Sanz Diéz, Socialist, is one of those whom we could describe as one of the Cordoba capital 'moles' who was saved after living in hiding for eleven years. If he had not, he would not have escaped the merciless brutality of the Cordova military. Martín Sanz was from Valladolid. He first worked as a tailor in Madrid, then in Cuba during the 1920s, where was politically active as a trade unionist, which is why Machado, the Cuban dictator, deported him back to Spain. Martín Sanz set up shop in Cordoba around 1930 and became totally involved in the political scene of the times, at the heart of the PSOE/UGT. He was elected Member of Parliament in 1931 but did not run for re-election in 1933. When the insurgents invaded Andalusia and took Cordoba, he found himself in such a perilous situation that during the first weeks he hid in some friend's house, whilst his wife said that he had gone over to the Republican zone. He later returned to his own house. His grandson Luís tells his remarkable story.

"My grandfather remained hidden in his own house, not only during the entire war but also, once it was over, until summer 1947 when he was discovered and denounced by a neighbour and arrested by the police. At that time, even though he was 74 years old and that eleven years had passed since the end of the war, he was held under Francoist law and imprisoned in Seville...

I don't know what happened during his trial, but he was sentenced and then sent to the Cordoba prison. By around 1949, he was already out of jail. From that moment on until he was almost 90 years old, he continued to work as a tailor in the workshop he set up in his house He was a personal friend of Pablo Iglesias, whose numerous correspondence with him he bequeathed to the PSOE.

The information that I remember regarding my grandfather Martin, I have, as you can imagine, because I know all this as my mother told it to my brothers and to me... I have not collected it all, for example the visits of the Francoist police to my grandfather's house during the war, and how, on one occasion, when they did not find him, they took my uncle Antonio who was only 13 or 14 years old at the time and kept him under arrest at the police station for some time... And how the family had to resort to a lot of ingenuity so that my grandfather could be operated on for a serious stomach ulcer; thanks to false documents and the generous complicity of several surgeons who were his friends.

Then there was the farce set up by the Guardia Civil the day they went to arrest him in 1947: the whole street was cut off and filled with uniformed guards, a huge number of armed vehicles, the amazement on the faces of the policemen who when they went to his house, instead of the 'dangerous Rojo' they had been sent to arrest, they were taking in an elderly 74-year-old who sported a long, white beard…"[lxxvi]

Father Gumersindo de Estella
Capuchin monk and Basque Nationalist
Navarra 11 November 1880 – Pamplona 7 November 1974
Prison Chaplain

AUTHOR'S ENDNOTES FOR CHAPTER III.

[i]   Rufino Ayuso Fernández. Field work given to Moreno Gómezz in 2002. An enthusiastic supporter of groups dedicated to recovering the historic memory; he was a policeman by profession who died very young. Moreno Gómez refers to this research in greater detail in "Huídos, guerrilleros, resistentes. La oposición armada a la dictadura" (Fugitives, volunteer militia and freedom fighters. Armed opposition to the dictatorship.) In *Morir, matar, sobrevivir. La violencia en la dictadura de Franco.* (Dying, killing and surviving. Violence during Franco's dictatorship). Collaborative book coordinated by Julián Casanova, Barcelona, Crítica, 2002, pp. 200 et al.

[ii] Vicente Fajardo. Short field work given to Moreno Gómez, a colleague of his at the Institute.

[iii] Ángel David Martín Rubio. *Paz, piedad, perdón… y verdad* (Peace, pity, forgivness… the truth). Madrid, Fénix, 1997, p. 249.

[iv] Jaacinta Gallardo Moreno, *Laa guerra civil en La Serena* (The civil war in La Serena). Badajoz, Diputación Provincial, 1994.

[v]   Agustina Merino Tena. *La represión franquista en Villenueva de la Serena, Badajoz* (The Francoist repression in Villanueva de la Serena, Badajoz). In *Memoria Antifranquista del Baix Llobregat. El genocidio franquista en Extremadura*, number 12, 2012, pp. 93 et al.

[vi] Letter from Nieves Casado Gómez, Madrid 5 April 1999, following the publication of Moreno Gómez' book *Victimas de la guerra civil* (Victims of the civil war). Madrid, Temas de Hoy, 1999.

[vii] Bartolomé Cabrera Peralbo. Written testimonial. Pozoblanco, 17 October 1986.

[viii] Letter shown to Moreno Gómez thanks to the kindness of Acisclo Romero Luque's family.

[ix] Gabrial García de Consuegra, Angel & Fernando López López. *La represión en Pozoblanco (guerra civil y posguerra)* (The repression in Pozoblanco. Civil war and postwar). Cordoba, F. Baena, 1989.

[x]   Documents that in 1987 were kept on file in the Pozoblanco Court archives.

[xi] Isabel González Jurado and Agustina González. *A dos voces* (Two voices). Cordoba, 2011, pp.37-38. In this book of memories, Isabel and her daughter describe all that happened, an event that is still well-remembered in the town today.

[xii] Antonio Gómez Cabello. Emails to Moreno Gómez January 6 and September 12, 2007. Notice posted on the *Todos Los Nombres* Internet Project, asking for information regarding the whereabouts of his uncle. Link: http://www.todoslosnombres.or/enlaces.

[xiii] Arthur Koestler. *Diálogo con la muerte (Un testamento español)* (Dialogue with death. A Spanish testimonial.) Madrid, Amaranto, 2004, p. 180.

[xiv] Moreno Gómez personally examined the application of the decree for his book *Victimas de la Guerra Civil* (Victims of the Civil War). Madrid, Temas de Hoy, 1999, pp. 407 et al. This calculation of 100,000 killed by Franco between 1936 and 1939 alone, is not an estimate, but a fully researched and documented total.

[xv] *Testimonios de mujeres en las cárceles franquisttas* (Testimonials of women in Francoist jails). Huesca, Instituto de Estudos Altoaragoneses, 2004.

[xvi] José Manuel Sabín Rodríguez. *La dictatura franquista. (1936-1975). Textos y documentos.* (The Francoist dictatorship. 1936-1975. Texts and documents.) Madrid, Akal, 1997.

[xvii] Arcángel Bedmar. Lucena, *Op. Cit.*, pp. 220-221.

[xviii] Case No. 128.712, Archives of Territorial Military Court I, Military Government, Madrid. Proceedings instructed by Col. Enrique Eymat. José A. Cepas *El Lobito*, and Alfonso Días *El Parrilleroi* were executed by firing squad 21 February 1946 in Madrid, in the same group as the famous Cristino García.

[xix] José Maria García Márquez and Miguel Guardado Rogriguez, *Op. Cit.*, p. 351.

xx José M. Solé i Sabaté. *La repressió franquista a Catalunya, 1938-1953*. (The Francoist repression in Catalonia.) Barcelona, Edicions 62, 1985, p. 61.

xxi Vicente Pascual Soler. Letter replying to Moreno Gómez from Cordoba, January 19, 1979, a short while before Vicente died.

xxii Arcángel Bedmar, *Op. Cit.*, p. 104.

xxiii Francisco Moreno Gómez. *Cordoba en la posguerra*, Op. Cit., *p. 139*. Moreno Gómez says that he had reported that 15-20% were accused of crimes, which has proven to be an overestimation, as it refers only to those who were actually executed and not to the total number of imprisoned.

xxiv *Causa General* documents, Cordoba Archives. National Historic Archives, Madrid. Only documents for some of the Cordoba city courts are found here.

xxv Emergency Proceeding of the Court, number 335 – Archives of the I Territorial Military Court, Military Governorship. Cited in Francisco Moreno Gómez, *Armed resistance against Franco.*, *Op. Cit.*, p. 42.

xxvi These and other tragic examples can be consulted in J. Subirats Piñana, *Pilatos 1939-1941. Prisión de Tarragona* (Pilate 1939-1941. Tarragona Prison.) Madrid, Ed Pablo Iglesias, 1993. Cited in Francisco Moreno Gómez, *La represión en la posguerra, Op. Cit.,* p. 312.

xxvii Francisco Copado Sánchez. Letter he wrote to his wife in prison in Valencia and shown to Moreno Gómez by his brother, Afonso.

xxviii Causa General (CG) Case file 11,395/39.

xxix Quoting Bravo Murillo, 19th century radical politician.

xxx Document on file at the Belalcázar Municipal Archives.

xxxi Raphael Lemkin, Polish Representative to the V International Conference on the Unification of Penal Law. Madrid, October 1933.

xxxii Baltasar Garzón and Vicente Romero. *El alma de los verdugos* (The soul of the executioners), Barcelona, RBA, 2008.

xxxiii Carolina Gil Fonce. Interview with Judge Baltasar Garzón, 12 May 2008, on Radio Argentina. Can be downloaded at https://www.inforam.nl.

xxxiv Arcángel Bedmar. *Baena, Op. Cit.,* p. 121.

xxxv *Idem.*

xxxvi José Manzanero. Testimony, reported by Moreno Gómez in his book *La resistencia armada contra Franco, Op. Cit.,* p. 190.

xxxvii Ana Marcos. *El País Semanal*, Madrid, 19 February 1984.

xxxviii Memories of Rafael Bedmara Guerrero, from Puente Genil, discussed with Moreno Gómez during an interview November 1981.

xxxix Arcángel Bedmar. Historian of the civil war in Cordoba province. Several publications.

xl Francisco Merino Cañaveras. *Castro del Río, del rojo al negro* (Castro del Río, from the red to the black). Terrassa, Barcelona, 1979, pp. 101 et al.

xli José Moreno Salazar. Letter to Moreno Gómez from Osa de la Veja, Cuenca, dated October 13 1985. Book of memoires: *El guerrillero que no pudo bailar – resistencia anarquista en la posguerra andaluza* (The guerrilla who could not dance – post-war anarchist resistance in Andausia. Silente, Ed. Victoriano Camas, 2004.

xlii Claro González. Interviewed by Moreno Gómez in Fuenteobejuna, August 1979.

xliii Told Moreno Gómez by Sebastián Gómez in Villanueva de Cordoba 7 August 1982.

xliv José Luna Mata. Unpublished memoires. Consulted by Moreno Gómez August 1982.

xlv Sebastián Gómez. Testimonial, Villanueva de Cordoba 1982.

xlvi Francisco Poyatos. Interviewed by Moreno Gómez in Cordoba, 1982.

xlvii Francisco Romero Cachinero. Interviewed by Moreno Gómez, August 1932, Villanueva de Cordoba.

xlviii Miguel Regalón Molinero. Letter sent to Moreno Gómez from Valencia, April 1986.

xlix Antonio Ramos Palomares. Oral and written testimonial, recorded by Moreno Gómez in Almodóvar del Río in 1982.

l Arcángel Bedmar. *Los puños y las pistolas. La repressión en Montilla (1936-1944)* (Fists and guns. The repression in Montilla). Montilla, Ayuntamiento, 2009, p. 105 & p. 143.

li Antonio D. Lóopez Rodriguez. *Cruz, bandera y caudillo El Campo de concentración de Castuera* (Cross, flag and Caudillo. Castuera concentration camp.) Badajoz, Ceder-L Serena, 2006, p. 232.

lii Casimiro Jambonero, *Diario del soldado republicano Casimiro Jambonero* (Diary of Casimiro Jambonero, a Republican soldier). Ed. Victor M. Santidrián Arias, La Coruña, 2004, p. 107.

liii Rafael Sánchez Guerra. *Mis prisiones* (My imprisonments). Buenos Aires, Alarid, 1946, p. 210.

liv Arthur Koestler, *Diálogo con la muerte. Un testamento español.* (Dialogue with death. A Spanish testament.) Madrid, Amaranto, 1937, pp. 87-88.

lv Ángela Cenarro. "La institucionalización del universo penitenciário franquista" (The institutionalization of the Francoist penitentiary universe), in *Una immensa prisión* (An immense prison), Barcelona, Crítica, 2003, p. 134.

lvi Antonio Muñoz Molina. "Guerreiros deseanados" (Apathetic Warriors). *El País, Cultura*, 13 October 2012.

lvii *Guerra de Gila.* Caustic monologue by the comedian Miguel Gila Cuestas, where he, as a soldier, appears to have a phone conversation with 'the enemy'. *Is that the enemy speaking?* 2014. YouTube 17/08/2009 https://www.youtube.com/watch?v=R7d4Aj4t-FA4.

lviii *La Vaquilla.* Comic movie by Luís García Berlanga, 1985. During the civil war, bored soldiers on either side of the Aragon front try to discover how to raise morale and depress the enemy. The Republican's idea of boycotting a local festival falls to pieces when both sides get involved in chasing a young cow (actually a bull) that had escaped from the village and had run into no-man's land...

lix Manuel Álvaro Dueñas, in *La gran repressión*, collective work, *Op. Cit.*, p. 103.

lx *Azul*, Cordoba, 21 May 1939.

lxi *Azul*, Cordoba, 23 May 1939.

lxii Catalina Cantador Romero. Written testimonial, Madrid 1980. Sent to Moreno Gómez.

lxiii Minutes of the 15 June 1939 court martial in Pozoblanco against Luís Romero Cortés and other accused from Torrecampo.

lxiv *Azul*, Cordoba, 23 May 1939.

lxv Adriano Romero. Undated, typed memoires, in his family's possession.

lxvi José Subirats. *Entre Vivéncies (*Memoires), Viena, 2003.

lxvii Ramón Alujas, June 12 1939, *Ibid., Op. Cit.*

lxviii Decree published in the *Boletín Oficial de la Provincia de Cordoba*, February 12 1940.

lxix Discussion based on Judge Juan José del Águila Torres' presentation at the IX Congress of Contemporary History, Murcia, September 17-20 2008, entitled "La jurisdicción militar de guerra en la represión política: Las Comisiones Provinciales (CPEP) y Central de Examen de Penas (CCEP), 11940-1947, 27 folios, Guadalajara General Military Archives.

lxx Alejandro Lerroux y García. Spanish politician, leader of the Radical Republican Party during the II Republic and Prime Minister of Spain three times from 1933 to 1935.

lxxi Eduardo Blanco Fernández, 1981. Interviewed several times by Moreno Gómez in his home in Madrid.

lxxii Eduardo Blanco's private archives, consulted by Moreno Gómez in Madrid in 1981.

lxxiii Adriano Romero's private family archives, consulted by Moreno Gómez thanks to the kindness of his brother Antonio, Also mentioned in his book *Eurocarrillisimo y oportunismo* (Eurocarrillism and opportunism). Bilbao, 1984, pp. 99 et al.

lxxiv Miguel Caballero. Interviewed by Moreno Gómez several times in Madrid from 1980 onwards, and consultation of personal papers belonging to him, by kind permission.

lxxv Brief witness accounts by his family who received Moreno Gómez at the Paseo de San Francisco de Sales, in Madrid, albeit somewhat unwillingly. Moreno Gómez published Blanco Pedraza's biography in number 17 of the newspaper *Villanueva*, Villanueva de Cordoba, September 1981.

lxxvi This testimonial regarding the 'mole' Martín Sanz, written by his grandson Luís 24 May 2003, was very kindly given to Moreno Gómez by José López Gavilan, another of the men who suffered the extreme cruelty and arbitrary actions of Francoism in his hoe town, Espiel, and in Quintana de la Sierra, where his father was murdered 18 September 1938, López Gavilán has published a small book of memories entitled *Aquellos duros tiempos, Anecdotario* (Those hard times. Anecdotal accounts.) Cordoba, 2004.

IV

CRUSHING THE VANQUISHED (C) REPRESSIVE LEGISLATION, DEATH RITUALS FIRING SQUADS

"Terror does not only when violence is being applied; it also exists when the violence has ceased and appears only as a constant threat that hangs over men's heads. The threat of terror creates an atmosphere, where terror is the determinant factor, an ambiance that poisons life even more than the actual root of the terror."
Oscar Steinberg, Buenos Aires
Quoted by Pablo Uriel 1988.

Introduction

Francoist regulations and repressive legislation stemmed from the illegality of the June 17 1936 military coup, as the dictates of the totalitarian, i.e., anti-democratic, system of governance on which Francoism was rooted. Francoist justice has been widely described as a farce, devised solely to provide quasi-legal cover the regime's criminality, both before and after the creation of the New State.

In addition to the innate vices of its origin, there immediately appeared a multitude of orders, decrees and laws unfettered by any due process guarantees of any kind and with utter contempt for the Rule of Law. As the latter was non-existent, the system understandably spawned hundreds of thousands of extra-judicial sentences. Such is the lingering influence of the Francoist Regime that, until October 20, 2023[1], Law 52 of 2007 only described these as being unlawful.

When Francoist criminality is looked at from the viewpoint of International Law, it becomes apparent that the Regime's crimes were considerably more grievous than one had imagined. Today, however, there still is a large sector of the community that pathetically insists on denying or at the very least, repudiating, this criminality.

---

[1] Date on which the Spanish Senate approved the second version of the Law of Democratic Memory which, in addition to declaring the Franco regime unlawful, repealed all the sentences handed down by the Francoist courts.

The oppressive maelstrom that began near the end of the civil war continued throughout the post-war period without any hesitation or pause in the Regime's repressive project. It was an ongoing program superimposed upon the contingencies of war, a continuum, beginning 17 July 1936 that lasted until the early 1950s, at the very earliest, occasionally lingering until Franco's death in 1975. Nothing less than 15 years of harsh, relentless punishments and extermination, unleashed against the social base of the Republic, its politicians and its illustrious elite. It was the ruthless annihilation of half of Spain, something that cannot continue to be excused or ignored.

The state of war declared by Franco to have begun 17 July 1936, officially remained in force until 1948. In other words, the first period of post-war repression that actually began before the 1939 Nationalist victory, ran concurrently with the civil war until the end of the 1940s. The transition of the Regime's methods and objectives provided a semblance of unity for a decade and a half of terror, extermination, punishment, ideological and economic repression, starvation, social exclusion, purging, annihilation and forced exile, a multifaceted repressive nightmare.

Franco's repressive legislation

All victims of Francoism were persecuted for the crime of military rebellion, a political crime that carried the death penalty. Justification for these assassinations was two-fold: a) sentences sanctioned by wartime edicts; or b) sentences handed down by courts martial, two faces of the same criminal coin.

This is the argument frequently advanced by the few magistrates who openly declare their observance of democratic principles, their real independence of thought and judicial understanding, whenever they are confronted with Francoist outrages against human rights. Magistrates such as Javier Moscoso, José A. Martín Pallin, the brothers Carlos and José Jiménez Villarejo and, of course, Judge Baltazar Garzón (recently ousted from office by a Supreme Court hidebound by inherited Francoist practices).

Although it is always worth turning to the Law when attempting to shed some light on historical events, the assumption that today there might be some academic consensus regarding Francoism is inconceivable. The theoretical *corpus* that issued from Academia regarding Francoist criminality is chaotic. For example, when speaking of Francoist criminality, Julio Ponce from Seville distinguishes between 'murdered' and 'executed', thereby attempting to spread the idea that among the victims, some of these acts were simply committed by common criminals. Today in Spain, right-wing members of the judiciary, especially those who prefer to ignore the advances in universal jurisprudence who disparage and rule in blatant contempt of international treaties (a typical feature of Francoism), are being forced to face reality in the form of occasional cases from outside Spain, such as the current cases arising from the famous 'Argentinian lawsuit' against Francoist crimes.

A noteworthy and instructive feature of the April 14 2010 lawsuit[i] before the National Criminal and Administrative Court, under the generic title of 'Genocide', is that it classified the crimes it claims were committed into five different groups:

- individuals who were seized from their homes or jails to be 'taken for a ride', or *paseo;*
- individuals sentenced to death and shot by firing squad following summary proceedings in which due process of law was notably absent;
- individuals sentenced to long prison sentences in trials equally devoid of due process of law and, furthermore, sometimes condemned to forced labour;
- mothers and fathers from whom children were stolen or made to disappear; and
- others who suffered directly from torture and arbitrary arrest.[ii]

Lastly, the most significant feature of Franco's great repressive project was the torrent of wartime edicts, orders, decrees and laws against Republican Spain, the toughest and the most forceful being those that were published in the immediate 1939 post-war period. Faced with this terrible reality, whose effects Moreno Gómez has and will continue to examine, no one can say that once the Nationalists claimed victory, Francoism acquired a moderate or benefactory nature, as this is totally false and contrary to the facts.

Francoist Jurisprudence

Franco's wartime jurisprudence began with the extra-legal publication of *bandos de guerra*, or wartime decrees, that the insurgent forces issued in profusion right from the very beginning of the civil war in 1936, that ushered in the wave of summary executions.

The initial summary executions were first 'legally' authorized with the publication of <u>National Defence Junta Decree of July 28 1936</u>, the point of reference for the beginning of the extermination. This and subsequent decrees gave real substance to all the so-called "Reserved Instructions" from 'Director' General Emilio Mola, with specific references to *'terror, cleansing, extermination and exemplary punishments'*, all of which are euphemisms for genocide.

This was followed by <u>National Defence Junta Decree 79 of August 31 1936</u>, published during the widespread atrocities committed by the entire Nationalist rear-guard, written in the same spirit as the Director's decrees and as legally farcical, based on directives such as the one that states: *'It is necessary at the present time (..) for speed to be the norm for military judiciary actions'*...

Franco's famous <u>Decree 55 of 1 November 1936</u> instituted the *emergency summary procedure* that during the first six months was only applied to a few military individuals loyal to the constitutional government, to Republican leaders and to anti-Franco persons of note. With few exceptions, the remainder of the repressed population was prosecuted on a large scale by extra-judicial summary executions, paseos and murder in situ, under the ex-judicial <u>Law of Escapes</u> during the entire war in all of Spain occupied by Nationalist troops. An excellent example of this are the martyrized towns in Badajoz province after the fall of the La Serena pocket of Republican resistance in 1938 and during the advance of the Italian-Francoist troops through Catalonia in the autumn of 1938.

Some historians believe, incorrectly, that the extra-judicial summary executions known as *paseos* stopped in March 1937 with the widespread resort to trial by courts martial, despite the fact that Francoists continued to resort to these as late as the early 1950s. Students of Francoism today show an alarming lack of general knowledge regarding this. With victory in 1939, during April and May, the period that Moreno Gómez calls the Black Spring, the paseos became a terribly reality in the entire Centre-South of Spain, the last region to fall to the insurgents, in 1941, there again was a marked increase in paseos.

Furthermore, during the period that Moreno Gómez calls the Triennium of Terror (1947-1949), the extermination programme rose to unexpected new heights with several thousand extra-legal summary executions under the Law of Escapes. During this period, Francoist terror fully penetrated the rural world throughout Spain under the leadership of Camilo Alfonso Vega, Director General of the Guardia Civil, who received his orders directly from Franco. In 1949, this terror spread without mercy throughout the countryside in the Seville mountains and in 1940, uncountable country people were assassinated in the region of Nerja, Malaga.

Decree 55 reappeared re-cast under the <u>July 12 1940 Ley de Securidad del Estado</u>[2] that created 'ordinary summary procedures', which only differed slightly from the 'emergency summary procedures' conceived by Franco in 1936, in that they included the prosecution of those Franco considered would be defeated after the fall of Madrid. Once again, the text of this law was accompanied by a so-called magnanimous slogan that these procedures would guarantee *'the indispensable speed and exemplary actions of military justice'*. The reader can well imagine what the concept of 'exemplary ations' implied....

The <u>November 21 1936 Circular from the High Court of Military Justice</u> issued in Valladolid made it almost impossible for anyone to appeal a revision of sentence because it excluded any appeal following a summary proceeding, which meant, almost every single case that was tried.

Lastly, Franco's <u>Decree 191 of January 26 1937</u> published in Salamanca, extended the application of Decree 55 to the newly occupied territory of Malaga.[iii] Decree 191 became the justification for the impending slaughter of the great number of Rojos the Nationalist troops soon expected to capture in Malaga.

Repressive Francoist Laws

Soon after the Nationalist victory, the above were followed by the enactment of major legislation under the jurisdiction of the Francoist

---

[2]   Law for the Security of the State.

military, directed at fine-tuning and redefining the crime of military rebellion, which encouraged the liberal application of the death penalty. The fact that this repressive legislation began to be forged as the end of the war approached, indicates that the Regime was not interested in lessening or even tempering the repression of the defeated. Quite the contrary. Its purpose was to assist with further planning and developing, from every possible angle, the multi-faceted program against the defeated. It was Franco's great all-embracing project.

Now began the great battery of judicial bazookas against the social base of the Republic and its political and illustrious elite who had already been beaten *in pectore* by the Law of Political Responsibilities (LPR) of February 9 1939 (later amended by Law of February 19 1942). The purpose of the LPR was to pull the plug on the defeated's financial system, an economy that had already been battered by multiple seizures of property during the war. This post-war act was both a fiscal punishment and a purifying and disqualifying measure to ensure that the defeated would be left without any means of survival. In actual fact, the total economic destruction of the defeated was not really a result of all these regulations, but of the plundering, the seizures and hands-on theft of property and assets belonging to Rojos during and after the war.

The January 25 1940 Order from the Presidency of the Government regarding the 'Examination of Sentences' by provincial committees, established extremely restrictive criteria and a detail description of those individuals who were prohibited from appealing under this legislation. All the excluded were persons of some political or trade unionist importance of one kind or another. In effect, only individuals of no importance whatsoever could benefit. This feature of the Order is important, if only to correct those persons who have misunderstood its purpose and who believe that this Order was a sign of the longed-for moderation of the Regime, which it definitely was not.

On the other hand, this Order is particularly significant as it reveals the importance of the extermination to Francoists, not so much regarding the extermination of individuals, as the destruction of the entire Republican system, or Marxism, according to the Nationalists, because of their belief that 'The Republic is guilty'. This purpose is clearly stated in the preamble that refers to the Regime's *'responsibilities regarding the criminal treason against the Fatherland that Marxism was guilty of when it opposed the Revolt of the Army and the National Cause'.*

What Francoist military jurisprudence did was initiate legal proceedings against the entire Republic as a political movement it described as bearing the effigy of Marxism, not so much against individual Marxists, but against all the citizens of the Spanish Republic who might have, in any way, participated in the democratic State's opposition to the military coup. This is a clear example of 'reverse justice' the Regime used to justify legal proceedings, whereby it re-qualified the democratic government's constitutional defence of the country as a military rebellion, an all-encompassing crime punishable under 19th century Military Law. In view of this, those who interpret those proceedings as a desire to revenge Nationalist martyrs are speaking nonsense, as from the onset, Franco had planned the extermination of his opponents, regardless of whether there were any Nationalist 'martyrs' or not.

According to Jiménez Villarejo, the 1940 Order embraced 83 different kinds of military rebellion, including being a Freemason, having held a public office, supporting the Rojo revolution and so forth. In other words, the 'Marxist rebellion' was guilty as charged.

The March 1 1949 Law for the Repression of Freemasonry and Communism had as a very specific objective, the ghosts raised by Francoism and the Church, because both Franco and Church had an obsessive hatred of Freemasonry. As part of the emergency summary procedures against Masons, with the publication of this law all Masons were forced to recant their association with Masonry, an atavistic procedure dating back to the Inquisition. An example of this was the trial of the Mayor of Posadas in 1931 who, as a member of the *Abril* Lodge, was forced to draft and sign the following recantation:

> "I, Rafael Marencio Muñoz, aged forty-nine years of age, married, profession porter and resident in Posadas, Cordoba, with a view to complying with the stipulations of Article 7 of the Law of March 1, 1940 for the repression of Freemasonry, and in conformity with Article 1 of the Order dated the said month, hereby declare..." [that he first joined a lodge in Seville, sponsored by Antonio Rueda Aguilar; that he

> was given the symbolic name of 'Galdós' and that he
> reached Grade Two; that he was assistant to the Grand
> Master of the Posadas lodge, Ángel Lara Muñoz]
> ...and "lastly, I formally declare in this recantation,
> that since the year one thousand nine hundred and
> twenty-eight, I have repudiated all my obligations to
> the sect, that I abjure my errors and endorse the above
> recantation. Cordoba, April 10, year one thousand
> nine hundred and forty."[iv]

Communism was the perceived threat and pretext for repressing all 20th century insurgents. The Court against Communism was based in Madrid and one of its most violent wielders of the judiciary gavel against heretics was the sadly infamous Colonel Enrique Eymar, a disabled veteran. Everyone captured during the repression of the guerrillas throughout Spain was tried by this Court.

The July 12 1940 amendment to the Law for the Security of the State qualified the crime of military rebellion under Article 237 of the Code of Military Justice, thus placing the absolute control of all these cases under military jurisdiction.

The March 29 1941 amendment to the Law for the Security of the State greatly toughened the repression, adding to the list of crimes punishable by death crimes such as ordinary theft of which those who had fled to the hills and their connections in the countryside could be accused. Under this law, numerous orders from the Guardia Civil were further directed at extermination. The punitive reaction was so swift that in 1941 there was a marked rise in the number of summary executions and in the application of the Law of Escapes. There was no decline in the repression during this year.

Law of March 2 1943 introduced amendments and extended the application of the crime of military rebellion, again adding fuel to the purifying fire, so that *"from now on, nobody will dare stray from a strict social discipline"*. Article 238 expanded the obsessive and neurotic definition of the crime of military rebellion. Military justice understood nothing else.

The new Code of Military Justice of July 17 1945, in which the crime of military rebellion was now brought under article 286, was a way of celebrating July 18 as a national holiday.

During the next couple of years, one could speak of a slight relaxation, or brief parenthesis, of the Francoist repressive furore, a loosening of the noose because of the international situation, the defeat of fascist Europe and the Regime's concern with the Allies' attitude towards Spain. However, seeing that the Allies were refraining from acting against Franco, the Regime gathered its forces and embarked on a new wave of oppression, the Triennium of Terror.

The <u>April 18 1947 Decree-Law on Military Rebellion, Banditry and Terrorism</u> was another of Franco's terrible laws, under which several thousand persons were sent to the grave.

This was not the last piece of Francoist repressive legislation. Once the guerrillas had been exterminated and the hills and the countryside had been subjugated, there remained a scattered clandestine resistance to the Regime that kept breaking out. With a view to suppressing this clandestine but stubborn resistance, particularly by members of the Communist Party, the Regime created the Court of Public Order (TOP – *Tribunal de Orden Publico*) under Law 15/1963.

A great many of these laws were rescinded by the first so-called Law of Historic Memory 52/2007, after thirty years of hesitant and timid democracy.

Now is a good time to close this section with a numeric allusion to the hyperbolic magnitude of the Francoist slaughter, enveloped as it was by the Regime in the glitter of extra-legal decrees and laws that were never more than extra-judicial coverage to a maelstrom of crimes against humanity. The data, updated in 2012 by Ángel Viñas[v], speaks for itself.

Before September 1939, during the year and a half that followed its coming to power, the III Reich had assassinated 473 individuals under the law and a thousand extra-legally. In comparison, in Seville General Queipo de Llano achieved this number in fifteen days in 1936 and by the end of the war in 1939, Franco had already assassinated 100,000 individuals. In fascist Italy, only 9 death sentences were handed out by 1939 and some 3,000 victims fell during the years of the regime. During the entire period of Italian fascism, 13,000 were exiled; in Spain, the number of exiled totalled almost half a million.

Until 1939, Franco was the most bloodthirsty of all the European fascists and he led the list in terms of criminality. Not only did he surpass his 'brother regimes' before that year, as he was also some kind of pioneer in setting standards of barbarity for totalitarian Europe. Franco only fell short in that he did not invent gas chambers nor set up soap factories from human fat. As Paul Preston said,[vi] Franco killed more Spaniards in Spain than Hitler killed Germans in Germany and Mussolini Italians in Italy. (German barbarity occurred outside its borders and the victims were predominantly foreign.) Extermination of fellow countrymen who opposed Franco's Regime was the real purpose of military jurisdiction during Francoism. (TABLE IV. Estimated Balance of Victims of Francoist Repression in Cordoba Province and Capital.)

Executions and the Death Ritual

"Most of us did not fear death, but we feared the act of dying... I was there when they died. They died with tears in their eyes, begging for help in vain, failing, as men should die. Because dying is a serious matter, and we must not make a drama of it. Pontius Pilate did not say: Ecce heroes; *he said:* Ecce homo."
Arthur Koestler[vii]

The culmination of the extermination ritual was execution by firing squad, but this was not the only way to die, very much not the only way. This was a chapter in the physical elimination of the defeated, essentially those classified as group C (Republican authorities, worker and trade union leaders, senior officers of the defeated Army and eminent individuals more or less those that are called politically significant people). Also eliminated were those in Group D (those accused of presumed 'crimes' that were rarely, if ever, proven). According to the classification of prisoners, at the end of 1937 only 2.13% of those executed belonged to group D. Also killed, a great many individuals from Group B (Republicans of no political significance, farmers from the country and ordinary townspeople, caught by the witch hunters and the furore for revenge killings during the first two post-war years).

Many thousands of people of all kinds and social classes died, in compliance with the criterion that a 'sufficient extermination' would ensure fear in the present and in the future and help in preventing even weak opposition in the present and the years to come. As the Seville war criminals said: "Thirty years from now, not a soul will be alive".

Regarding the professional occupation of the eliminated, logically the majority belonged to the working class, especially tenant farmers and farmhands. Employing pure fascist logic, rural fascism drew its sustenance from the numerous and demanding peasantry as did Italian fascist theoreticians who defined war as 'the hygiene of the world'. Francoists believed that the time had come for 'cleansing and exterminating as a fundamental principle of the Movement', when they justified the bloodshed they initiated on July 18 1936.

All Moreno Gómez' lists of victims in Cordoba capital and province, indicate their profession, when known, which enable him to confirm that the overwhelming majority were rural labourers and farmhands. Several historians have compiled some percentages elsewhere. In Toledo, 52% of all executed were workers, according to José Manuel Sabín.[viii] In Albacete, of a total 72% of executed individuals, 40.2% were farmhands and 31.9% were general working class, according to Ortíz Heras. In Malaga, 64.27% of those executed in the provincial prison 1939-1942, were farmhands. The percentages are similar for all of Spain, especially in the south.[ix] Members of the working class and, most specifically, the rural proletariat, were scapegoats for the fascist extermination, which was the whole purpose of the military coup – to cleanse and force the rebellious proletariat to its knees.

A particularly important feature that is barely touched upon in many studies, is the tactical use that Francoism made of the great number of death sentences handed down during the first two years after victory, when so many suffered the terror of that period, although later the 'magnanimous Caudillo' accepted *motu proprio* to commute a number of sentences, at his discretion.

What the Regime immediately wanted was a program of great terror that would destroy the dignity of the individual, based on the abundance of death sentences, so that the greatest possible number of defeated would have to suffer the dreadful experience of having the death penalty hanging over their heads for several months, as long as a year in many cases.

Someone who has endured the anguish of expecting that each night his name might be called for the last trip, who cannot rest, who has lived each day as if it were his last, will be forever damaged. After such a terrible experience, the inner psyche of a prisoner can never be the same as before, even though at the end his death sentence is commuted. Many of those whose sentences were commuted remained seriously damaged in terms of their will to fight.

The percentage of sentences that were commuted varies and is difficult to compute. Some studies have come close. According to Matilde Eiroa, of the 800 death sentences handed down in Malaga capital 1939-1942, 710 detainees were executed and only 90 sentences (11.25%) were commuted.[x] In Cartagena, according to Pedro María Egea, there were 176 executions 1939-1945, but he only discovered 20 commutations in 1939. In Monóvar (Alicante), according to Sánchez Reio, of 324 sentences, 38 were death sentences; of these, 21 were executed and 17 were commuted. The figures vary greatly.

The possibility of a commutation increased greatly after 1939, inasmuch as the revision of sentences became possible after changes to the law in 1940. In summary, in addition to those who were executed, may prisoners suffered the long-term threat of the death sentence and, once that terror was removed, their minds were irretrievably damaged in one way or another. An example of that is the playwright Antonio Buero Vallejo (whose dossier Moreno Gómez examined), who lost the fighting spirit he was known for before 1939, as he became totally apathetic from the suffering he endured in captivity. Francoism had its methods.

The *enterado*

The last step in the humiliation consisted in the manner by which a sentence was confirmed. The annotation *enterado* (meaning 'noted') written by Franco himself on a sentenced individual's dossier was the ultimate life or death decision. Once a case was tried by a military court, whether permanent or temporary, the dossiers for prisoners who had been condemned to death were sent to the Judge Advocate's office, first in Cordoba and later in Seville. The Judge Advocate for the Army of the South's mission (two judges in 1939: Ignacio Cuervo and Francisco Bohórquez) was to examine each dossier, approve the death sentence, order the execution of the sentence and send his recommendation to Franco – "His Excellency, the Generalissimo" – for confirmation.[3] More specificallyy, the Judge Advocate was

---

[3]   Rafael Sánchez Guerra, an eminent Cordovan victim of reprisals, sarcastically nicknamed

responsible for confirming the final ruling on the sentence proposed by the military court.

The case remained with the Advocate's Office until the Legal Advisor to the Ministry of the Army, the woefully infamous Lot. Colonel Lorenzo Martínez Fuset, returned the individual's dossier with Franco's decision. The Legal Advisor would meet daily with Franco and present him with a huge file filled with death sentence recommendations from numerous towns in Spain. It is said that when time came for the coffee break, the dictator would indicate his approval by scribbling *enterado* on death sentence after death sentence, whilst continuously joking with Martínez Fuset. There is a famously shameful jocose moment when Franco's chaplain, Rev. José Maria Bulart, called out in his usual macabre voice: "What? Do you mean buried?" playing on the Spanish meaning of *enteRado* – noted – and *enteRRado* – buried. The priest's behaviour reflected National Catholicism's self-satisfied and exultant approach to the 'purging' of the 'Godless'.

Every dossier arrived done and dusted on Franco's desk and he would rarely make any changes unless it was to underline the word *garrotte* or *garrote and crush* or write *C* for *commutado* – commuted. Some proposals for the commutation of a sentence already bore the notice *Ojo! Take note*! some high-ranking individual's remark. For example, General Varela wrote this on the dossier of Republican Artillery Major Francisco Blanco Pedraza, who had served under Pérez Salas on the Cordoba front. This was the result of the efforts of the Major's mother, who was able to soften Varela's hard soul, pleading that as the Rojos had killed one of her sons who was serving in Madrid, she couldn't let the Nationalist kill her other son. Franco commuted the sentence.

In 1939, when the Regime launched itself with special urgency on its wave of punishments, revenge and ruthless repression, Franco's *enterado* arrived swiftly by teletype.[4] For example, Luís Romero Cortés, from Torrecampo, was sentenced to death in Pozoblanco June 15 1939. June 27, the Seville Judge Advocate approved the sentence,

---

Franco 'His Excellency the *Criminalissimo*', meaning Supreme Criminal, a play on words for the Spanish 'Generalissimo' meaning Supreme General.

[4] The Advocates always confirmed and approved the sentences, never disagreeing. At least, Moreno Gómez knows of no such case where they disagreed.

the *enterado* arrived quickly by teletype from Burgos where Franco was living at the time, and Luís was executed August 5. In a single month, the death sentences that rained incessantly upon the defeated during the second half of 1939 were processed without appeals or revisions of any kind. Every case was judged to be most summary and most urgent and besides, before January 1940, there were few regulations allowing for the possibility of appealing or reviewing sentences.

In 1940, there was a kind of slowing down in the processing of Francoist justice, not because the 'Criminalissimo' wished to show any generosity of spirit as such altruism would be unthinkable, but because the overwhelming volume of summary trial actions in all of Spain resulted in a monumental bureaucratic gridlock. This slowdown, which enabled a small minority to escape death, prolonged the suffering of the many thousands of prisoners who had already been condemned and had to endure a year or a year and a half of fear of the pending death threat, in the knowledge that there was no possible escape from the final solution.

When the exterminating bureaucracy came to a virtual standstill whilst Franco moved his headquarters from Burgos to Madrid on October 18 1939, his Generals took matters in their own hands and, from 1940 onwards, dealt directly with the great volume of proposed commutations and appeals to sentences. Orders for sentencing piled up in the War Courts as they waited for these proposals and appeals to be processed. In Cordoba capital, the military bureaucracy struggled even further after the summer as the military began emptying the provincial prisons of inmates and sending them all to Cordoba capital where they would be interned at the end of September and throughout October.[5]

Whilst requests for commutations and appeals in 1940-41 were examined and in most cases, denied, inmates in Cordoba who had been condemned to death were transferred to Burgos prison, where they remained for about one year, until they were again brought back in the so-called 'train of death', for immediate execution by firing squads in Cordoba cemeteries. Few unfortunate individuals, knowing of their fatal destination, were occasionally successful in escaping during that journey.

---

[5]   There were two prisons in Cordoba capital. The main one, the *Prisión Provincial*, Plaza del Alcázar, and a new prison, the *Prisión Habilitada*, on the road to Pedroches.

A case in point illustrating the1940-41 delays in carrying out the sentences, is that of Socialist Antonio Baena Moreno, from Pozoblanco, condemned to death by garrotte, April 22 1940. November 1941 the dictator had still not given his *enterado* to the order nor had the Military Governor for the 2nd Region approved it. Baena Moreno was finally executed ten days later, November 17. In all, nineteen months waiting in anguish. The following was written at the end of his dossier:

> "...We thereby pronounce, order and sign our sentence on 05-1940. I, Judge Advocate for the Army of the South hereby rule that the proposed sentence should be approved, that the ruling is final and must be executed, and order that H.E. the Head of State is informed of this capital punishment. H.E. the Commander General of the 2nd Region has approved the sentence and stated that it should not be carried out until the appeal for commutation of the sentence is examined and decided upon at the highest level.
>
> November 7 1941, I, Cirílo Genovés, Head Advocate of the Ministry of the Army, hereby certify that His Excellency was notified of the sentence pronounced by the Pozoblanco Court Martial against Antonio Baena Moreno – case 27.505, and that H.E. gave his approval..."

In yet another case, Pedro Torralbo Gómez from Villanueva de Cordoba was judged and sentenced to death May 1 1940. May 11, the Judge Advocate confirmed his sentence and April 23 1941, the Commander General of the 2nd Region approved it. June 3, 1941, thirteen months after he had been condemned, Pedro Torralbo was executed. His dossier shows:

> ...11 May 1940, the Judge Advocate for the Army of the South approved the sentence, declaring that his ruling was final and must be executed. He further ordered that H.E. the Head of State should be informed of this capital punishment and that this case

should remain with his Office until His Excellency's reply was received.

April 23 1941, the Commander General of the 2nd Region, in view of the aforementioned sentence and the Judge Advocate's opinion, approved the capital punishment and ordered that this case be sent to the Military Governor for Cordoba so that the latter could appoint a judge for purposes of the notification, carrying out of the sentence, and publishing all other necessary information...

A few more examples of the long wait between sentencing and execution. Juan F. Chuán Soto sentenced to death May 1 1940 was executed November 6 1941; Juan Lorenzo Cantador tried on the same date was executed September 12 1941; Juan Escoriza Segura, sentenced April 25 1940 was executed November 6 1941. There are many more examples of men who were condemned at the beginning of 1940 but not executed until well into the following year. On the other hand, in 1941 now that all courts martial were operating in Cordoba capital, death sentences were carried out much more swiftly, the same year.

Special note is made that the approval, or *enterado* could be two-fold. On the one hand, there was Franco's *enterado* issued through the Judge Advocate for the Ministry of War, and on the other hand, the enterado proclaimed by the Commander General of the 2nd Region. In Seville, this post was filled by General Queipo de Llano from July 18 to 20 1939, when he was replaced by General Andrés Saliquet. At the height of the enterados in 1941, the Commander General in Seville was General Miguel Ponte y Manso de Zúñiga.

The final step for the hundreds of enterados that were processed daily by the Seville Judge Advocate and returned to the mot remote towns and villages in Andalusia, was the provincial Military Commander's Office, responsible for appointing the judge who would inform the prisoner of the final decision and for carrying out the sentence. However, as the Francoist repressive bureaucracy was so overwhelmed during these first years of the Regime, that as many witnesses and data will testify, numerous notices of the commutation of a sentence were received long after the condemned inmate had been executed. A few diaries and letters sent from the jails speak of the dramatic nightmare that each prisoner had to face night after night, month after month, as he waited to hear whether his name was next on the list of those to be executed or not. The real number of those affected by this terrible situation will never be really known.

Sacas or illegal removals of prisoners for execution

The first step to the gallows was the *saca*, meaning the illegal removal of a prisoner whose sentence had not been commuted, from prison for execution. The sound of metal doors and bolts at midnight, of wardens reading out the names of the doomed; indescribable tension and heart-breaking silence from all the condemned men in every cell. The networks of signs and unusual information that existed in the jails meant that the prisoners often knew several hours beforehand when a saca was about to take place. With the whiff of tragedy, panic swiftly spread from cell to cell. There was a restlessness and continuous suspicion that totally affected the daily lives of the condemned.

An indication that a saca was near, was when the daily lists of commuted sentences arrived at the prison and the 'fortunate' were called to sign their receipt of the news. That night there inevitably was a saca from among those whose sentences had not been commuted. The condemned only had a break, a slight pause in the tension on Sundays and holy days, because there never were any sacas on those days. Antonio Baena's unpublished diary, an excerpt of which is given herein, gives one an idea of what it was like, an experience that Moreno Gómez described as a *spreading wave of terror*.[xi]

The Francoist repressive program was not just directed at exterminating or physically eliminating the defeated, nor just at tormenting, mistreating and starving them in the jails. Above all, the Francoist repression was a program of terror, both past and present, terror that provided the essential basis for the survival of the Regime. This terror sprawled over all like a wave. In other words, it not only affected the victim, but it spread outwards to affect all his cellmates and also, logically, his fellow prisoners throughout the jail. As we read in Antonio Baena's diary, the next day in the patio, everyone 'looked as if he was just recovering from a serious illness'.

The wave of terror did not stop there: it oozed out into the street. First affecting the victim's family, then, all his neighbours, friends and acquaintances and lastly, to all the public (in the defeated sector, of course), as everyone, everywhere, spoke with horror at what had happened to Tom, Dick or Harry. Each execution terrorized a great many people both inside and outside the jail, in every town and village. That terror was swallowed like a bitter pill by everyone and grew like a painful cyst that silenced all, a silence that continues to this day as hundreds of times relatives of the victims still refuse to speak. It was a Regime of terror that survived by feeding on the hidden panic of its victims. As Koestler wrote:

> "I never thought that the dictatorship of a minority could remain in power through terror alone. I ignored the extent to which those primeval forces that paralyze the majority from within, are living and real."[xii]

Preventive terror also was Franco's strategic instrument for inhibiting any possible domestic opposition and for nipping it in the bud if necessary. In Zaragoza, in a Paseo de Ruiseñores bungalow that housed the local Military Police headquarters, the Francoist colonel in charge decided on the lives of hundreds of Republican soldiers. These men were culled and removed from the San Gregorio military prison in Fall 1936 when 400 of those soldiers and some civilians as well were executed.[xiii] A military chaplain who was present, Rev. Gómez, made an unprecedented request – he asked the colonel to show some clemency. The Nationalist colonel's reply was as peculiar as it was political. He stated that he was willing to forgive a poor worker for becoming an anarchist, but Justice must be ruthless with these men, with intellectuals, public servants, all those who betray their class. That it was the Regime's painful legal duty to be implacable in exterminating these vermin. As it upheld the necessary background terror... [xiv]

Terror would manifest itself in many other ways, the most insidious perhaps being the fear it instilled in inmates condemned to thirty years in prison as they contemplated what the future might bring for their families. This is reflected in a fragment of the diary that Sebastian Blanco Copado, an alderman for the Izquierda Republicans in Pozoblanco, wrote in prison in Puerto de Santa María, Cadiz, after he was condemned to 30 years in prison. This excerpt from a letter he wrote to his wife and family illustrates the immense pain of ordinary people who felt the weight of the destruction created by the military coup.

> "Faced with such a cruel and prolonged separation I could die and never again see you, and that idea made me tremble; to leave you alone in the intemperate world, freezes my blood and I remember our humble home, our home that harboured so many memories.
>
> That home has now been destroyed and broken, left without direction or order, like a fragile sailing ship floating rudderless at the mercy of the waves. This brought tears to my eyes, and I thought of you, poor innocent victim of a time that should never have existed, of you, shipwrecked by a storm of human passion, buffered by a whirlwind of hate such as Man has never known. I do not want to die without leaving you some memories..."[xv]

This excerpt from Sebastian's letter to his wife, published September 1985 in an article entitled 'The Nationalist repression in Pozoblanco' in the *Revista de Feria*, erupted in the form of a major scandal among the right-wing residents of Pozoblanco. They showered the municipal newsletter with letters, accusing the author of being a fanatic, of poisoning the townspeople and demanding that City Hall ban the sale of the said magazine, describing it as unwise and lacking in objectivity that would only reopen old wounds.

This was typical apocalyptic right-wing ranting, after almost ten years of the return to democracy, clearly reflecting the difference between the French right-wing Gaullists and their clear anti-fascist tradition and the Spanish Right who haver had such a tradition.

Pablo Uriel, in his prison diary, described how the life of the prisoner with the threat of death hanging over his head, suffering the anguish of the nightly sacas, the fear that his name will be called for execution, changed his brother's life forever.[xvi] When his brother was released from the Zaragoza Provincial Prison, the family found him greatly changed, much quieter and withdrawn. He had witnessed the last hours of thousands of persons and because of that he had become totally indifferent as to his own future.

One of the effects of the terror was the mindless resignation that the carnage produced in the victims. As this bacchanalian bloodletting convinced the prisoners of the inevitably fatal outcome of their situation, the men allowed themselves to be led to sacrifice with a disconcerting apathy. This was always such an almost total lack of resistance that the assassins found it much easier to do their odious work. The passivity with which those men accepted that they were going to die, whether they were civilians or had served as soldiers on the battlefield, is impossible to comprehend today. In truth, their minds had been irretrievably damaged.

The expanding wave of terror resulted in what Pablo Uriel, citing Steinberg, described as an *atmosphere of terror*, which had an impact as effective as the terror itself.[xvii]

Another important factor in the atmosphere of terror is its unpredictability, when feeling terror is so arbitrary and so irrational that no citizen feels safe, when anybody can become a victim, as it was in Spain during the 1947-1949 Triennium of Terror when the regime concentrated its repressive effort on persecuting the guerrillas in the mountains and in the countryside.

Returning to the ritual of the sacas, in the prisons, the condemned men survived without really living, as they awaited death. The bureaucracy of the execution began with a telegram from the Military Governor to the Prison Director, listing the names of those who were to be executed and other relevant information. The eve of the execution, the Director would receive another telegram from the regional military commander, instructing him to 'send the condemned to the chapel'. Next, the Military Governor would send another telegram to the Prison Director, instructing him to hand the condemned over to the Guardia Civil. The head guard on duty at the prison had to sign the telegram himself in confirmation that he would take charge of the condemned men whose names were on the list. It took three or four days for all these telegrams to go back and forth.

In Catalonia, the Military Governor's fatal list would be delivered to the prison by car between nine and ten at night. On the San Simón, Pontevedra, penal island, the enterado would arrive at night, in a motorboat, with a prison officer and the chaplain, Rev. Nieto. When the prisoners heard the motorboat, they froze with fear. In Pozoblanco, the sign that a saca was about to be carried out and that there would be executions, was the presence of the Salesian chaplain, Rev. Antonio Do Muiño, whom the prisoners had nicknamed the 'bird of death'. In Villanueva de Cordoba, it was Rev. Marcial, the parish priest's arrival at the jail the afternoon before an execution to administer the last rites to the condemned. His arrival shook the prisoners like an earthquake.

There are few witness accounts, diaries or memoires referring to the moment of the saca from Cordoba. One such account, however, Arthur Koestler' diary from when he as imprisoned in Malaga and Seville, excerpts of which are published herein, describes in greater detail those terrible experiences the weeks during which he lived in constant panic and his fear that he, too, would be taken to be executed. His descriptions begin in Malaga, February 1937, a couple of days after the fall of that city, and they continued in April, by which time he had been transferred to Seville, as if he were another Anne Frank, describing that which those individuals who persist in erasing the events of the past would like us to ignore.

Each page in Koestler's diary from the Seville prison is from April 1937 and each one makes shivers run down your back. The truth of the matter was that the terror that afflicted the men and women of defeated Spain was not the fear of death, which many who were tortured may well have welcomed.

What terrorized them was the dying itself, the pain, the shots, the blood. Koestler himself said so "Most of were not afraid of death; we were afraid of the act of dying."[xviii]

Moreno Gómez was unable to find more diaries like Baena's and Koestler's, but he was able to obtain a few eyewitness accounts referring to events in the Villanueva de Cordoba prison. April 2 1940, three inmates were taken by saca, one of whom was Manuel Salazar Vilches, a very active Communist, who was accused of having used the Guardia Civil horses after the barracks surrendered. One of his fellow prisoners told Moreno Gómez that before leaving his cell, like the rest of the inmates, he refused to confess. He left shouting Viva the Republic! in a most courageous manner, even in the street, as most of those who were imprisoned were very apathetic when they were taken out.[xix]

Juan Escribano Fernández, aged 23, an active member of the JSU was executed at 5 p.m. April 6 1940 in a saca in Villanueva de Cordoba. His family were able to see him before he died, and his sister gave Moreno Gómez her eyewitness account at the beginning of the 1980s. (She has since died.)

"Emilio *El del Lunar* came to our house to arrest Juan. He was accused of having attacked the barracks. They beat him so severely that the clothes he sent home were covered in blood, with patches of skin sticking to them. He had hoped that his sentence would be revised, but the terrible Judge Calero intervened, and his appeal was not heard.

The day that he was shot (in the afternoon), his clean clothes and the spoon with his name on it were returned to us, as souvenirs. That is how we knew that he was going to be executed. Moments before he was taken out, his brothers and sister arrived to see him. I hugged him non-stop, and the sergeant told them to bring me something to drink. I replied that I didn't even want the smell of that place. My brother Juan tried to console us, telling us not to cry, that 'he was not born to live for eternity'. He was rather inattentive and lethargic.

The next day, our father managed to get permission from Judge Calero to exhume him and place his body in a niche in the cemetery. He had been hit by five bullets in the chest and he had a large hole in his forehead."

Judge Calero would frequently grant favours such as allowing a prisoner's family to say their farewells, in exchange for substantial rewards. Juan Escribano's family complained that he accepted many presents, in exchange for saving Juan, which he did not.

One of the first executions carried out in Villanueva de Cordoba November 7 1939 at 5 p.m., was the saca that included Pedro Juan Martínez from the Fuente Vieja prison. It was a theatrical affair, with armed Guardia Civiles posted on every corner outside the prison. The condemned men were loaded on a truck whilst vast crowds of right-wingers gathered around to see the show. Tearful relatives of the prisoners remained at a distance, behind the guards on the street corners. As Pedro Juan climbed onto the truck, he went mad. Seeing so many people come to watch him die, he lost his head and began shouting: *"Let's go to the bullfight! Come on everybody! Let's party! Let's have some fun!"*

The truck then drove down the Calle Alta, the Cruz de Piedra and the street that led to the cemetery, where a huge Falangist crowd had gathered, singing *Cara al sol* and applauding wildly as the victims were mowed down by the gunfire, the last humiliating sounds they took with them to the other world. It is extremely hard to die, but so much harder to die under absolute humiliation, to the applause and singing of the enemy.

One witness in Cordoba told of the story of a melon and a saca. An unknown farmer from Villa del Río was saving a melon that his family had sent him: an absolute banquet to stave the hunger in the prison. When his name was called, he asked permission to share the melon with his cellmates. The banquet over, he went calmly to his death. Someone should tackle the impossible task of writing a treatise on the courageous death of Spanish farmers, victims of National Catholicism.

The multiple circumstances of the sacas in the Cordoba prison will remain unknown forever. The fury and the cruelty were such that special dates were set for sacas of working men. This was the case in Cordoba capital in 1941, when 34 men were executed om Labour Day, May 1st. According to Pedro Molinero's eyewitness account, Blas Gómez Medina, a leading Communist from Villanueva de Cordoba, was the least prepared of the lot for what was about to happen to them.

June 22 1940, a saca of 22 condemned men from the Cordoba prison included the famous Paco Dios, from Villafranca, who had commanded a company in the Villafranca Battalion and later served as a Militia Major in the 74th Brigade. He was a great leader, born of the people and to whom Pedro Garfias, the poet, dedicated a poem in his wartime anthology *Héroes del Sur.*[xx] Joe Monks identifies him as Capitán Paco, an Irish International Brigade volunteer, in his book on the Spanish civil war.[xxi]

Neither author knew how their hero died, but it did not follow the usual ritual. He was a hero in life, a hero in commanding his men, a hero in death. Moreno Gómez was able to interview his brother Juan, who has since died, in Cordoba capital.

"The night of the saca, he changed his clothes and shared his belongings with his cellmates. Later, there was an incident with the priest who insisted on pressing a crucifix against his chest. Paco pushed him away angrily, which earned him a slap in the face from a Guardia Civil. While the condemned men waited, talking to each other in front of the La Salud cemetery wall, Paco asked for a cigarette and a last wish: he wanted to say a few words, before everybody, before he died.

They say that he spoke valiantly, censuring fascism and rebuking the Regime's executioners, ending with a resounding *Viva la Republica! Long Live the Republic!* When his brother-in-law Antonio Gómez Torres collapsed, in tears, Paco consoled him with words of courage for what was about to happen. The priest, impressed by Paco Dios' speech, asked him what his profession was, and he was surprised with his answer – a simple bricklayer."

The ritual and anguish of the sacas is described in Pablo Uriel's excellent book Moreno Gómez quoted earlier. This young doctor was arrested as a prisoner-of-war in the Autumn of 1936 and imprisoned in the General Military Academy prison at San Gregorio, Zaragoza, where he came close to death but survived. When you read his account in the Appendix, it is important that you remember, however, that he is reporting on a military prison where most inmates were soldiers who had been sent there from the front lines, based on unproven accusations from Falangistas, mainly members of the Spanish fascist party in Zaragoza. Apparently, the Nationalist Army also had to be 'cleansed'.

In that prison, the person immediately responsible for determining who would be executed was the Military Police Colonel at the Paseo de Ruiseñores headquarters in Zaragoza. Above him, the Francoist repressive machinery.

In the case of those who disappeared unrecorded, the Machiavellian nature of these disappearances was apparent in a short directive, sent to the head of the Brigade of the San Gregorio military prison, instructing him to 'Release imprisoned soldier (name) if he has not been requested by any other authority'.

The last part of the instruction was the key word for the *paseo*[6] and the soldier's subsequent disappearance. It was in this extraordinary manner that 400 soldiers and two sergeants 'disappeared' in the Autumn of 1936 from that Zaragoza prison.

The Chapel

The ritual of the saca was followed, usually but not always, by a period of waiting before the execution, known as «The Chapel». Also, after a condemned man was removed from his cell, he was taken elsewhere, also referred to as the chapel and where, as he waited, he could write farewell letters to his family, confess to the priest – which many refused to do and resulted in heated arguments, attend mass and/or take communion. Mass and communion were offered to the condemned in Zaragoza, but not in Cordoba nor in other provincial capitals.

The bureaucratic procedure was simple. The condemned person's dossier was returned to the military court by the Judge Advocate or Commander General, with the annotation enterado as confirmation of his sentence. In the prison, the saca was put into action and the prisoner was taken to an office where he was read the ruling of the Court, informed of the sentence, then taken to the chapel.

In Zaragoza prison, because it was a military prison and the condemned were taken directly to the cemetery for execution, there was no chapel. What usually happened in other prisons was that when the saca was scheduled for dawn, the condemned had most of the night to let the sentence sink in or to write to their families. In 1939, the time in the chapel was much shorter as the entire process was much quicker in many townships (although not in the capital cities) where sentences were read in the middle of the afternoon and the condemned were executed at 8 p.m. Other times, the executions

---

6   As a reminder, the *paseo* was the practice of the Guardia Civil showing up at an individual's house in the middle of the night, loading him onto a truck a taking him 'for a ride' from which he never returned. Also, once a soldier was released from jail, he went home as an ordinary citizen, only to be taken on a paseo from his home, never to be heard from again, hence he just 'disappeared'.

were carried out even earlier, at 5 p.m. There could be so little time in the chapel that the victims did not even have enough time to write their last letter. Later, however, in 1940, most executions were carried out at dawn both in the capital cities and the townships.

The letters that were written in the chapel were very difficult to write and today, continue to make extremely painful reading. They were the last thoughts of an entire generation of Republicans that was exterminated. They express the pain of the Andalusian farmers who were sacrificed by the fascism of the barracks, the Casino and the Church vestry. Moreno Gómez has quite a few of these letters, handwritten by those who wrote them before they died, several of which are reproduced herein.

One of the most moving letters written by the condemned men who were executed in Cordoba capital, is the letter that Captain Paco Dios wrote to his wife and daughter when he was in the chapel awaiting his imminent execution:

"Cordoba Provincial Prison
June 22, 1940

Dear Lucía: keep these lines until our dear daughter is old enough to understand them. Dear daughter. Just a few words, as if this were my Last Will and Testament, because as I have nothing else to bequeath you. I am going to tell you why they are killing me.

Ever since I was able to think with reason, I became aware of the injustice that affects a large part of Mankind, and of the inequality that exists between it and those who control the money. Ever since, I fought to correct that evil and, like me, so did many working men. The more united we were, the more the number of grievances we were able to get corrected, as we defended ourselves from the misery that capitalists impressed on us.

However, those gentlemen, aware of the danger that our union represented to their rich living, rather than share with us part of that which they had leftover, chose to launch the most horrible tragedy ever recorded by History, with a view to robbing us of the few rights that we had obtained.

We defended ourselves and, briefly, were defeated. Immediately, the victors decided that I must die, as they did thousands and thousands of my comrades. That is all. Many died on both sides, although you will be told differently.

As I am certain that, when you understand all this, you will benefit from something of that which I fought for, others will take care to inform you of what happened, both to you and to all those who, like you, lost their father.

But whatever happens, I have a word of advice that is, at the same time, a mandate: follow the road that I took and if as you travel it, the same thing happens to you that happened to me, how proud you shall be to give your life when you have the satisfaction of having done your duty.

Nothing more. Your father, Paco Dios."

As we saw earlier when Moreno Gómez described Paco Dios' execution, his morale and personal dignity remained intact when he faced death, unbroken by the punishments he had suffered, as did others, although not all desired to be heroes. As a military commander, Paco inspired his men and at his death, he caused consternation among his enemies, such as the priest who was astounded that this 'evil man' could express such profound thoughts.

From Villanueva de Cordoba, there are several letters written in the chapel with quite different overtones, including several in which the authors clearly ask their relatives to take revenge upon the executioners when the state of affairs changes. May 17 1940 there was a saca of 5 condemned men, the first one scheduled for 6 a.m., a novelty as during the previous six months, these were held at 5 p.m. to provide a show for the townspeople. One of the executed was a renowned local politician who belonged to the Izquierda Republicana, a moderate man, who had been targeted for extermination like all political leaders: Francisco Sánchez Muñoz, known as *Curro Beatas*, past alderman for the Frente Popular. He left two letters written in the chapel, the first one of which was personal. He wrote a second letter which he most certainly had delivered by a different means, addressed to his family, undated but also written in chapel. In this letter, he denounces the many who tortured him in Villanueva, adding the wish that someday they shall suffer the revenge they deserve.

On their own, publication of the above letters that the family gave Moreno Gómez is a form of historic redress as well as the condemnation of his Francoist torturers. Nine days later, May 16 1940, another eighteen men were executed, among whom Eugenio Palmera Jurado Pozuelo, renowned leader in Villanueva, considered a Communist stronghold. Eugenio was horribly tortured and beaten. His death was truly a merciful release. He also wrote a letter in the chapel in which he asks his family to revenge him. This letter, also reproduced herein, is now made public for the first time. It was given to Moreno Gómez by his nephew, Miguel Jurado Tintorero, who was visibly moved. All the papers he had came from his uncle José Palmera's archives, Mayor of Villanueva in 1935 representing the Leroux political party, Eugenio's brother.

Several residents of Villanueva de Cordoba were executed elsewhere in Spain, some of their letters of farewell are also recorded herein.

May 19 1943, Rafael Porras Caballero from Pozoblanco, son of Antonio Porras the Republican poet and politician who at the same time was living in exile in France, fell victim to Francoism in Madrid. That morning, Rafael wrote two letters whilst he was in the chapel: one to his siblings and another to his parents and brothers. Although only the second letter is reproduced here, attention is drawn to the extremely unusual note that he wrote on the first one: *"I have been allowed to send these last letters of farewell by the kind favour of the prison priest"*. (This remark confirms other information according to which many condemned men agreed to say confession only to ensure that their last letters would be sent to their families. Some of those who refused confession were even denied the possibility of writing a last letter.)

Rafael's second farewell letter to the whole family ends with a moving postscript in which he asks his parents to help the widow and children of a fellow prisoner, who were left destitute:

"José María San Ildefonso dies with me. He leaves a widow and children. Please do as much as possible for them. They live at Carretera de Aragón number 15, Ventas."[xxii]

As farmers and workers all over Spain and Republicans of all kinds continued to be eliminated by the Regime, they bequeathed short and simple letters of farewell to their families. In his book of memoires, Galo Vierge leaves us some heartrending examples from Pamplona, massacred by Falangistas and Carlistas, which just shows how the extermination program reached every corner of the country.

"Elías Sesma, from Sartuguda[7] was a thirty-year-old man whose great worry when he was incarcerated was, who would harvest the little bit of farmland that he owned in his village? As he did not know how to read or write, I wrote his letters to his wife Teresa Arpón, and when he dictated what he wanted to say, he always expressed his great concern regarding his beloved vegetables: 'Teresa, go to uncle Vicente's house and ask to borrow the cart and go to the field and pick the green peppers and the tomatoes, before they rot'. Again and again, in every single letter. The day that he was executed, 16 November 1936, with sixteen other prisoners from his village, Elías left his cell happily, naïvely believing that he was being set free. In his hands, he carried a drawstring bag for bread where he kept some trinkets for his daughter Teresita. That man, in his supreme ignorance and good faith, had the noble and simple soul of those who work in the fields, and even though they had shot two of his brothers, it never entered his mind that men could go as far as to criminalize innocent victims such as he in order to sow terror among the country folk."[xxiii]

---

[7] Known as *village of widows*.

That was poor Elias' last wish: to harvest his green peppers. There was a great rush to exterminate the defeated during the last months of 1936. There was no time for the chapel, no time to write letters, no time for anything. Prisoners went from their cell to the grave, without further ado. The prolonged imprisonments of the post-war period were almost a privilege.

A prisoner's stay in the chapel reached its nadir with the priest. Books of memoires frequently refer to how shocked the condemned men were with the priests' impertinence. They mention cases of irate priests who would go so far as to beat an unrepentant inmate and of some less fanatic priests.

The most descriptive book of memoires so far, one of the most terrible documents regarding the extermination, is the one written by a Capuchin monk and Basque nationalist, Gumersindo de Estella, who was charged with giving religious solace to the condemned in Zaragoza during the war years.[xxiv] Although there is no such book for Cordoba, the best documents regarding the Francoist death ritual are by two Zaragoza prison diarists: aforementioned Pablo Uriel for the San Gregório Military Prison and Father Gumersindo for the Provincial Prison. In the latter prison, the chapel was in the judge's chambers, where an improvised altar had been set up so that mass could be celebrated during the terrible fatal hours. There is no evidence that any chapel masses were ever said in the Cordoba prison.

An ominous group of individuals gathered in these chambers at dawn. The Judge Executioner, his secretary, the doctor, one or more priests, several persons dressed in black (members of the Brotherhood of the Blood of Christ, elsewhere known as the Brotherhood of Peace and Charity) who collected the bodies, and some others. The last to arrive at the chapel were the condemned whose hearts must have sunk at the sight of this melodramatic gathering.

The first thing that enters one's mind is that to most of the Republican condemned, having to face a priest at such a terrible moment must have been a tremendous psychological, moral and intellectual shock. Logic would suggest – although we know it was impossible – that at such a crucial moment in their lives, they might have been comforted by the representative of an ideology they believed in; not by one they did not. Why not a political commissar who might have praised the great honour of giving their lives for freedom and democracy, or some such? Another blow to their ideals, the recognition that the ideology at the base of the military coup that they had fought against, was now responsible for their deaths. Apparently, no authors have commented on this latter aspect, one that surely meant a great deal to the Republicans who were about to die.

Another feature of note is that in addition to the San Gregorio Military Prison, numerous Nationalist soldiers were also executed at the Zaragoza Provincial Prison. This would indicate that the Francoist cleansing of its own ranks was likewise terrible and widespread. Something like this, so minutely planned, so systematic and devastating, did not occur in the Republican zone, at least not to any extent.

Father Gumersindo's assistance to the condemned at the Zaragoza Provincial Prison, began June 22 1937. On his first day at work, he began by attempting to give spiritual guidance to a very cultured and highly educated individual, a Socialist Secretary of the Escarrón Municipality, near Caspe. Father Gumersindo's diary entries, many of which are reproduced herein, are descriptive, informative, heart-breaking, and profoundly empathetic. In his diary, Father Gumersindo tells how in February 1938, the Zaragoza Provincial Prison alone housed 5,200 men and 800 women, in a prison that was built for 250 inmates. The overcrowding was evident. In July of the same year, a prison doctor's report stated that 2,000 inmates were suffering from scabies.

Two high-ranking Republican prisoners were executed March 17 1938: General José María Enciso and Colonel José María González Tablas, both of whom were taken prisoner in Escarrón March 12, during the major Francoist offensive in the Lower Aragon province. The morning of their execution, the General Commander of Zaragoza, Francisco Rañoy Carvajal, was present in the chapel as he did not want to miss the great event. General Enciso refused spiritual comfort, whilst Colonel González turned himself with fervour to confession, mass and communion. Father Gumersindo said that the Colonel burst into tears after confessing and that he sobbed as he received communion.

April 20 1038, a mutilated man was shot in Zaragoza. Thirty-five-year-old Antonio Botela from Gerafe's leg had been amputated above the knees because of wounds he had received during a Republican bombing on the prison November 5 1937. Not even this amputation saved his life. In the chapel, when he finished confessing, he sobbed incessantly: "My poor children, what is going to happen to them in this world?... My poor children." He then attended mass and took communion. Several soldiers had to carry him to his execution. There was a similar case in Cordoba of a mutilated prisoner who was executed by garrotte. One cannot deny that bad things occurred in the Republican zone, but Spanish fascists were solely to blame for this kind of barbarous behaviour.

The May 12 1938 entry in Father Gumersindo's diary refers to nine prisoners who were brought to the chapel, one of whom a woman, Maria de Assis Figueras. They were from Alcañiz, possibly prisoners of war. That morning, the hour in the chapel was an absolute scandal. The woman refused to go to the chapel from the identification room where the sentences were read out. At that moment, a female prison guard shamefully mocked her: "Ay, María, María. Look how the blessed CNT has betrayed you!" Father Gumersindo immediately rebuked the guard. He then noticed that the prisoner appeared to be pregnant, so he approached the Judge Executioner for clemency, but the Judge refused his request, and she was tied up and executed.

Some say that we historians lose ourselves in the angst of the facts and that we neglect the overall legal picture. Not so. If we choose to resurrect these hidden pages of history, it is because we have to make an elemental reconstruction of events that have, until now, been intentionally forgotten. There is always time to analyse the legal aspect. For now, we need to give expression, without excuses or distinctions of any kind, to how the victims of Francoism faced death. We must make known their state of mind, their protestations of innocence, their pain, the way they faced the priests' harassment, as well as the series of cruel punishments to which they were subjected during this colossal genocide. This is not losing oneself in miseries or an obsession with the facts. Historians are well aware of the task that is before them, a job that has been avoided for far too long.

Father Gumersindo's diary shows us that many prisoners-of-war from the Teruel front were continuously being shot. One of those prisoners, from Vich, Isídro Franquisa, denied all spiritual comfort. The great majority of the condemned who arrived at Father Gumersindo's chapel in Zaragoza Provincial Prison protested their innocence and denounced the injustice – remember most of them were civil prisoners-of-war.

This brings us back to the farce of the court martial. It is as if, with these summary rulings, we are looking for a non-existing legality, an illusion that innocence would somehow shine through and that only the truly guilty would be condemned. To even suggest such a hope, would be to encourage a fantasy and a false mirroring of the truth, to presume that there actually was a state of law in the Francoist Regime and an independent application of justice, something that proved to be totally impossible.

Everything that took place in the so-called Zaragoza 'chapel' and in all of Spain was 'standard' Francoist 'justice', i.e., non-existent justice. This shows, once again, that the courts martial were a farce, a travesty of justice, an absurdity that only served to provide genocide with a respectable image. The judges, Jesuit chaplains, prison directors, advocates, General Commanders and the *Criminalissimo* himself, were not interested in justice. They just wanted to kill, and kill, and kill some more, the more the better. The distinction between innocent and guilty was impertinent, anecdotal and unthinkable.

Ingenuous Father Gumersindo appeared to believe the fantasy and he trusted the presumed existence of a real justice. It was the rare dawn that he was not taken in. October 19, 1938 he attended to four condemned men, one of whom was from Monforte, Teruel:

> "At first, he expressed his indignation saying that the accusations against him were totally false.
> - But man, I said. Why didn't you say all this when you were interrogated...?
> - Because I was never interrogated at my trial. The Prosecutor made many false accusations almost all of which were lies. I wanted to protest but I was ignored. The defence attorney never visited me."

Nine months later, July 14 1939, when eight prisoners from Alcañiz were taken to the chapel, nothing had changed. Francoist 'justice' continued as usual: falsehoods, lack of guarantees of all kinds, lack of evidence, powerless accused, manifest injustices, disrespect for innocence... One of the prisoners, a young, well-educated man, told Father Gumersindo:

"I believe that what they are doing to me is a mistake. I was tried in Alcañiz, was absolved and set free. I was later again arrested. I have not been interrogated; they have not spoken a single word to me. When I expected, with some reason, that I was going to be set free, I find that they are going to kill me. Who can explain this to me? A mistake or injustice."

Francoist military justice was injustice. Injustice is the *per se* definition of a revolutionary, totalitarian and anti-democratic regime. The opposite is the rule of law of a democracy. Moreover, the genocide of an entire social class in Spain could not be perpetrated with justice, only by injustice. It is therefore clear, that all the Francoist extermination practices were likewise unjust and judicially repugnant, both under the 19396-1937 wartime decrees and those ordered by the courts martial from 1937 onwards, such as under the Law of Escapes at various stages of Francoism (1939, 1947, 1948, 1949, 1950 and later). Sadly, we are again reminded that until very recently, [November 2022], today's democratic government had neither dared nor been capable of passing laws to rescind the rulings of the Dictator's courts martial.

A last vivid entry in Father Gumersindo's diary in Zaragoza that could apply to all the Spanish prisons, is dated December 13 1937 when a young woman, Nicolasa Aguirrezagala, aged 22, from San Sebastián, was brought into the chapel. One of so many Spanish *Roses*, she cried non-stop: *"Ay, mama! Ay, mama!* How far they have brought me to kill me!" Father Gumersindo begged her to calm down.

– "How can I calm down when everything they read out is false. I didn't denounce anybody. I didn't know the person called Portolés; I never knew him and never heard his name before they read it out in Court. It is all a lie. I cannot resign myself to die because of a lie. He was denounced by a waitress. I was a cook."

Father Gumersindo continues. When I heard these reasons, I looked at the Public Defender, who had entered the chapel behind me, and I asked him:

- "Are you listening to what she is saying? How can this be possible?"

The Public Defender replied, lowering his head as if in shame:

- "The country courts can make mistakes...! I thought to myself that they should err by pardoning, not by condemning. We do not have the right to play with people's lives.... I did everything that I could..." he added.

Nicolasa looked him in the eye and replied: "But you did nothing...." and she continued to protest her innocence.

- "But my daughter, didn't you say all this when you were interrogated?"
- "I was not interrogated, nor was I allowed to speak."
- "But didn't you explain all this to the Public Defender?"
- "I never spoke with the Public Defender. I didn't even know who he was, nor which one he was."

I was astonished to hear this and felt confused; I didn't know what to think, how to explain what had happened."

Enough. Having once and for all described the scene for everything that occurred in the chapel when death was imminent, there is no need to go back to the Francoist court martial. The reality is clear from the convictions of the multitude of innocents who were led to the grave. Moreno Gómez again insists that they were not 'mistakes'. It was the implementation of the extermination plan developed by General Mola, months before the military coup actually took place. There were no mistakes.

The firing squad

Execution before the firing squad marked the end of the ritual of death. Well, almost the end as the Regime would still attack the condemned by post-mortem enforcement of the Law of Political Responsibilities that resulted in the economic ruin of their families.

There is little information regarding the final moment in Cordoba, although Moreno Gómez was able to obtain several testimonies for 1936 in the capital's cemeteries.[xxv] He could add some more for the post-war period, but this information will never be as complete and detailed as that described for Zaragoza in Father Gumersindo's remarkable diary.

The Morning of October 22, 1936, 41 prisoners were executed in Seville under the war decrees by order and direction from General Gonzalo Queipo de Llano, who until 2022 was still publicly honoured in Macarena Basilica in Seville, where he was buried[8], as the Church continued to lend a deaf ear to the 5th Commandment. An exchange of letters with relatives of those condemned with José Garcia Márquez, shed light on how the firing squads acted in Seville on those dates.

Márquez' information begins with an examination of the files kept at Seville Military Archives, when investigating the wounds that a corporal received from a bullet that ricocheted during an execution 25 October, three days after the previous one and in which 35 prisoners were executed. At the time, the condemned were shot against the righthand wall of San Fernando cemetery, by firing squads of Moroccan soldiers, who held their rifles under their armpits and were disastrous marksmen. The executions were held between 2 a.m. and 3:30 a.m. The condemned were grouped two by two and shot in pairs, while the remaining prisoners watched and waited their turn. This was such a slow affair that the entire procedure lasted about an hour and a half.

The wounded corporal reported as follows:

> "...Early in the morning of the 25th, as he was on duty at the Commissariat of the Watch, Calle Jesus del Gran Poder, he left with the sergeant and five militia as part of the guard for the convoy of trucks bearing the prisoners and the firing squad to the cemetery walls.
>
> In preparation for the execution of the prisoners, these were lined up as usual, two by two, in front of the said wall and the firing squad, and on their

---

[8] Among other measures, the Democratic Memory Law of 21 October 2022, ordered the removal of the remains of Franco and Queipo de Llano from the Valle de los Caídos and the Macareno Basilica in which they were publicly honoured as heroes of Spain.

right side the truck whose lights shone on the place of execution. The guard stood a bit in front of the truck and to the right, to prevent any of the prisoners from escaping on that side.

They had been executing prisoners for some fifty minutes when the declarant felt that he had been hit in the stomach and that the bullet was in his body, which is why he was immediately taken to the Hospital in one of the cars that was there."

At this point we need to return to Father Gumersindo's diary, as it contains the most correct description, we have so far regarding the Francoist method of execution by firing squad.

The day (22 June 1937) that Don Tregidio, Secretary of the Escarrón Municipality, near Caspe, was shot in Torrero (Zaragoza) cemetery, Father Gumersindo describes the great cruelty towards the two prisoners who were about to be shot:

"... We had arrived at the cemetery. We drove along the wall to get to that part of the wall at the front, facing the city and neighbourhood of Torrero, from where we had just come. And... we found a detachment of some one hundred soldiers. They were formed in rows facing the wall, but about fifty metres distant. Sixteen of them were closest to the wall... As we arrived, our truck stopped. We got off and began to walk towards the soldiers. They looked at us with curiosity. I walked next to Don Tregido. The other prisoner was accompanied by Rev. Victor, who could not contain his roughness. We walked past a Red Cross van, almost touching it... Next to the van, two gurneys that had been prepared to receive the bodies of both men. And they could see it all. What a terribly sad walk! Sixty or seventy bitterly difficult steps for the condemned men and for anyone who had been born with a bit of heart...

Nobody asked the prisoners whether they wanted a blindfold. I still did not abandon my friend. I stood at his side, stroking his right arm and neck with my hand, and I repeated the prayer: 'Merciful Jesus, save my soul'. He repeated it and he kissed the crucifix. I offered it to the other prisoner for him to kiss, but he shook his head. The silence was deafening. I realized that the officer who had to give the sign to fire was waiting for me to leave. I walked away and stood behind the advance squad of soldiers. The officer shouted: 'Aim!' Don Tregidio shouted: 'Long live God and Socialism!' The officer shouted again: 'Fire!' The fatal shots rang out. Each body was riddled by eight bullets. They fell backwards, onto the ground...

Some Guardia Civiles approached to remove the metal handcuffs that bound their hands. I approached to administer extreme unction to one and absolution and a prayer. Both bodies were lying in a large pool of blood that had run down their legs and was mixing with the dew... A lieutenant shot them each twice in the head. The doctor approached to confirm their death. The members of the Brotherhood of the Blood of Christ picked them up and placed them on the gurneys..."[xxvi]

22 September 1937, three women and one man were executed. When they arrived at the place of the execution and saw all the soldiers who had been assembled – an entire company and a squad of soldiers (six for each person). Father Gumersindo writes:

"... We began the slow walk towards the place of execution. It was the most horrible walk of my life. The three women wobbled as they walked, their hands had been tied, their clothes were in disarray, their hair was a mess (the babies of two of the women had been torn from their breasts as they entered the chapel) ...One of the women shouted: 'So many men just to kill three women...!"[xxvii]

Something that occurred when the condemned were shot must be addressed painful as it is: may of the executions were bugled, the soldiers did not aim properly, the condemned did not die at once, only after horrendous suffering, as Father Gumersindo tells us. On 5 October 1937, when eight prisoners were shot in Zaragoza, two of whom were French, the execution was a horrible disaster.

> "Four soldiers were assigned to each prisoner. They stood in a row, two metres apart from each other, between the cemetery wall and the firing squad... I moved away. The silence was deafening. It was 6 a.m. and dawn had just broken. The fatal shots rang out and they all fell to the round, but the majority were only slightly wounded. The soldiers always shot reluctantly. Someone shot at the wall. On the ground, almost all the condemned were moving around, all painfully crying out.
>
> I began moving down the line, administering extreme unction and absolving their sins. The second was young Doñate. When I reached him, he began to rise very slowly, he looked straight at me and raised his bound wrists in supplication. When he was on his knees and I was holding him up with my left arm as I absolved him with my right, the lieutenant came up and shot him in the head, a couple of inches from my face. Poor Doñate's blood stained my sleeve. When I saw him given the mercy shot, my stomach turned in protest and anger... But I had to continue absolving the remaining prisoners, before they too, were killed off.
>
> ‒ How badly you shot ‒ I told the soldiers. You made them suffer terribly..."[xxviii]

1 February 1938, Father Gumersindo records the execution of six condemned men, among whom Joaquim Laguna, 18 years old, who was taken prisoner in Sigüenza cathedral. The execution was another great fiasco.

> "Of the six condemned men, only one died instantly from the shots received from the eight rifles aimed

at each prisoner. The others, who had fallen to the ground, still lived and screamed with pain. I gave them absolution and the extreme unction to all but one, the most rebellious."

It would be particularly useful today if there were death statistics for the Cordoba provincial cemeteries and for the prison in Cordoba capital. The carnage in San Rafael or in La Salud, dozens and dozens of men, shot one after the other, tragically waiting their end. The specific cases, what they shouted, their protestations of innocence, firing squads that aimed badly, and so forth. Sadly, there are no formal records either from chaplains, from soldiers (these would never dare document their own crimes), from Guardia Civiles (even less likely to do so), or prison guards. We lack this information and the details, as for example, how did Federico García Lorca die, what might he have said before he was shot, how much might he have suffered? This is not anecdotal information, but very important data if we wish to reconstruct the life and death of those men and women.[9]

Another disaster of suffering was the execution 12 May 1938 of nine condemned from Alcañiz, one of whom was María Figueras whose passage in the Zaragoza prison chapel was also recorded earlier by Father Gumersindo.

"When we arrived at the place of the execution, we placed ourselves between the wall ad the firing squad. I suggested that they turn themselves to the wall so that they would not suffer from looking at the soldiers.

Old man Andrel[10] seeing the soldiers arrive, spoke to them:

'Men! You are about to kill sons of the people...!'
I stood between the wall and the condemned, unceasingly begging them, one by one, to have faith in God... Suddenly, I heard the Prison Director's voice shouting: "Father Gumersindo! Get out of the

---

[9] And if we wish to retrieve the Historic Memory of these fallen throughout Spain, so many of whom are unidentified, i.e., 'disappeared'.

[10] 61-year-old Miguel Andrel.

way!" He wanted to give the order to fire, and I hadn't realized that. In fact, the officer hadn't noticed that I was still talking to the prisoners. There still was little daylight.

Young Maria was twisting and turning in place. She continued to sob and say: *Ay! Padre mio! Ay, padre! Why are they killing me...?*

The soldiers were excited, and I saw their indifference and disgust mirrored in their faces.

No sooner did I get out of the way that the shots rang out. Four rifles for each prisoner. None of them died instantly. They were not mortally wounded. One of the wounded, rolling about on the ground, cried out: 'They have finished us; we only have a minute left to live!' They all cried out with pain, and some begged for the mercy shot. I went up to each one and absolved them. An officer administered the mercy shot, sometimes repeatedly shooting the dying three times in their heads."[xxix]

14 July 1938, another eight condemned from Alcañiz and another disgraceful execution. On this occasion, the firing squad forgot their ammunition.

"As we approached the soldiers, the truck stopped but we were ordered not to get off. A quarter of an hour later and we still had not been ordered to get off. I jumped down and asked a soldier what was happening and what was the reason for the delay. He told me that the soldiers had not brought their ammunition. Fifteen minutes later, some soldiers arrived in a car...

That delay was very harmful to the condemned. Some lost patience and began complaining: 'Why are they keeping us suffering here? How pleased they are to make us suffer! Then they dare say that it is the Rojos who are cruel...! Hurry up and kill us! Please, kill us and be done...!'

But that morning, something else added to the suffering. Father Gumersindo continues:

> "Finally, we got there. They made another mistake. After lining the eight condemned in a single row, they decided that they would execute four first, then the other four. Right there, in full view of the last four, the officer gave the order to fire against their companions. They saw the soldiers take aim and fire and they heard the shots. They then watched them lying on the ground, rolling in pools of blood and crying out in pain.
>
> Then the same soldiers took their positions in front of the surviving prisoners, who fell next to the first. Looking at the eight wounded lying on the ground, one got the impression that this was a battlefield."[xxx]

When faced with such terrible scenes, the historian attempts to determine what occurred in the numerous executions in the cemeteries in Cordoba capital, such as when 1 May 1941, 34 were shot one after the other in the La Salud cemetery. What did the condemned do during the pre-dawn hours; how was the saca done; how long did they have in the chapel; did they protest on the way to their death; what were the circumstances surrounding the final act? Knowledge of these details is important to the history of the men and women involved and when one is attempting to reconstruct documents regarding the Francoist genocide against the Republicans. Another 34 fell in the San Rafael cemetery 3 May of the same year; 28 fell 3 June. All of this was a spectacular waste to promote the expansion of the wave of terror.

Also, one must not forget the executions in Cordoba capital in Summer 1936, when on several occasions more than one hundred victims were executed each time. Very few details of these massacres have come to light and been collected by historians; sadly, they are doomed to oblivion.

The Church's unsupportable harassment of the condemned at the supreme moment has been commented on in a proliferate manner. In the general plan of religious propaganda written by the Jesuit priest, Rev. Pérez del Pulgar, S.J., among other things he states:

"...In several executions, we have seen the impressive contrast that, whilst a group of men died kissing the crucifix and saying Viva! Spain, another waited to be executed singing the International, blaspheming, and swearing." [xxxi]

Two last reports regarding the execution debacles and the unerring Calvary of the condemned, as Father Gumersindo records in his diary 26 July 1938, marked the execution of 7 victims against the wall of Torrero cemetery, almost all prisoners of war, several from Gelsa and one from Belchite. The procedure was the same as everywhere else.

"The victims of the 26th suffered a great deal. The soldiers shot badly. They were apathetic. One of the executed fell to the ground shouting 'Hurry up and kill me!' All cried and screamed with pain. A deplorable and sorry spectacle...!" [xxxii]

Finally, a few words regarding the four victims who were put to death in Torrero 18 October 1938:

"The soldiers who on that day ad the misfortune of being the executioners or part of the firing squad, shot badly. The unfortunate prisoners rolled around on the ground, with such pitiful cries and screams that tore at my soul. The unfortunate Martín raised his feet and legs, whilst harsh, deep *Ayes!* from his chest sounded like a death rattle.

The two men who refused confession, exclaimed just before they were shot: *Viva the Republic! You will soon suffer the same fate!* "[xxxiii]

With these notes from the Zaragoza prison chaplain, we get an approximate idea of that which occurred in the prisons and cemeteries all over Spain, both during and after the post-war period, regarding the mechanics of the ritual of death and the extermination of all Republicans by the Francoist 'New State' - the National Catholicism.

All the paraphernalia of death responded, without ado, to the brutality of European fascism of the times. Everything obeyed the strict characteristics of fascism, without any doubt. Nothing of such cruelty can be explained without determining the coordinates of the Nazi-fascism-Francoism cult of violence, with all the variations that one must consider, but always within this triangle. Without this, one cannot begin to clearly explain such violence and such extermination.

ENDNOTES FOR CHAPTER IV

i   Lawsuit filed 14 April 2010 before the National Criminal and Administrative Court, Federal Criminal Correctional Court 1, Case 4591/19, under the generic title of 'Genocide'.

ii   Ana Messuti. La querella argentina: La aplicación del principio de justicia universal al caso de las desapariciones forzadas. (The Argentine lawsuit. The application of the principle of universal justice to the case of the forced disappearances). In Rafael Escudero Alday and Carmen Pérez González Desapariciones forzadas, represión política y crímenes del franquismo) Forced disappearances, political repression and crimes of Francoism). Madrid, Trotta, 2013, p. 129.

iii   Carlos Jiménez Villarejo and Antonio Doñate Martín. Jueces, pero parciales. La pervivencia del franquismo en el poder judicial. (Judges, but partly so. The survival of Francoism in the Judiciary.) Pasado & Presente, Barcelona, 2012.

iv   Summary Emergency Procedure No. 12.720/1939 against Rafael Marencio Muñoz, Territorial Military Archives II, Seville. Courtesy of Joaquín Casado.

v   Ángel Viñas. En el combate por la historia. Op. Cit., p. 20, citing Richard J. Evans and Bosworth.

vi   Paul Preston. El gran manipulador La mentira quotidiana de Franco. (The great manipulator. Franco's daily lies), Barcelona, Ediciones B., 2008.

vii   Arthur Koestler. Diálogo con la muerte. Un testamento español. (Dialogue with death. A Spanish testimony.) Madrid, Amarante, 2004.

viii   José Manuel Sabín Rodríguez. La dictatura franquista (1936-1975). Textos y documentos. (The Francoist dictatorship (1936-1975). Texts and documents). Madrid, Alcal 1997.

ix   Francisco Moreno Gómez. La repression en la posguerra (The post-war repression), in Victimas de la guerra civil (Victims of the civil war). Op. Cit., p. 330.

x   Matilde Eiroa. Recent article in the online newspaper Hispania Nova, number 10, 2021, available at http://hispanianova.rediris.es/

xi   Antonio Baena. Pozoblanco. Unpublished personal Diary.

xii   Arthur Koestler. Op. Cit., p. 183.

xiii   Pablo Uriel. Mi guerra civil (My civil war). Self-published. Valencia, 1988, pp. 83 et al.

xiv   Ibid., p. 94.

xv   Sebastian Blanco Copado. Letter to his wife, excerpt published by García de Consuegara in an article entitled "The Nationalist repression in Pozoblanco", Revista de Feria, Pozoblanco, September 1985.

xvi   Pablo Uriel, Op. Cit., p. 97.

xvii   Ibid., p. 30.

xviii   Arthur Koestler. Op. Cit., p. 261.

xix   Juan Gutierrez Romero Bruno. Interviewed several times by Moreno Gómez in Villanueva de Cordoba.

xx   Pedro Garfias. Hérois del Sur (Heroes of the South), Mexico, 1941.

xxi   Joe Monks. Con los Rojosen Andalucía (With the Rojos in Andalucía). Seville, Renacimiento, 2012, pp. 110 et al.

xxii   Letter kindly given to Moreno Gómez by Floriano Sánchez Bermuda from Pozoblanco.

xxiii   Galo Vierge. Los culpables. Pamplona 1936 (The guilty. Pamplona 1936). Pamiela, Pamplona, 2006, p. 101.

xxiv   Gumersindo de Estella. Fusilados en Zaragoza, 1938-1939 (Executed in Zaragoza, 1938-1939). Mira Editores, Zaragoza, 2003.

xxv   Francisco Moreno Gómez. El genocidio franquista en Cordoba (The Francoist genocide in Cordoba), Crítica, Barcelona, 2008.

xxvi   Gumersindo de Estella, Op. Cit., pp. 57-58.

xxvii   Ibid., p. 65.

xxviii   Ibid., p. 70.

xxix   Ibid., p. 121.

xxx   Ibid., p. 143.

xxxi   José Manuel Sabín, Op. Cit., p. 116.

xxxii   Gumersindo de Estrela, Op. Cit., p. 144.

xxxiii   Ibid., p. 151.

1. Adamuz	28. Fuente La Lancha	53. Posadas
2. Aguilar	29. Fuenteobejuna	54. Pozoblanco
3. Alcaracejos	30. Fuente Palmera	55. Priego
4. Almadinília	31. Fuente Tójar	56. Puente Genil
5. Amodóvar del Río	32. La Granjuela	57. La Rambla
6. Añora	33. Guadalcázar	58. Rute
7. Baena	34. El Guijo	59. San Sebastián de los
8. Belalcázar	35. Hinojosa del Duque	Ballesteros
9. Belmez	36. Hornachuelos	60. Santaella
10. Benamejí	37. Iznájar	61. Santa Eufemia
11. Los Blázquez	38. Lucena	62. Torrecampo
12. Bujalance	39. Luque	63. Valenzuela
13. Cabra	40. Montalbán	64. Valsequillo
14. Cañete de las Torres	41. Montemayor	65. La Victoria
15. Carcabuey	42. Montilla	66. Villa del Río
16. Cardeña	43. Montoro	67. Villafranca
17. La Carlota	44. Monturque	68. Villaharta
18. El Carpio	45. Moriles	69. Villanueva de
19. Castro del Río	46. Nueva Carteya	Córdoba
20. Conquista	47. Obejo	70. Villanueva del
21. Córdoba	48. Palenciana	Duque
22. Doña Mencía	49. Palma del Río	71. Villanueva del Rey
23. Dos Torres	50. Pedro Abad	72. Villaralto
24. Encinas Reales	51. Pedroche	73. Villaviciosa
25. Espejo	52. Peñarroya-	74. El Viso
26. Espiel	Pueblonuevo	75. Zuheros
27. Fernán Núñez		

MAP OF THE 75 MUNICIPALITIES
IN CORDOBA PROVINCE

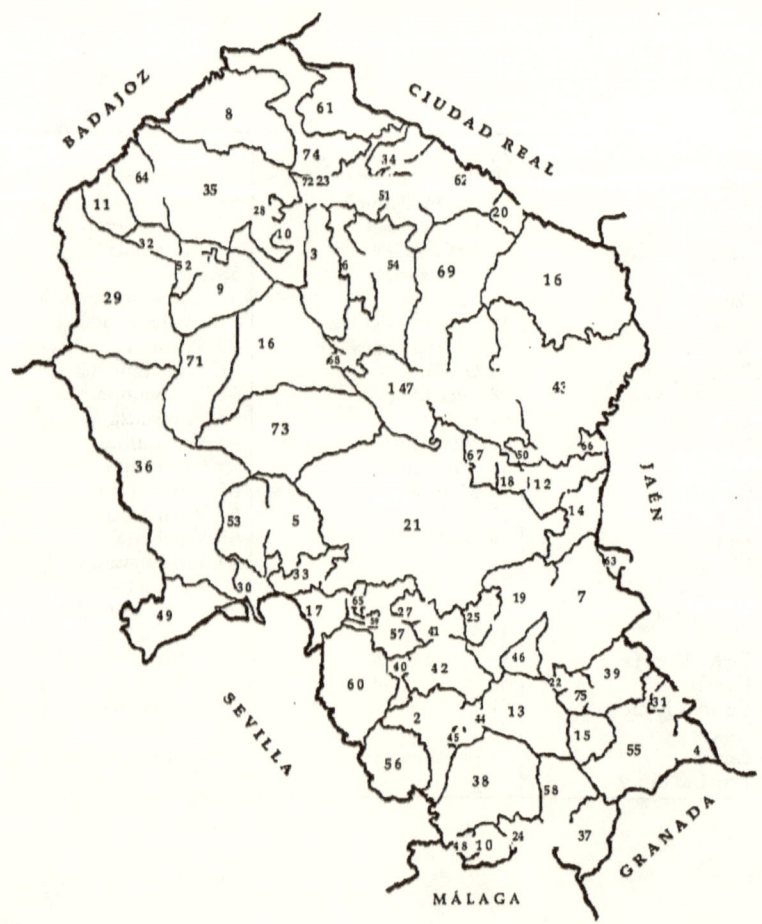

V

CRUSHING THE VANQUISHED (D) WIDESPREAD EXECUTIONS AND OTHER FORMS OF TERROR

"Terror debases everyone... Terror, whenever it is applied with callousness, will produce an abject acceptance by the community, a humiliating feeling of gratitude on the part of those who managed to escape the punishment... The bloody orgies dishonour everyone involved, executioners and victims. Only thus, can one explain many of the things that occurred during those months."
Pablo Uriel 1988.[i]

Introduction

Executions by firing squad were not carried out in every one of the 75 municipalities in Cordoba province. The slaughter was concentrated in the main towns of the province, in whose jails prisoners from neighbouring townships were interned.

These post-war executions were not carried out with the same intensity in the regions of the Sierra as in the countryside, as the latter region had been under Franco's rule since 1936 and continuously subject to the extermination from the start of the war. For all that, the repression continued to hit hard on Puente Genil, Montilla, Castro del Río, Bujalance, Palma del Río and other important towns. Nonetheless, the most terrible extermination was carried out in the Sierra, which remained Republican until the very end.

One's attention is drawn to the fact that Franco planned the first stage of the post-war extermination in the hometowns of the defeated prisoners. The entire repression during 1939 and 1940 was carried out in townships although it was centered in the more important towns. It is not difficult to discern the reasons for concentrating the repressive strategy in the Centre-South region of Spain as this was the last area of the country to be occupied by the Nationalists.

1. First, the Regime implemented the usual 'cleaning operation' in the recently conquered townships. That is how the Francoists

themselves described it: the «military operations» of the occupation, they said, were followed by «cleaning operations». You can imagine what they were referring to with that tragic euphemism one they bandied during the entire war. This was no 'improvised' repression, as some scholars have suggested, merely devised to 'shock'. It was planed, widespread and devastating. It is well known that in Franco's opinion, dominance of the land was much less important than the domination of the people.

2. By staging the first step of the post-war repression in the townships, Franco guaranteed that the most extensive, more profound and greatest amount of violence would be felt in those places where the disaffected defeated were to the Regime'. This context guaranteed the perfect targeting of reprisals and acts of revenge. As a result, all the principal towns in the Centre-South, in our case Cordoba province, became centres of imprisonment with mass incarcerations, beatings and systematic torture, and the venue for the courts martial and the first wave of executions.

3. This rattling of rifles in the provincial towns was a matter of some satisfaction for the winners, and a source of fear for the anonymous mass of disaffected. An atmosphere of terror, exemplariness, and general punishment that would not have been felt in the countryside had the executions been limited to the provincial capital cities.

4. Lastly, in this ways the Regime could ensure the complicity and involvement of its entire social and political base in the great repressive task. Loyalty and connivance from the victorious half of Spain, encouraged and programmed for their unwavering support of the Dictator and his punitive undertakings.

The principal townships of Cordoba province, where Franco concentrated the 1939-1940 repression, were as follows.[1] **Baena**, natives of this town and from Albendín, Valenzuela ad Luque. **Belalcázar**, only this town. **Belmez**, natives of this town and from Villanueva del Rei, Espiel and some from outside the province who

---

[1]   In **bold** the centre of imprisonment; in lower case, details of neighbouring townships and villages whose residents with either tried and/or imprisoned and/or executed here.

were miners and railroad workers. **Bujalance**, this town only. **Castro del Río**, in addition to natives of this town, prisoners from Baena, Espejo, Luque and some from Doña Mencía, Priego, Bujalance and Valenzuela. **Espiel**, overseen by the genocide Corporal *Pepinillo* during the 1941 slaughter; disaffected from Espiel, Pozoblanco, Villaralto and one from Peñarroya, were executed here. **Fuenteobejuna**, in addition to natives of neighbouring villages, some from Hinojosa and Villaviciosa. **Hinojosa**, location of a great many victims, in addition to natives of this town, many from Belalcázar, Santa Eufemia, El Viso, Villaralto and some from Fuente La Lancha. **Montilla**, natives from this and neighbouring villages, others from Fernán Núñez, Nueva Carteya and Doña Mencía. **Montoro**, in addition to natives of this and neighbouring towns, some from Azuel and Venta del Charco. **Peñarroya**, in addition to natives of that town, the final resting place of disaffected from a great many places: Villaralto, La Granjuela, Fuenteobejuna, Alcaracejos, Añora, Belmez, some from Hinojosa, Los Blázquez, Pozoblanco and a great many from outside the province (such as Badajoz), such was the influence of the mines and the railway on local employment. **Pozadas**, natives of this town and from Frente Palmera and Hornachuelos. **Pozoblanco**, a patchwork of local townships plus El Viso, Villaharta, Villaralto, Torrecampo, Villanueva del Duque, Dos Torres, Alaracejos, Añora and some from La Granjuela. **Puente Genil,** almost purely local. **Villafranca**, natives of the town and some from Adamuz and Villa del Río. **Villanueva de Córdoba**, in addition to natives of the town, there was a special concentration of prisoners from Pedroche, Torrecampo, Adamuz, El Horajo and some from Conquista and Villaralto.

The towns that were established as centres of imprisonment quickly became poles of attraction for two different groups of Spaniards. On the one land, Falangistas from small towns flocked there with the only purpose of wielding rods and beating prisoners from their towns. These were the emissaries of the torture. The scene repeated itself in every centre of imprisonment in the province.

The other group of travellers from small towns were the relatives of those who were imprisoned. A multitude of women, mothers, wives and sisters, plus old people and children, walked tirelessly carrying small bundles of food for their imprisoned relatives when they visited them. The country roads, especially in the Sierra, were a constant coming and going of people with bundles and sacks to help their kin. The fact of the matter was that the Regime was totally disorganized when it came to feeding its prisoners, so that any prisoners unable to get food from their families had an unbelievably bad time of it, often with fatal consequences. During 1939 and 940, sustenance of the prisoners was, in the most part, left to their close relatives.

The first great food catastrophe came at the end of September and beginning of October 1940 when the mass of prisoners was transferred out of the smaller provincial towns and concentrated in the capitals, well out of the reach of their families and the food that was vital to their survival. This led to that which is known as the «1941 Spanish Auschwitz», when the death rate in Francoist prisons rose to scandalous heights, something that has barely been studied. 1941 was also the year of uncountable mass executions by firing squad in the provincial capital cities and a year of the massive death toll in Francoist prisons for all reasons. It was not a 'year of modulation' for the repression as one uninformed individual wrote elsewhere.

TABLE V
Number of individuals executed by firing squad after the end of the war in the towns and villages of Córdoba province.
Data from the Civil Registry offices

Location	Number executed	Location	Number executed
Adamuz	3	Hornachuelos	8
Aguilar de la Frontera	1	Montalbán	1
Alcaracejos	1	Montilla	22
Almodóvar del Rio	6	Montoro	21
Baena	32	Palma del Río	40
Belalcázar	30	Pedro Abad	40
Belmez	36	PeñarroyaPueblonuevo	88
Bujalance	55	Posadas	19
Cañete de las Torres	3	Pozoblanco	209
Cardeña	5	Priego	2
Castro del Rio	167	Puente Genil	29
Conquista	1	Santa Eufemia	21
Dos Torres	6	Torrecampo	3
Espejo	6	Villafranca	8
Espiel	16	Villanueva de Córdoba	102
Fernán Nuñez	4	Villaralto	(8) 1
Fuenteobejuna	40	Villaviciosa	3
Hinojosa del Duque	67	El Viso	(8) 6
TOTAL post-war executions by firing squad: 1,102			

We need to keep in mind that the great post-war slaughter was part of Franco's repressive project from 1936 until 1950, at the very least, always subservient to the demands of the war. The executions were not a side effect of a collateral project, they formed a mission in its own right, a structural and strategic project that was characteristic of the Regime. Fifteen years of programmed bloodshed, with different modules, but always directed at the greatest efficiency. If there were times when there appeared to be some lessening of those crimes, it is because other repressive instruments were being implemented more intensely (hunger, social exclusion, humiliating practices, beating in the barracks, forced labour, economic repression, cleansings, death rate in the prisons, repression and brainwashing of children, imprisonment with banishment, etc.)

Do not forget that *Francoism was a land with two rivers: a river of blood for the dead and a river of tears for the living.* If sometimes the flow of blood lessened, the flow of tears increased. It is time to stop glossing over and publishing syrupy sweet interpretations of the events of these years, in respect for the historic truth and the victims. There is another adage: *Negation is inversely proportional to the work of historical research.* In other words, denial and ignorance of the facts go hand in hand. The field work necessary to investigate and document the events in the townships of Cordoba province goes to prove that adage.

Post-war executions in Cordoba towns and villages (Part I)

Respect for the historic truth and the victims.

In **Almodóvar del Río**, there is an unusual post-war case whereby the 6 prisoners who were shot November 12 1939, presumably so sentenced by a court martial, were not recorded in the Civil Registry. This information was obtained from an oral source. The six were leaders of the CNT trade union and some also belonged to the local War Committee.

Baena. Blood again flowed in 1939 in this martyred city of Cordoba in 1936. The Civil Registry records 32 victims in the immediate post-war period (and more in Castro del Río and the capital). The executions began at dawn June 22 1939. The bourgeois olive growers and landowners were in a hurry to carry out the killing that had already began terribly in 1936 with the execution of an additional 700 victims.[2] 32 more died in the post-war period (and another 10 in the capital). Since then, there has been not a hint of any kind of collective bargaining in Baena. The Regime got what it wanted.

June 22 1939, 13 post-war victims were executed. The leader of the group was the famous peasant leader José Joaquín Gómez Tienda – *El Transio*, of the CNT. Seven were natives of Baena, one from Albendín, another from Valenzela, another from Castillo de Locubín and three from Luque. They were all court martialled in Baena with great pomp and circumstance, one month earlier, 20 May. The court was presided by Guardia Civil Colonel Evaristo Peñalver. The Prosecutor was the flamboyant fanatic, José Ramón de La Lastra y Hoces, Marquis of Ugena and a grandson of the Duke of Hornachuelos. A slimy magistrate from Rute, Bernabé Andrés Pérez Jiménez also participated in the trial.

It was all very symbolic: the Guardia Civil, the old landholding aristocracy and the Francoist judiciary, against a handful of peasants, perhaps a couple of revolutionaries, with calloused hands, shirtless and wearing rope sandals. «Peace» had to be restored to the fields by chopping off a few heads. Rural fascism, without a shadow of a doubt. Another Guardia Civil commander also presided over court martials in Baena and Castro del Rei: Lt. Colonel Rafael Herrera Doblas. The Guardia Civil had been sent to bring order to Spanish fields.

José Joaquín Gómez Tienda – *El Transio*, was accused of participating in the killing of right-wingers in San Francisco Sanctuary on the night of July 28 1936, although it was proven that on that day, he was in Castro del Río trying to get reinforcements. He lived the last six months of his life fully aware that the victors wanted his head. At his trial – a great public event attended by all Baena's finest – when the Prosecutor read out the charges against him, he said: "My hands leave this world clean, whilst yours are dripping with blood, especially Don José Cubillo's." [ii] Joaquim was condemned to three

---

[2] Arcángel Bedmar has managed to describe 403 victims in Baena as follows: killed in the war 327 (officially recorded, only 206); post-war, 43; in prison and forced labour battalions, 17; in Nazi camps, 15; and guerrilla fighting, 1. *Baena, roja y negra. Guerra civil y represión (1936-1943)*. Juan de Maitena, Lucena, 2008, p. 274.

death sentences, by garrotte, to make certain that the penalty would not be commuted. The charges against his companions did not go beyond a simple arrest.[iii]

According to what Antonio Gómez, *El Transio*'s brother, told Moreno Gómez, Joaquim and his twelve companions left at dawn June 22 from the prison that had been opened on the Plaza de Francisco Valverde, in Baena.

> "The day of his execution, my mother found out beforehand from the lady who cleaned the jail. As the execution party had to walk down one street, my mother spent all night in vigil, next to the window. At dawn, my mother heard them approach and heard my brother say, as he walked past our house, *Salud, padres míos*. My mother stayed there until she heard the shot of mercy, then she broke down unconsolably. Before he died, Joaquim chastised his executioners and others present, including some rural policemen and the great landowner, Manuel Trujillo."

In Baena, executions continued throughout Summer 1939. There was a saca of 10 victims November 8. On the 11th of the same month, a young farmer, Manuel Cañete Tarifa, was killed during the transfer of prisoners, also according to Antonio Gómez Tienda.

> "Prisoners were being transferred in the town when the prison guard, an animal called Palomero, came in and began calling out names. One asked him whether he needed to take his backpack and he replied that he wouldn't need his backpack where he was going. This caused some concern among the prisoners who were being tied together, two by two. Soon, they heard a shot. One of the prisoners, so-called Cañete, who was tied to *El Mota*, had cut the rope with a shaving stick and ran away. Shot in the leg, they caught and finished him off in the Marbella arroyo.
>
> The prisoners who had remained inside refused to leave, believing that they would be killed in the

street. They said that if they were going to kill them, they might as well do it indoors. The prison guard explained what had happened, but they did not believe him. They then had to bring the prisoners who were already out in the street, back into the prison so that those who were inside could see for themselves that the others were still alive. The transfer of prisoners was then accomplished without further difficulties. The prisoners were marched down the street, forced to sing *Cara al Sol* on the way."

Although there was no violence in the Albedin district of Baena in 1936, Diogo Cantero Morales sent Moreno Gómez the following account of what he witnessed July 1 1939:

"...those who returned bit by bit from the Republican zone at the end of the war, were arrested and put into prison but as there were no serious accusations against them, the Falangistas invented the following ploy. They printed out several handwritten flyers with the following saying: 'Death to Franco! Death to the Guardia Civil! Log Live the Passionaria!' and other similar declarations. These were distributed July 1 1939. It was a clear provocation by the Falange, to provide it with an excuse to round up some 40 prisoners whom they tortured. The following August 14, they took five to Baena. Four were executed and one was saved."[iv]

The idea of printing false anti-Regime leaflets was a type of provocation Francoists frequently used in other towns and villages in the province. Summer 1940, another raid was carried out in Villanueva de Córdoba, the excuse being some posters saying: "Viva Negrín, bring him back to Spain" written by a venal municipal policeman nicknamed *Berenger*. Those who were caught in the raid were cruelly tortured and given severe sentences. The same ploy continued to be used in other townships in the Sierra in 1948, together with another provocation such as lighting firecrackers, as a pretext for the ruthless application of the Law of Escapes.

Prisoners from Baena, Albendín, Luque and Valenzuela were transferred February 1940 to the Castro del Río prison in the Convent of Nuns, which meant that executions would continue to be carried out in Castro. Later, September 1940, there was another transfer of prisoners but this time to Cordoba capital where they were executed. The last execution by firing squad in Baena was April 11 1940, Rafael Herenas Espartero, a native of Albendín. His was a strange, theatrical judgement (one of his cousins had denounced him), in which he was accused of having shot at Guardia Civiles in Cartagena, other guards in Valencia and against right-wingers in El Grao (Castellón), and for having participated in the attacks against the cathedral in Jaén. In other words, Rafael was supposed to committed crimes all over Spain, all of which he denied during his trial in Cordoba, October 1939. A witness interviewed by Arcángel Bedmar in 2008, told that Herena's confession to the Baena Guardia Civil crime was obtained under torture (all post-war declarations, i.e.., confessions, were obtained under torture) and he had been made to kneel on a board with spikes.[v]

Belalcázar remained a Republican zone to the bitter end. A town of unionists and very active country people during the Republic, in 1936 it immediately took up arms and fiercely fought the Nationalists. The Republican militia last went to battle August 13 and 14, killing 170 Nationalists, the largest number in the entire province. They were to pay dearly for their bravery as during the post-war, the victors entered the town like starving dogs. A division of the Spanish Legión under General Salvador Bañuls Navarro was quartered in the town in 1940 with the purpose of sowing terror in the region, which they did to the full as they spread out throughout the Los Pedroches district (Belalcázar, Santa Eufemia, Pozoblanco, Villanueva de Córdoba and Cardeña). Such was the terror, that all the guerrillas in the region, led by *El Francés*, retreated in a long walk towards Cáceres, which they entered during the first days of December 1940 through Alma, Cañamero and Guadalupe.

The deaths of 30 residents executed during the immediate post-war period (as well as others in Hinojosa and in the capital) are recorded in the Civil Registry. Executions by firing squad in Belalcázar began early 16 May, because of the work of Military Court No. 4. Death sentences fell heavily on inmates in the local prison in the Divina Pastora School, and the executions followed immediately.

In addition to these executions, there were several sacas June 4 and 20. Because of this, the prisoners planned a great escape August 4 1939. Fifteen condemned to death escaped, of which 3 died in the attempt: Benedicto Cabanillas, José Paredes - *El Boche* and Alejandro Gómez - *El Chefe*. The remainder escaped into the hills where they played major roles as guerrillas (*Palomo, Huevero, El Portugués, Quivicán, El Fiscal, El de la Carmela*) as did others who fled a year later from the Hinojosa del Duque prison: *El León, Bellota* and others.

Unusually, two women were executed in Belalcázar: Carmen Rubio Cáceres and Matilde Medina Pizarro, August 6 and 30 1939, taken to their death separately from the men, as was customary in post-war Spain. An example of this was the Thirteen Roses tragedy in Madrid. These young women left Ventas prison hoping they would re-join their companions so they could die together, but when they arrived at Este cemetery, the men had already been shot.

Francisco Mesa Paredes - *Mesilla* died November 1. He had served as a Lieutenant in the Militia and assistant to Aldo Morandi, an Italian International Brigade Commander. They fought on the Peñarroya front and were quartered with the 63rd Division in Villanueva de Córdoba Fall and Winter 1937. *Mesililla* was accused of presiding over the local War Committee.

From Belalcázar, prisoners were sent to Hinojosa, where the firing squads continued to operate, and then later to Cordoba capital. An eyewitness told Moreno Gómez of some natives of Belalcázar who were executed in Hinojosa, apparently at the end of 1939, although their names are not recorded in any Civil Registry. This was the case of young Luís Prat Blanco, local Secretary of the PCE, who had lost a leg from gangrene, having been wounded on the Madrid front. His niece Guadalupe, kindly allowed Moreno Gómez to examine the two letters of farewell he wrote to his family – one to his sisters and the other to his parents.

"To my dear sisters, nephews and brothers-in-law:
It is with the greatest pain and with tears in my eyes, that I dedicate these words to you so that you may remember me. Sisters, at the last moments, as I am about to be executed, my heart beats with the

memory of such dear sisters of mine alone, as you have had no other brother.

You who have always been my greatest joy and hopes, you who have been the mirror of my soul, you who were everything to me, my dreams and my joy, who on this tragic day, fate will take me from you. I ask you to be always good, that you look after the girl of my dreams (Guadalupe), that you forgive everyone as I forgive them, that you educate my little niece well, to be truthful and when she grows up, that you tell her that she had and she still has, albeit underground, an uncle who loves her with all his heart and soul, that she was my only joy.

Sisters, live in peace knowing that your brother was never bad; you know that I always was, as far as I could be, an honest worker. Farewell dear sisters and brothers-in-law, forever. Farewell with all my soul, farewell until eternity, your brother, Luís Prat Branco, 32 years old."

"To my dear parents, from prison:

With pain in my heart and sadness for myself, in the last moments of my life, I dedicate this letter to you, so that you always remember with fondness, the advice from your son Luís.

Dear parents, bad luck now that my life is ending, despite the misfortune of my leg, the best time of my life. Father, keep in mind what you [*sic* both]] must do in the years left to you, resign yourselves to the calamity and do all the good that you can for everyone, most especially for my sisters and nephews. Be truthful and good, forgive everyone as I forgive them in my last hours. I hope that you will live for many years, considering that I in my last and best years of my life am being separated forever from you.

Although it pains me, I ask that you do not go into a corner with pity for me, that you try to dry up as much of the pain as you can, in the hope that you may live life as well as possible, because I am going to rest forever, and my sole and only concern is for you, for how much you will suffer for me. Be certain that your son always believed that he was as good to you as you always were to me. I beg you to forgive me for anything I might have done to offend you, as I myself I leave without ever having been offended.

Farewell dear parents, farewell forever and for eternity. Farewell with all my soul. Your son, Luís Prat Blanco, 32 years old, Belalcázar, 1939."

Belmez. The Civil Registry records 36 executed by firing squad during the immediate post-war period (some more in the capital). Executions by firing squad occurred rather late in this village of the Peñarroya mining district. Only one in 1939; the remainder in 1940, with one exception. 16 of the victims were taken to the Hinojosa cemetery to be shot and the largest saca of the lot was 25 May with 15 inmates. A few days before they also took a woman to be shot there, Leonor Expósito Palomo (one of the post-war '21 Roses of Cordoba'. In Belmez, the victims were not farm workers, but miners, railway employees, production workers or craftsmen.

Bujalance. One of the great fiefs of Cordoban anarchism since the beginning of the 20th century, undoubtedly why the local notary, Juan Díaz del Moral, became an icon of social historiography with his book *Historia de las agitaciones compesinos andaluzas*[3], a unique book of its kind. This notary's celebrity led to his being elected Deputy to the 1931 Constitutional Courts where, as a member of the Group at the Service of the Republic, he became a great Agricultural Reform expert.

The eight first post-war executed in Bujalance fell June 5 1939. On of these, Pedro García Cano - *El Cojo*, was a member of the UGT and President of the local War Committee in 1936. Another, Pedro Martínez Ortiz, was a member of the same union and committee. The remainder of the victims belonged to the immensely popular CNT. Executions took place at wide intervals throughout 1939. One of the Haro Manzano brothers, noted local union activist, Manuel, was shot November 18. Two other brothers, Luís and Francisco, were

---

[3] *History of the rural uprisings in Andalusia.*

executed several months later. December 7, a historic member of the CNT, Ildefonso Coca Chocero - *El Viejo* aged 65, was shot.

In Bujalance, Francoism specifically targeted the energetic conscience of the Bujalance peasants, parts of the Regime's program for cleansing and mercilessly subjugating the Spanish working class. That is why the military coup was planned and what fascism was invented for: to abort, preventively the alleged revolution of the proletariat. "There is a phantom over Europe" and it was on the altar of that phantom that Mussolini, Hitler and Franco officiated and immolated their opponents.

Because of the great repression, Bujalance was one of the Cordoba townships that sent many men (twenty) to the mountains to join the guerrillas commanded by the famous *Los Jubiles* freedom fighters which included men from Belalcázar, Hinojosa, Villanueva de Córdoba, Adamuz and other towns.

To the 21 executions in Bujalance in 1939, we must add another 28 in 1940, in small, frequently individual, sacas. Four men were executed March 30 1940, among whom another famous local anarchist, Francisco García Cabello - *El Niño del Aceite*, aged 56 years, whose name was made popular by the press following the successful worker uprisings in Bujalance December 1933. Also fell with him, the young Luís Haro Manzano. *Niño Aceite's* brother Manuel was executed 20 May. The last executions in Bujalance, before the new wave of terror in 1941, took place September 12 1940. Beginning October 1940, the Regime began to concentrate prisoners from the town prisons in the Cordoba Provincial Prison. The great bloodshed that followed this was what one would expect, as were the ensuing tragic consequences of those deaths.

Castro del Río. Perhaps the first fief of Cordoban anarchism, this town was subjected to one of the bloodiest punishment in post-war Cordoba province. Cradle of Benito Cordobés, one of the old master lay preachers who published the *La Idea Libre* magazine, Castro del Río had earned the unusual fame of having defeated Nationalist General Varela August 7 1936. Such 'audacity' so infuriated General Queipo de Llano that he exploded with typical gross verbosity and publicly threatened the town, telling it to "Begin preparing graves". So, it happened.

In truth, it was not only natives of the town who were killed here, but also others from Baena and Espejo before everyone was sent to the capital to be slaughtered. Between 1939 and 1940, 167 victims ended up in the mass graves of the Castro del Río cemetery. A usual feature was the time set for the executions, not all at dawn but more often in the afternoon, especially in 1939 and 940, both in Castro as in other neighbouring towns (Pozoblanco, Villanueva de Córdoba and others). Moreno Gómez presumes that the *enterado* may have arrived in the morning and the authorities wanted to get the business over as soon as possible. This, aside from the fact that this enabled the Francoists to take advantage of teaching the townspeople a lesson, spreading fear as the executions were carried out in the daytime, while they put on a free show for the Falangistas.

Thus, the first fatal shots rang out at 3 p.m. and at 4 p.m., the first on June 19 with 11 victims. Earlier, 11 May, Felipe Aguilera committed suicide in the latrines as he could no longer bear the torture. August 3, Francisco Recio Roja, a member of the 1936 War Committee and José Criado – *Taraje*, another local peasant leader, were both executed. The sacas were more numerous as of November 1939 and included some farm workers from Espejo. Six were executed November 19, among them Antonio Márques Bello – *El Chino*, age 22 years, a fan of flamenco who, it is said, attempted to raise his companions' spirits on the road to the cemetery by singing the tragic verses: *Madre, coge tu pañuelo / y vete para la Audiencia / y dile al señor fiscal / que te lea la sentencia, que a mí me van a matar.*[4]

Ten were executed November 24, including young José Cañasveras Villatoro, who fulfilled his wish of being married *in artículo mortis* in his cell, before he died. The last execution of the year, seven victims, was December 5. The leader of this group was the eminent Unión Republicana attorney, Manuel Castro Merino, aged 53 years, a typical example of the so-called attorneys of the poor. He had defended workers at trials with mixed juries and even donated land that he owned, for social housing. He had cultivated a comfortable living as a member of the typical Republican bourgeoisie, with its strict customs and considerable knowledge of altruistic ideals. A graduate

---

[4] "Mother, gather your kerchief / and go to the Court / and ask the Prosecutor / to read you the sentence, / because they are going to kill me."

of the Seminary in Cordoba, he studied Law in Granada, practicing in Granada, in Castro del Río and in other courts. He was active in Madrid and Cordoban circles. A friend of Fernando de los Ríos, he was associated with the *Institución de Libre Enseñanza* and as President of the Cordoba Alliance of Farmers, was an advisor to small farmers. Castro Merino participated actively in the Republican build-up of 1931. He ran for office in 1933 and soon joined the Unión Republicana party, from which he organized the Castro del Río Frente Popular.

Because of his age, Manuel Castro Merino only carried out administrative tasks with the 24th Mobile Brigade and wrote newspaper articles before and during the war. Returning to his hometown in May 1939, he was immediately arrested and badly beaten during the interrogations. Once again, the rancid right-wing rural population, Señoritos and Catholic fascists showed the least respect either for the poor or for the illustrious individuals who had placed their knowledge and expertise at the service of the underprivileged. Even worse, to have done so was considered an aggravating factor and these men, both attorneys of the poor and doctors of the poor such as Cayetano Bolivar, who was also assassinated, were considered traitors to their class.

All these victims were accused, as almost everyone was, of belonging to a so-called Revolutionary Committee, which in reality was the Frente Popular democratic political movement. Especially accused of belonging to this committee were Antonio Elias Herencia, Alfonso Nieves Núñez, José Díaz Criado, Juan Gómez Gutiérrez, Manuel Castro Merino, Pedro Calvo García, Francisco Recio Rojano and José Porcel Rivas, although it has proved impossible to confirm these accusations. Besides, as a rule the Regime labelled everyone they wanted to exterminate as belonging to that Committee. Manuel Castro, because of his intellectual importance and his philanthropy, had been sentenced beforehand.[vi]

Like Goebbels, Spanish fascists despised culture, books and intellectuals with a radical and visceral hatred. It was that fury that deprived Spain of the brilliant García Lorca.

Two farmhands from Espejo were shot at the same time as the lawyer of the poor: Luís Cordoba Jiménez and Francisco Cordoba Lucena, regarding whose lives we have no information. As least we were able to save their names from oblivion. All the executed ended up in the same large mass grave that had been dug in a side path in the Castro del Río cemetery. That same morning of December 5, Francisco Torronterras García, aged 37 years, appeared dead in the middle of a pool of blood, we do not know whether he was in the chapel or not. He had taken his own life during the night by cutting his wrists with a razor.

The number and size of the sacas increased in 1940, and there were also more in Baena, Espejo and elsewhere. Ten victims were executed March 18. One of the largest sacas of eighteen victims, April 10, included Diego Prado Bracero, a young farmer who also was married in prison before he died (it is not known whether this was of his own free will or because he was pressured to do so by the priest, a frequent occurrence). 13 May, one of the four executed was Alfonso Criado Garrido, son of the union leader *Taraje* who had been shot several months earlier. 18 May, the group of seven included the man who had been the soul of the organization of the local militias in 1936, Rafael Moreno Herencia – *Maruca*, who commanded a Cavalry Squadron in the Andalusia-Extremadura Unit. He died aged 66 years, blinded by a war wound and condemned to sevenl death sentences. Also shot at the same time, Alfonso Camargo Ortega whose elder brother José had been executed a month earlier.

28 May, another thirteen victims. June 4, they killed José Porcel – *El Sastre* and José Sánchez Alcántara. It is said that the latter left for the cemetery barely able to walk and that *El Sastre* had to hold him up. Andrés López Luque – *El Colorin*, a 25-year-old farmer who had prepared his escape but was caught, was waiting several days for the right time to escape, but he was caught. Another version says that he gave himself up because he was starving. All the inmates were forced to walk past his body as it lay in the patio. This was the typical barracks discipline that Franco had determined would apply to all of Spain.

June 29, Eugenio Rodríguez – *El Pavero*, from Baena, who had served as Captain with the 88th Mobile Brigade and was wounded on the Pozoblanco front, was executed with two other prisoners of war.[vii] August 28, there was another suicide in the Castro del Río prison: José Sánchez, aged 30 years, threw himself down a well in the prison patio. August 31, another famous anarchist leader, Manuel Mármol Algaba – *Loreto*, was executed. September 7, 11 executed by firing squad.

The last saca in post-war Castro del Río was September 12, again with 18 victims. The slaughter then moved on to the capital.[5]

Espejo. The repression in Espejo was especially handled in Castro del Río and later in Cordoba capital. Despite everything, the repressive tyrants wanted the people of Espejo to hear the bursts of shots from the machine guns, as a reminder of who was in control. August 3 1939, they shot five men in the cemetery. The settling of accounts then continued in Castro del Río and in the capital. Nonetheless, Francisco Jiménez García, aged 40, who had served as Mayor for the Frente Popular and was being kept as a spoil of war in the municipal jail, was sacrificed in the town. He was barbarously tortured, and we do not know whether he attempted to kill himself or whether he 'was suicided' as the cause of death entry in the Civil Registry is most peculiar: «suffocated due to incomplete hanging». Just what did that mean? The exact circumstances of his death are unknown, but there is no doubt that he was a victim of torture. This was February 27 1940.

Espejo, like Castro del Río and Bujalance had anarchist origins during the so-called Bolshevik triennium (1918-1920), but little by little, the townspeople turned towards the Communist Party, particularly around 1930 following the preaching of Adriano Romero, a political activist from Villanueva de Córdoba who served as PCE Deputy for Pontevedra in 1936. The great readership of the *La Idea Libre* magazine in this town had something to do with a charismatic rationalist teacher, Clodoaldo García (like Benito Cordobés in Castro del Río and Montemayor). Don Clodoaldo, already an old man, was saved the firing squad as he was condemned to thirty years. Not Benito Cordobés of whom we only know that he was exterminated with a

---

[5] Moreno Gómez owes some of the details of the events in Castro del Río to the pioneering work of a grassroots diarist, Francisco Merino Cañasveras – *Castro del Río, del rojo al negro*, 2nd edition 1989. On the other hand, a booklet published by the University of Cordoba by Francisco López Villatoro, *Cambios políticos y sociales en Castro del Río 1923-1979*, Cordoba, 1999, has a fancy title but it says nothing of social changes. The great social landmark in Castro del Río, summer of 1936, is dispatched by the author with only a few lines. As to his 'objectivity', he only speaks of right-wing victims and not a single word about leftists who also fell. A useless and inane piece of work. Merino Cañasveras' pioneering study, by a truck driver who emigrated, has given us all lessons as to how to approach the social aspect of the townships. Moreno Gómez often met with Merino Cañasveras in Castro del Río.

batch of prisoners in front of Castro, August 7 1936, on the orders of General Varela, one of the New State's war criminals.[6]

August 4 1936, Robert Capa and Gerda Taro arrived in Espejo, where he took his iconic photograph *Death of a militiaman* in the outskirts of Espejo, not Cerro Muriano as has been reported. It has also been said that Capa followed the road of anarchism, which is equally incorrect as Espejo was a PCE stronghold during the entire Republic. Likewise, there is no basis to associate Capa with the anarchists. When August 5 Capa was in Cerro Muriano, this town was not on any anarchist route, and when in May and June they accompanied the International Brigade in La Granjuela, there was even less anarchism there.

Fuenteobejuna was another of the martyrized cities of Cordoba as it had already lost some 500 residents in 1936. Franco's victory would unleash the final solution on its townspeople who were socialists and active UGT trade unionists with a longstanding tradition of workers and sharecroppers. In reality, Fuenteobejuna never ceased being punished throughout the entire war. Post-war, executions continued to be carried out in 1938, when there were at least 20 victims, the last one 18 December.

As the bells of victory rang out, the townspeople eagerly awaited the return of their relatives, whether civil employees who had been evacuated or returning soldiers, all who had been away for three years because of the war. The Municipal jail was immediately filled to burst with captured Rojos, one of whom, Agustín León, had been Mayor for the Frente Popular. Courts martial were quickly organized and the first to fall before the firing squad was Obdulio Romero, June 2, whose file Moreno Gómez was unable to consult. Three more fell August 4, including Felisario Cidoncha who did not even make it to the cemetery as he was shot at the church door. It is said that he was totally broken by the barbarous torture he had suffered. He was accused that in 1936 someone had seen him during the saca or on the truck taking Fuenteobejuna right-wingers to La Granja and Azuaga (Badajoz), when 57 were highjacked and killed by anarchist militia from Alanís (Seville), it was supposed in collaboration with

---

6   General Varela's bloodshed reached its height during the march from Toledo to Madrid October 1936, when he ordered the execution of more than 4,000 Republican prisoners of war.

some from Fuenteobejuna. Anyhow, the determination of guilt is a vain exercise because at the courts martial the prisoners always paid as the accusations were considered "fair because they were sinners", whilst those above them and responsible almost always found a way to get out of the way.

The serious business of killing began in Fall 1939. October 14, Santiago Rodríguez, a 55-year-old worker, was shot. His cellmate, who survived, Ildefonso Sedano, gave his eyewitness report: [viii]

> "I remember a night that I was walking with Santiago Rodríguez. A Guardia Civil entered the patio and asked him if he was Santiago Rodríguez. He replied that he was, and the guard continued: 'There is no point in beating around the bush. I have come to get you to shoot you.' And he was taken away. This companion belonged to the CNT from La Cardenchosa."

October 24 at 8 p.m., there was a saca of four inmates, one of whom Tomás Gallardo Habas, from a family that had been sorely punished by Francoism, particularly during the persecution of escapees. One of his sons, Tomás Gallardo Medina, escaped to the hills during the post-war and was never heard of again. The Habas family were from La Cardenchosa, and they suffered many lost in the guerrillas (brothers Nemesio, Eugenio and Dionisio Habas Rodriguez, the first two killed in 1947 and Dionisio in 1951, in Seville). The Gallardo family also lost relatives among the people from the hills through the Law of Escapes during the triennium of terror. In the same saca, Socialist Alejandro Cuadrado, second Assistant Mayor during the last Republican municipality.

October 30, five executed, among which brothers José and Aurélio Romero Agredando. For the rest of the year, the executions were held at 8 p.m. or 9 p.m. The custom of executing prisoners in the afternoon, at the usual time for the bullfights, seemed to be the usual practice in 1939 and during the first quarter of 1940, in most Cordoba province towns. There is no apparent explanation for the abandonment of dawn as the time for the executions.

A heavy dose of killing was planned for November 6, with six victims, led by aforementioned Agustín Sánchez, last Frente Popular Mayor in Fuenteobejuna, who had attempted to live incognito in a village in Almeria at the end of the war. He was, however, identified four months afterwards and taken to Fuenteobejuna and incarcerated there until he ended up in the cemetery. The officer in charge was a Lieutenant Flores. Several of Agustin's friends, accused of belonging to the War Committee in 1936, were executed at the same time: Teófilo Mateos Rivera, Antonio Murillo, Francisco Zurita and Rafael Gómez Ríos.

A survivor, Ángel Horrillo from the village of Ojuelos Altos, describes the tragic moments that were lived then in the Municipal prison:[ix]

> "When I was returned to my village in May, even before I entered it, I was surrounded by armed men and taken to the Falange. I could easily deny the first accusations, because when they met in their committee where they used to determine the accusations, they came up with such a number of false accusations, with the sole purpose of eliminating me. I was tried by summary court martial with another nine men, and there were ten death sentences.
>
> There were some 80 of us in jail during 1939 and the beginning of 1940. From the patio, we could see the arrival of those who came to sign the list of those who would be executed that night. In this terrible situation, we found ourselves at 8 or 9 at night and, arranged in circles, the jailor – *Don Manuel* he was called – made us sing *Cara al Sol*, once or twice. He then gave us five seconds for all to go through the 70-centimetre door, but before doing so, hit us left, right and centre. A little later, speaking very slowly, he would read out the list and after each name, sarcastically say: 'To the chop'.
>
> Four of my group were shot on four different nights: Antonio Ruíz, Luís Romero, Antonio Múñez and Juan Pedro Hidalgo, the last of whom was

my uncle. He and his wife were kept in a separate room, but of the same lot of prisoners, and when his name was called, she let out a terrible cry. I found it spectacular as I could not control either my heart or my nerves."

The Civil Registry recorded 40 Fuenteobejuna victims during this first stage of the repression, 1939-1940. The prisoners were transferred to Peñarroya later, where the shootings continued until later, in Cordoba capital, where all the prisoners in the province were gathered at the beginning of Fall 1941.

Hinojosa del Duque was another of the Sierra townships that were greatly punished. Its Civil Registry records 66 victims, not just from Hinojosa, but also from Belalcázer and some from Santa Eufemia, El Viso and Villarelto. In Fall 1940, all prisoners left there were transferred with the rest from the province, to Cordoba capital.

April 19 1939, the Law of Escapes was applied to three individuals at the La Gutierra farm: Sebastián Martínez, Prudencio García Gómez (a retired soldier for whom there are no details) and 18-year-old Pablo Gómez Leal, brother of the famous *Vidal* (three Gómez brothers were executed: Pablo, Antonio and José).

June 22 1939 there was a burst of shootings with a saca of four led by the man who had been Mayor during the war, Eloy Pizarro – *El Barón*. Together with his companions, they were accused of belonging to the 1936 War Committee, rightly or wrongly (Fermín Muñoz, Ramón Nava and Isidro Barbancho). September 1 1939 there was a three-man saca, which included Luís Ramírez – *El Montillano*, a wealthy individual of a certain social standing. This man, who owned property, who had money in the bank and even a typewriter at home, was killed so they could rob him. He belonged to an interclass ideology and created a so-called Committee Pro Peace in 1936, to try and get arbitration and agreement between contending parties. This work alone and the wish to steal from him, led him to his grave.

August 14 a single woman was taken to be shot, Carmen Aranda Caballero, who left 6 orphan children. This family's history is terrible. Her husband, Lázaro Leal Martínez – *El Perdigón*, fled to the hills where he died on an unknown date. One of their sons, Francisco Leal Aranda also fled to the hills where he died in 1949, between Belmez and Hinojosa. There were 30 executions in Hinojosa up to the end of 1939, plus two who were not recorded in the Civil Registry (Luís Prats Blanco, local secretary of the PCE and Alfonso Vélez, both from Belalcázar). The largest saca of the war was 7 victims on November 1.

There were executions in Hinojosa throughout 1940, but rather irregularly. The March 8 saca of 5 victims included Pedro Rubio Cáceres, a 28-year-old baker from Belalcázar, whose sister Carmen had been shot the year before in their hometown. May 8, they killed Dionisio Blázquez Forgas, accused of having belonged to the War Committee, as was Alfonso Arellano Muñoz, shot May 25. The great number of condemned to death can be explained by the plan for a great escape from the Concepcionista Convent-Prison, at dawn between 31 August 31 and September 1, the last market day. Master bricklayer, Lázaro Leal Martínez - *El Perdigón*, whose wife Carmen Aranda was shot the previous August 14, made a hole in the wall that led to the church vestry and to the convent patio. Some twenty men ran out through that hole in the middle of the night, chased by the guards through the streets of the town. 15 men managed to make it to the open fields, and they became famous in the guerrilla fighting (Pedro Díaz Monje – *El Francés* (who commanded the guerrilla in Cáceres), Manuel Hidalgo Medina – *Bellota*, Lázaro Leal – *El Perdigón* and his son Francisco, Demetrio Morales – *Cuatete*, Francisco Vigara Mesa – *El León* (from Belalcázar), Francisco Corchado Silveira – *Lazarete* (from El Viso), among others. Of those who died when escaping through the streets, four are recorded in the Civil Registry at Hinojosa (Antonio Ramos and Miguel Jiménez of Hinojosa; Pedro Molero and Rafael Herrero, of Belalcázar). Another two from Belalcázar are not recorded: Manuel Paredes and Francisco Cáceres Calderón. There were many such great fugitives from prisons all over Spain, such was the great despair caused by the tremendous repression that was launched, beyond anything that all the defeated could ever have imagined. The last person shot in Hinojosa, before the general transfer of prisoners to Cordoba, was Manuel García Blanco, September 18 1940, a native of Villaralto. He was accused by the Causa General as an «eminent revolutionary» in this village, which does not appear to be quite correct as he was elderly.

Lastly, 1941 the Hinojosa Civil Registry records two peculiar unidentified deaths due to 'hanging' or 'suicide' in the Guardia Civil barracks (an odd place to commit suicide), which is no more than a euphemism for death due to torture, under the aegis of the very hard 29 March 1941 Law of Safety of the State that led to a great rise in Francoist terror that year in a dozen townships. That same year, more than 500 inmates died in Cordoba city prison.

Montilla was sorely punished, without explanation, because the only thing residents of its extensive Republican neighbourhood, more Socialist than Communist, did was to flee the town in 1936. Already in this year of the military coup, Francoists eliminated at least 200 persons.[7] Despite this genocidal punishment, the Regime looked for another punishment in the post-war, shooting 15 residents and 8 out-of-towners (from Fernán Nuñez, Castro del Río, Nueva Carteya and Doña Mencia) in the Montilla cemetery.

The slaughter in post-war Montilla began 7 November 1939, with a saca of 7 victims, headed by the local leader Antonio Córdoba Gálvez and both Taper Ruíz brothers – Antonio and Francisco. Some of Montilla's most eminent Republicans fell in the 1940 sacas. 16 May, with some out-of-towners, a noted representative of Montillano antifascism, Juan Cordoba Zafra, who reached the rank of Militia Commander in 1938, a member of the board of the *Casa del Pueblo* community centre, secretary of the Socialist Youth Movement, alderman for the Frente Popular and, following the coup, organizer of a military unit in Bujalance, of which he was the Captain, later integrated in the Jaén Militias and in the 92nd Mobile Brigade. In 1939, he attempted to reach safety in the port of Alicante, at the head of his unit, but was captured there and brought to Montilla[x], to the great delight of the local landowners. His great qualifications as a politician and as a Republican soldier were responsible for his death.

Another heavyweight from Montilla fell with the May 25 saca, Manuel García Espejo – *Chicuelo*, aged 30 years, member of the board of the Casa del Pueblo, secretary of the Young Socialists in 1934, and Militia Captain, founder of the Montilla Company Summer 1936, integrated in the Garcés Battalion and in the 73rd Mobile Brigade,

---

[7]   This figure was estimated by Moreno Gómez in his book *1936: El genocidio franquista en Cordoba*, Crítica, Barcelona, 2008, p. 282. Arcángel Bedmar also arrived at the same value in his book *Los puños y las pistolas,* on the repression in Montilla, published by the Montilla Municipality in 2009, p. 204. Both authors, however, were only able to identify 116 of the dead.

under the command of Antonio Ortiz from Espejo. They fought for the Republic to the bitter end and there could be no other end to this undertaking than the firing squad. *Chicuelo*, a brilliant and good-looking man, went to his death May 25 1940 in his native Montilla, the *Munda* for which Caesar and Pompey fought two millennia earlier. He was accompanied at the end by a young local Communist, José de la Torre Requena, whose appeal for mercy arrived a few days later.

The Montilla and neighbouring prison inmates were transferred to the Cordoba Provincial Prison between October 17 and 19 1940. The executions continued in Cordoba, and it is there the Frente Popular Mayor of Montilla, Manuel Sánchez Ruíz – *El Perla*, and National Vice President of the FNTT, died. Yet another a great Montilla personality sacrificed by a mad, reactionary and tridentate Spain as already done in 1936 to another historic leader of Cordoban socialism from Montilla: Francisco Zafra Contreras, member of Parliament and first Republican Mayor, one of the founders of the FNTT, who the troops under Sáenz de Buruaga dragged to Baena where he was murdered by pistol shots in the public square July 28 1936.

The only ones left in Montilla were elderly and the senile individuals represented by the Carlist group of the Count of La Cortina Francisco de Alvear, president of the National Confederation of Catholic Farmers. All very Catholic, such as archpriest Luís Fernández Casado, responsible for 'saving the souls' of those murdered, but not their bodies.

Montoro. In addition to those executed later in the capital, 11 were executed in Montoro in 1939 and 10 in 1940. There were executions on Christmas Day 1936 when the Nationalists occupied the town, although there is no mention of them in the local Civil Registry. 9 of the executed belonged to the International Brigade and several were public servants. Moreno Gómez estimated the number of victims in 1936 at 40.[xi] When the post-war arrived, the Francoists of Montoro received the returnees with loaded rifles. This was a continuation of the same repressive program that lasted fifteen years (936-1950), at its harshest and with various modifications conceived to ensure the greatest effectiveness and application.

The slaughter began with the application of the 29 April 1939 Law of Escapes, against Miguel Cañuelo Lozano, a baker. November 28 was the first formal execution, with 10 victims:one from Azuel (a village in which nothing had happened in 1936), four others from the village of Venta del Charo (Cardeña) whom the Causa General accused of two deaths in 1936, on what basis is unknown: José Antonio Romero, Falangist, and Bartolomé Coleto, landowner from Villanueva de Córdoba.

From Montoro, Fernando Ruíz Zorro and Francisco Cepas, accused of attacking a prison full of right-wingers with axes, weapons and dynamite, July 22 1936, causing 50 victims, a popular uprising in turn motivated by another mass killing that the Francoists had just committed in neighbouring Pedro Abad. The accusation of the attack on the Montoro prison was generally applied to everyone they wanted to eliminate, both in the countryside and in the capital. It is now impossible to determine which of these accusations were well based and which were not. Whatever, Francoists systematically eliminated everyone with any standing in the labour movement, trade unions, Republican army (first of all, officers and non-commissioned officers) and Republican authorities (Members of Parliament, Mayors, etc., without exception), whether there had been right-wing victims in a town or not. If there is any doubt, look at the killings in provinces such as Navarra, Zamora, Galicia, Castilla-León, Canarias, Ceuta, Melilla, and so forth.

The 1940 executions in Montoro were mainly carried out in September. They began the 4th with the shooting of a local Communist: Juan Barbado Zamora, bricklayer, aged 52 years, father of Francisco Barbado, national PCE leader, commander of the 5th Regiment and who played an important role in the restoration of Democracy in Andalusia. As he told Moreno Gómez in1983:

> "My father was a Communist and they accused him of having participated in the famous attack on the prison in 1936. He was imprisoned in Montoro for a year and the infamous Sergeant Arenas, of such sad memories in the town, destroyed him with beatings. Arenas would have breakfast at the Casino with the Señoritos and when he finished said: 'OK. I am now off to give them their breakfast' and the beatings would begin in the jail. On another occasion, the Señoritos in the Casino, referring to my father, told him: "Either you

kill him, or we will". He was shot. He had become blind from the torture he was given."

Young Francisco Caballero Majuelos fell September 7. He had been accused by the Causa General of killing a retired soldier, Pedro Arroyo, July 1936. Again, it was impossible to confirm the veracity of this. September 11 there was a larger saca of 6 victims, also accused of the assault on the jail in 1936. Finally, September 20 they executed a woman, Patrocinio Purificación Juárez Pareja, aged 39 years, regarding whom Moreno Gómez was unable to get any information.

Palma del Río, another town which suffered terribly in 1936, with several hundred assassinated in Don Félix's ranch at the hands of this landowner and breeder of fighting bulls. When Moreno Gómez fist visited Palma del Río around 1980, he went to look at the tragic yard, which was very rundown and overgrown. Today, it has been renovated and fenced in, to keep the curious away. This is without a doubt, one of the Historic Memory Locations in Andalusia.

During the post-war, there were 32 victims in 1939, 8 in 1940, and more in the capital. The executions began November 7[8] and right away, November 1939 was most tragic, both for Cordoba province and for all Spain. On this occasion, in Palma del Río, 16 workers and farmers were taken in a saca (most of them anarchists and members of the JSU). November 16, another saca at 2 a.m. with 15 victims, among which the Díaz León brothers – Julio and José, workers.

March 29 1940 there was a peculiar execution of Gumersindo Santiago Páez, a 59-year-old worker, member of the *Luz y Prosperidad* Masonic lodge that existed in Palma del Río from 1913 to 1936, of which he was a founding member and in which he held important positions. It appears that the 'crime of being a Mason' was what led to his death, or perhaps because the Palma del Río Francoists were unable to capture the Grand Master of that lodge, Antonio España Ocaña, father of the young member of the JSU, José España Algarrada, who organized the militia from this town.

Beginning Spring, the executions in Palma were held at 8 p.m. March 30, 6 more victims, most of them workers. The last execution

---

[8] November 7 is an interesting date as it was chosen in many Cordoba townships for the first executions, because it was the anniversary of the Russian Revolution and the defence of Madrid.

in Palma was May 15 at 11 p.m. Francisco Jiménez Ordóñez, a 24-year-old chauffeur. The remaining natives of Palma inmates killed were executed in Cordoba capital.

Pedro Abad, on the banks of the Guadalquivir, a small farming town (pop. 4,143 in 1926), suffered an equally merciless repression. There is nothing more enlightening than reading local writings to understand what Francoism was, particularly during the first stage of its development (1936-1943) when the Regime itself classified itself as totalitarian and adopted the Nazi-type salute that even Bishops displayed with remarkable dexterity.

The infernal microhistory of how the townspeople suffered in Pedro Abad (as in countless unknown towns and villages) we owe to a local essay by the Adán Gaitán brothers: Félix and Juan Manuel.[xii] The following sections that are now reproduced, are truly astounding. The torture, the humiliations, the suicides, aggressions of all kinds, even cases of lynching by the «victorious Catholics», were the order of the day.

According to these authors, 150 were condemned to various sentences by the courts martial in Pedro Abad, of which one third, 150, were executed.

Implementation of the extermination program began early during the war, July 22 1936, when a column of Nationalist fighters from Cordoba (artillery soldiers, Falangistas and some peasants) appeared that morning in the town with a view to carrying out the typical 'punishment sortie', as they called it then. They arrived, ready to kill with a vengeance. As they got off their trucks, the first people they saw were several farmworkers who were working a vegetable patch, the Teviño Nieto brothers, whom they shot and left where they fell. They did the same with another young man who was tending his goats, Salvador Aguiar. As they arrived at the first houses, they saw and shot Nicolás Alonso who was cleaning something; they did the same to José Escobar – *El Gorrilla*, who was selling sunflower seeds on the streets. Seeing what was going on, Francisco Arena, a 50-year-old, tried to save himself by climbing up a tree. They saw him and shot him down. His sister Ana (mother of *Gato Negro*), finding out about the death of her brother, killed herself by throwing herself down a well.

They saw a car on the street being driven by a young disabled man and ordered him to Stop! At that point, Juan Lora Escobar, the village schoolmaster, a man of order who had nothing to do with anything political, called out from the door to his house where he had gone to see what was happening. "Leave him alone; can't you see he is mentally disabled?" A local Falangista turned around and, without a word, shot the schoolmaster in the head and left him for dead. That is how the «Saviours of Spain» behaved.

(A similar mad crime was committed about the same dates in Aguilar de la Frontera. A poor farmer was leading his donkey laden with water casks, when down the road he met up with a truck full of young Fascists, *Señoritos* from Aguilar. For fun, they shot and killed the poor man, shouting 'For God and Spain', as they drove off roaring with laughter. Their targets were farmworkers, shepherds, shirtless and poor people wearing rope sandals. This is how these people enjoyed themselves. Pure and hard fascism, very much like the Italian blackshirts during the 1920s, as depicted by Bertolucci in the first part of his movie *Nine hundred*.)

The tragic incursion of the Fascists from Cordoba in Pedro Abad left bitter memories. They went from house to house asking, 'Where is the owner of the house?', randomly killing townspeople right left and centre. One of those who opened his door was Pedro Parilla, a 70-year-old, whom they shot dead in his entrance hall. This was the same tactic of entering and leaving the houses we saw in Villafranca (when the Moorish troops were leading the fighting), in Baena and elsewhere.[xiii] That was 1936 in the villages in Andalusia.

The consequence of the schizophrenic incursions in Pedro Abad, was the hasty retreat of the Nationalist militia who, as they left the town, shot into the windows of houses as they went, killing 12 and wounding 17 civilians, in protest against the arrest of right-wingers in Casa Olaya. To these, we add another 5 or 6 (including a 17-year-old) dead the Fascists had already killed on the streets, plus several more townspeople who, before the Fascists returned to Cordoba, they took with them as they left town, killed in the outskirts, then burnt their bodies with gasoline. That morning ended with another 16 murdered, plus another 11 whom the Fascists took with them to Cordoba and later killed. Moreno Gómez was able to confirm at least 30 wartime victims in Pedro Abad.

There were 39 executions in Pedro Abad during the post-war period, plus an additional 14 natives of the town executed in the capital and 8 who died from starvation in the Provincial Prison, a total 61.

Seven of those executed during the post-war period fell November 2. The Military Judge was Francisco Iglesias. The firing squad consisted of an officer and soldiers from the 3rd Falange Brigade of Cordoba. Taken in this saca: Antonio Arenas – *Graniaino*, who had been a Socialist Mayor of the town and Francisco García Lara who was killed because he was on guard at the Casa Olaya when right-wing prisoners were shot. In reality, they accused everyone in the town of this crime, indiscriminately. Juan Pulido Canales – *El Zorro* was accused of one death, something that he forcefully denied during his trial; Juan Ruíz López, a 45-year-old tradesman, father of 7 children, only acknowledged at his trial that he had taken prisoners to Casa Olaya but that he did nothing else; Rafael Jurado Castillan, a member of the Libertarian Youth, was also accused of belonging to the Self-Defence Group, as were almost everybody, of the attack on the barracks and the Casa Olaya episode; José Fernández of the CNT was accused of being a member of the War Committee and other crimes including that 'he bore arms in 1938; lastly, Juan Gaitán Valderramas – *Maura*, a tavern owner belonging to the Izquierda Republicana, accused of nothing specifically except that during the war he was living in Villafrance de Cordoba, where he was Treasurer of the Committee for Refugees. From the onset, the Dark World of the accusations under Francoist 'justice displays its clearly apparent nature as a sovereign, totally untrustworthy farce.

November 5 (the Adán brothers say it was the 6th), there was a three-man saca in Pedro Abad: Alfonso Carcelé Galán, a Socialist, who after seeing the cadavers of the priest Antonio Pérez and the retired Guardia Civil Manuel Ortega who had been executed earlier, thought that he should take them to be buried. This act of compassion was turned against him, he was arrested and accused of having taken parts of the statue of Christ when statues of the saints were damaged in the church; Juan Aguilar Cailla who was accused of presumably having belonged to a Self-Defence Group (earlier, Moreno Gómez described how his brother Salvador, who had a pronounced limp, was shot July 22 1936 when he was watching the goats outside the town limits); lastly, Matías Prieto Castilla, father of five, also accused of belonging to a Self-Defence Group and of being a Communist.

November 8 1939, a 4-man saca including Antonio Rojas Arenas, accused of having shot against his cousin, the priest Alfonso Canales.

They said he handed him to the War Committee of Villa del Río, who ordered that he should be taken for a walk, or *paseo*. It is impossible to determine the authenticity of all of this. Regarding Fernando Triviño – *El Gordito*, there was no accusation, unless that he was a member of the Farming Committee and that he was put in charge of the right-wing women who were made to work in the fields and that he belonged to the UGT. There were no accusations against Juan Martín Carcelé Rojas. There was a surprising circumstance indicative of the revengeful post-war rage and the Fascists ability to invest wicked acts, in the case of Juan's brothers who played their drums in processions and were forced to be present when their brother was executed and made to play the drums. There was no specific accusation against Rodrigo Arenas Durán, other than that he belonged to the UGT and to a Self-Defence Group.

November 20 a single individual was executed: Rafael García Cambronero - *El del Centro*, who had been a Municipal Judge. As a member of the War Committee, he was a negotiator in the surrender of the barracks and did not advocate any excessive intervention, just attempted to put some order into the middle of all the chaos. Before he was killed, he was so brutally tortured and beaten so badly that they had to take him urgently to hospital in Cordoba to heal his wounds until he was well enough to be executed. He was visited by María Muñoz, his wife, in hospital and he told her: "I will soon be taken back to Pedro Abad so they can finish their work". He was 52 years old and the father of Rafael García Contreras, who had been a PCE Senator under the Republic. The latter writes, in his book of memoires:

> "What they did in the barracks to my father was unbelievable. Beating after beating, he was tortured until he had a large wound in his back... On one of the 'walks' from the barracks to the prison, my aunt carried me in her arms, so that I could see my father... He was dripping blood all over him."[xiv]

November 24 1939 was the last execution of the year, with the death of Alfonso Valcarteras Arenas who was repeatedly accused of belonging to a Self-Defence Group and of having shot against two right-wingers. Frequently, the accused signed confessions through torture, which is why we cannot always trust what is written in their dossiers.

Executions in Pedro Abad recommenced March 7 1940, this time against a single person, Andrés Lara Gaitlán, aged 49. He had a long experience as an alderman and was famous as a speaker in meetings. He belonged to the War Committee, and he participated in arrests. When the town was evacuated at the end of December 1936, he moved to Villanueva de Córdoba where he worked with the Committee for Refugees.

April 5, three men were executed: Domingo Ruíz López, a member of the CNT who was killed because he knocked the image of the Sacred Heart of Jesus that was attached to the façade of the City Hall building, to the ground. When these men were being taken away in the truck, on the way to the cemetery, they drove past the prison that housed their companions. Domingo called out to them to keep their spirits up, while the soldiers kicked him in the butt. His brother-in-law, Nieto, the Mayor, was executed some days later; Juan Cerdá Moreno was accused of almost the same crimes as all the others, i.e., of belonging to a Self-Defence Group and of having been a guard at the Casa Olaya. Two more of his brothers (Miguel and Melchor) were later also executed and another, Bartolomé, condemned to a long prison sentence. Regarding Rafael Arjona Hernández, there were no accusations of importance, only that he was a Communist and that he bore arms against the «Movement to Save Spain».

This brings us to the great May 11 1940 scandal in Pedro Abad, the so-called Casa Barcos Case, the description of which is a bit confused in the Adán Gaitán brothers' book (pp. 524 et al.). There was some kind of stickup May 5 by several individuals from Morente, apparently members of the *Los Jubiles* guerrillas. Pistols in hands, they entered the house of the wealthy Fernando Carezo Barcos and accidently a gun went off and mortally wounded 14-year-old Francisco Carezo. Six townspeople from Pedro Abad were arrested as charged by Manuel Albenden Ribas, Captain of the Guardia Civil from Montoro. Charges were urgently filed against all six, although it was not clear that they were all accomplices to the robbery as they were accused of carrying (non-existent) pistols. Among those arrested was the presumed head of the gang, Juan Gallardo, from Morente, who was beaten to death May 9. That same day, another prisoner, Jerónimo Grando – *Jeromo*, 19 years old, from Pedro Abad, unable to bear the torture, committed suicide by throwing himself down a well.⁹ The four remaining prisoners were executed on the 11th.

The Emergency Summary Court was set up in two days and it ended with the statement: "The execution of the condemned will take place at 7 a.m. tomorrow", which was incorrect. The execution of the four presumed accomplices to the robbery, which was not at all proven, took place in full daylight, at noon, and it was a public execution, much more like a lynching.

The Adán Gaitán author say that the four prisoners were taken from the jail at 12:30 a.m. 11 May, to be executed in the cemetery. The Francoist population prepared itself for a great party. The prisoners were taken from the Old City Hall building with great excitement from the townspeople, while calls were made to invite all the neighbours into the street, including the school children, to watch the prisoners march down the street to the cemetery outside the town.

The executioners cut rods as they reached the outskirts, with which they poked and beat the condemned men, pulled them by their hair and their moustaches. The poor victims begged: *Kill us once and for all. Stop this!* No sooner said then done, right there, before they even got to the cemetery, they were placed before a public firing squad where earlier, the Francoists had brought two of Alfonso López Salinas' young brothers to watch the shooting. Their bodies were left where they fell for 24 hours, for the general public to jeer and laugh at.

Juan Martinez' mother and fiancée, who were from Morente, came to see him. A soldier told the victim's mother that she would be allowed to give him a kiss, on the condition that she did not cry. If she cried, she would be arrested. The two women somehow held back their tears and complied with the ritual, but when they moved away amongst the olive trees, one could hear their cries echo throughout the Cordoban countryside. Only one of the detained was temporarily saved; although condemned to twenty years in prison, Bartolomé Regalón died of starvation in Cordoba Provincial Prison, February 25 1941.

The bloody events in Pedro Abad were not over, not in the very least. 16 May, five days after the terrible Casa Barcos episode, there remained a last coup, a saca of 6 presumed prisoners who were to

---

⁹ The Grande family was another massacred by Francoism: two Jerónimos brothers were executed in 1936 and his father was imprisoned.

be shot that day at dawn in the cemetery, an event the proved to be even more tragic than usual. The prisoners were put in the middle of a truck, with soldiers on either side – a corporal and a sergeant. However, two of the *morituri* who were tied to each other, had a desperate plan in mind: José García Yeste – *El Porcelano*, 23-year-old anarchist, and Francisco Cambronero Rodríguez – *El Fraile*, aged 41 years. At a given moment on their way, having managed to free themselves from each other, they jumped off the truck, one on either side. The soldiers, in the middle of the following confusion, started shooting wildly and one of them, shot corporal José Alonso in the head. As to the fugitives, *El Fraile* was caught on a street, and he died there. *El Porcelano* managed to reach the Guadalquivir River, which he swam across but was unable to move from the other bank as he was paralyzed with cold because of the freezing water. Half of the soldiers found him there (the other half remained guarding the prisoners on the truck). The unfortunate *Porcelano* was beaten, kicked in the butt and punched in an authentic *via crucis*, then taken to a place called Santo Cristo, where they finished him off in a true lynching, kicking and hitting him with machetes; a horrendous death.

The remaining four prisoners were executed by firing squad in the cemetery. Rafael Mejias Rivera, a 30-year-old bricklayer, was accused of nothing special other than that he had arrested right-wingers, which he denied, had belonged to the UGT, had attended leftist meetings and had belonged to a Self-Defence Group. José Gallego Olanda, a Communist, who had belonged to a Self-Defence Group was discovered to have killed his cousin Bartolomé Olanda and gone to the cemetery to bury him. This discovery sent off a furious persecution by his relatives on his mother's side, who did not rest until they had killed him. As regards Francisco Arenas, it is only known that his brother Rodrigo had also been executed the previous November.

The Civil Registry for May 24 1940 records a death due to Pedro Alcaraz' 'collapse' in prison, difficult to clear up. The Adán Gaitán authors mention someone nicknamed *Paradito* who died of a heart attack when he heard his name called for execution.

Another past Republican Mayor, Francisco Nieto Romero, a Socialist and a very moderate person, was executed May 24. A member of the War Committee, he was savagely tortured – they pulled his teeth out with pliers, tore off his moustache and beat him so much that he could not take off his shirt because it was stuck to his bloody body. His companion in this saca, Antonio Castilla Canalejo, had belonged to the UGT and was an active recruiter for the guerrillas in neighbouring villages.

June 4 they only shot Francisco Luque Arenas, of the CNT, accused of belonging to a Self-Defence Group and of signing sentences as a member of the War Committee, which was false, because he belonged to the Committee in November 1936 when there was no longer any violence against right-wingers. There were so many accusations, it was impossible to determine which were true and which were false. Besides, as the Francoist plan was total cleansing, it did not matter that they used all means to this end.

The last men were executed in Pedro Abad June 13, with three victims: Diego Arenas García, 50 years old and father of 7 children, was accused of seizures and guarding. He had belonged to the War Committee and to make matters worse, was a Communist, everything necessary to ensure that he would be exterminated; Miguel Cerdá Moreno was another longstanding Communist, member of the *Nueva Aurora* Society who, during the war, had served as Provincial Delegate for Agricultural Reform in Ciudad Real. There was little foundation to his accusations. It was his longstanding militancy that led him to the cemetery wall, just like two of his brothers – Juan and Melchor –mentioned earlier., yet another example of Francoism exterminating entire families without second thoughts; Pedro Gomáriz Castilla, also a Communist, was accused of the usual: belonging to the War Committee and a Self-Defence Group. Although it is believed that War Committees everywhere usually consisted of only five or six people, the Francoists used this as an accusation right, left and centre, the most certain accusation for doing away with somebody.

Later, the local Pedro Abad Francoists had prepared a last major coup before all the prisoners were transferred to Cordoba capital. This was the execution of a woman, Josefa Ortega Igea, aged 37 years, a widow with five children. When at the end of 1936 Pedro Abad was evacuated, she got separated from her husband, Manuel Abajo, whom she never again saw because he died in a bombing. When she returned to the village, she had to face the accusations of her neighbour across the street, a fanatic right-winger, who never stopped until she got Josefa killed, which she did 3 October 1940, after waiting for her to give birth to a child. Josefa Ortega was the typical woman who does not shut up before anything or anybody. One could say that her crime was a crime of the tongue and nothing else. Already under arrest, her accuser, no doubt a 'person of good standing' in Francoist eyes, arranged for her to get one beating after another, even though she was pregnant, whilst they called out: "Hurry up and give birth because we have to kill you". One of the accusations against her was as stupid as the following "She was noted for her Communist ideas, and she took part in public acts and in spreading revolutionary propaganda. During the Rojo regime, she led groups of women who persecuted people belonging to the Right". These and other absurd accusations were accepted by the Pedro Abad Military Judge, Francisco Iglesias Sánchez, and she ended up before the firing squad.

This last judicial aberration did not mark the end of the repression in Pedro Abad, just the end of firing squad executions in the town, as the second half of the repression was still to come in the Cordoba cemeteries. Fortunately, Moreno Gómez was able to consult the microhistory of this village written by the Adán Gaitán brothers, which helped him descend into the inferno of the Francoist repression in the countryside. Everywhere in Spain, it was the same thing, an inhuman and unbearable punishment, but there are very few detailed studies of this. Thus, when investigations and research are oversimplified, the sweeping profiles of the repression become thinner and get lost, giving food to present-day negationists who, in generalizing everything, have found their perfect habitat.

Post-war executions in Cordoba towns and villages (Part II)

Major targets of the repression in the Cordoba mountains.

Peñarroya-Pueblonuevo, the virtual capital of a large mining basin in the northern part of Cordoba province, served as a Nationalist Headquarters during the war. In the early days of the Republic, Peñarroya stood out as an important Socialist stronghold although with time, its members gradually became more subdued, as opposed to Puente Genil and Palma del Río, where the JSU was increasingly more forceful and active.

With the outbreak of war, the Peñarroya Socialists, led by the moderate Eduardo Blanco Fernández, Member of Parliament for the Frente Popular, formed a Militia Battalion known as the *Batallón del Terrible*, but they were unable to launch a defence of Cordoba down the Pedroches road. The miners from Peñarroya were not sufficiently powerful to defend their district which was captured by the Nationalists October 13 1936. All future attempts to recover their homeland were unfruitful and cost them rivers of blood throughout the war. Apparently the Peñarroya miners may have been less combative than others from Linares, La Carolina or Puertollano who were also more active from a syndicalist viewpoint. The Terrible Battalion eventually merged with the Garcés Battalion.

When the time came for the Nationalists to celebrate the «Victory of Revenge», the Francoists filled Peñarroya with prisoners from all over North Cordoba and many other places in Spain. If you look at the list of victims, you see that few were natives of Peñarroya-Pueblonuevo. Most those murdered by Francoist 'justice' beginning in 1939 were prisoners of war – soldiers and militia – who had been imprisoned in Peñarroya or had been sent there with the declaration of victory from the neighbouring concentration camps of Valsequillo, La Granjuela and Los Blázquez, or just before, following the Republic's last battle January 1939.

The first sentenced to death by Courts Martial 1, 2 and 3 in Peñarroya, was shot by firing squad June 20 at the unusual hour of 9 p.m. Hermenegildo Estévez Ferrer, a soldier from Valencia, was shot August 2. Emilio Soligo Molino, a sergeant from Lerida, was shot September 5. Little or nothing is known of all those who fell, soldiers in the service of the Army of the Republic, the «Distinguished Army».

These executions were soon followed by those of prisoners from villages in the Los Pedroches district (Villaralto, Pedroche, La Granjuela, Los Blásquez) wo had been imprisoned in Peñarroya. Florentino Cubero Aranda from La Granjuela, whose brother Rafael *El Manco de La Granjuela* became a famous guerrilla fighter in the region, was executed November 8. December 11, two men and one woman, Martina Alcántara, aged 53, were shot. A few days later, according to the Judge Advocate, another woman, Dionisia Alcántara Calvo was said to have 'collapsed and died', under suspicious circumstances.

During this first stage of the post-war repression in the provincial towns and villages – the 1939 and 1940 biennial – 88 residents of

Peñarroya were executed in a first bloody tribute to the victorious Nationalists.

Julio Perea Peña, whose brother Felipe had been a leading Socialist in the mining region before the Republic and an alderman in 1931, was executed January 1 1940. Another active Republican, Fermín Pradera Nieto, was shot February 27 at the same time as the Mayor of Los Blázquez, Honorio Esquinas Benavente, a peaceful man who had sought to prevent any kind of violence in his town. They were followed by another active Republican, Alfonso Rodríguez Imbernón, from Alcoy, April 8. There was a large saca of 12 victims March 16, most of them from Alcaracejos, such as brothers Nereo and José Mansilla Moreno. A third brother, Juan de Dios, was executed elsewhere. All the prisoners from Alcaracejos were accused, en bloc, by the Causa General, in a totally bigoted indictment of minimal historiographic value.

October 30 1939 marked a saca of six prisoners in Peñarroya-Pueblonuevo, one of whom was Gabriel González Godoy, famous Commander of the Stalin Battalion and of the 25th Mixed Brigade (MB). A miner from Linares, he was the local secretary of the UGT. When war broke out, he was charged with organizing the Militia from Linares, with the rank of sergeant. Later, in Baeza, he created the famous Stalin Battalion that became part of the 25th MB as the 99th Battalion. He held the rank of Captain on the Pozoblanco front March 1937, with the 25th MB and his invaluable Battalion, as he led the Republican attack against the Nationalist troops in Villanueva del Duque. He was promoted to Commander[10] in May and spent the rest of the war on the North Cordoba frontlines.

Like many other Republican officers, Gabriel González was taken prisoner in Peñarroya-Pueblonuevo. His trial was based on a mountain of accusations sent to the Court by telegram from the Linares Guardia Civil, describing him as of 'bad past behaviour and having participated in numerous outrages'. He submitted appeal after appeal disputing these accusations, to no avail as a militia Commander could not expect clemency from a Francoist court His death sentence, dated July 6 1939, was approved by the Judge Advocate for the Army of the South in Seville on August 8 and his dossier with Franco's *enterado* arrived in

---

[10] The rank of Commander in the Spanish Army is equivalent to the rank of Major in the British or American army.

Peñarroya on the very day of the saca. The Court met at 5 p.m. to read out the death sentences to the sentenced men, after which they were taken to the chapel where they remained for a very short time as the execution was set for 8p.m. According to the little his family knows, Gabriel's body ended up in the mass grave in Pueblonuevo cemetery. He wrote no letter of farewell, or if he did his family never received it.[xv]

Some political leaders and ex-Mayors managed to evade the death penalty as they fought for survival through the great labyrinth of death. The ex-Mayor of Peñarroya, Fernando Cartión, also Republican Governor for a short time in August 1936, was sentenced to death but some benevolent hand saved him. The same occurred with the Socialist ex-Member of Parliament, Eduardo Blanco Fernández, who had also served in the past as Civil Governor during the Republic. He was arrested in Ciudad Real and in April 1940 was brought to the improvised Peñarroya prison that had been created in the old UGT Trade Union Building built by the miners. He appeared before a court martial in Cordoba June 11 1943 and was sentenced to *only* thirty years in jail, because "despite his extremely leftist ideas, he always appeared to have protected persons belonging to the right". His salvation, like that of others, was that he was tried four years after the end of the war. Had he been tried in 1939, at the height of the Francoist thirst for vengeance, he would never have escaped the firing squad. Others like him, who had protected right-wing individuals, were less fortunate: Antonio Baena from Pozoblanco was sentenced to death by garrotte; Joaquim Pérez Salas, a well-known community benefactor, was executed in Murcia, as were the aforementioned ex-Mayor of Los Blázquez and so many others.

April 1940, both Peñarroya prisons (Municipal and Trade Union Building) received inmates from the entire Fuenteobejuna region. From there, they executed Manuel Caballero, accused of belonging to the War Committee. As mentioned earlier, the Causa General regularly accused 15 to 20 persons of belonging to this Committee to ensure that nobody escaped its grasp, regardless that these local Committees never consisted of more than four, five or at the most, six individuals.

The post-war slaughter was just beginning. When Constanza de la Mora (wife of Hidalgo de Cisneros) wrote the finale to her book of memoires in New York, July 1939, she concluded with these words:

"Franco has murdered thousands of Spaniards. As I write these words, the firing squads continue to kill men and women... Hundreds of thousands more are living lives of continuous torture and humiliation in Franco's prisons and concentration camps."[xvi]

Pozoblanco was another Cordoba district martyred by Francoism. The thirst for blood here was one of the worst in the entire province, as we shall also see in Villanueva de Córdoba, which had been the regional Republican capital during the war. Both townships were symbols of resistance to the coup on the frontlines in the North of the province. Pozoblanco was particularly vulnerable to Nationalist violence because in early Summer of 1936, the local right-wing and all the Guardia Civiles from a dozen towns in the Los Pedroches district gathered in the town to stage an armed uprising against the Republic. The Republican Government had to send part of the Miaja Column to take control of the town. The insurgents surrendered August 15 with the understanding that no militia would enter the town, only regular army, and that the Guardia Civiles and right-wingers would be evacuated to Valencia. That same day, two trains filled with the Pozoblanco insurgents left for Valencia. They were interned in the harbour on the steamship *Legazpi*. The People's Court began working in September, whilst Largo Caballero's government refused to accept any commutation of the sentences. Consequently, more than 300 Pozoblanco insurgents, civilians and Guardia Civiles were executed in Paterra cemetery. To make matters worse, when the militia later entered the town August 15-16, they took another 50 right-wingers. It was to be expected that the Pozoblanco right wing would wait until the time was right for them to take their revenge. That opportunity came with Franco's victory.

Even though the residents of Pozoblanco had little or nothing to do with the People's Court in Valencia, nor with the activities of local militia groups. Be it as it may, there was no excuse for the imminent bloodbath that could be expected with the victory celebrations. In the previous chapter, Moreno Gómez described the explosion of burning terror that fell upon the town during April and May 1939, in the form of summary executions and the Law of Escapes here and in neighbouring towns and villages where at least 101 victims were 'taken for walks', the infamous *paseos*. This phenomenon of immediate terror, devoid of all legality, was applied to the entire recently conquered Centre-South of the country, a terrible reality barely addressed in all the studies of the war.

Now follows Moreno Gómez' study of the application of Francoist 'military justice' properly speaking, in Pozoblanco, at the hands of Courts Martial 9 and 13, whose sentences began to be executed June 22. That day, they only shot Alfonso de la Cruz (María *La Posadera*'s son).

This marked the first frequent use of the euphemisms entered as the cause of death in official death certificates to cover up the real reason for so many 'disappearances': traumatic shock, internal haemorrhage, etc. The Judge Advocate sent the death certificates to the Municipal Judge who then sent them to the Civil Registry office where they were recorded.

When examining the death certificates, researchers have to be especially astute in determining which expressed causes of death are actually factual or simply euphemistic, especially when it is known that the prisoners were tortured considering the widespread use of torture and its frequently fatal outcome.

The following certificate from the Pozoblanco Civil Registry, for example, leads on to suspect death under torture:

> "War Judge Advocate – Military Court No. 13 – Pozoblanco.
>
> Please record the death of JUAN ÁLVREZ POZO, resident of Villanueva del Duque, born in Hinojosa... age 37 years, married to Valeriana Rubio, with two children, who died yesterday from cardiac asystole, as certified by the Coroner. Pozoblanco, 15 October 1939."

In this case, that the prisoner, a healthy young man, died from torture, is very plausible. However, unless the historian speaks to relatives of the deceased or to witnesses, he cannot confirm his suspicions.

The courts martial in Pozoblanco began May 26 1939, according to Bartolomé *Pérez Salas* Cabrera Peralbo[xvii], one of the accused. The first trial of eight accused resulted in twenty death sentences. Only

Miguel Arroyo García, secretary of the Socialist Party and secretary of the Workers Agricultural Cooperative, escaped execution; the remainder were shot. The eight accused tried at the second trial May 27, were sentenced to 18 death sentences.

Only Francisco Rodríguez Arroyo and Bartolomé Pérez Salas were saved at the second trial. Leaving the Court, Mario Cabrera Amor looked straight at him as he told the Public Defender of the terrible situations of some of the men who were being tried. The Defender replied: 'Of this lot the only one who will be saved is the one they call *Pérez Salas!*' And so, it was. All the others were shot.

Executions in Pozoblanco continued to be held in full daylight at the end of the day, around teatime, and some Señoritos went to the cemetery to watch. Almost nobody escaped the death sentence in the 1939 trials. Franco's approval, the *enterado*, arrived within a few days of the trial, by teletype, almost by return mail.

During the first two years of the repression (1939 and 1940), there were 209 victims in Pozoblanco (not including guerrillas and *paseados*) the highest number post-war in the province alone, followed by Castro del Rio (167) and Villanueva de Córdoba (102). None of these and other data include the number of residents of the provincial towns and villages who later died in Cordoba capital.

Six prisoners were executed August 5 1939, among whom Luís Romero Cortés, from Torrecampo, whose dossier Moreno Gómez was able to obtain. He was accused of the usual crimes: he was a Communist, had participated in the attack on the town, belonged to the War Committee, was a political leader in the Republican zone and had arrested members of the right wing. Despite the lack of an accusation of a blood crime, he was sentenced to death June 15.

One of the executions that had a major impact on the residents of Pozoblanco occurred October 28 when at 7 p.m., Tomasa Díaz Moreno was taken out to be shot. Aged only 21 years and in the flower of youth, she was a Los Pedroches 'Rose' and a leader in the local JSU, just like the Madrid Roses[11]. Poor Tomasa was taken to her death both physically and morally crushed by the National Catholic

---

[11]   *Treze Rosas*, or thirteen roses, was the name given to thirteen very young women, seven who were still teenagers, members of the left-wing JSU, who were arrested in Madrid at the end of the civil war, imprisoned, tortured and shot. One of the most tragic early examples of Francoist repression of the defeated.

victors. She was tortured, they shaved her head, she was forced to drink castor oil and to walk up and down the town streets with a signing hanging from her neck.

When the Catholics decided to kill, they did so in a refined manner, with the torture and fires they learned to perfection during the Inquisition and many other momentous occasions throughout History. Tomasa died full of emotional pain as her father, Antonio Díaz Jurado, had been executed the day before. Her fiancé, Gaspar Jiménez Cebrián was executed earlier, April 11. In Tomasa's case, the victorious tripartite (barracks casino and church vestry) treated her family with special cruelty. Why? For the sole reason that they owned a newspaper kiosque on the main street where her family sold copies of the workers trade union newspaper. This was a case of killing the messengers.

Cesáreo Romero, a local Socialist leader and Mayor of Torrecampo for the Frente Popular, was executed November 3 in Pozoblanco. It so happened that he was out of town July 25 when the militia entered Torrecampo and he was only able to return many weeks later. Nevertheless, the records of his trial that Moreno Gómez consulted show that he was sentenced to death on clearly futile charges. His wife went to Pozoblanco to visit him the day he was shot, without knowing that this was about to happen, but as there had been a saca, she was not allowed to enter the prison. As the truck with the sentenced drove away, she recognized her husband among them. She ran to the cemetery. When she got there, the firing squad had done its job; still she was allowed to go to his body and wipe the blood from his face She was refused his body and they threw him into the mass grave with the others.

Socialist Cesáreo's death is important in that it shows how Franco did not shoot his victims as punishment for their crimes, but because of their ideas. The beliefs that were at the base of the proletarian *IDEA* philosophy that representatives of the working classes preached the length and breadth of the Andalusian countryside during the 19th and 20th centuries. The IDEA that Franco's rural and Catholic Fascists put before the firing squads.

As he wrote this, Moreno Gómez was reminded of an occasion from his childhood. One day, he asked his mother why someone in the town had been killed. "Because he had ideas", she replied. In retrospect, this was an unforgettable answer from an apolitical country woman, who was able to summarize, without knowing it, the essence of Fascism: *to eliminate all proletarian and syndicalist IDEAs*, by means of a campaign of enraged destruction that was nothing more than preventive violence against the ghost of a 'revolution', a theoretical nightmare that in the 1930s haunted the Right more than the Left.

More words of wisdom from his mother: "Never sign up for anything". Another magnificent motto: never sign up, never get involved, in other words, do not become an activist. She knew that those who were executed since 1936 were those whose names were on Lists. That was Francoism: elimination of all those who were committed to an ideal, those who stood out, who formed the social and political base of the Republic. His mother never told him that people were being killed because of their crimes; only because *they had ideas*. As a child, he never paid much attention to her sayings, but today, he finds that they were brilliant words from a simple woman who never belonged to a trade union, never held a job, never worked for any public department. The fact is that ordinary people are prone to giving magnificent lessons that put fancy-pants scholars who promote violence to shame.

The elimination of all Republicans with positions of authority of any kind at every level of society, was the Eleventh Francoist Commandment. Acisclo *El Fraile* García Dueñas, an ex-seminarian Secondary School Latin teacher, was executed November 21 1939. Nothing more political is known of him other than that he served as the Health Officer for the 7th Mixed Brigade. Francisco Díaz Pastor, ex-alderman for the Frente Popular and Manuel Fernández Contreras, member of the Izquierda Republicana, were also taken in that saca.

The last one of these was involved in a most unfortunate incident as it was he who helped the notorious Juan Calero Rubio one of the most bloodthirsty Francoist judges in post-war Los Pedroches, to escape imprisonment by the Republicans before the war. Judge Calero acknowledged Manuel's help but he 'could not forgive him for not having also helped his son-in-law to escape'. At the end of the war, Manuel Fernández who had refused to go into exile because he considered that he was innocent of any blood crime, was interned in the Valsequillo concentration camp, followed by an additional seven months in Pozoblanco prison. At his trial, several individuals falsely alleged that he organized crimes and looting, and he was sentenced to death. His wife begged Judge Calero to be merciful in recognition for past favours, to no avail. Manuel was executed the next day.

We must not forget the thousands who died from torture, those who 'were left in our hands' (Francoist slang for prisoners), a terrible iceberg of which only the tip is known. These were such as Juan Álvarez Pozo, from Hinojosa but resident near Villanueva del Duque, mentioned earlier, a healthy 37-year-od who died October 14 1939 from 'cardiac asystole', according to the Coroner. Another one whose heart failed him, David Cuello Amadeo, from Alcaracejos, died from 'acute endocarditis' December 17 1939, according to the Judge Advocate. Others who 'were left in their hands' include Gervasio Martínez Hidalgo, who died from 'epileptic seizures' September 1939, according to the note from the Military Police. November 13 1939, Rafael Bueno Roldán, a 53-year-old physician, was allowed to die in prison from 'septicaemia', considered a suicide apparently because he refused medication for a wound on his knee. One of those executed November 29, Severo García Gonzalbo, arrived before the firing squad on his knees, unable to stand because of the beatings and torture he had been subjected to. How many more cases like these were there?

Antonio Márquez Jurado, ex-Socialist Mayor of Pozoblanco during the last part of the war and who had absolutely nothing at all to do with the fighting, was executed November 29. His political sympathies, as was the case with so many others, led to his death.

The great sacas in Pozoblanco took place during 1940, especially in May. The executed came from this and neighbouring towns: Añora, Dos Torres, Alcaracejos and El Viso, Aril 11 there was a particularly tragic execution, that of the Head Postmaster, Enrique Ramirez Dópido. It appears that his great misfortune was that the Pozoblanco Falangistas hated his father-in-law, Antonio Varo Granados, with a passion. Varo, a Public Prosecutor from Aguilar de la Frontera, was in Madrid when war broke out. Even so, he was accused by omission, because 'instead of coming back to the town to prevent excesses, he remained in Madrid'. This, of course, was a farce, but Francoist 'justice' was like that. Varo was also accused of being 'a prominent member of the Left and effective propagandist of its ideas', also absurd but because of all this, he was considered 'the supreme responsible for all the crimes committed in Pozoblanco'. The rage directed against this man was uncontrollable. He was sentenced to death by garrotte but, fortunately for Varo, a Nationalist Captain that Varo had hidden in his house in Madrid, spoke in his defence (which was most unusual) and his sentence was commuted.

Not so his unfortunate son-in-law, Enrique Ramírez Dópido, who was executed April 11 1940, on the false grounds that he was the link between those who besieged the town, when during the whole time that Pozoblanco was under siege, Enrique was imprisoned in the barracks and was only released after the town had surrendered. To add to both these families' miseries, there were ulterior motives to their arrests, namely that father and son-in-law both owned a flour factory in Fuentes de Andalucía that was seized by Falangists. Enrique Dópido also owned a valuable market garden at the outskirts of the town that was expropriated by the local leader of the Falange.

There is another interesting case where a prisoner was sentenced to death although his sentence was later commuted, José Madueño, a veteran Socialist, who, when war began, travelled to Elche, to Santa Pula and then to Madrid. Nevertheless, when Moreno Gómez was able to consult his dossier, he noticed that José was charged with: 'by leaving town he did not avoid excesses and crimes' and also was 'considered to be conceited'. Such a list of absurdities was used to send Republicans to their deaths. Madueño escaped on this occasion, but very few others did.

As mentioned earlier, almost nobody accused of belonging to the local War Committee escaped. For this reason, May 11 they executed one of the few leading Communists in Pozoblanco, Bautista Herruzo de la Cruz, ex-Captain of the Pedroches Battalion, ex-alderman for the Frente Popular. The calvary suffered by this man is still spoken of in Pozoblanco because he was savagely tortured and almost beaten to death. He was accused of having sent information to the People's Court in Valencia.

Another member of the Pozoblanco War Committee, Samuel Romero Estrella, was executed May 28 1940 and another, Miguel Justo *El Policial* Sánchez Garrido, June 12. It is said that he was despatched with three *coups de grâce*. Miguel was captured in the mountains where he had fled at the end of the war with his father, Justo Sánchez, who was also executed. His brother-in-law Juan Guijo died in Mauthausen Nazi concentration camp, his sister Josefa and her husband Juan Serrano suffered many years imprisonment and all their property was seized. Another family destroyed by Francoism.

On previous pages, Moreno Gómez spoke of the expansive wave of terror, that is, the impact that the executions of inmates had, first on their fellow prisoners in their cells, then on their families, then on left-wing supporters and finally, on the townspeople as a whole, and he cited some extracts from the diary of one of the prisoners from Pozoblanco, Antonio Baena (later executed in Cordoba capital).

Repression by summary execution ceased in Pozoblanco after September 20 1940 when all the prisoners in the province were transferred from the local jails to the Cordoba Provincial Prison in the capital, where military justice continued to implement Franco's program of extermination.

There was a curious incident in Pozoblanco in 1986, regarding recognition of the historic memory of the victims of the repression, when Gabriel García de Consuegra, the local historian, published his extensive study of the Francoist repression in Pozoblanco, in the September issue of *Revista da Feira*, an annual magazine celebrating the principal town fair. All of Pozoblanco rushed to read his article which included a list of 220 Republican victims (not including guerrillas). The ensuing scandal exploded with right-wing protests raining on City Hall, to be published in the Municipal Bulletin, demanding the seizure of the issue that dared to address the 'other memory'.

The local assembly of the Alianza Popular political party and several private individuals wrote letters of protest:

> "We totally condemn this, because it is neither the time nor the place to set the populace to fighting over four useless and mediocre individuals, who furthermore knew nothing about anything. Pozoblanco does not deserve what has been done to it, on such an important occasion, the celebration of our annual fair."
>
> "How is it possible that in these times and days, so much is said about peace, which is what all of us want, that a nobody has thought to waste his time writing this article, unless he is an authentic fanatic, in which case he can be considered guilty of everything, except of being a democrat (...) with articles such as this one

(....) the only thing that the author has achieved is an attempt to rekindle the fire and upset the peace."

The October 1986 issue of the Municipal Bulletin published a petition signed by 40 right-wing individuals containing incredible statements such as:

"...the unfortunate article's lack of objectivity has caused great discomfort and indignation amongst most of the population of Pozoblanco. What will the people think of us? Is Pozoblanco reviving a subject that has already been forgotten? It is revitalizing an aftermath, whose fire began some fifty years ago and has been extinguished for most of the inhabitants."

The above group of 40 did not let matters rest. Since it was rumoured that what was published was just an extract of García de Contrera's work, the group of 40 vociferously demanded that that the rest of his work should not be published and that he should be stripped of the coveted municipal Ginés de Sepúlveda[12] prize he had received for this article. Furthermore, they insinuated that as the author belonged to the monied class in Pozoblanco, he was betraying his class, pointing out that relatives of his had also been victims of the Rojos in 1936. All in all, a reactionary hecatomb in Pozoblanco, in September 1986, not long before the 50th anniversary of the beginning of the civil war. All because a list of left-wing victims had been made public.

The great scandal did not end there. The Socialist Municipality lost its nerve and rather than standing firm and remaining consistent with its party ideals, it issued a public apology on the local Radio and ordered the seizure of the relevant issue of the *Revista da Feira*, even though it had already been distributed and had sold out. Such anxiety on the part of the local Socialists was neither more nor less

---

[12] Juan Ginés de Sepúlveda (1490-1573) was born in Pozoblanco. A great humanist, philosopher and Greek scholar, he was the royal chronicler of Charles V and tutor to Philip II. He wrote numerous books on philosophy and ethics, theology, law, history and political theory.

than a reflection of the position of the Spanish Left during the period of transition.

After this skirmish with the Right in Pozoblanco in 1986, García de Consuegra, a secondary school teacher with whom Moreno Gómez had exchanged data and information, asked to be transferred and he disappeared from Pozoblanco. The proposal to build a monument to the victims of Francoism in Pozoblanco, hopefully in the cemetery where so many were executed, disappeared.

The Right felt that they were the only ones with the right to publish the names of their fallen and that those of the Left had no right to a memorial of any kind, an opinion that is still reflected today. [13]

A similar initiative in Villanueva de Córdoba, long before the one in Pozoblanco, had a totally different result. In April 1981, Moreno Gómez had already spoken in public and published the list of 'the other fallen' (provisional number 14) in the town, in the local press.[xviii] Nobody protested, although Moreno Gómez knew that the opposition could not get over their amazement at his temerity. However, this was when the UCD was in power and the right-wing was still marching to a different tune. Not content with the List, he and the other members of the Historic Memory Association were encouraged to push for a monument to the defeated in the local cemetery. UCD City Hall, much more open-minded than others that followed it, did not deny their request. There was, however, a caveat. The inscription on the monument must read: 'To those who died for their ideals'. The organizing committee refused that. It had to read: 'To those who died for freedom'.

There followed several months of wrangling, during which the Historic Memory Association published a manifesto and collected the signatures of notable and distinguished individuals from Madrid. Today, as he re-reads their manifesto, Moreno Gómez is amazed at the list of signatories: José Luís Aranguren, Buero Vallejo Castilla del Pino Carmen Conde, Antonio Gala, Gibson, Bishop Iniesta, Rev. Gómez Cafarena and Rev. Llanos, Laín Entralgo, Daniel Sueiro, Antonio

---

[13] Known as the pact of silence, this reflected the stubborn official denial of the existence of Republican victims, a condition essential to the restoration of democracy in 1977. Today, the Spanish Historic Memory Movement is actively engaged in putting an end to the victors' uncompromising refusal to pay the minimum attention to the memory of the defeated Republicans.

Tovar, Tuñón de Lara, Francisco Umbral and a few more.[xix] The effect of the manifesto was to demolish the ancestral administrative immobility. The monument 'To those who died for Freedom' was duly inaugurated in Villanueva cemetery on All Saint's Day, November 1, 1982. It may even be the first such monument in the province. The local Marching Band entertained the townspeople by playing *The International*, which they rehearsed from the score Moreno Gómez previously obtained from the Madrid Marching Band. Small successes in the fight for the Democratic Memory, before this concept had become a national by-word.

Puente Genil is next in our retrieval of tragic events. After Cordoba capital, it was here that the province was subjected to the most extensive genocide, almost a thousand civilian victims after General Castejón's troops entered the town August 1 1936. The slaughter in Puente Genil reached historic levels throughout the remainder of the year and in 1937. This town was one of the birthplaces of several Cordoban trade union movements, boasting a youthful Socialism that began in 1936 with very dynamic JSU groups. In 1921, Puente Genil also was one of the places at the heart of the birth of the PCE in the province, at the same time as Villanueva de Córdoba, which explains why the rebel military decided to crush this stronghold of workers' movements.

The post-war eradication was led by Jaime García del Val, who arrived in Montilla as Judge Advocate. He was notorious for his cruelty, which was written in blood, and for his licentious behaviour with a courtesan whom the people nicknamed *the Judge's Anita*. Many of the courts martial in Puente Genil and in many other towns in the province were presided by Guardia Civil Lt. Evaristo Peñalver, with the infamous Ramón de La Lastra y Hoces as Prosecutor. On one of the Court's visits to Puente Genil, Pentalver's Court tried the eminent elderly Republican, Antonio Romero Jiménez, a highly respected individual, past Republican Mayor and Freemason. Despite the fact that he held the title of Honorary Captain of the Army of Africa from when he served with Franco in Morocco in the past, this did not work in his favour as an Order came from the highest to ensure that the strictest sanctions were applied to him, the only person to Moreno Gómez' knowledge who was executed by garrotte, October 24 1939. One of his sons was taken on a paseo in 1936.

A similar case of Francoist cruelty affected two sets of father and son, November 2 1939. The reason for their execution is unknown but, on that day, they executed Francisco Villar and his son José, and Francisco Reyes and his son, also named José. Meanwhile, torture was being liberally applied in the several Puente Genil jails. Francisco *Jesús* Preso Delgado was executed November 6. Broken by torture, he attempted to take his life in prison by slashing his wrists, but he was treated then shot.

The terror reached such heights in Puente Genil that November 16 there occurred an incident that could have been taken from a movie. When two municipal policemen (José Palos Ramírez and Juan Mendonza), accompanied by Jesús Aguilar, the Falangista Mayor when t to Francisco Palos Gálvez' house to arrest him, he was waiting for them in his living room. He lashed out against the policemen with a steel bar, killing them both, and only stopped when the mayor shot him.

At this time, many notable individuals fled forever in exile, from this village of berserk genocides, such as the poet Juan Rejano (a celebrated intellectual who died in Mexico) and the great Socialist leader and Freemason Gabriel Morón Díaz (who joined the PCE at the end of his life and who also died in Mexico). The April 24 1940 saca of six prisoners included the socialist railway worker José Mora Valencia who appeared to have been President of the War Committee, had been so cruelly tortured that he almost did not survive for his execution. Also, Rafael Reina Hidalgo whose brother Juan was executed a few months earlier.

The last execution in Puente Genil was held June 11 1940 (October 20, the remaining prisoners were also transferred to Cordoba where many were executed.) Of the three victims of this saca, Emilio *Palomo* López Arisbal was executed despite having been promised his safety in exchange for rendering some confidential services. It is said that when he was taken from prison a lieutenant called him 'criminal' and Emilio shouted back: "You are the criminals. Murdering wolves! Who, in reprisal for a hundred of yours killed by the Left, have already slaughtered several thousand patriots".[xx] In all, twenty-nine were executed during the immediate post-war period. If to this number, we add those who later died in Cordoba capital and those who were executed in 1936, we note that this town sacrificed 900 victims in homage to Franco.

Santa Eufemia is a town in the extreme north of Cordoba province, on the road to Almadén, in the Los Pedroches district. The town's resistance to the military coup was organized by the famous Dr. Pedro Vallina, whose anarchist beliefs were shared by most of the townspeople. Santa Eufemia suffered its own dose of terror during the post-war, with 21 executed in the town, other natives of the town shot in Hinojosa del Duque and in a neighbouring village, and still more later, in Cordoba capital. In 1940, the Spanish legion, under the command of General Salvador Bañuls, arrived in Los Pedroches to begin its reign of terror in the region, centring its operations in Pozoblanco as well as Belalcázer, Villanueva de Córdoba, Cardeña and Santa Eufemia.

With the onset of this terror, large numbers of men (guerrillas) from the North of Cordoba began to flee towards Cáceres which was becoming a guerrilla centre. In Santa Eufemia, a dozen townspeople fled to the hills, such as Manuel *El Secretario* Fernández and Norberto *Veneno* Castillejo. The Legion was merciless in its pursuit of escapees in the Los Pedroches district throughout the Summer and Fall of 1940. In Santa Eufemia, legionnaires poured boiling oil into the ears of Norberto Catillejo, the fugitive guerrilla's father, to force him to tell them where his son was hiding. In Villanueva, they beat four men to death with sticks, another one in Belalcázer and yet another in Cardeña.

The notorious Causa General accused all the remaining Santa Eufemia left-wing sympathizers en bloc, of crimes and of belonging to the War Committee. Seven men were first executed July 4 1939 and eight more on the 24th of the same month. Later, they executed a few more. It is said that the Francoists also executed a group of townspeople in the La Parraga mine, Agudo, Ciudad Real, supposedly because it was there that a band of Republican militia had executed 35 right-wingers from Santa Eufemia earlier during the war. Moreno Gómez was unable to confirm this.

Villafranca. A few words regarding a small town that, contrary to what usually happened in the villages, suffered eight post-war executions. Villafranca had been a strong supporter of workers' rights since the dawn of the Republic. It was one of the only four towns in the province (with Villa del Rio, Doña Mencia and Villanueva de Córdoba) that elected Communist aldermen in 1931. A typical town of farmers and labourers, it always was a source of trade union leaders. In 1936, it became the general headquarters of the Andalusian Militia, later the famous Villafranca Battalion, under the command of Francisco del Castillo and his assistant Pedro Garfias. One of the Militia Captains of the Battalion, Pacio Díos, was executed in Cordoaba.

Francisco Jurado Fernández, who had gained much fame for his bravery as a member of the Villafranca Battalion, was killed June 9 1939. After returning to his hometown at the end of the war, he was arrested and forced to present himself to the military headquarters every day, where he was sorely beaten. The day he died, he was again taken to the barracks, and he flatly refused to go. He ran away to force them to apply the Law of Escapes and end his torment, which they did at 11 a.m. November 25 they executed António Serrohe Ramos, whose brother Juan had been the first Communist alderman in 1931. The last execution in the town that year was December 12 1939, when they shot four men – three from Adamuz and one from Villa del Río.

This was not the end of the bloodshed in Villafranca. November 1948, the Guardia Civil applied the Law of Escapes to two sexagenarian brothers, Andrés and Diego González Fernández, simply because they were the brothers of a guerrilla fighter, Juan *El Álvarez* González Fernández, who had been hiding in the mountains since 1941 when he escaped from the 'train of death' that was taking the condemned from Burgos to Cordoba to be shot. Ripping off some slats in the side of the cattle wagon in which the prisoners were travelling, Juan jumped off the moving train. He survived eight years in the mountains until he was again caught, and later executed in Montoro, November 1949.

Villanueva de Córdoba. We now enter another of the iconic towns of Cordovan workers' movements, to which Moreno Gómez dedicated an earlier book that is frequently quoted herein. In 1931, the Republican Ministry of Government declared that this town was one of the greatest centres of Communism in Andalusia. The townspeople fought heart and soul against the 1936 military coup, so it is no surprise that the post-war repression fell on the town with all the force of a hurricane, with 102 executions. To this number we must add 36 executed later in Cordoba capital, 20 guerrillas killed, 22 *paseados* in 1948, 23 townspeople who starved to death in Cordoba prison, and 4 who died in Mauthausen concentration camp to which they were sent.

To this is added the widespread application of torture, the 1940 wave of terror by the Legion, starvation, exclusion, repression of all kind, the Law of Escapes in 1948, military justice under the most cruel of Franco's executioners such as Judge Juan Calero, Lieutenant *Pepinillo* (Juan Moreno Sevillano, from Osuna) Captain Fernández (from Bailén), Commander Felipe Martínez Machado and Captain Aznar, all who followed the Nero-like dictates of the Regime. Today, all the energetic and forceful syndicalism of the 1930s in this town, one of the main reasons for the military coup, has been erased from the map. The following is a brief step-by-step reconstruction of the post-war extermination from different viewpoints.

The slaughter in Villanueva de Córdoba began in the early hours of April 19-20 1939 when the Law of Escapes was applied to six prisoners near the municipal water tanks at the beginning of the road to Conquista. They were buried in the fields where they fell, and to this day, there bodies remain there. During April, several temporary jails in the town were packed with prisoners (the Refugio, Romo's House on Calle Herradores, the Municipal Depot and Juan Herrero's house on Calle Conquista). Later, they were all taken to the large, improvised Fuente Vieja Schools jail. The centres of torture were the Military Headquarters in Los Laurianos house on the main square and the Guardia Civil temporary barracks and SIPM (Military Police) headquarters in Malagón house on La Preturilla square. Everyone who was arrested passed through both these centres; some left their skin, others, their lives.

The murderous master of ceremonies of the great local repression was Judge Juan Calero Rubio, from Pozoblanco, who presided over Military Court No. 11 and whose inflicted cruelty and moral torment could fill an entire encyclopaedia. Mentally imbalanced, he ended his homicidal career by committing suicide in his home in Pozoblanco at the end of August 1940. The court martials over which he presided were held in the Torres Hall, Calle Concejo, on the days that the military court arrived in town from Cordoba city.

The Francoists launched the bloodbath in Villanueva de Córdoba on All Soul's Day, November 2 1939. That day, at 5 p.m., a truck went to the door of the Fuente Vieja Schools jail to collect the first five victims, all from Pedroche. As the executions were to be carried out in daylight, many supporters of the regime or simply curious public, crowded in front of the Fuente Vieja. Others went to the cemetery: Falangistas, some upper-class bourgeois ladies and several well-known accusers, who vociferously insulted the condemned, calling them criminals and other names. As soon as the first round of shots was heard, the Francoist public, as usual in the rest of Spain, burst out with applause and sang the *Cara al Sol*, as the condemned men breathed their last.

A few days later, November 7, the anniversary of the Russian revolution and of the first day of the battle of Madrid, a second execution of three victims: Marcial Cobos, from Torrecampo and a member of the Committee on Communities, and two from Villanueva: Pedro *El Chunga* Juan Martínez Capitán and Francisco *Villaralteño* Rubio Gómez. As reported earlier, the latter's family was almost totally exterminated, as several months later they executed his father-in-law (Juan Gómez, aged 67) and two brothers-in-law (Manuel and Juan Gómez Luna, aged 27). A third brother-in-law, Hilário Gómez, committed suicide when, upon his release from ail in 1946 he discovered the family tragedy. All the men in the family were eliminated and only two women whom Moreno Gómez visited in Torrecampo at the beginning of the 1980s, were left to mourn.

As to El Chunga, who was very well-known in Villanueva, not because of his politics but because he was some kind of local hero, he is still remembered for inviting everyone to "Come to the party in the bullring" as he left for the cemetery. The infamous bloodthirsty Lieutenant Pepinillo, who ended up taking his own life in Espiel July 18 1941 after killing fifteen farmworkers, was officer in charge of the firing squad and of administering the coup de grâce. Still today, the general public knows nothing of these and similar barbaric acts that too place throughout victorious Spain.

Zacarías Romero Regalón, from Pedroche, was executed November 30. According to eyewitnesses, he was arrested for the simple reason that they had found a razor in his backpack, for which he was also cruelly beaten.

1939 ended with the first 20 victims; more than 80 would fall the following year. As 1940 dawned, returned townspeople continued to be forced to make declarations, to receive beatings and be tried by summary courts martial, ending with the shooting of those who had been 'duly sentenced'.

30 January was the date of the execution of a prisoner who had been sentenced to death by Judge Juan Calero: Misael López, the Head Postmaster, whose reprieve obtained by a relative of his, a high-ranking Nationalist military officer, the Judge had hidden in a drawer. Although it cannot be proven, this may have been the 'serious administrative mistake' that led the Judge to commit suicide that same year August 28.

Leap year was celebrated with a numerous saca of 14 men at 5 pm. February 29, almost all from Pedroche and Torrecampo, and one from Villanueva: Lope *El Diamitero* Ibáñez. A native of the mining region of El Horcajo, he was an explosives expert and apparently fought the coup in 1936. He was arrested in Almodóvar del Campo in 1939, from where he was taken on orders from members of the Villanueva Casino who received him with beating after beating, until he was executed. Several individuals from Torrecampo died with him, namely Sebastián Luque (whose son Juan had been taken for a paseo right after the Francoist victory), Cesário Serrano (because he had belonged to the Rojo Red Cross), young Manuel Luque Herrero, who was cruelly tortured by El Colodro, the Torrecampo gravedigger. Another four were from Pedroche, among whom the famous Francisco *E Pindolo* Tirado, a farmer who had been very active in the revolutionary days of Pedroche, and others from that town whose charges are unknown: (*El Princeso*, Pedro *El Asentero*, Francisco *El de la Rosa*, David El *Cabezón*, Diego *El de la Zorra*, etc.). El Pindolo was tried in Villanueva September 7 1939 at the court martial presided by Commander Ramóm Navarro, on unknown charges, and sentenced to death by garrotte – he was actually shot.

Ten more victims were executed at 5 p.m. March 5 1940. One, Pedro Amil Cuadrado, had been Socialist Mayor of Adamuz, took refuge in Villanueva in 1937 and served as leader of the Committee of the Evacuated from Adamuz. He held various positions behind the Republican lines. Arrested after the end of the war, he was unable to recover from the usual terrible beatings and torture he endured.

Fructuoso Prieto Arévalo was part of that same saca. The *ABC* newspaper of 1931 printed his photograph as President of a committee in the famous October 1931 strike in Villanueva that was repressed in a scandalous manner. He was arrested by Falangista Diego El Chunga, who was famous for torturing Pepe Delgado, Juan Lucio and others.

Another leading activist for workers' rights died that same day, Communist Diego Ranchal Plazuelo, brother of the well-known Socialist Miguel Rachal who had served as Mayor of Villanueva del Duque and was executed in Barcelona by the infamous Campo da Bota firing squad. Diego's 'crime' was that he had been a member of a street band during Carnival 1936, whose jingles strongly criticized the local landowners: *Vote for Toico, a very reputable man, who buys votes with cuddles from...* they sang. Several members of the street band who participated in that event ended up facing the firing squad. Diego was forced to sing that ditty as he was tortured. Diego Majuelos was also part of that saca, falsely accused by the widow of a Guardia Civil for something he denied. One of his brothers, Mateo, was later executed in Cordoba.

The March 26 saca included Antonio Rubio Cobos, from Pedroche, father of Moreno Gómez' secondary school teacher, Manuel Rubio. Many of the children of those who were shot by the National Catholics were made to attend religious schools and encouraged to follow a religious career, as a form of penitence and reparation for the 'sins of their fathers'. Manuel Rubio was one of those children who was sent to the Salesian seminary. As a young man, he abandoned his religious studies in the early 1950s and moved to Villanueva where he turned to teaching.

One of the four executed April 2, the elderly Socialist Ramón Ruíz Hernán, had frequently represented his party in agricultural meetings and committees, which is why landowners hated him. He was accused of having insulted Franco in a letter he wrote during the war and which they found. Also executed that day, Manuel Salazar Vilches, was a very active Communist. Both men were also accused of having taken the horses belonging to the Guardia Civil after they surrendered their barracks in July 1936. Vilches died shouting *Viva! La Republica!* "It was one of the most valiant deaths that I witnessed".[xxi]

Juan Escribano Fermández, son of Juan *El Pedrocheño*, a member of the JSU, was executed at 5 p.m. April 6. He was accused of having been present at the unproven execution of 'the 21' right-wingers August 6 1936. Judge Juan Calero accepted numerous gifts from his family in exchange for saving their son, but he kept the gifts and sentenced the young man to death and allowed his family to say farewell to him.

The first early morning execution took place 6 a.m. May 17, with five victims. Prisciliano Orellana was accused of having gone to arrest a right-winger, Victoriano Muñoz, who was later executed by a person called Rojas. Despite the fact that everyone in the town knew this, all who were in the arresting party were exterminated, whilst the only truly guilty person escaped into exile. Also executed at the same time, Matías Villarreal (whose brother Basílio, member of the War Committee, was in the mountains with the guerrillas) and Francisco *Curro Beatas* Sánchez Muñoz, whose letters from the chapel are reproduced herein. Curro Beatas was a member of the War Committee, a leader of the Izquierda Republicana, First Assistant Mayor in 1936, and during the war, joined the PCE. He was always a moderate leader and avoided all possible excesses. Two individuals he 'saved' from Republican action, the Pedraza brothers (Matías and Gregorio, the Francoist Mayor), were both landowners and his principal accusers.

The most infamous execution in the town during those days occurred May 26, when the saca took the largest number of victims – 18. It was exceptionally held on a Sunday, precisely on Holy Sunday, when the religious procession took twice to the streets. It is said that each time that the procession walked past the Fuente Vieja Schools prison, the hooded members of the religious brotherhood shouted

"Kill them all! Don't leave a single man alive!"

Historically, religions have been at the root of the most violent movements worldwide. The largest bloodbaths have always occurred in the name of religion and Catholics have always been leaders when it came to exterminating heretics. As prof that nothing in this book is an exaggeration of this, here is a brief extract of that which a priest from a church in Rota, Cádiz, had to say in 1936:

> "The most guilty and ungodly have already answered to God for their actions. They are now paying for their guilt in having infiltrated the town with the Marxist poison, thus distancing themselves from God. But there still remain some who intend to deceive us. We shall discover them all. Everyone shall get what he deserves; no-one shall escape. Mark my words. Nobody! We must go even deeper with the cleansing

until we have destroyed all the rot that Russia has introduced into this town..."ˣˣⁱⁱ

That was the Church under Franco. Moreno Gómez recently received a letter from Jaén informing him that a local priest, Tomás de la Torre Lendínez, was currently travelling all around the province collecting signatures (he already had more than a thousand) for a petition to abolish or revoke the Historic Memory Law. This is an example of what he writes in his blog:

> "During the many years of the Zapatero Government, we were forced to swallow some disgusting legislation. One of those filthy laws is the ill-named law for the historic memory, the source of vindictive conflicts and hatred..."ˣˣⁱⁱⁱ

The Church continues to go its way without pain in the heart, nor any desire to change, without confessing its own sins or doing any penance. *Plus ça change, plus c'est la même chose.*

Leading the victims in the unholy Easter saca, was a moderate Communist leader, Eugenio *Palmera* Jurado Pozuelo, an educated man, a violinist with the Municipal Orchestra, who had represented the local PCE at several meetings in Andalusia during the 1930s. Occasionally in the jail, other prisoners asked him to play his violin, but those sessions were always interrupted by a drunken subaltern from the regular army, with kicks and slaps and shouts of: "Obviously, wild animals also like music."

All the members of the Palmera family were involved in politics. His brother José who had served as Mayor in 1935 for the Leroux Party, did what he could for Eugenio, writing letters and trying to pull influential strings, to no avail. To make matters worse, another brother, Zacharias, was a rabid Falangista who wanted nothing to do with his brother Eugenio. The latter's main accuser was Pepe Delgado whose brother had been tried in Jaén and he accused Eugenio and his wife Florentina of having gone to the People's Court in Jaén to depose against him. His letter from the Chapel is reproduced herein.

Others included in the same saca were: father and son Juan Gómez and his son Juan, of the aforementioned Los Villaralteños family; Alfonso Leal, a member of the Zamora family, secretary of the Committee of Communities, whom the casino Señoritos accused of having stolen their land; Avelino *Campana* Ayala, alleged that he had been present at the execution of some right-wingers; Antonio *Casillas* Pedraza, saddle maker, Socialist, hated by Pedro *Sargento Chicorro* Muñoz; Fernando Fernandez de Haro, for being a Socialist, private tutor and seller of trade union newspapers, and also because it was alleged that he had 'entered the town at the head of two thousand riflemen', which was absurd. (His family insisted that he was executed after he received a reprieve but that it was hidden by Judge Juan Calero); Sebastián *Figurillas* Santofimia, involved in the Carnaval marches; Francisco López Justos, uncle of the famous Mojea (María Josefa López Garrido, a member of Julián Caballero's guerrilla group; and Blás *Catalán* Arévalo, where everything conspired against his survival.

Blás Arévalo's brother-in-law was the great Communist leader, Nemesio *El Floro* Pozuelo, of Jaén, where he was Secretary of the PCE, Civil Governor during the war and a member of the party's National Board of Directors. Nemesio went into exile in the USSR where he became quite famous.[xxiv] Blás married Nemesio's sister Juana *La Flora*, who was sentenced to 20 years in jail for having sheltered the wives and children of the Guardia Civiles who had surrendered their barracks, in her house on Calle Cañada Baja, an act of Good Samaritan kindness that was, nonetheless, held against her. One of their sons-in-law, Antonio Pérez de la Riva, was beaten to death in Burgos prison. To make matters worse, when Blás Arévalo was on guard duty at the famous Guerrilla Headquarters at Calle Conquista 14, Villanueva, at the end of 1937, he did not prevent a commando from removing a right-winger who had been arrested, Francisco Díaz, and made him disappear.

Pedro Amil Ruíz, from Adamuz, was executed in the May 26 saca and his brother Juan, the following saca, June 3. Both were elderly. Of the four men executed June 3, one was a past Mayor of Pedroche, Tomás Rodríguez de la Fuente. From Villanueva, they executed Juan J. *E Conejero* Mohedano Sánchez, a well-known Socialist accused only of standing guard over right-wing prisoners at the Regajito Schools. He was a past alderman for the Frente Popular and held several positions during the war, including Mayor during the last few days, a position that was cut short at the end of the war because of the Casado coup.

The June 12 dawn saca took three victims, one of whom was the Communist Juan Antonio Bustos Casado, son of Gaspar *Legañas* Bustos both of whom worked for the great landowner Torrico. Juan Antonio Bustos, a member of the PCE, was an alderman during the war and Treasurer of the local Trade Union Federation. He played no part in the anti-coup fighting in July 1936 because some miners from Puertollano mistakenly shot him in the arm and he spent considerable tine in hospital. His father, Gaspar, went to Torrico, his employer, begging for his son's life, to no avail. Worse, Juan Antonio and his wife Antonia had owned a tavern across the street from City Hall that was a favourite haunt of left-wing townspeople and the Communist Mayor, Julián Caballero.

Juan Antonio Bustos was one of the prisoners who confessed in chapel before he died, perhaps hoping to ask the parish priest, Rev. Don Marcial, to look after his two sons. One of them, Gaspar, was pushed towards the church and he became a priest in Cordoba, where he remains to this date. Moreno Gómez attended classes he taught at the Seminary.

Also executed at the same saca, Juan Luna Enríquez, Communist, brother of the first Republican Mayor in 1931, Andrés Luna, a member of the Leroux Party. His last letter to his family, written in the Chapel, is reproduced herein.

There were only two more executions in Villanueva. June 15, they executed 55-year-oldd Socialist Cecilio *El Vinatero* Ruíz, accused only of having stood guard over right-wing prisoners. The last individual executed, September 12, was Alfonso *El de La Loma* Sánchez Pozuelo, a Communist, because Diego López alleged, on no apparent grounds, that he took part in the 1936 revolt. One of his brothers was later shot in Cordoba capital. Both were sons of María La Loma, head of another local family that was almost exterminated.

September 26, all the prisoners in the town were transferred to Cordoba Provincial Prison where they would join the hundreds of men who would die during 1941, at the height of the mass Francoist extermination that curiously appeared to coincide with the exterminations in the Nazi concentration camps.

This tragic description of so much that Villanueva de Córdoba suffered during the post-war period ends with a brief reference to three eminent natives of the town who were executed by Francoists in Valencia and in Barcelona.

Francisco Copado Sánchez was first cousin of the notorious Jesuit Rev. Bernabé Copado S.J., chaplain to the Carlist unit under Colonel Luís Redondo, whose humane and Christian qualities are best described by the fact that he let his cousin Francisco be executed on the basis of Juan Fernández and Manuel Rodríguez Moreno's clearly false allegations. Both swore that Francisco was a dynamiter and leader of numerous mobs when he actually was a truck driver and belonged to the War Committee. Everything else was false. The end of the war found him in Valencia where, with others, he took refuge in the Panamanian Embassy. The Embassy, however, was attacked by Francoists who arrested all the occupants. Copado was sentenced to death with other Republican officers, June 12 1939. Taken at dawn from Modelo Prison in Valencia November 1, he was executed by firing squad in Paterna cemetery. In his cell, he left a lock of his hair for his wife María and several farewell letters, extracts of which are quoted earlier or reproduced herein.

Another tragic story was that of Miguel Ranchal Plazuelo, also resident in Villanueva de Córdoba and later in Villanueva del Duque, where he was a famous Socialist Mayor during the Republic and part of the war. He was married to Maria Josefa Luna, sister of the aforementioned Bartolomé Luna, Socialist Mayor of Villanueva de Córdoba in 1932. When in in 1931, the PCE was founded in Villanueva and joined by all the members of the Socialist Youth and a great many local working men, he and his brother-in-law strove to maintain the essence of the PSOE in the town. Miguel Ranchal was frequently mentioned in the trade union press during the Republic because of his defence of the miners in the El Soldado, Villanueva del Duque, mines, as a truly dedicated Mayor should. Within the PSOE, Ranchal, a *Prietista*,[14] opposed Francisco Largo Caballero's faction.

When war broke out July 18, the party was profoundly divided. Ranchal, as Mayor, did his utmost to check the outbreak of violence in his town, which he was generally successful at achieving. For example, several times he prevented the Jaén Militia from attacking the local prison where right-wingers were being held. During the war, he held leading positions in the Cordoba PSOE, and he argued strongly with

---

[14] Supporter of Indalecio Prieto, right-wing leader of the Spanish Socialist Party during the Republic.

the PCE against unifying the two parties; on the frontlines, he served as a political commissar. At the end of the war, he was arrested in Albatera Alicante. From there, he was taken first to Valencia, then to Barcelona, where he was executed June 13 1940 at the tragic Campo de la Bota[15] despite numerous appeals to leading Francoists whose lives Ranchal had saved. One, Demetrio Carvajal Arrieta, a Judge Advocate, cynically told Ranchal's wife not to worry that at the utmost her husband would only receive a more or less lengthy prison sentence.[xxv]

José Cantador Huertos, another eminent Socialist from Villanueva de Córdoba, was executed in Paterna cemetery, Valencia, August 29 1940. A bank employee, in 1933 he moved to Játiva where he was an alderman for the Frente Popular, director of the UGT and the local branch of the PSOE. He previously served as Municipal Secretary in La Carlota, Cordoba. At the end of the war, he was taken prisoner in Albatera, then taken to the Játiva jail where he was sentenced to death by Military Judge Eusebio González on false allegations by a fellow bank worker, Ricardo Diego Ruíz, for whom he had done some favours in 1936. This was a situation that repeated itself time and time again, when members of the right-wing who had been favoured in one way or another by Republicans, became furious accusers against Republicans after the end of the war. Cantador's letter of farewell to his family, reproduced herein, was found hidden in his pillow in his cell.

Miscellaneous executions in other small towns and villages

Natives of **Villaralto** were executed in various places in the Los Pedroches district (Pozoblanco, Peñarroya, Hinojosa and others) and, after September 1940, in Cordoba capital. The Civil Register lists eight executed. Six of these obits also indicate Pozoblanco as the place of death and one Peñarroya, which is why only one name is listed for this town. Likewise, in **El Viso**, where of eight dead, two are recorded as

---

[15] The *Campo de la Bota*, a shooting range for French troops under Napoleon, was a Barcelona neighbourhood and military installation on the outskirts of the town, by the sea. Between 1939 and 1952, Francoists shot and killed 1717 persons sentenced by summary trials and executions. Eyewitnesses describe truck after truck laden with prisoners and so many killed as they fled towards the shore, attempting to escape the firing squads, that the sea was stained red with blood. Description downloaded from http://www.barcelonarutas.com/el-campo-de-la-bota-de-triste-recuerdo/.

having died in Pozoblanco, which is why only six names are listed. In **Almodóvar del Río,** there is no entry for death by firing squad in the Civil Register, although eyewitness accounts confirm that six residents were shot there November 1 1939, which is why six names are listed for that town.

In **Cañete de las Torres**, there were only three executions: one October 18 and two November 7 1939. Other residents died in Baena, Castro del Río and Cordoba capital. The **Espejo** Civil Register only records five victims in the cemetery, August 3 1939. Others from Espejo died in Castro del Río and in Cordoba capital. It is also here that the record for Francisco Jiménez García states that he died February 27 1940 by 'incomplete hanging', a euphemism for torture. Just how is one to interpret what is described as 'incomplete'?

There are eight victims in **Hornachuelos**, two in 1939, two in 1940 and four in 1941, for whom there is no additional information. Five more natives of Hornachuelos died in Posadas and more in Cordoba capital. There was a single execution in **Montalbán,** Juan Río Jiménez, a shoemaker aged 33, August 10 1939, with no further information. This village appears to have been overlooked by the bloody storm that devastated the region beginning in 1936, although recent information from an eyewitness, Alfonso Vaquero Zamorano, indicates there are quite a few more than are included therein.[xxvi]

Moreno Gómez reminds the reader that the data regarding victims that are given in the Tables, town by town, refer only to the official entries in the respective Civil Registry Offices, and that we cannot lose sight of the possibility that the number of victims for each town and village may be considerably greater as many unrecorded executions were held in the more important towns.

The terror of the Spanish Legion in the District of Pedroches

"No Government has any right to refute a fundamental right of the victims...The truth! There has been no truth. If this is contained in the archives, it still remains hidden from us. During the period of transition, it was not for nothing that Francoists committed crimes. These have been forgotten. There has not been even a single debate... And what about the reconciliation? Where is that reconciliation? In Spain, suffice to even bring up the subject of the civil war for people

to begin to panic. Where is the reconciliation? Lastly, when is Justice going to ask forgiveness for never having done anything more..."
Baltazar Garzón
University of Seville, 13 January 2014

None of the many historians who write about the Francoist repression, appear to have addressed the dispersal of the Spanish Foreign Legion at different places in Spain with express repressive intentions, especially in 1940.

In 1940, the New State used the African Foreign Legion troops to add an extra touch of terror in the regions that sheltered those who had fled to the mountains, which is a large section of the country. In 1939, part of that mission was entrusted to several battalions (*tabores*) of regular troops, but the Moroccan mercenaries who had been useful during the war in the forefront of Franco's army proved unable to act in the rear-guard during 'peacetime'. The Cordoban district of Los Pedroches, the last one to fall into the hands of the Nationalists and a very mountainous region that sheltered fugitives in abundance, was especially suited for repression by the Legion.

The legionnaires were sent to the north of Cordoba province at the beginning of April 1940 in the shape of the 3rd *Bandera* of the First *Tercio*[16] of the Legion, under General Salvador Bañuls Navarro who, at the same time, was appointed local military commander, first in Villanueva and then in Pozoblanco. The Legion replaced the Military Police (SIPM) units that had been the first to act in the recently conquered villages, indiscriminately applying the Law of Escapes,[17] setting up headquarters in all the buildings that the SIPM had occupied in the towns and villages. In Villanueva de Córdoba, this was in the Los Laurianos house on the main square, where many of the defeated had already been cruelly beaten and some killed.

Commanding sections under General Bañuls' orders, we find Lieutenant San Antón, 2nd Lieutenant Barcieiro, Physician Lieutenant Segovia, and others. One section in **Santa Eufemia** was infamous for

---

[16] Spanish *Tercios* were the infantry unit most used during the time in which the Habsburg ruled in Spain. Tercios were composed of about 3000 men divided into 10 or 12 smaller units, called *compañías* or *banderas*. They consisted of three types of soldiers: pikemen, swordsmen and harquebusiers (riflemen).

[17] SIPM Lieutenant Leopoldo Mena was notorious for the amount of blood he shed in this manner in Villanueva de Córdoba.

the atrocity of its actions, such as the boiling oil incident, a form of torture that so horrified the people in Los Pedroches, that many years later Moreno Gómez heard reports of this in several places in the district.

In Autumn of 1940, legionnaires were particularly vicious in **Belalcázar** under Lieutenant Ortega and Sergeants Escobar and Rodríguez.[xxvii] The presence of a unit of Legion troops in this town was surely due to the considerable fears the Francoists had of the great many fugitives who had fled to the mountains (prison escapes in Hinojosa 1940 and in Belalcázar 1939). There were so many groups of these fugitives that there were frequent skirmishes in the surrounding farms (Pellejeros, Ochavo, La Encinilla, and more). Some 40 fugitives participated in one such skirmish September 1940. One of the most terrifying legionnaire officers, Lieutenant Juan Tamayo Vián arrived in Belalcázar December 1940. He was so crazy, so mad, that even the Francoist Mayor Justo Rivallo was afraid of him, as he wrote to the Provincial Governor in Cordoba:

> "When this Legion Lieutenant came to this town on the 13th instant, to take up his command, he declared that he intended to beat a lot of people and that is just what he did. Yesterday evening, at 11 p.m., he went to an establishment where several parishioners were meeting in a peaceful manner, and he began beating them with rods, ordering that the establishment be immediately closed. He did the same at a dance that he also cancelled, but not before he had rained beatings on several people who were present..."[xxviii]

From this, one can deduce that there were aspects of Francoism that went beyond a dedication to the overwhelming repression of the defeated, to spreading terror everywhere, creating a general atmosphere of terror that distressed both the disaffected and the affected as well. If these barbarities were committed by the legionnaires in 'times of peace', one can just imagine what these troops had been capable of in times of war, as they ruthlessly conquered towns and villages in their assaults on Badajoz, Talavera de la Reina and everywhere they went.

The legionnaires went about the towns as if they were the owners of the peoples' lives and property, exceeding their mandate to go after fugitives (which is what they did the least), imposing a reign of terror in the manner of barracks discipline, the only obedience they knew. It was the logical supremacy of the military over the civil society that came with the New State. Lieutenant Tamayo ridiculed the Mayor with his "I order, therefore you obey" attitude. He began by ordering that the streets had to be swept clean as if they were patios in a barracks, then he set closing times for the shops and curfews for other activities, as if everyone came under martial law once *Taps* was played.

The power of the military as the essence of Spanish fascism, liberally sprinkled with the Holy Water of National Catholicism, explains all kinds of controls, excesses and abuses that occurred. Nevertheless, after all that, there are uninformed individuals who still insist that Francoism in 1941 had already become a sort of paradise. This, despite the fact that if executions decreased in number, not in 1941, but in 1942, it partly was because there was practically nobody left to shoot. Besides, why did they want or need to execute more people as society lived in terror, cowed by the many other types of repression: exclusion, hunger, blacklists, beatings, continuous raids for prisoners, shaven heads, castor oil treatments, humiliations of all kinds, rationing, daily signing-ins at the barracks under a fragile parole, not just by those who had been freed from camps or prison, but also many left-wing sympathizers including adolescents[xxix], police abuses in the capital cities, in addition to slave labour all over Spain, prisons bursting with inmates and so much more, *ad infinitum*. In other words, one cannot speak of a softening of the Regime, either in 1941, or in 1942, or in 1947, or in 1948, or even later (the Law of Escapes was applied with full force in Nerja, Malaga, in 1950). To speak of a softening of the regime is to belittle historical rigor.

Returning to the terror in the North of Cordoba province, the persecution was so great that those who had escaped to the maquis in the region, some 50 individuals, organized the so-called Great March to Caceres. There were fugitives from Hinojosa (*El Francés*), Belalcázar (*Loro, Sincolor*), Santa Eufemia and some villages of Badajoz. They entered Las Villuercas, Caceres, December 6 1940. Caceres Guardia Civil Commander, Lieutenant Colonel Manuel Gómez Cantos, got wind of this but, unable to find the fugitives, used the civil population to teach the escapees a horrendous lesson. He organized a raid of peaceful villages, chosen totally at random, and applied a *paseo* to 12 residents of neighbouring Cañamero and 16 townspeople of Logrosán. Gómez Cantos got tired of murdering people in Badajoz, where he had been appointed Head of Public Order by Queipo in Marbella, (after Malaga fell) and in the hills around Huelva, where he chased fugitives in 1937. Responsible for the bloodbath in the Badajoz bullring, Gómez Cantos proved to be so out of control that the Guardia Civil finally arrested him in 1945, stripping him of his duties. He is not totally forgotten as to this day, a popular insult in the region is to accuse an individual of being 'more evil than Gómez Cantos'.

Once the Legion abandoned Los Pedroches, the Cordoban fugitives who had taken refuge in Cáceres at the end of 1940, returned to the North of Cordoba beginning Summer 194. (*El Francés* from Hinojosa, remained in Cáceres where he led a group of guerrillas.)

Let us now centre our attention on the Legion's activities in **Villanueva de Córdoba**. Beatings that were given to each and any townsperson were the norm, as in other places, part of the strategy of spreading terror throughout the rural world. Several eyewitness reports collected by Moreno Gómez at the beginning of the period of transition (later, this was no longer possible as the government restricted access to archives), revealed an unsustainable hardship. Bartolomé Caballero Coleto (His brother Juan *El Herero* Caballero had escaped to the hills) told Moreno Gómez:

> "April 1940, I was making charcoal in the Los Marines farm in Barranco de los Pobos (between Villanueva and Adamuz) and, because I had a brother who had fled to the mountains, the Legionnaires went there to get me, and they took me to Pedro Luis' mill. They gave me a terrible beating and then set me free. I finally made it back to the farm, with great difficulty, because of my wounds."

There is something more to the above story. In the same farm, a young man who belonged to the Las Marines family, was apparently friendly with *El Hebrero*, the guerrilla. He left food for him at night, in a hiding place where the guerrilla would collect it later. Whether the repressor forces knew of this or not, the fact is that they beat Miguel Gutiérrez so badly that this young man could not stop crying and he lost his mind for good. Moreno Gómez was unable to determine whether it was the Guardia Civil or the Legion who beat him, nor whether the beating was given on the farm, at the landowner's mill, or in the town barracks. The fact is, they beat him so badly they drove him mad, and he never recovered his wits.

Still in Villanueva, July 1940 the Legion implemented a series of menacing actions against the country folk, most certainly with a view to sowing preventive terror against anyone thinking of helping the maquis. The legionnaires organized a warped field trip, taking with them the men and women they found on the farms. It was an exhausting expedition lasting four days and three nights, in which they gathered up some fifty country folk, including some women such as Manuela Illescas, who had to leave new-born children behind.[xxx] "We will not stop until we find the fugitives," said the legionnaires. They travelled as far as the olive grove and the oil press belonging to Don Dionisio, close to the village of Obejo. The ill-treatment of their captives was constant. Finally, they took everyone with them to Villanueva, ending the odyssey in the Guardia Civil barracks on the Plaza del Carmen. The prize given to each for four days of deportation was a beating, especially a poor old man named Claudio (whose son Pedro Coleto Díaz was shot under the Law of Escapes June 17 1948).

The Legion's terror in Los Pedroches grew in strength under General Bañuls as Summer 1940 advanced. In Villanueva de Córdoba, 54-year-old Juan Fernández Moreno (father of Sebastián *El del Banco*), an apolitical and simple man, was arrested as he worked on Navalazarza farm, August 24. They took him to headquarters and, in the presence of General Bañuls, they hung him from the rafters and beat him brutally. As he lay dying, he was taken to Romo's House on Calle Herradores that served as an improvised prison and torture chamber. According to oral testimonies from his family, the omnipresent Pepe Higuera (son of Miguel Higuera the builder) and several other notorious local torturers who helped the Legion during those terrible months, were present. The Civil Registry records his cause of death as 'died at 8 p.m. in a deserted area from gunshot wounds'. This agrees with his family's version that as Juan Fernández was clearly dying, they took him out to the fields at nightfall and gave him the coup de grâce, to simulate an attempt at escaping. They took his body away and his family was never able to find out where they buried him. Yet another 'disappeared'.

Two days later, another crime at the hands of the legionnaires: a 43-year-old apolitical man, Juan *Horozco* Cantador Cachinero, who was arrested as he worked at José Cámara's farm, also in Navalazarza. His family recalls that some days earlier, several Rojos had been seen in the Dehesilla farm, a neighbouring farm. For no known reason, the legionnaires went to Cámara's farm and took him off to the deadly location: General Bañul's headquarters. The session of torture was so brutal that when Juan Cantador was removed to Romo's House, he was in such a sorry state that he died soon afterwards. The Civil Registry states 'died p.m. from a heart attack'. If it had not been for Moreno Gómez' research and interviews with witnesses, we would have never known that this so-called 'cardiac syncope' was covering up the death of a man who had been beaten to death, a crime of *lèse humanité*.

These are the death records that never appear in the statistics. Like those, how many more hidden crimes and 'disappearances' that History will never unveil? Only oral eyewitness accounts and oral testimonies, so despised by so-called academics, can offer access to the hidden nooks and crannies of Francoist criminality.

The Legion continued its wave of crime in Los Pedroches. In Villanueva, under the watchful eye of the Mayor, the Falangista Gregorio Pedraza Cámara, October 16 1940. Bañul's men organized another of their atypical field trips into the countryside, arresting all the country folk they found along the way, both men and women, on the excuse that they were hunting down Rojos. Amongst the arrested was elderly José Huertas Valverde, aged 68, and his son-in-law Francisco Valle.[18] They took them from Antonio Herrero's La Sierra farm, down to the town, beating them along the way. There,

---

[18] Moreno Gómez frequently interviewed the victim's daughter, Antonia Huertas, in her house Calle Pelayo, Villanueva de Córdoba, in 1979. An extremely kindly woman, she talked to him and gave him many details of those terrible days. In addition to the tragedy of her father, she told him how one day when she was working in the fields, her children came to tell her that there were some men in the hills. Taking care, she got a loaf of bread and went to see them: it was the group led by Julián Caballera, past Communist Mayor of Villanueva de Córdoba. "The Guardia Civil found out about this contact, arrested the whole family and took us all to the Cordoba prison. I was pregnant at the time and gave birth there. One year later, in 1941, when they felt like it, they set us free".

they tortured the elderly man, making him swallow buckets of water and pushing a tube down his throat. His throat ruptured and he died that night in the Hospital, from 'cardiac arrest' according to the Civil Registry.

Some people may be confused in thinking that during the war and the post-war genocide, the dirty killing work was somehow organized by the Falangistas, but that was definitely not the case. Falangistas could not kill or execute anyone on their own accord. Francoist genocide was always committed under the orders of and directed by the military police (including the Guardia Civil). When victory was declared, it was a military police unit that sowed the terror in Pozoblanco (100 victims of the Law of Escapes). The 2nd Military Police Company, under the orders of notorious Lieutenant Leopoldo Mena, was the one that sowed the terror in Villanueva de Córdoba and was responsible for the first deaths under torture. The arrival of the much worse and bloodier Legion substituted and 'improved' on what the Military Police had started.

The Legion also acted at will in **Cardeña**, a Villanueva neighbourhood. Moreno Gómez was able to discover at least one crime, the case of a humble stonemason, Pedro *Perico el de la Aurora* Gutiérrez Díaz, aged 46 years, married to Juana Cano and father of 6, whom they beat to death. According to the Civil Registry, he died from 'septicaemia, from infected wounds'. He died in the Municipal Depot, at the end of the day, August 1 1940. Another crime that would have remained uncovered had it not been for the family's oral testimony.

The 3rd Battalion of the 1st Tercio of the Legion set up its first command post in Villanueva de Córdoba Summer 1940, and that Fall moved its headquarters to Pozoblanco. These mercenary troops were garrisoned throughout the district; in addition to the Pozoblanco headquarters, the Legion established command posts in Villanueva (November), Belalcázar (December), and elsewhere. There is a record of the death 'from a gunshot wound' of a legionnaire in the Villanueva de Córdoba barracks New Year's Eve 1940-41. The reason for his death is not given, but it could have been either suicide or the result of a disciplinary procedure gone wrong. There are eyewitness accounts that legionnaires were punished by being made to sweep the Fuente Vieja square in Villanueva, with a bare torso and sacks of sand tied to their backs with wires.

When the Legion left Los Pedroches in Spring 1941, it was replaced by the 3rd Mobile Guardia Civil Company from Seville, under Captain Sebastián Carmona y Pérez de Vera, who began his bloody apprenticeship in Aguilar de la Frontera in 1936. Their task was to chase and capture fugitives but most especially, to keep up the terror and beatings of the population, both urban and rural. These Mobile Units served in the so-called Fugitives and Bandits Districts, and they answered to the Guardia Civil Headquarters in Cordoba capital.

We know the date on which the Legion left **Pozoblanco**, May 12 1941, when they were bid a 'warm farewell' from the municipality after having been there for 7 months.[xxxi] Most curious is that the bloody commander Salvador Bañuls left Pozoblanco either engaged or already married. The fortunate bride was a member of the Vizcáino family. As a 'love' present for his adopted marital homeland, when he was promoted to General and appointed Captain General of Catalonia, he donated his red sash of office to the Virgin of Luna, patron saint of Pozoblanco. On procession days, the statue of the Virgin is still taken through the streets, adorned with this war criminal's blood red sash.[19]

Faced with this and other examples of the still apparent alliance of the cross and the sword, as embodied by the Causa General, it is a wonder that today deniers of the Francoist repression have not promoted the General Canonization of all the Francoist war criminals. There are many precedents for this, as in the movie *Àgora*, the Church blesses the canonization of Saint Cyril of Alexandria. Why not also the Francos, the Queipos, the Don Brunos, the Díaz Criados, Gómez Cantos and so on, and so on, without forgetting the Bañuls of those times?[20] A magnificent flowering of war criminals, smelling of sainthood. Would Moses ever have thought of writing the 5th Commandment on the tablets if he could have known that the above unholy alliance was to honour that Commandment even less than it did the 6th?

---

[19] Although Moreno Gómez has protested against this practice several times in the social media, photos of the 2017 procession show the red sash still tied around the waist of the statue of the Virgin. *Vide* http://miralospedroches.es/la -virgen-de-luna/.

[20] Thanks to the approval of the revised History Memory Legislation in 2022, the declaration of the illegality of the military coup, the removal of the 'sainted public memorials' to Franco and Queipo, this would be most unlikely today.

**Republican inmates in Seville being taken from the prison
to the cemetery where they were to be executed.**[21]

[21] Photograph courtesy of ICAS-SAHP, Fototeca Municipal de Sevilla, Fondo Serrano.
https://www.nytimes.com/2012/05/13/books/review/

ENDNOTES FOR CHAPTER V

[i] Pablo Uriel, Op. Cit., p. 27.

[ii] Antonio Gómez Tienda. Account regarding his brother. Interviewed by Moreno Gómez in Baena, 6 to 10 January 1982.

[iii] Further details of this case can be found in Arcángl Bedmar, Op. Cit., p. 258.

[iv] Arcángel Bedmar, Op. Cit., p. 260.

[v] Ibid.

[vi] Manuel Castro Delgado, Information provided to Moreno Gómez by Manuel Castro Merino's grandson, in several letters, February 2014.

[vii] Arcángel Bedmar, Op. Cit., p. 249.

[viii] Ildefonso Sedano, from La Posadilla, Fuenteobejuna. Letter to Moreno Gómez from Cordoba, dated 19 November 1985.

[ix] Ángel Horrillo, eyewitness report sent to Moreno Gómez from Peñarroya-Pueblonuevo 1 July 1983. He was a native of Ojuelos Altos, Fuenteobejuna.

[x] Ángel Bedmar, Op. Cit., pp. 109 et al.

[xi] Francisco Moreno Gómez, El genocidio franquista en Cordoba, Op. Cit., p. 582.

[xii] Félix and Juan Manuel Adán Gaitán. Mártires de una esperanza. República, guerra civil y represión en Pedro Abad.(Martyrs of hope. Republic, civil war and repression in Pedro Abad). Lopeta, Jaén, 2009.

[xiii] Francisco Moreno Gómez, El genocidio franquista en Cordoba, Idem.

[xiv] Rafael García Contreras. Susurros de libertad. Memorias. (Whispers of Freedom. Memories).El Páramo, Cordoba, 2009, p. 22.

[xv] Moreno Gómez owes much of this information to the kindness of Juan Peralta, the victim's grandson, who sent him Gabriel González' dossier February 24, 2014.

[xvi] Constanza de la Mota Maura, Doble spendor (Double splendor), Gadir, Madrid, 2004, p. 550.

[xvii] Bartolomé Cabrera Peralbo, eyewitness reports, interviewed by Moreno Gómez in Pozoblanco, November 21 1985 and October 17 1986.

[xviii] Francisco Moreno Gómez. Villanueva newspaper, Villanueva de Córdoba, number 12, April 1981.

[xix] En solidaridad con la Comissión y Proyecto de Monumento (In support of the Monument Committee and Project), Villanueva newspaper, Villanueva de Córdoba, number 19, November 1981.

[xx] Manuel Bedmar, eyewitness accounts during several interviews with Moreno Gómez in Puente Genil.

[xxi] Juan Gutiérrez Romero, a companion of Vilches in prison. Eyewitness report.

[xxii] Antonio Bahamonde. Un año con Queipo de Llano. Memorias de un Nacionalista. (A year with Queipo de Llano. Memoires of a Nationalist.). Espuela de Plata, Seville, 2005, p. 113.

[xxiii] Letter to Moreno Gómez from the Todos Los Nombres de Porcuna (All the Names from Porcuna) Internet project, dated January 30 2013. Blog Religión en Libertad (Religion in Freedom): http://www.religionenlibertad.com/blog/tomas-de-la-torre-lendinez-297.html.

[xxiv] Francisco Moreno Gómez. Hombres que dejan huella. Nemesio Pozuelo. (Men who have made their mark. Nemesio Pozuelo). Biography of this leader in the Villanueva newspaper, Villanueva de Córdoba, issue 23, March 1982.

[xxv] For an ore complete biography of Miguel Ranchal, see Francisco Moreno Gómez, Villanueva newspaper, Villanueva de Córdoba, issue 23, March 1982.

[xxvi] Alfonso Vaquero Zamorano, interviewed by Moreno Gómez 16 October 2013.

[xxvii] Azul newspaper, Cordoba, 16 October 1940.

[xxviii] Letter from the Mayor of Belalcázar to the Civil Governor in Cordoba 16 December 1940. Belalcázar Municipal Archives.

[xxix] Ernesto Caballero, Vivir con Memoria (Living with Memories). Planeta, Barcelona, 1979.

[xxx] Manuela Illescas, one of the women who had to leave a new-born child behind. Eyewitness account to Moreno Gómez

[xxxi] Pozoblanco Municipal Archives, Book of Minutes of the Sessions, Box Number 246.

VI

CRUSHING THE VANQUISHED (E) LAW OF FUGITIVES AND THE UPSURGE OF EXTRA-LEGAL SUMMARY EXECUTIONS IN 1941 CAUSA GENERAL RAIDS, FIRING SQUAD EXECUTIONS IN CORDOBA CAPITAL

> *"[Saturated with memories] ... we should qualify what kind of saturation we are talking about. Who is so saturated? Is it really absurd to continue speaking of forgetfulness and silence? ... In all cases, there appears to be no doubt that the intended saturation, in case it existed, is there but basically only in one direction. If there is a saturation of memory, it is the memory of the victors, not that of the defeated. The saturated memory (of the former) and the satiation with the silence (of the latter)."*
>
> **Alberto Reig Tapa**
> *Memoria de la Guerra Civil*

The Law of Fugitives and the upsurge of extra-legal summary executions in 1941

Nobody could imagine the upsurge in the application of the Law of Fugitives in 1941. Certainly, none of those who minimize the humanitarian catastrophe of Francoism, those who believe that all was 'legal' during the post-war period, who have written about the Law of Fugitives applied during the Black Spring (April-May 1939) in the entire Centre-South region of Spain and even less regarding how and why there was an upsurge in the Law of Fugitives and other forms of extra-legal summary executions that were brought back in 1941. Nor has any one of those so-called historians written about how hundreds were taken for paseos during the period that Moreno Gómez calls the Triennium of Terror (1947-1949) in all of Spain, from Asturias to Málaga, the golden years of the Law of Fugitives. In the many writings about post-war Francoist government repression that have been published, despite so much talk about political violence and exemplary repression, there is little mention of the paralyzing terror, genocide, widespread extra-judicial extermination... the multiple crimes against humanity.

Something strange occurred in 1941 in quite a few towns and villages in Cordoba province when there was a series of executions that could not have been 'legal' because all the 'legally condemned' prisoners had been transferred to Cordoba Provincial Prison at the end of September 1940. These 1941 executions were summary executions devoid of all due process of law, an extra-legal, calculated surge of the repression in 1941, as the Law of Fugitives was applied in Cordoba towns and villages.

The so-called legal basis for this 'pull back' as some ignoramus has called it, was the Law for the Safety of the State of March 20 1941, as published in the Official Bulletin of the Province on the orders of the Captain General of the 2nd Military Region, General Miguel Ponte y Manso de Zuñiga. This law contained a new list of crimes punishable by death, (with the activities of the maquis in mind) including armed robbery. August 26 of the same year, an Order signed by General Emilio Álvarez Areces called, in no uncertain terms, for the 'elimination' of everybody who had escaped to the hills and their accomplices in the plains. Accordingly, all the deaths that occurred in the Cordoba countryside in 1941 were a consequence of these new extermination measures.

In Adamuz, there was a usual execution in the cemetery of a young farmer, Ricardo Molina Pastor, August 8, in a village when there were no longer any 'lawful' prisoners left as they had all been sent to Cordoba city. Without a doubt, this was a summary execution. In Aguilar, the Guardia Civil killed José Vega Castellón at the railway station, November 25.

In Bujalance, three young men were summarily executed in the cemetery, without a trial. They were not fugitives but, motivated by hunger and the post-war shortages, they would go out at night to scavenge for food whilst going about their usual lives in the town during the day. They were surprised during one of these raids by the manager of an outlying farm, who shot and wounded one of them, which is how they came to be caught. They were taken to the cemetery and shot on the spot.

In Dos Torres, 6 young men were summarily executed during the Summer. Moreno Gómez was unable to determine the exact reason, but presumably it was related either to the fugitives or to some kind of robbery. Eyewitnesses with whom he spoke in the village told him of a 2nd Lieutenant who was posted in the village and who, helped by some low-life accusers, invented a false Communist plot and some supposed contacts with the fugitives in the hills, as a pretext to summarily execute 6 disaffected individuals. The Law of Escapes was applied at dawn July 27, to Antonio Jurado and Genaro Cazorla; July 30, to Pedro Romero, José Talero and Sebastián Lunar; and August 1 to José Romero Iglesias. These executions were recorded as the 'elimination of subversive elements', in Special Military Court for Fugitives Number 8 – a foretaste of that which would happen in 1947-49 during the Triennium of Terror.

The nadir of the criminality of this period occurred in Espiel, at the hands of Carlist Militia 2nd Lieutenant José Moreno Sevillano, the notorious genocide *Teniente Pepinillo*. Also nicknamed 'the red beret', he sowed terror when he was stationed in Villanueva de Cordoba and continued to do so after he was transferred to Espiel in 1941. Obsessed with the idea that there were Rojos all over the neighbouring fields and countryside, he dedicated himself to arresting farmworkers like crazy. He was assisted by perverted accusers such as Teodoro Valero from Pozoblanco, and a so-called Francisco from Espiel (brother-in-law to olive oil producer Emilio González). Pepinillo and his henchmen were responsible for one of the most criminal post-war episodes in Cordoba province, with arbitrary arrests, tortures and beatings, right left and centre to everyone they lay their hands on, without distinction.

The apotheosis of a crime of *lèse humanité* marked the July 18 1941 celebrations in Espiel, as recently described by an eyewitness, the relative of two of the victims: his father – Antonio Arévalo, aged 53, and his brother – Adrián, aged 19[i]. He would strip the inmates' chests, handcuff them and tie their hands to their legs, and then beat the unfortunate men on their backs. He would send their clothes for washing to their families, soaked in blood and with strips of skin sticking to them. One inmate, Adrián Arévalo, our witness' brother, wrote his fiancée: *"I don't know whether we will ever see each other again, We are being made to pay for something we did not do."*

One day, several inmates asked to confess to the village priest. Some took this opportunity to tell the priest of the unfair treatment they were being subjected to, especially as they had not committed any crime and were only humble working farmhands with no involvement in anything political or with the guerrillas in the mountains. The priest, believing the truth of their declarations, went to Pepinillo telling him that he was mistaken regarding those individuals. Pepinillo was absolutely furious: he slapped the priest hard across his face and accused him of being a Rojo. The next day, the priest went to Seville to complain at military headquarters that he had been slapped by Pepinillo and he did not return to Espiel. The military authorities acted swiftly, and an order arrived from Cordoba Military Headquarters ordering Pepinillo to present himself immediately to Headquarters, within a maximum of four hours, unarmed. This was late on the eve of July 18.

Pepinillo's revenge was swift and terrible. The very next night, July 18, he ordered a truck to collect all the inmates in the town jail and take them to the cemetery, whilst he went with another truck to the Municipal Depot where there were another 15 prisoners. Arriving at the cemetery, Pepinillo unloaded his truck and had all 15 prisoners executed. Meanwhile, the driver of the first truck, upon arriving at the cemetery and terrified with what he saw and what about to happen to the men in his truck, said "I am going to risk my life", floored the gas pedal and drove the truck and the prisoners post-haste to Cordoba.

No sooner were the executions over, Pepinillo went to a town dance where the townspeople were celebrating the date and asked a young lady for a dance. Soon afterwards, in the middle of the dance, he took his revolver out and shot himself in the head, thus ending his life. The young woman was so shaken by his action that she suffered mental problems for the rest of her life.

Among those executed on that date: 7 were from Pozoblanco, 4 from Villaralto, 3 from Espiel and 1 from Pueblonuevo. Bláz Muñoz Márquez, brother to Baudilio Muñoz Márquez, one of the victims from Pozoblanco, provided a background account of the events that led to his brother's death. At the end of the war, he said, there were many deplorable and painful events. There was a man in Pozoblanco, Teodoro Valero, who owned an olive grove in Peñon de Lazarillo, Espiel, who saw a potential Communist in every man in the village. This man spent his time accusing men of proven honesty, they were arrested and taken to Pepinillo's headquarters in Espiel.

It is useless to describe the scenes of what they did to these men there for whom there was no specific accusation and whose fabricated confessions were obtained by the most tremendous beatings. Teodoro and a so-called Francisco dedicated themselves to going to the farms in the mountains and, on Pepinillo's orders, capturing men they accused of being Rojos or Communists. One notable case was that of a Falangista, also named Teodoro, who owned some money to the Pozoblanco judge, Antonio García Ruíz. The latter forgave Teodoro his debt on the condition that they stop beating his shepherd. Even the Espiel parish priest began to complain of Pepinillo and his gang's behaviour.[ii]

Pepinillo's activities included the economic destruction of his victims and their families, which was usual Francoist behaviour in those days. Farmhands who were arrested in the fields near Espiel also suffered the loss of their belonging and, occasionally their homesteads. A case in point is that of the Arévalo Bajo family. Pepinillo seized their herd of goats, a mare and other belongings which ended up in a military barracks in Cerro Muriano. After death had decimated the family, his widow and her surviving sons went to Cerro Muriano to retrieve their cattle. The soldier in charge understood her outrage and, taking pity on the family, told her: "Most of the goats have been killed, but the mare is over there. Take them" The family that had been plundered, returned to Espiel with the few animals they had left.

This was not a case of an occasional excess nor a Francoist officer's individual's fit of madness, but yet another example of the fact that Francoism consisted precisely of excesses, arbitrary actions and unlimited cruelty. With the onset of the military coup, bloodthirsty individuals popped up like poison mushrooms all over Spain, such as a policeman called Naranjo, one of Pepinillo's more diligent assistants. Under his orders, women's heads were shaved, they were stripped naked and forced to drink castor oil. Everyone was terrorized. One day, a travelling musician arrived in the town with his goat, but he did not know how to play *Cara al Sol*. Pepinillo had him arrested and beaten until he learnt how to play the tune and then made him spend an entire day playing the fascist hymn all over town.

In Fuenteobejuna, Jesús Franje Carlos, a young working man of whom nothing else is known, was summarily executed by firing squad February 9.

In Hinojosa del Duque, two men were executed in Summer, perhaps in relation to robberies on farms, supposedly by suicide, possibly due to torture. A man whose name is unknown is recorded as having died, supposedly by hanging in the Guardia Civil barracks, June 22. Likewise, July 2, the cause of another unknown prisoner's death is recorded as 'wounded himself'.

In Hornachuelos, 4 victims were summarily executed in 1941: Julio Ramos, February 13; Antonio Sevilla and Francisco Moyano, February 17; José

In Villaviciosa, Francisco Martínez Aguilera is officially presumed as having committed 'suicide by hanging' in the municipal jail, August 10. Even if this were a true case of suicide, the disproportionately high number of suicides in Franco's prisons is, on its own, an indication of Francoist methods and criminality.

In Santa Eufemia prison, July 13, another peculiar cause of death in the jail, officially due to 'alcoholic coma' (were alcoholic drinks allowed in jail?). Moreno Gómez' informants in the town, however, assured him that he died from a beating.

Lastly, a complicated case in Torrecampo where another 2nd Lieutenant posted to this town was busy doing his own thing. July 31 in the evening, two men who had been arrested were being viciously beaten. When the tormentors broke one of Florencio Rísquez Andújar's arms, he exploded like a caged wild animal, attacked the Lieutenant and ran out of the jail. As one might expect, he was caught in the Torrecampo main square where he was shot and killed. The Lieutenant also ordered the execution of another prisoner, Sebastián Pastor Romero, who had also attempted to escape from the jail but was hit on the head with a stone and brought back, where he sat quietly in his cell. Sebastián's wife was also beaten on the Lieutenant's orders. All this, that evening.

It appears that in all the above cases, this was related to presumed contacts between these men and those who had fled to the hills. Men working in the fields were continuously subject to raid after raid and tempestuous beatings, always because of malicious allegations. As the post-war period progressed, the climate of terror spread from the urban centre to the rural world as official interest in repressive activities acquired a new impetus. By 1941, the New State was no longer centering its attention on the wartime causes of anti-Francoist activities, but on all sources of post-war clandestine and reorganizational political activities, including populace support to fugitives. The punishments the authorities meted were terrible indeed.

Causa General Raids

The *Causa General de la Guerra Civil Española, la dominación roja en España* (General Cause of the Spanish Civil War, the Red rule over Spain), was a procedure instituted by Supreme Court Decree April 16 1940, retroactive to April 1 1939, endowing Franco's Ministry of 'Justice' with the means to investigate and punish the nature and extent of the 'criminal activities of the subversive elements who openly opposed the existence of the essential values of the Fatherland that were saved, at the last moment, by the Liberating Movement".

The Causa General was essentially an archive containing evidence with which Francoism could prosecute any remaining Republicans and all those it considered disaffected, for crimes against Spain. Public prosecutors were sent into the provinces to collect documents and compile witness statements, although legal standards of proof were not always observed. The more than 1500 files that were compiled were used to complement the military and political repression and as a justification for swooping down on towns and villages to collect prisoners and later, as evidence, for the trials that were organized in the provinces before 1943-45.[iii] This triggered thousands of raids and arrests throughout the country and thousands of summary proceedings that ended up in thousands of executions.

One particularly scandalous trial, or *processo*, the *Processo de La Parrala* or *de La Centena* – the Parrala or Centena Trial – Causa General Case L546/41, was the outcome of the great November 221 raid in several towns and villages in Cordoba province, especially in Villanueva de Cordoba, with consequences in Jaén and Seville, after which over a hundred were arrested.

The cause of the November raid was the activities of an extensive network of wives and relatives created to help the prisoners who had been taken to Cordoba Provincial Prison and who were dying by the dozens from hunger and starvation, typical Nazi prison extermination practices. Despite the support to the organization to provide assistance for the imprisoned and which, under Victoria Fernández, undoubtedly saved many lives, more than 500 inmates died in 1941 alone (a total 756 died during the entire decade). During the post-war period, more men died in Cordoba capital from hunger and other causes than by firing squads.

403

Early 1980s, Moreno Gómez obtained a great many witness accounts of this raid. In the Vallecas neighbourhood of Madrid, he interviewed one of the accused, Isabel *La Chata* Gutiérrez Romero from Villanueva. He interviewed María *La Loba* Muñoz in Villanueva de Cordoba and then went to Puertollano, the hometown of another of the Villanueva accused, Juan Reyes Gómez (brother-in-law to *La Moejea* of the maquis), who gave him many more details.

Isabel *La Chata*, María *La Loba* and other women who were prominent members of the local PCE, were already targeted because they fled to the mountains in the first days of the Nationalist victory. They had gone to the La Garganta area but only stayed away for fifteen days before returning home because they found that kind of life unbearable. Upon their arrival, they were picked up by the authorities, soundly beaten, subjected to the castor oil and shaven head treatment, and labelled 'the Mayor's whores'. As Rojas, their names were on a list. This brings us to 1941 and the scandalous high mortality from starvation of the Cordoba prison inmates. Many of the aforementioned Rojas, as well as Victoria Fernández (Adriano Romero's wife) and his sisters Dolores and Isabel, began to gather food to try and save their lives.

Victoria Fernández was the first one arrested 21 November and with this, the link to Seville and Villanueva de Cordoba was established. The Jaén connection was established through José Lupiáñez from Jaén, who was already imprisoned with Adriano Romero in Cordoba.

The raid began in Villanueva de Cordoba November 21 with the arrest of María *La Loba*, Isabel *La Chata*, and several other women, as well as quite a few men: Juan Reyes Gómez, Miguel *El de la Fragua* Cabrera, José Vioque García, Juan Capitán Gutiérrez, Román *El Carbonero*, Diego García *El Mosico* García Cachinero, Manuel *Mazo* Torralbo Cantador, the schoolmaster, and several others.

Under Francoism, arrests were not simple conventional judiciary procedures but the first step for terrible beatings. First came the beatings and the rest followed. In this case, under the orders of a rabid Guardia Civil Captain, all the arrested were subjected to torture, beatings and humiliations in the Fuente Vieja Schools improvised jail. Juan Reyes testified that he was cruelly beaten on the grounds that he was a 'Communist accomplice of those who had fled to the mountains'. Manuel *Mazo* the schoolmaster, was accused of being a Communist leader. The women were not beaten.

All these charges were idiotic as the only thing that brought them together was a simple organization to help the Cordoba inmates. Victoria Fernández and Dolores Romero were imprisoned in Cordoba. Rosa Alcaraz was imprisoned in Jaén and several Sevillians, José Muñoz, Faustino García Marín, Salvador Galiana Serra, Antonia and Carmen Navarro, Rosarito Navarro and others, as well as José Merino Campos, were captured in Cordoba.

A few days later, the hundred prisoners who had been caught in the raid were tied two by two and taken to the railway station and from there to Cordoba capital prison, then onto Seville where they spent a year in prison awaiting trial. Cause 1,546/41 was tried in Seville by a Council of War presided by Judge Carlos Ollero y Sierra. Considering that the charges were inconsistent with each other - assistance to rebellion, plotting, clandestine reorganization of the Communist Party, etc. - he avoided handing out any death sentences.

By October 1942, many of the accused were out on parole, with all that this implied: daily checking in at the local barracks and regular doses of 'syrup of the rod'. Some, such as schoolmaster Manuel *Mazo* and Juan Reyes, although allowed out on parole, were not formally tried until October 20 1945, when they received a light sentence, already served with their first year in prison. Mazo, the schoolmaster, was re-arrested in June 1948 and killed under the Law of Fugitives.

These raids, based on Causa General information and carried out without any pretext, served from the onset to keep those disaffected with Francoism on their toes and make their daily lives unbearable. After many raids, the military authorities did not even bother with arrests or any legal procedures, they just held the persons they caught as prisoners of the local government. Even when there was a semblance of due process, the result was the same: several months in jail, children abandoned, jobs lost, or work remained undone and fields untended. The purpose: 'Do not let the disaffected rest and make it clear who are the bosses in Spain and who holds the upper hand'. This was so characteristic of Spain during the 1940s, that the jails became a kind of unholy social meeting place for supporters of the Left, and when some left, others arrived. All were tortured and beaten to some extent.

These comings and goings went on for an entire decade, with another aggravating factor: all those who were released, were released on parole, that is, under supervision. What did this mean? That every afternoon they had to present themselves at the barracks and have their names checked off a list, a humiliating ceremony. Everyday, some were lectured, and others taken into the barracks to be 'warmed up'. As a result, quite a few fled to the hills. The famous guerrilla Juanin from Cantabria, fled to the mountains because of the daily beating he received in his hometown. All this extreme behaviour is ignored by Julius Ruíz and the deniers who are still determined to believe that the repression consisted only of shooting a handful of disaffected criminals. They have no idea or still choose to ignore what was going on in the depths of the country, where exclusion, hunger, beatings, humiliations, constant persecution, slave labour, blacklists, castor oil, and shavings were the order of the day.

Generally speaking, most people today have very little idea of just how chancy life was for the defeated during the first years of the 1940s, as the relentless persecution of fugitives in most of the Spanish provinces as a pretext for stoking the fire of terror, not only in the mountains but also in the countryside, in the rural world. Just reading the list of individuals executed does not tell us anything of this or of the multitude of raids that were launched for the most banal reasons. The following witness account by one who was caught in such a raid, Juan Gutiérrez Romero, a native of Villanueva de Cordoba but resident in Villafranca, is witness to this:

> "In November of the same year (1939), after the maquis' attack on Pozoblanco, I was arrested for the first time. We were accused by Blas *El Sillonero* Carbonero and held in Villanueva jail until September 1940 when we were taken to prison in Cordoba, together with all the other prisoners in the province. One year later, the accusation was cancelled, and we were allowed to return to Villanueva. January 1942, I was free, but on parole. I could not go far out of town because every evening I had to present myself at the barracks to mark my presence.
>
> At the beginning of 1943, several men from the Villanueva maquis (Diego *El Chato* and others) went to the mountains to join Julián's maquis in Las Umbrias... A few days later, all the farmworkers from the region were ordered to Torrico farm in La Loma del Caballero

and they kept us all under arrest for three months, first in Montoro and later in Cordoba capital. I was lucky and they did not discover my contacts with Julián.

December 1944, I was arrested for a third time. They caught Isidoro *El Lobo* and because of that, they caught me also, I was taken to the Villanueva Mobile Headquarters in the Fuente Vieja Schools jail where I was beaten then released.

At the beginning of 1945 I was again taken and made to take part in a raid looking for Rojos. By then, other friends such as Pablo Agenjo and Diego *El Mosico* had also been released from prison. The latter was made to go along several raids with me. A sergeant told us: "In war, when you catch a confident, you liquidate him, and we are at war."[iv]

Another example from the 1941 year of terror occurred Fall that year, when November 25 a mixed patrol of Guardia Civiles and Falangistas, under the orders of 2nd Lieutenant Eduardo Solar, was carrying out a routine inspection in the countryside. The patrol was searching to the left of the Villanueva-Adamuz road, when the Lieutenant walked off towards some shrubs where a group of Julián Caballero's guerrillas were hiding. The guerrillas had no choice but to shoot the Lieutenant on the spot and to run for their lives. The next day, the Falangista Diego *El Chunga* and his usual group of Falangista civilians and soldiers rounded up all the farmworkers in the region and took them under arrest. They beat a family they nicknamed "Los Quemados' to death. The system was always the same and repeated itself constantly: incident in the mountains > general raid and roundup of farmworkers > beatings all round.

The 21 Roses of Cordoba

Francoists exterminated 21 women in Cordoba province during the post-war period. Many more fell during the war, especially in 1936, both in the capital and the province, as happened all over Spain. The following Table gives the name of these *21 Roses*, divided into three groups: those who were summarily executed during the first two post-war years, those who were killed in the fight against the guerrillas, and those who were shot under the Law of Fugitives in the countryside.

The designation *21 Roses* is applied here as an honour and as being similar to the case of the *13 Roses of Madrid*, the name given to a group of thirteen young women summarily executed in Madrid just after the conclusion of the civil war. Their execution was part of a massive execution campaign known as the *saca de agosto*, which included 43 young men, among them a fourteen-year-old.

TABLE VI. The *Twenty-One Roses of Cordoba*

Location	Name	Date
Shot by firing squad during 1939 and 1940		
Belalcázar	Carmen Rubio Cáceres, age 32	August 6 1939
	Matilde Medina Pizarro, age 50	August 20 1939
Belmez	Leonor Expósito Palomo, age 38	May 9 1940
Hinojosa	Carmen Aranda Caballero, age 47	August 13 1939
Montoro	Patrocinia Juárez Pareja, age 39	September 9 1940
Pedro Abad	Josefa Ortega Egea, age 37	October 3 1940
Peñarroya	Martina Alcántara, age 53	December 11 1939
	Dionisia Alcántara Calvo, age 60[1]	December 16, 1939
Pozoblanco	Tomasa Díaz Moreno, age 21	October 28 1939
Killed while fighting with the guerrillas		
Fuenteobejuna	Isidora Merino Merino (from Esparragosa)	February 27 1947
Belalcázar	Soledad Moreno García (from Guadalmez)	June 2 1947
Villaviciosa	Luísa Lina Montero (from La Granja)	June 11 1947
El Viso	Maria Josefa López Garrido (from Villanueva de Cordoba)	March 5 1948
	Sergia Flores Sanz, in Ciudad Real	"
Assassinated in the countryside under the Law of Fugitives		
Fuente Tójar	Josefa Briones Molina, age 58	December 27 1946
Villanueva de Cordoba	Catalina Colero Muñoz, age 52	June 8 1948
	Amelia Rodríguez López, age 49	September 10 1948
Pozoblanco	Amelia García Rodríguez, age 18	"
	Isabel Tejada López, age 60	"
Cardeña	Brígida Muñoz Días, age 60 (from Obejo)	September 14 1948
Belmez	Teresa Molina Sánchez, age 26 (from Espiel)	February 27, 1949

---

[1] Not shot. Recorded as having collapsed and died, probably after being tortured.

Firing squad executions in Cordoba capital 1939-1942

"Everything that is told here is true, because I believe that it will be useful for History, and I like it when History tells the truth."
Rosario La Dinamatera, in
Tomasa Cuevas, Barcelona, 2004.

"The day that one truly knows the history of the repression in these villages of Toledo, Extremadura, in all of Spain, there will be those who believe that it is not possible that so much happened, and that human beings,because we have to call them something, were capable of taking their sadism and their hatred to such tremendous extremes."
Carmen Machado, in
Tomasa Cuevas, Barcelona, 2004.

The genocide in Cordoba capital was shocking. During the war, Francoism had ended the life of at least 4,000 persons in the capital, not all of them from that city but also from many towns and villages in the province who were taken to Cordoba to be shot, together with prisoners whom they captured on the Los Pedroches front lines and members of the International Brigade they captured on the Lopera-Porcuna front.

The terror was supervised by Commander Zurdo and the bloodbath was at the hands of Lt. Colonel Bruno Ibáñez, both of the Guardia Civil, and other bloodthirsty repressors who emptied Cordoba of Republicans.[v] Rafael Castejón y Martínez de Arizala, past-President of the Real Academia of Cordoba, assured Moreno Gómez, during an interview July 13 1983, that according to Spanish Red Cross archives, the number of victims in Cordoba capital totalled 7,770. In Moreno Gómez' opinion, this number is excessive but there are no means by which he can confirm or correct it.

To make matters more difficult, in 2013, the current Director of the Spanish Red Cross in Madrid, Carmen Flórez Pérez, refused to allow Moreno Gómez access to the Red Cross archives. As Moreno Gómez already had reliable confirmation that up to the middle of 1938, 3,495 prisoners had already been executed in Cordoba capital, he wish to further his research at the Red Cross. Ms. Flórez Pérez denied him access on the grounds that: "We are subject to Law 15/1999, Data Protection, and consequently, personal data will only and exclusively be made available to the interested parties themselves and/or their relatives".[vi]

It was quite clear: the freedom to do historical research in Spanish archives still shines by its absence. The right to privacy after death, over the rights of historical science.

Winter 1939

The list of victims executed in San Rafael and the La Salud cemeteries in Cordoba capital during the post-war period, was again outrageous. Executions by firing squad began anew in Cordoba capital with the onset of the courts martial. At first, there were few sacas, mostly involving people from outside the province – from Jaén, Málaga and Badajoz. Basically, the Regime was liquidating the last groups of prisoners taken during the war.

November 8

The largest saca that year, 11 victims, among whom Ricardo Rubio Calero, aged 27, son of *El Calor*, a well-known elderly Socialist from Pozoblanco. The young man was a prisoner of war who had been captured in Vinaroz, Castellón, in 1938. There were two major strikes against him: on the one hand, he was the son of a time-honoured Socialist leader, whilst on the other hand, he had served as a political commissar in the Republican Army. Although some attributes and occupations were immediately fatal in the great Francoist cleansing operation, the executioners usually looked for additional, albeit false, specific accusations with which to advance ulterior motives. Accordingly, Ricardo Rubio was also accused of having belonged to the local War Committee, a totally false charge, but the reality was that the Regime had decided that all of El Calor's family had to be exterminated. Several other prisoners executed in the same saca were from Puente Genil, the Cordoba town that, after the capital, sacrificed the greatest number of its residents.

December 6

6 more executed, a few more from Puente Genil, among them one of the heavy-weights of Cordoban Socialism, Justo Deza Montero. Actually, he belonged to the Young Socialists, and as a member of the War Committee was one of the angry townspeople who rose

against the military coup in Puente Genil and managed to thrash the equally furious local Fascists. Indeed, the Republicans enjoyed a brief victory when August 1, they repelled an attack by General Castejón's Legionnaires who in this town later perpetrated one of the largest acts of genocide in Spain. During the war, Justo lived in Pozoblanco where he presided over one of the Community Councils. It is no surprise that there was a price on his head. He is reported to have faced death valiantly shouting: "Compañeros, you have to revenge so much bloodshed!"

Winter 1940

At the beginning of 1940, the Regime began executing a number of professional soldiers, members of the Republican Army. In Cordoba, professional soldiers were executed in a killing field of their own, the Polígono de Casillas. Considered by Franco as war criminals, in their dossiers they are only described with the rank they held before July 18, not the rank they rose to in the Republican Army during the war. First, however, the officers were stripped of their highest wartime rank and reduced to the rank they held before July 18.[vii]

January 17
Two high-ranking Republican officers, prisoners of a defeated army and therefore subject to the Geneva Convention which Franco always ignored, were executed on this date, thus marking the beginning of a new series of war crimes at the hands of the Regime. Executed were Lieutenant-Colonel Narciso Sánchez Aparicio, 50-year-old Artillery Commander, a member of the Central Army Chiefs of Staff and at the end of the war, Head of the Chiefs of Staff of the XVII Army Group and then of the XXIII Army Group. Executed with him: Artillery Commander, Lieutenant Esteban Rodríguez Domingo from Valencia. Captain in 1936 and Comandante[2] in 1938.

January 27
Two more Republican soldiers executed in the Polígono de Casillas. Captain José Bueno Queijo, a valiant Basque, who defended the Republic in the North of Spain. He commanded the 2nd Division of

---

2   Equivalent to Major in US and UK armies.

the Santander Army Group, where he rose to the rank of Comandante. When the North fell, he went to France and from there returned to the Centre of Spain to defend the legal government. At the end of the war, José Bueno was Head of the Chiefs of Staff of the 22nd Division of the Army of Andalusia. Defeated and taken prisoner, he was denied his rights under the Geneva Convention and was executed. Furthermore, also charged under the Francoist Law of Political Responsibilities, he was fined 5,000 pesetas that his family had to pay. Executed with him, Lieutenant Luís Soler Espianba from Cartagena, who although he had retired from the army in 1936, re-enlisted to defend the Republic when war broke out. He rose to the rank of Captain, then to Comandante (1938) and he fought in the Granada sector. He ended the war in the 54th Battalion of the 23rd Division of the XXII Army of Andalusia.

February

That February, they executed a great loyal soldier, Lieutenant Roberto García Domenech, who had come to Cordoba in 1936 with the Alcoy Unit, under the command of Colonel Giralf. When this unit was divided into two sections and sent one to Cerro Muriano and another to Espejo, Lieutenant García was sent to Espejo where September 23-25 he fought with great valour under Colonel Jesús Perez Salas.

March 18

The wholesale killing of civilians in 1940 Cordoba capital began March 18, in this case with a saca of 19 victims from several towns in Cordoba province – Hornachuelos, Villa del Río, Obejo, Almodóvar and more. Among the executed: Gonzalo Obrero Duque, an eminent local trade union leader from Villafranca; Francisco Haro Manzano, belonged to a Bujalance family that was all but exterminated – 4 brothers were executed, and another spent many years in prison in Puerto Santa Maria – only one sister survived.[3] The workers' movement

---

[3] Moreno Gómez recently discovered that a nephew of the Haro Manzanos lived near him in Getafe (Madrid). "I would have liked to hear what he had to say about these events, but I never did because every tine I brought up the subject, he burst into tears. This is why I repeat that Francoism was a river of blood for the dead and a river of tears for the living. These features continue to be ignored by all those who neither investigate or show any interest in the victims and their families."

in Bujalance was wiped from the map when the last activists, the Jubiles brothers, died as guerrillas in the mountains.

Spring

April 5

6 Republican military prisoners of war were executed: Lieutenant Lorenzo Almaraz de Pedro, who had fled from Badajoz to Portugal in 1936, then on the ship *Nyassa*[4] to Tarragona where he re-joined the Army. Promoted to Captain, he fought in Catalonia and with the XXII Army Corps, Army Group of Levante, when he was taken prisoner in 1938; Lieutenant Damián Contreras began in the Jaén Volunteer Battalion. Promoted to Captain October 1936, he commanded the 148 MB of the 37th Division of the XII Army of Extremadura, and June 1938, promoted to Comandante; 2nd Lieutenant Felipe Gallardo Linares, retired in 1936, re-enlisted and rose to the rank of Lieutenant, then Captain, then Comandante in 1938, always with the Army of Andalusia; Lieutenant Eugenio Muñoz Hoyela, a *Guarda de Asalto*[5], was posted to Linares in 1936. He has been described as very active during the entire agitated Summer 1936, especially in creating the Jaén Militia where he became famous as the soul of the resistance to the coup in the entire Upper River Guadalquivir region. He was promoted Captain in October 1936 and Comandante in 1938 when he was posted to the Army of the Centre. Artillery 2nd Lieutenant Antonio Fernández Sánchez was always posted in Cartagena. He was promoted Lieutenant October 1936 and Captain in March 1938; last, Captain Enrique Medina Vega, retired in 1936, re-enlisted and rose to the rank of Comandante, serving with a Machine Gun Unit in Almería until the end of the war.

---

[4] Sailing under the Portuguese flag, in 1944, a large number of Jewish refugees from the German holocaust sailed on this "refugee ship" to safety in what was to become Israel.

[5] An elite army corps similar to a Pretorian Guard.

April 8

Another huge saca of 24 victims. Little or nothing is known regarding these men other than their names. Quite a few were from Villa del Río, Hornachuelos, Adamuz, Villafranca and other provincial towns. Although there is no information regarding their professions, the men themselves or their lives, at least their names were recorded for History.

April 20

Civilian executions in Cordoba usually took place in the City Prison but April 20, for some unknown reason, 5 men were executed in the Provincial Prison: 3 from Posadas, one from Villa del Río and another from Adamuz.

June 4

16 executed, several from Palma del Río, Hornachuelos, Posadas, Espiel and other towns. This month there was a saca practically every day.

June 6

17 executed. The sacas were getting larger and larger and the panic in the jails was terrible. The men shuffled off to their deaths like robots, resigned to their fate, at the same time that Francoists were shooting their countrymen extra-legally in all the towns and capitals of Spain. This was the great cleansing of Spain, to the rhythm of rural and Catholic fascism.

June 8

14 men from Palma del Río, Montoro, Villa del Río and other Cordoba towns executed. Included in this saca, Rafael Polonio Delgado, a native of Montilla, resident in Palma del Río. He belonged to the JSU where he was very active, as were his brothers. His brother Francisco also imprisoned in Cordoba, had come from a French concentration camp. He was permitted to accompany Rafael in the chapel to say farewell. Another brother, Antonio, had already given his life for the Republic on the Peñarroya front.

June 22

18 executed. The most notable individual in this saca was Paco Dios. Francisco Dios Muñoz, *El Capitán Paco* as he was known, was a

young bricklayer from Villafranca, the heart and soul of the Villafranca Battalion and later of the 74MB, and another great fighter of that irreplaceable generation of the 1930s. At the beginning of 1980, Moreno Gómez interviewed his brother Juan, in Cordoba, who told him of his brother's valiant end, as reported earlier. Capitán Paco encouraged his companions at all times and when facing the firing squad, behaved like a hero. Asking for and being granted permission to say a few words, he spoke with such inspiration and high ideals, that all present were amazed, including the executioners and the chaplain. Pedro Garfias dedicated a poem to this great man from Villafrance in his book *Héroes del Sur* and he truly was such a hero. Joe Monks, a member of the International Brigade, also mentions him in his book *With the Reds in Andalusia*.[viii] Today in Villafranca, despite the fact that since the war all the Mayors have been either Socialist or Communist, there is nothing to celebrate the memory of this exemplary leader, not a street, or any kind of memorial plaque named after him, or after Comandante Castillo, nor the great poet Pedro Garfias who sang of Villafranca and died in exile in Mexico. It would appear that all these politicians are latter-day accomplices of those who preach historic amnesia.

Summer-Fall 1940

July 20
6 executed, almost all from Montoro.

September 2
11 executed, many from Villaviciosa, among them a local trade union leader: Tomás de la Torre Barbero.

September 20
7 executed

September 30
6 executed

November 12
Execution of Juan *El Gato Negro* Rojas Arenas, an anarchist from Pedro Abad. He had belonged to the War Committee and was a great

fighter. Imprisoned in San Miguel de los Reyes, Valencia, for his participation in the revolutionary strike in Bujalance in December 1933, he was released after the Frente Popular elections.[ix] Co-founder of the *Nueva Aurora* workers association. The Prosecutor asked for 10 executions by garrotte. His brother Antonio was executed in Pedro Abad the previous year.

December 18
4 executed

December 20
3 from Santiago de Calatrava, Jaén, executed.

December 27
Francoists celebrated the end of the year with bloody fireworks - 34 executions, the largest saca so far in post-war Cordoba capital. Many were from Montoro, Belalcázar, Puente Genil, Alcaracejos, Castro del Río and other towns. In other words, a sampling of the entire province.

1941

January 31
1941 began with a great saca of 25 victims, mostly men from Los Pedroches district: 6 from Belalcázar, 7 from Hinojosa, 4 from Villaralto, 2 from Fuenteobejuna and other towns. The most notable prisoner in the saca was Féliz Chaves Caballero, past Socialist Mayor of Fuente La Lancha. From Belalcázar, they took the elderly Socialist Antonio *El Sabio* Vigara Regidor, a peaceful, vulnerable man, a typical nineteenth century individual and past President of the *Casa del Pueblo*. He had lived in hiding for a year until discovered by Fernando Ballester, a policeman who denounced him to the Legion when it arrived in Belalcázar October 16, 1940. His son Agustín was executed 3 months later, in May. One of the victims from El Viso, Miguel Ruíz Fernández, was accused by Causa General of belonging to the War Committee.

This particular saca had a huge impact both inside and outside the prison, which is exactly what the Francoists wanted. Never before had so many been killed in this manner, either in Cordoba or in all of Spain.

Still, it appears that although the people had become inured to the lists of so many executed, to see so many killed at one go, one after the other, was much more criminal than anything else that the cold mentality of today can grasp, "The bourgeois aloofness", as Reyes Mate calls it.[x]

Winter 1940 – Spring 1941

In Cordoba, the rest of the winter and much of Spring was undisturbed by the terror of the sacas because the sentenced to death from Cordoba and other Andalusian provinces had been sent to Burgos where they would be tortured by the bitterly cold winter weather until Francoism decided to kill them once and for all. To give the reader an idea of how harsh the 1940-41 Winter was, the thermometer on Christmas Day registered -4°C (24.8°F).

Cordoba capital, however, was not free from the effects of the extermination during this Winter and Spring. For the moment, although there may not have been many sacas, the prisoners were still dying like flies in the Provincial prison. In the new Cordoba Prison in 1941, more than 500 inmates died of starvation, almost as many as were executed by firing squads in all of post-war Cordoba – 584 in all. This number continued to rise during the entire 1940s, during which 756 inmates in the new Cordoba Prison, the Spanish Auschwitz, died of hunger.

April & May

Once Winter was over, in April and May 1941, the prisoners who had been sent to Burgos for the winter were returned to Cordoba, in several trains known as 'trains of death', for the sole purpose of carrying out their sentences. From the onset, the condemned men knew their fate and there were a few desperate attempts to escape the cattle wagons in which they were travelling.

There were no limits to Francoist cruelty as in all of Spain, it became customary to celebrate executions on festive dates and public holidays such as February 16, April 14, May 1, and November 7. There could be no more cruel way of teaching townspeople a lesson than scheduling a saca to coincide with local festivities, as on the traditional May 1 working men's holiday in Cordoba.

May 1

The train of death with the condemned had arrived from Burgos a few days earlier and the 34 men on board knew why. They came from many towns and villages in the province: 4 from Villanueva de Cordoba, 3 from Villaralto, 3 from Hinojosa, 3 from Bujalance, others from La Rambla, Cañete, Montoro, etc. The most famous member of this saca, Manuel _El Perla_ Sánchez Ruíz, aged 33, past Socialist Mayor of Montilla, was an outstanding person who had held leading positions in the JSU and the local Workers Society, was Secretary of the provincial ENTT and promoted National Vice President in 1938, in Valencia. He was captured in Alicante and sent to Albatera concentration camp. His wife and two sons were fortunately able to get on a ship to Oran, Algeria. Manuel was taken from the Montilla jail to Cordoba.

Blás Gómez Medica, native of Villanueva de Cordoba, had been a leading member of the PCE for many years, was alderman for the Frente Popular, presided over a Community Council and served as an early volunteer with the Garcés Battalion. He was about to flee on a ship at the end of the war when he was denounced by a fellow countryman, a so-called Aplicos, who handed him over to the Nationalists with the words: "Take him. You have caught yourselves a great Communist leader". After being sent from one prison to another, he ended up in Villanueva where he was exhibited to the Señoritos at the Casino, Later, it was the _via crucis_ of all the prisoners including the train of death from Burgos. According to Pedro Molinero, a cellmate, when Blás Gómez' name was called out for the saca, 'he became totally downcast, shuffling to his death like a sleepwalker'.

Another native of Villanueva executed on this day, José María Sánchez Jurado, a solicitor and clerk, local director of the Unión Republicana, alderman and member of the Frente Popular. An inoffensive person, he never imagined that he would be killed. Harking to his experience with the Law, he attempted to console his companions saying: "Don't worry; the only thing they can take from us is the right to vote". Sadly, he appeared to ignore that Fascism has no laws. A cellmate, Manuel Pascual Soler, told Moreno Gómez that when they called his name, he shared his personal objects among his cellmates, in such a way that they were all very moved. "That was one of the most terrible nights in Cordoba Prison: "you heard nothing, but nobody slept".

Pedro Padilla Moreno, another native of Villanueva, was executed for no 'revolutionary' reason, just because he had committed 'an act of war'. When Padilla was fighting on the Belalcázer front at the end of 1938, he was sent on patrol for the nightly relief of the sentinels. They left one at his post, but as they left, this guy attacked them behind their backs as he intended to go over to the enemy. As the patrol threw itself to the ground, Padilla shot and killed the traitorous sentinel, When the family of the dead man found out what had happened, they filed a suit in Villanueva against Padilla, Afonso Militos Torralbo, a solider from Villanueva, and another soldier from El Horcajo. All three men were arrested, and Padilla assumed full responsibility, justifying his actions as legitimate self-defence in an obvious act of war. The authorities refused to accept his defence.

Lastly, of the four natives of Villanueva who were executed, José Telesforo Torralbo Expósito was denounced by the family of José Fernández Martos who had died in battle in 1936. He left a widow, María *La Lavandera*, who later became stepmother to the well-known Rev. Pedro León Moreno, a high-ranking Dominican monk, currently posted in Seville.

Other victims of that May 1 saca included Bartolomé Parrado Serrano of the Bujalance War Committee; Francisco Casado Pedrajas from Pozoblanco, who had been photographed smoking a cigarette whilst sitting on the body of a dead Nationalist; Agustín Viagara from Belalcázar, son of Antonio *El Sabio* Vigara who had been executed three months earlier; Antonio Guerra from Pedro Abad, member of the UGT, accused of belonging to the War Committee but no other specific charges. He had been imprisoned in the Granada bullring at the end of the war, then in Padul concentration camp (so-called Camp of Glorification, by the Francoists), and finally transferred to Cordoba capital.

May 3

Another great saca, 34 victims executed in San Rafael cemetery. It was a matter of expediting all the passengers from the first train of death that had arrived from Burgos some days earlier, but not executed in the May 1 saca: 5 victims from Pedro Abad, 4 from Santa Eufemia, 3 each from Hinojosa, Belmez and Castro del Río; several more from Montoro, Villaralto and other towns.

Vicente Blanco García, a Communist leader from Belmez, ex-Militia Captain, was accused of having organized a People's Court from the City Hall balcony in 1936, to try a group of right-wingers who ended up being shot. Vicente Blanco was the scapegoat for all these events, and he was cruelly tortured, including by Falangista women. In Cordoba Prison he attempted suicide by cutting his wrists, but his wounds healed, and he was executed.

In the same saca, Alfonso Gómez Gutiérrez from Castro del Río, whose brother Juan had been a member of the War Committee; Tomás Pizarro Rodriguez from Belalcázar, whose brother Juan had also been a member of the War Committee; Francisco Muñoz Gutiérrez from Montoro, accused of the death of some right-wingers in Ventorrillo de la Lola, Montoro, July 1936. A similar accusation was made against Manuel Ocón Fleitas from Adamuz, and Antonio José Calero Tirado from Pedroche. The latter was accused by the family of José Manosalvas, a Falangista who was killed inside a church in 1936. All of these Causa General accusations were accompanied without a shred of proof of any kind, therefore unreliable.

June 3

Yet another great saca in Cordoba Prison following the arrival of another train of death from Burgos, this time with 28 victims. These trains full of Andalusians sentenced to death, were sent not only to Cordoba but also to Jaén and other locations. It was the terrible Winter during which the Francoists 'entertained themselves' at the expense of the Andalusians in Burgos.

Claro González Sánchez, a native of Fuenteobejuna, who managed to escape from one of the trains of death, describes their despair:[xi]

> "We arrived in a convoy of prisoners from Burgos, destination Cordoba, straight to the cemetery. The doctor bade us farewell in Burgos with these words: "You know where you are going".
>
> Before we arrived at Arévalo Station in Avila, a group of 18 of us escaped, with the idea of going to Portugal where there was a British ship and for which a teacher from Villa del Río had some documents.

Unfortunately, we were all immediately recaptured. A Lieutenant Colonel took me and four other men and indicated that he was putting us before a firing squad. If it had not been the intervention of a sergeant, I truly believe that he would have killed us all there and then. We were imprisoned in Avila for three months and between one thing and another, by the time we were sent to Cordoba it was May 1942 and we were able to save ourselves."

Another time, Santiago Cepas Romero, a native of Villanueva de Cordoba, told Moreno Gómez that on another occasion, when the train slowed down to go through Despeñaperros station, a group of prisoners tore strips off the side of the cattle wagon and escaped. Some of them were immediately recaptured, but others managed to disappear. Antonio Ramos from Almodóvar del Río told him of Juan Pato Velázquez, also from Almodóvar, who like others made a hole in the side of the goods wagon that was taking them to Cordoba to be executed, and escaped, but his footprints in the snow led the guards directly to him and he was recaptured. Then there was the convoy of prisoners from Madrid to Cordoba at the beginning of Summer 1941, when a prisoner from Badalarosa, Seville, escaped when the train stopped in the Linares-Baeza station. He was immediately surrounded on the platform by Guardia Civiles who shot him on the spot.[xii]

Of the 28 who were executed June 3, 11 were natives of Villanueva de Cordoba and some well-known in their community. The most famous of the lot was Pedro *Cuadrado* Torralbo Gómez, Communist alderman n 1931 and 1936, provincial Member of Parliament in 1936 and Militia Captain in the Garcés Battalion. His adventures when he returned to his hometown in 1939, on foot from Jaén, and his capture, have been described herein earlier. Pedro Torralbo was tried as Causa General Case 27.404/39 and suffered considerable torture before he was sent to spend the winter in Burgos. He was tried in Villanueva May 1 1940 in the Torres hall, Francoist Pedro Luengo Benítez as the Presiding Judge of the Military Court and sentenced to death. He immediately appealed his sentence to no avail. The Captain General of the II Military Region approved the sentence April 23 1941.

Also from Villanueva de Cordoba, the Castro family alleged that Bartolomé Viveros Torralbo had been present when Miguel *El Fresco*'s house was examined by the police in July 1939 and where Falangista Juan A. Castro Díaz died. Manuel Orellana Gómez and Avelino Nevado Asencio were denounced by the widow of Victoriano Muñoz, whom they went to arrest in the countryside in August 1936, when a man called Rojas shot and killed her husband. Avelino Nevado and his brother Alfonso were members of the PCE, served as volunteers in the defence of Madrid, and belonged to a family that was well aware of its social standing and was greatly persecuted by Francoists. Alfonso Bujalance Gallego, also from Villanueva and executed in this saca, had been a Militia Commander and was related to the above Falangista, Victoriano Muñoz. Francisco Illescas Palomo from Villanueva, the especially tragic case of the man whose wife and three children died when the Las Navas powder magazine exploded February 28, 1939, was also executed in this saca.

29 executed were from several other towns in the province, including Ángel Trujillo Medina from Villanueva del Duque, accused of belonging to the War Committee and Manuel Madueño Navarro from Montoro, accused by the Causa General and the FAL Militia, of being responsible for several deaths during the evacuation of the village, Christmas 1936.

One of the most notable victims of these deaths was Eduardo Bujalance López, a socialist from Hornachelos, whose brother Antonio, Member of Parliament for the Frente Popular, was murdered in Cordoba by the Nationalists soon after the coup, July 30 1936. Eduardo was executed solely because of who his brother was. It was a purely political elimination, like so many thousands in Spain, as there was nothing of any importance of which he could be accused. The landowners in his town accused him of totally false crimes, with the sole purpose of erasing the Bujalance family from the map. When Moreno Gómez visited Hornachuelos, he was given a copy of Eduardo's last letter, written in the chapel, that clearly illustrates the political nature of his execution.[xiii] This letter is also reproduced herein.

June 9
9 executed: Blás Gajete López from Villafranca, 26-year-old public school teacher about whom Moreno Gómez has written; more men

from Pedro Abad, a small village that was terribly punished during the post-war when 150 men were imprisoned and more than 50 executed; Francisco Olanda Garrido, Socialist who served with the 108MB on the Madrid front, accused of participating in the negotiations for the surrender of the Guardia Civil barracks in the town and of several other unproven acts. From Palenciana, a neighbouring village to Pedro Abad: Joaquín Antequera Gómez, who according to the Causa 'took up arms against the Glorious National Movement' and engaged in guarding and arming patrols (the so-called Self-Defence Groups); Rafael Rojas Navarro, member of the UGT and the same Self-Défense Groups. Apparently, he was on guard in the jail August 10 1936 when the Militia from Alcoy arrived in the town and took a saca of seven right-wingers whom they killed in exactly the same place that Francoists had shot and burnt a group of townspeople July 22 1936.

June 28

9 executed in Cordoba, including Francisco Moya Gómez from Pedroche, accused by the family of landowner José T. Gutiérrez Rane, executed in July 1936 (Causa General data).

Also, a few individual executions on different days that month.

July 15

Next large saca with 14 victims from different townships including Francisco Pedragosa Velasco, a native of Valenzuela, accused of belonging to a War Committee. Moreno Gómez regrets that he was unable to obtain more information regarding the history of the multiple and tragic personal activities for so many of these victims, so that he could honour their memories.

August 20

8 executed, including Rafael Deza Montero, a native of Puente Genil, brother of Justo, a great trade union leader, member of the JSU, executed November 9 1939. A third brother, Marcos, was also executed. Another family totally eradicated by the Regime.

Fall and Winter

September 12

11 executed, 5 from Villanueva de Cordoba, including Juan Cantador Zamora, whose story Moreno Gómez told earlier as a typical example of Francoist 'justice', when referring to the riot and violent treatment of the prisoner that followed a case of mistaken identity at his trial under the notorious Judge Juan Calero Rubio. Executed on the same day, Juan Lorenzo *Cucharas* Cantador, ex-Militia Captain. It was alleged, as all the others were, that he participated in the attack on the Villanueva barracks, belonged to the War Committee and other such activities, all without any proof whatsoever. All the following prisoners arrived from Burgos in another train of death September 8: Miguel Campos Toledo, accused of persecuting right-wingers in Villanueva July 24 1936; Faustino *El Peno* García Calero, well-known native of Pozoblanco, ex-Militia Lieutenant, accused, probably without cause as usual, by the families of two right-wingers executed in 1936 (José Alcaíde Dueñas and Antonio Cañuelo).

September 29

One of the 3 executed on this day, Juan Flores López, past Mayor of Espiel, and José Palomo Huertas, a native of Villanueva de Cordoba and another of the many who fell because he had taken part in a carnival prank involving a ditty ridiculing a well-known landowner who was running for office in 1936.[xiv] José Palomo Huertas and his wife Francisca Gómez Cuevas belonged to a large Communist working-class family. His wife, together with other brave women in the town, stood in front of the Guardia Civil horses to prevent them from charging against their husbands.

José Paloma was sent to a concentration camp in Valladolid until June 1939 when he was released to return to Villanueva. No sooner had he arrived home that he was arrested on totally false charges June 22 1939, beaten and tortured until daybreak the next day when they took him to the Fuente Vieja Schools prison. They had placed slivers of wood under his toenails and beaten his arms to bits and his body so badly that he was left lying on the floor. His cellmates took care of him, and he recovered. The charges against him were totally false. He was sentenced to death by court martial at the end of 1939. September

26 1940 he was transferred to Cordoba New Prison with all the other local prisoners and then sent to Burgos where he spent 11 months, from where he returned to Cordoba shortly before he was executed. As to his wife, Francisca Gómez Cuevas, she was arrested with all the other Rojo women in the town and imprisoned in Juan Herrero's house on Calle Conquista, where she gave birth to a son. The child fell and died without ever knowing his father who was imprisoned in the Fuente Vieja Schools prison at the time.

In Cordoba, men were sent to the new prison and women to the old one. Francisca suffered many years in jail, like Juana *La Flora* and other front-line Rojas. The tragedy of the Palomo family did not end there as another brother, José Antonio, was killed in 1948 under the Law of Fugitives after being caught listening to the *La Pirenaica* radio station.[6]

October 11

7 executed in San Rafael cemetery, including a well-known individual from Villaviciosa, Nicomedes de la Fuente, an anarchist who had become known for his role in the October 1934 revolutionary uprisings, which is why he fled the town and took refuge in Bujalance. He was now being executed on the anniversary of those events. Tomás Cçuadrado Ruíz was another activist involved in the Adamuz uprisings.

November 6

8 executed, including several from Villanueva de Cordoba. Juan Escoriza Segura, a native of Almeria, one of the famous *materos* whose job it was to keep the bushes and shrubs on the hillsides under control. He was accused of 'participating in the attack on the village and in a shooting in Fuente Vieja', which he denied. As his family told Moreno Gómez, Judge Juan Calero propositioned Juan's daughter, promising to save her father in exchange for her 'services'. As she snubbed him,

---

[6] *Radio España Independente* was a clandestine radio station broadcasting from Bucharest from July 21 1941 to July 14 1977, nicknamed *La Pirenaica* because it was believed to be broadcasting from somewhere in the Pyrenee mountains. Run from Moscow by the Spanish Communist Party as the voice of the victims of Franco's regime. *Vide* article by Armand Balsebrea and Rosario Fontova at http://www.davidpublisher.org/Public/uploads/Contributer/55079243cc777.pdf.

he refused to process Juan's appeal and had him placed in solitary confinement where he suffered all the usual mistreatments until he was shipped to Burgos.

Juan F. Chuán Soso, from the same town, was condemned under the usual accusations of 'attack on the town and shooting in Fuente Vieja' that the authorities indiscriminately levied against all Almeria townspeople they captured, It is said that Chuán refused to confess so forcefully, that he began to shout, and they had to muzzle him. José Domenech Martínez, native of Montoro, was a member of another of the exterminated families whose sons Valeriano, Lucio and José were executed. In Montoro, the charges were the same: everyone was accused of attacking the Falangista inmates of the local jail July 22 1936.

November 17

This saca of 3 is remembered as when one of the three victims, Antonio Baena Moreno, schoolmaster from Pozoblanco, who had presided over the War Committee, was executed. An influential Socialist, somewhat of a visionary and very kind, who truly never participated in any kind of blood event, quite the contrary. He avoided every conflict he could. To make matters worse for him, he had served as political commissar for the Republican 8th Army Group and later General Political Commissar for the Army of Andalusia. Throughout this imprisonment, he continuously attempted to demonstrate his innocence, with detailed appeals, to no avail. The fact of the matter was that the great 'crime' in Franco's eyes was the 'political significance' of an individual, not whether he had been involved in bloodshed or not.

Antonio Baena wrote a lengthy diary in prison, that his family kindly allowed Moreno Gómez to consult and copy, a perfect description of the disgrace, constant humiliations, shortages, hunger and terror that everyone inside a Francoist prison suffered.

Baena was tried by court martial in Pozoblanco April 22 1940. He was accused by several local Falangistas: the Bosch Caballero family, relatives of Lázaro Delgado Cabrera, Juan García Tirado, Bartolomé Caballero, Domingo Márquez, José Delgado Dueñas and others. In other words, the majority of the Pozoblanco bourgeoisie, most of whom wanted to revenge relatives of theirs who had fallen victim to the People's Court in Valencia. With all these accusations, there was no way out for Antonio Baena. September 19 1940 he was sent to Cordoba with the local prisoners then onto Burgos for the winter. November 2 1941, after a year and a half with a death sentence hanging over his head, he was returned to Cordoba on the train of death, for execution. That same day, he concluded his diary with the following words: "I leave Burgos for Cordoba, having lost all hope. How I think of you, my children!"

Two weeks later, he was listed for the saca. A cellmate of his from Dos Torres, Carlos Menéndez, told Moreno Gómez that he bade farewell to Antonio Baena, when they took him. He went, tied with wires to another prisoner from Pozoblanco. Baena was depressed and crying. He didn't say anything. Carlos Menéndez was assigned to work in the prison office and that night he was made to stand guard and accompany the Head of Services to confirm the departure of the sentenced.

Jerónimo Jurado Carrillo, Baena's companion, had been accused by Falangista Francisco Peralbo's family. Jerónimo had served as a guard over several Falangistas imprisoned in the Teatro jail and they accused him of having been present at the November 20 1936 executions.

November 25

Only 1 execution that day, Miguel Lindo Serrano from Adamuz, a member of the famous Lindo family, two of whom, the Luque Lindo brothers, had escaped to the mountains because they were tired of being so badly and frequently beaten in their town. They joined Claudio Romera's Socialist fighters and survived as guerrillas until around 1949.

December 10

10 executed from several towns: Villa del Río, Posadas, Castro del Río, Pedroche and Pedro Abad. Francisco Carrillo Cobos, from Pedroche, was accused by two families whose relatives had died while serving in Franco's army. Julián Claudio Carrillo, also from Pedroche, was accused by the family of José Tirado Vaquero who had been killed by a fugitive but, so they said, because he had been called Falangista by Julián. He was also accused by relatives of being responsible for the death of Antonio Rodriguez, a right-wing student who died in 1936.

The accusations against Juan Castilla Rivera from Pedro Abad, was a novel one: that he was responsible for organizing the 'Children of the Night' who put petards on the railway line at night, as well as of belonging to the typical local Self-Defence Groups, of course.

December 20
The last execution of the year in Cordoba capital, with 6 victims.

1942

Fewer sacas during the year, but with more victims each time.

January 10
9 victims

March 10
2 executed. Moreno Gómez considered that this execution was especially significant because of the extremely moving letter that Joaquim Moreno Muñoz from Baena, wrote his family from prison, a copy of which he obtained from Arcángel Bedmar and reproduced herin. Dated February 4 1942, it reflects the enormity of the suffering of the prisoners under Francoism.

> "Mother, regarding what you sent me, I received everything, and you have no idea how much good it did me as I have improved from the debility I suffered, so you can continue bringing or sending me clothes and what food you can, as it will all help me feel better. If you cannot, without it I will find it impossible to continue... A couple of handkerchiefs to wipe my nose, some size 38 cord sandals, and tobacco.
>
> Dear brother, you have no idea how happy I was to read your letter and... don't forget me. You have no idea the satisfaction that it is for a prisoner to know that he is remembered, especially under the circumstances in which I find myself, because nobody knows it better than he who has the misfortune to go through such situations: this eats me up. Tomás of

my heart, do what you can, both for my health and for taking this weight off me, as it is unbearable..."

Everything ended for this unfortunate man March 10, but at least Moreno Gómez was able to record the suffering he endured, for History. Regarding so many others it has been impossible to get any information but at least we have most of their names, which is not a little.

March 25
5 executed: Juan Sánchez Pozuelo, second son of María *La Loma*'s, from Villanueva de Cordoba, whose brother Alfonso was executed September 12 1940. Juan had been hallmarked, as were another 2,000 militiamen, for having participated in the great anti-coup uprising July 1936, with the assistance of 20 miners from Puertollano. It was a technically perfect operation in which the Republicans fought the Nationalists by breaking through the walls of adjoining houses as they moved through the town, instead of fighting openly along the streets, a most unusual military operation. The Republicans were ultimately defeated and thereafter, all the townspeople were considered guilty of 'the attack on the town'.

For some unknown reason, the Francoists waited until March 1943 before executing a third of María's family. Juan *El de La Loma* was denounced by *El Cuco* who lived on Calle Pozoblanco. Juan's family later told Moreno Gómez that he caught tuberculosis in Cordoba prison and that he attempted suicide by throwing himself onto the prison patio from the third-floor gallery but did not die. As customary, his guards probably waited until he recovered before they executed him.

June 25
11 victims

August 8
7 victims

November 7
8 victims. This date, the anniversary of the Russian revolution and the first offensive against Madrid, was always celebrated with one or more sacas.

1943

Executions by firing squads occurred less frequently during this year and most often, only one prisoner at a time. Frankly, there were not many left to kill, at least for the time being, because it was not yet time for the 1945-1947 Triennium of Terror. Meanwhile, other assorted features of the 'multi-repression' program began to be applied. (Reyes Serroche, executed February 6, is not included in the list because unlike the others, he was a common criminal.)

The first stage of the Francoist repression, in terms of executions by firing squad for reasons related to the civil war itself, can for all intents and purposes be considered ended in 1943. If on the one hand there were fewer prison inmates sentenced to death for this kind of crimes, left to kill, there were other means by which the repression could be enforced. Resort to these was rising to new heights: forced labour, beatings in the barracks, blacklists, the humiliation of being freed on parole with the attending restrictions, the continuing drama of those serving long prison terms, not to mention the Law of Fugitives in itself a budding catastrophe. Typical causes of post-war crimes were those associated with the clandestine reorganization of political parties, partisan support to the maquis and guerrillas, and so forth. These would be typically resolved with large sacas in the Fall of 1944.

We have had an indication of the forces of death that enraged and asphyxiated the provincial towns and villages during this first stage of Francoist repression, and Cordoba capital was no exception to the overall severity of the Regime. This, because martyrized Cordoba capital had already been enduring an unbearable genocide since July 1936. How many shootings, how many bloody dawns, how much bloodshed in the cemeteries...

In Cordoba, there were many nights in which so many were executed by firing squad that the bloodbath often lasted three hours, as groups of 8 or 10 men fell together, one after the other, onto the bodies of those who preceded them.

In Cordoba capital, as in all other towns and villages in Spain, the people continued to live the same nightmare of terror that they had endured since the terrible days of 1936, even though in the post-war period executions were more directed at those who had already been sentenced to death and were incarcerated. Few were left to execute in the capital, but the harassing and the insistent climate of terror persisted in the neighbourhoods mostly populated by the working classes. The Guardia Civil, the Municipal Police, and the Falange took it upon themselves to ensure that the population would continue to be subdued by the threats, the beatings, the castor oil treatments, the registrations, detentions, denouncements, accusations, the apprehension of children by the Social Welfare program, exclusion, pursuit of the black market and hunger in the most literal sense of the words.

Guardia Civil Police and Falange multiplied their efforts to sacar Rojos whom they said were hiding under stones everywhere. Many persons who had been in the Republican zone in both capital and townships, tried to live incognito in the city, without success. The Cordoba capital *vox populi* had not forgotten the group of Fascist policemen who were forerunners of the New State's agents of repression and who would kindle tragic memories in the 1939 immediate post-war period: Commissar José González de Lara and policemen Aparicio Romero, Ricardo de la Fuente, Heredia Espinosa, Ruíz Molina, López Linares, Rafael del Olmo García and others. In 1942, Commissar Aurelio Cortegero de la Cuerda and an extensive list of especially active political policemen.

Amongst the Francoist Guardia Civil, another body devoted to the political repression, we note the tragic notoriety of Corporal Payán (José Payán Porcel?), Corporal Arenas and Corporal Manolo *El Colorao*. The latter, from the La Magdalena barracks, took pleasure in torture and in applying enormous beatings for the most trivial reasons such as the theft of a cabbage or a head of lettuce. If he caught someone hunting birds in the countryside, he made them eat them raw; if he caught someone picking olives, he made them eat the green, unripe, bitter ones, to all of which he added the well-known beatings. Another Guardia Civil nicknamed *El Bizco* was feared for the savagery of his torture, even before the October 1934 uprisings. Another one was known as *El Dino*, specialized in torturing and applying the castor oil purge to women who were caught with goods they had purchased on the black market. Others such as Corporal de Asalto Ballesteros, municipal policeman Moreno, Guardia Civiles Baltasar *El Maño*, Rafael *El de los Bigotes,* and a so-called Urbano Cantizano who had served as a Legionnaire during the war and immediately re-enlisted.

Likewise, engaged at their side there was an entire cohort of Falangista admirers, social-climbers and servants to the Army of Evil, in uncountable numbers. The names of some of the more notorious are still remembered in Cordoba: Juan *El Gitano* a Falangista sergeant, and his local contacts Isidoro *El Comandante* and Antonio *El Loco*, suppliers of the barracks and rabid hunters of women black marketeers whom they beat viciously when caught.

Powerful Falangista women were no less evil than the men. One, Conchita Costa, Director of the Prison for Women, was particularly notorious for her ruthless and inhumane discipline. Another, María Campos y Carmelita *Caraquemá*, a Falange leader, would send her centurions down the streets of Cordoba, arrest women in the working-class neighbourhoods and then take them to the Falange barracks where they shaved their heads and purged them with castor oil, a treatment invented and widely applied by Italian Fascists. Other groups of puritanical Falangistas organized raids of women and took them to the Roman baths behind the Alcázar and there, on the pretext of hygiene, shaved and mistreated them.

The Falange paramilitary barracks were located in the Rinconcito, in front of the Isabel la Católica movie theatre, another in the Calle Jesús María, near the Góngora theatre, another in Calle Carbonell y Morand, near Capuchinos square. All of this and much more, formed the climate of harassment, abasement and humiliation of those who were disaffected with the Regime. Together with the firing squads, they formed the essence of the Francoist repression.

Some clandestine reorganization 1944-1945
Causa General Case 94/44.

"Our party has never thought that the solution to this war might be the establishment of a Communist regime. If the mass of the working classes, the farmworkers and the urban bourgeoisie follow us and support us, it is because they know that we are the most resolute defenders of national independence, liberty and the Republican Constitution."
José Díaz Letter to
Mundo Obrero newspaper 19 March 1938

Despite the defeat of the Republic, the workers' political parties, especially the PCE, never ceased attempting to reorganize themselves even in the face of the relentless Francoist repression, with its torture, imprisonments and executions, both in Cordoba and throughout Spain. In Madrid, the provincial reconstruction of the PCE involved the well-known playwright Antonio Buero Vallejo, who produced false documents and was almost shot. Curiously, the most reliable witnesses state that the principal leaders of the political reorganization were interned together in the same prisons. Considering that the best-known leaders were under arrest, as were the most combative members of the workers' movements, it is not surprising that the important base at the origin of the clandestine fight against Francoism could be found in the prisons.

During Moreno Gómez' research into the clandestine reorganization in Cordoba, he recorded observations regarding the spontaneous homegrown nature of the reorganization attempts, regardless of any directions from outside the country (almost impossible to apply, at any rate), as well as the lack of homogeneity in the dynamism of any kind of political opposition. At the same time, while there is information regarding the Communists who were clearly active and the Anarchists who were less consistently so, there is practically no available data regarding the clandestine activities of Cordoban Socialists who virtually drop out of sight in the province in 1939.

On the other hand, the harshness of the Francoist repression against any attempt at reorganization was a powerful deterrent, as to be found with clandestine propaganda or being listed in the flowchart of a local or provincial party committee, was punished by death. Already in 1941, political activists such as Heriberto Quiñones and Enrique Sánchez in Madrid, began to fall like flies before the firing squads, not because they had committed a crime of war but because they had engaged in crimes of clandestine activities. Matilde Landa and Buero Vallejo escaped by miracle. Still, regardless of the underlying reason for their arrest, when tried, all offenders were charged with "a crime of military rebellion', the only charge allowed under Military Law.

Moreno Gómez therefore extended his research into mass executions after 1942, beyond the repression of defeated Republicans and those considered guilty of crimes of war, to include aspects of the government's recent muti-repression policy directed at a new civilian target – everyone involved in some form of 'clandestine' activity – meaning almost everyone who opposed Francoism in Spain. Beginning with an intent to behead the nascent clandestine organizations and to maintain the climate of terror by which Francoism controlled the civil population, the government refined and updated its tried-and-true methods of repression. As the Causa General became increasingly active in its investigations, the number of raids of suspect organizations rose exponentially, as did the number of individuals arrested and tried by the courts and condemned to death.

Causa General Case 94/44 that was instructed and tried in Cordoba capital July 27 1944,[xv] provided details of the clandestine organization of the Communists since 1943, beginning with the creation of a local committee in Cordoba capital, then a Provincial Committee, followed by several branch committees in several towns. (The Communist Party was doing the same in Galicia and in Madrid, at about the same time.) All this organization came crashing down at the end of 1943, following a massive raid in which 145 individuals were arrested (primarily in Cordoba, La Rambla, Castro del Río, Belmez and Peñarroya-Pueblonuevo), 68 of whom were formally tried. Of these, 8 were executed by firing squad October 19 1944.

The events that led to this disaster began August 1943, primarily due to the initiatives of three Communist leaders: Antonio Guerrero Lebrón, recently released from Málaga prison, and two other activists: Adriano Romero from Villanova de Cordoba, past member of Parliament for the Frente Popular, and José Lupiáñez from Jaén, both of whom were imprisoned in Cordoba capital. This was a matter of divulging the politics of the Unión Nacional, a new populist creation of the PCE, already being disseminated throughout Spain by Jesús Monzón, from France.

Summer 1943, one of the first local PCE committees was created in Cordoba capital, with the following officers:

José Molero Berlanga, Local Secretary
Alfonso Cerezo Regalón, Financial Secretary
Antonio Fernández Cuenca, Uprisings Secretary
Juan Vélez López, Organization Secretary
Celestino Lara Ruíz, Propaganda Secretary.

This local committee's immediate objective was to provide assistance to the prisoners (the same thing that was the basis for Causa General Case 1,546/41), contact with like-minded organizations and to attract members. Contact with the Communist organization in Madrid was managed quickly thanks to José Molera Berlanga's efforts. October 5, 1943, Madrid sent a delegate to Cordoba, Manuel Santuagi Álvarez Aguado, who had to confirm his identity before the Cordoba Committee with the password *on behalf of Maruja*. Manuel Álvarez had been exiled in France but he returned to Spain via Catalonia with false documents. He had served as a political commissar with the military during the war. Under his direction and assistance, the local officers intended to expand the Communist organization in the region. Cordoba province was divided into four sectors (La Rambla, Belmez, Villafranca and Castro del Río).

The Provincial Committee was created, and the following were appointed as officers:

Manuel Álvarez Aguado, Political Avisor, nicknamed *Santiago*
José Molero Berlanga, General Secretary
Francisco Medina Rodríguez, Organization Secretary
Juan Vélez López, Military Secretary
Alfonso Cerezo Regalón, Financial Secretary
Antonio Fernández Cuenca, Uprisings Secretary
Celestino Lara Ruíz, Propaganda Secretary

Contact with the Party Regional Committee in Seville was soon established. One of its members, Manuel *Bartolo y Luísa* Castro Campos proposed two Secretaries for Cordoba: Manuel Álvarez and José Molero. The latter did not accept and was substituted by Emilio Jiménez Rascón. Manuel Álvarez made another change to the first Provincial Committee: Juan Vélez, Celestino Lara and Francisco Medina Rodríguez were charged with the Cordoba capital's activities creating cells in several capital neighbourhoods.

It is important to keep in mind that these clandestine activities in Cordoba city and province abided by the guidelines of the new National Union policy disseminated by the PCE and Jesús Monzón from France, that had already begun being infiltrated throughout Spain. In Cordoba, clandestine National Union pamphlets and handouts were distributed, with the enormous risks that this entailed. The handouts were written on Rogelio Díaz García's typewriter. The police discovered another in Julio Priego Ordonez's home.[xvi] All the above were indicted under Causa General Case 94.

Local committees were created in the three provincial towns: La Rambla, Belmez and Peñarroya-Pueblonuevo. Shoemaker Bartolomé Mendoza Caballero tried to do so in Castro del Río, but without success.

The Belmez Committee consisted of José Békar Luque, Daniel Gallardo Gallego, Andrés Díaz Bonilla and José Fernández Braga (the latter was a representative on the Provincial Committee and senior officer in Belmez).

The La Rambla Committee was directed by Andrés Ruiz Urbano, followed by Martín Peinado Alcaide, both of whom had been attracted by Juan Vélez's activism and in whose boarding house in the capital, Manuel Álvarez the link with Madrid, was staying.

The Peñarroya-Pueblonuevo Committee was directed by Pablo González Calvo, Secretary General, whom Moreno Gómez interviewed years later in Madrid. The remaining Secretaries were Cipriano Tapía Quinta (Organization), Antonio Igualador Gómez (Information), Mariano Fernández Romero (Finances) and Juan Manuel Muñoz Mansilla (Raising funds for prisoners).

All the above committee members were arrested in the great raid at the end of 1943, as were many supporters or subscribers from each town. 32 were arrested in the capital, including Benito del Puerto Baltanas (head of a radio station and several cells in the capital) and Blás Herencia Burgos (who was in contact with Castro del Río) and many more.

In Belmez, in addition to the Committee, another 7 activists who, according to the Causa General, acted in three-by-three cells, were arrested. In La Rambla, the Committee and another 9 members fell into the clutches of the Regime's political police. 8 activists, in addition to the local Committee, were arrested in the Peñarroya raid. One, Juan Gallardo Sánchez, was arrested on the grounds that he had donated 15 pesetas to a fund for prisoners. The raid was carried out by police From Madrid and from Cordoba.

All the oral witnesses stated that at a first stage, 145 were arrested in 1943 and 1944. Many remained in prison as 'guests' of the government, without trial; 68 were tried under Causa General Case 94/44.

The great raid was disastrous for the organization that had worked so hard to get it going from prison as early as 1941. When Adriano Romero was transferred to prison in Granada July 1942, he had to stop being the soul of the clandestine organization in Cordoba. Adriano has recorded their actions in his memoires:[xvii]

> "When I arrived in Cordoba, I ran into Alfredo Caballero and many other Party leaders. José Lupiáñez arrived later and with Guerrero Lebrón, also of the Cordoba Provincial Committee, the three of us did our utmost to help the Party leaders in the prison and on the street, without ever taking over their jobs. The main work involved information, preparing leaders, and legal and economic assistance, as much as possible. Alfredo Caballero was in Sevilla, where there were also many but less important members, and Lupiáñez in Madrid."

Captain Manuel Álvares Núñez was the Presiding Judge for Causa General Case 94/44; Captain Fructuoso Delgado Hernández prosecuted the July 27 1944 court martial. He had been instructed to be very harsh and so he was. He asked for 45 death sentences, 18 indicted were sentenced to death, 5 to life imprisonment, 1 to twenty years in jail, 8 to thirteen years in jail, 28 to six years in jail and 7 were absolved. The court martial was held in the same building as the Provincial Prison.

August 5 1944, the Seville Judge Advocate approved the 18 death sentences, and they were confirmed August 10 by decree from Captain General Miguel Ponte y Manso de Zúñiga, General Headquarters for the 2nd Region. Nevertheless, the Government's *enterado* of October 3 1944 only sent the following 8 to the firing squad, to be shot at dawn October 19:

- Manuel Álvarez Aguado, 29, single, bank employee, from Madrid
- Juan Vélez López, 32, married, manual labourer, from Posadas, resident in Cordoba
- Celestino Lara Ruíz, 29, married, manual labourer, from Cazorla, resident in Cordoba

- Francisco Medina Rodríguez, 34, married, metalworker, from Cordoba
- Antonio Fernández Cuenca, 29 single, salesman, from Brazil, resident in Cordoba
- Alfonso Cerezo Regalón, 40, married, bricklayer, from Adamuz, resident in Cordoba
- Emilio Jiménez Rascón, 29, married, employed, from Cordoba
- José Molero Berlanga, 33, single, telegraph operator, from Espiel, resident in Cordoba.

Another five activists who were also caught in that raid were sentenced to death in a different Causa General trial, details of which Moreno Gómez was unable to find. One, Corporal Antonio Cobos León, from El Carpio, had been found with propaganda for the Unión Nacional. Moreno Gómez only knows that name of one of the four other victims: Sebastián *El Niño del Dinero* Caravaca Martínez, one of the Los Jubiles guerrillas, who had dared to come down from the mountains in Bujalance to see his family, and there was either captured or he turned himself in. They beat him, made him talk, then executed him.

Julio Priego Ordóñez who was caught in the great trial and sentenced to 30 years in prison, told Moreno Gómez the following in an interview some years later:[xviii]

"I remember the words of those unforgettable companions, when October 19 1944, they were taken in a saca from the Cordoba Provincial Prison to face the firing squad. They shouted Viva la Republica! and they asked us to ensure that the blood they were about to shed was not forgotten.

We were arrested by Falangista police who shut us up in the Puerta del Rincón Falange barracks in Cordoba. The repression was so severe that it continued through January and February 1944, with 145 arrests. The raid began in Cordoba capital and later spread to the towns of Peñarroya-Pueblonuevo, Belmez, Villafranca, Pedro Abad, El Carpio, La Rambla and Castro del Río. They leaned on us so harshly that for the supposed crime of military rebellion they asked for 45 death sentences; 9 were shot outright without any accusation. This was the most criminal trial in History."

The repression was undoubtedly out of control, considering that the only proof the authorities found and the only 'subversive' material they were able to present to the Court was twenty flimsy eight-line leaflets from the Unión Nacional. The Regime did not appear to be at all magnanimous, as some swear it was.

There is some information regarding the clandestine activities of the PCE during 1945, based primarily on an undated letter from Bartolomé Fernández Sánchez, Socialist from Pozoblanco, who served as a Major in the Militia, was captured and at this time of writing, was imprisoned in Cordoba. In his letter, Bartolomé refers to a December 30 1945 report from imprisoned Communists, in which they reflect on the evolution of their Party's political aspirations during that year. According to Bartolomé, the report consisted of six items that the Party proposed for discussion by the bases:[xix]

1. Difficulties and hardships that the prison committee identified, following the Causa 94 prosecution [assistance provided to prisoners].
2. The authority that today the leaders of the prison committee share with the Provincial Committee of the PCE, thanks to the strength of its base.
3. It is said that the union of Republicans, Socialists and themselves is a certainty as there exist committees of relationships between them, although many are highly critical of their confederates.
4. Preparation of union leaders for the future: directing them along the Communist lines. Knowing how, for this purpose, to take advantage of the masses at the right time, as the people are aware of the PSOE's reformist position and its revolutionary passivity throughout the world.
5. The work of reorganizing the UGT involves accepting as a transitory fact, all that connects it with obtaining the desired relationships, and that the work of the PCE within the UGT tends towards creating a single Central Committee of the

Proletariat, ceasing with factory, company, workshop and general field committees by uniting workers and farmers in the implantation of the Supreme Soviet.

6. Give strength of purpose and action to all Communists who, as members of the PCE, can express their opinion and elect, in a democratic manner, those comrades who are best prepared to give political guidance within the Party. If this is not being done today, they say it is because of the current Regime's repression.

7. This Agenda shall be discussed by all the Communists in the prison and later sent, with the opinion of all the members, to the Party leaders.

From this document, one deduces that the clandestine political activity in the prison was active and hegemonically Communist. The usual Communist pretension of its being the supreme authority is quite clear, despite its unspoken recognition that the Socialists and especially, the Anarchists, were reticent when faced with such unitary policies. Nevertheless, some Socialist prisoners such as Bartolomé Fernández, imprisoned in Cordoba, became involved in the PCE's deliberations and activities.

The Communist base in Cordoba replied to the above report as follows:

"All the Communist cells, having discussed the six points of the report via the committees of radios, sectors and subsectors, approve the report as it describes the work of the Communist Party for one year and because it complies with true Communist guidelines and doctrine. Although it reveals defects and great omissions that today are irreparable because we are engulfed by a dark cloud, if we are to ensure the success of our efforts, we must strive to bring about a unity with the other political sectors, especially the CNT/UGT and generally speaking, avoid an in-depth attack on confederates, nor censure the few young, imprisoned Socialists, as this would seriously harm unity.

We also wish that this leadership were more democratic when appointing the comrades that will have to fill the different positions, that it disseminate all the propaganda material it receives, that it allow every Communist to associate with anyone who he believes good for the Party and that it cease all censorship of this kind.

We ask that you inform the Party base, as soon as possible and in general terms, which political line is to be followed at these times of repression of the Spanish people, whether the U.N.A. or the Democratic Alliance. Communists are not much concerned with the colour of the politics as long as all the anti-Francoist forces are united."

Bartolomé Fernández attached an interesting commentary on the report, that he appears to have sent to his organization:

"As you can see, this is a summary of the comments, but they are sincere and revolutionary and do not admit any partisan criticism, simply a desire to break the yolk that oppresses us today; tomorrow will be something else and shall stir the attention of the mass of working men. The political movement inside the prison is, as you can see, an agreement of criteria within the Marxist parties and an alliance of combative political forces..."

Outside the prison, capturing clandestine Communists was the Regime's order of the day. There was another great raid in Cordoba capital, tried as Causa General Case 443/1945. This court martial of 24 accused, two of whom were women, was held inside the Provincial Prison July 24 1946. The Socio-Political Brigade of the Cordoba government police arrested them on charges of attempting to reorganize the Communist Party.

Accused under Causa General Case 443/1945	
Cordoba residents	
Domingo Escola Verde	Antonio Carrasco Expósito
Antonio Sánchez Alcaide	Juan Conde Navarrete
Sebastián Fernández Martínez	Antonio Rivas García
Rafael Obrero López	Juan Sánchez Cabrera
José Días Bueno	Francisco Catalán Higuera
Joaquin Cabello Labrador	Manuel Jiménez Escribano
Pedro Moreno Pino	Jaime Cuello González
Benito Núñez Vélez	Antonio Rabadán Carmona
Fausto Contreras Hervás	Francisco Páez Ortega
Rafael Camargo Montes	Manuel Jiménez Sánchez
Seville residents	
Ildefonso Becerra Galindo	Ángeles Vargas *[Ildefonso Becerra's wife]*
Josefa Arévalo Pozuelo	Ana Ponce Barneto

The crime for which they were indicted was summarized by Francoist military jurisprudence in a single expression: military rebellion. As evidence, the police presented some brochures entitled the "Booklover's Guide" which consisted of a whole bunch of news clippings from *Mundo Obrero* regarding the Unión Nacional policies, dated 1944.

They were accused of creating the provincial organization of the PCE, as successors of those who had fallen in Causa General Case 94/43. Domingo *Antonio el Misterioso* Escola Verde, considered one of the most important accused, had returned covertly to Spain from France, April 1944, where he was living in exile. He arrived in Cordoba in May and January 1945 he contacted the Regional Committee in Seville, to where he went. Upon his return to Cordoba, he dedicated himself to organizing the so-called Young Communist Fighters of Cordoba, along Jesús Monzón's guidelines. At an assembly in March, they agreed to merge with the Communist Party. Domingo Escola Verde suffered terribly after he was arrested. To prevent saying anything under torture that could damage the organization, he attempted suicide by throwing himself off the top floor of the Police Headquarters to the patio below. He was seriously injured but they treated his wounds and court-martialled him with the other accused.

Another eminent accused, Antonio Carrasco Expósito, was an officer of the Cordoba Regional Committee. A representative of the last Regional Committee of Seville where he held the position of Assistant Secretary for Organization and Finance, he also was a regional instructor for the Party with the mission of reorganizing the Cordoba City Committee. At this time, there was a so-called First Committee that functioned until March 1945 when the Young Communist Fighters and another group that had been created independently, merged with the Cordoba Committee, hereinafter known as the Second Committee.

Identification of the members of the First Committee is difficult, because some of its members were sentenced and executed under Causa General Case 98/44, regarding whom there is no data. Also tried as belonging to the First Committee under Causa General Case 443/45: Arturo Sánchez Alcaide, Secretary General; Juan Conde Navarrete, Organization and Finance; and Juan Sánchez Cabrera, Uprisings and Propaganda), were not executed.

After the 1945 merger, the following are known to have been officers of the Second Committee: both Antonio Rivas García, Uprisings and Propaganda, and Rafael Obrero López, Organization and Finance, had earlier belonged to the independent activist group; Fausto Contreras Hervás had been a member of the Young Fighters.

In 1945, Cordoba province was divided into three sectorial or provincial subcommittees: the First Sector (North), the Second Sector Ç(South), and the Third Sector (possibly Cordoba capital).

The Committee for the First Sector was comprised of, among others, Manuel Jiménez Escribano, Secretary General; Joaquin Cabello Labrador, Uprisings and Propaganda; and Sebastián Fernández Martínez, Organization and Finance, in the name of whose wife, Barnarda Gárate, the Regional Committee sent its correspondence to Cordoba.

Three members of the Second Sector Committee are known: Francisco Catalán Higuera, Secretary General; José Díaz Bueno, Organization and Finance and Juan Conde Navarrete. Only one member of the Third Sector Committee could be identified under Causa General Case 443/45: Juan Sánchez Cabrera.

Of those indicted under Causa General Case 443/45, those who were resident in Seville were tried in that city because they received

Communist correspondence in their homes: Josefa Arévalo, Ana Ponce Barnero and Ángeles Vargas.

Only one of the accused, Domingo Escola Verde, was sentenced to death by this court martial July 24 1946, even though the Prosecutor had asked for seven death sentences. Six accused were sentenced to 39 years in jail, one to 20 years in jail, three to 12 years in jail, one to 6 years in jail and three to 3 years in jail. The death sentence was confirmed by the Seville Judge Advocate September 6 1946 and by the General Capitania on the 17th of that same month. The Government commuted Domingo Escola Verde's sentence January 29 1947.

It has been extremely difficult to reconstruct the labyrinth of raids and arrests during the 1940s. Much information comes from the reports of oral witnesses who have spoken of many of these *caídas*, or nettings, such as the *Caída del Horno* because it left from a brick factory in the Campo de la Verdad, Cordoba capital. A great many number of Communists were netted on this occasion, such as Melchor Ranchal Rísquez, a native of Añora. When at the end of 1945, five months later the police were again looking for him in another raid, he chose to flee to the hills where he joined the guerrillas with the nickname of *Curro de Añona*.[7]

The members of clandestine organizations who were netted and arrested in the caídas, were subjected to the traditional beatings and tortures of all prisoners under Franco. We cannot forget that an outstanding feature of Francoism was that nobody who was arrested escaped a beating. For many of the political prisoners, the arrest alone would have been punishment enough. Torture obeyed various ends and where the clandestine were concerned, it was as a means of obtaining the names of Party members and details of their links. In the case of Melchor Ranchal for example, his wife went to see him in prison after he had suffered a non-stop 24-hour torture session. She said she did not recognize him. That is why he fled to the hills, not because of a whim or any desire to become a guerrilla fighter.

The repressive terror reached such heights during 1945-46 that more and more fled to the hills, which undoubtedly explains the rapid increase in the number of guerrilla groups at that time. Many

---

[7] Melchor died in the Umbria de la Huesa, Villaviciosa, 1947 tragedy, when all the leaders of Julián Caballero's 3rd Guerrilla Group were executed.

of those who discovered that they thought they were to be named or identified as clandestine, had no other choice than to flee to the hills. José *Felipe* Merino Campos is an example of a young man who, believing that he may have been discovered for having set off some firecrackers in Cordoba to celebrate July 18, fled to the hills and joined Julián Caballero's guerrillas.

During the entire decade of the 1940s, every time there was a raid and netting of country folk, both men and women, in the neighbourhood of a town on the grounds that they were giving support to the guerrillas, there followed an outbreak of extra-legal shootings, deaths and imprisonments. Whenever a guerrilla turned himself in to the authorities, it was the same. There was a constant in and out of imprisoned individuals, justified by the authorities as anti-guerrilla repression. Of the many thousands incarcerated during this decade, according to an official source at least 60,000 individuals were imprisoned for this reason. No less than 5,349 were arrested in Toledo province between 1941-1948 and there were other 'hot' regions such as Galicia, Asturias, León, Levante, Málaga, Granada, Cordoba and more. To those numbers we must add the thousands who suffered the great repression of clandestine activities in major cities such as Madrid, Barcelona and Seville. The great penitentiary world of Francoist prisons was so hyperbolic that one could be forgiven for thinking that any understanding of it is almost beyond one's reach.

ENDNOTES FOR CHAPTER VI

[i] Señor Arévalo Bajo, personal testimony and eyewitness account to Moreno Gómez at the entrance to the Pozoblanco Old People's Home, Pozoblanco, December 12, 2013.

[ii] Blas Muñoz Márquez. Information sent to Moreno Gómez, Pozoblanco, 21 November 1985.

[iii] Mary Vincent. *The Splintering of Spain: Cultural History and the Spanish Civil War*. Ed. Chris Earlham and Michael Richards, Cambridge University Press, 2005, p. 75.

[iv] Juan Gutiérrez Romero from Villanueva but resident in Villafranca. Interviewed several times by Moreno Gómez during the 1980s.

[v] Francisco Moreno Gómez. *1936: El genocídio franquista en Córdoba* (The Francoist Genocide in Cordoba). Crítica, Barcelona, 2008.

[vi] Carmen Flórez Pérez, letter to Moreno Gómez denying him access to the Red Cross archives, 13 November 2013.

[vii] Pedro Gafias. *Héroes del sur* (Heroes of the South). Madrid-Barcelona, 1938.

[viii] With the Reds in Andalusia. London, 1985. (pp.110 et al, in the Spanish edition).

[ix] Data about Pedro Abad provided by the Gaitán brothers – Adán, Félix and Juan Manuel. *Mártires de una esperanza* (Martyrs to Hope), Lopera, Jaén, 2009, p. 495.

[x] Reyes Mate. *Memoria de Auschwitz* (Memory of Auschwitz), Trotta, Madrid, 2003.

[xi] Claro González Sánchez, interviewed by Moreno Gómez in Fuenteobejuna, August 1979.

[xii] Miguel Navarro González, oral testimonial to Moreno Gómez in Obejo, 24 August 1985.

[xiii] Copy of a letter given Moreno Gómez by his sister, Salud Bujalance. Hornachuelos, April 1981.

[xiv] One of the stanzas of that ditty, which his mother *taught* him, was based on the idea that the right-wing did not win the 1933 elections because women were allowed to vote for the first time. "If women had not been given the vote, nor a say, the perfidious Right would have beaten us."

[xv] True copy of the Case 94/44 dossier was provided Moreno Gómez by one of the indicted, Pablo González Calvo, from Peñarroya.

[xvi] Moreno Gómez interviewed Julio Priego several times and he still has his address in Cordoba.

[xvii] Adriano Romero Cachinero, *Eurocarrillismo y oportunismo* (Eurocarrilism and opportunism). Bilbao, 1984, p. 102.

[xviii] Julio Priego Ordóñez, interviewed by Moreno Gómez, January 2 1983.

[xix] This document is kept with Bartolomé Fernández's family archives in Pozoblanco.

VII

DESCENT INTO THE BLUE HELL

Mass deprivation of Physical Liberty. Mass transfer of prisoners to Cordoba prisons and Burgos Summer-Fall 1940 1941 - Franco's Auschwitz

"One of the things that I truly desire, which is why I have agreed to speak of everything that we have endured, is because above all, I want this to serve as an example to others, that they might know just what a Fascist Regime and Dictatorship such as we have suffered, is like. As I bear witness, I describe and explain all that we suffered and how we lived during the eleven years that I was held in jail, every day under the threat of death and what this did to the children and what it meant for my family. I am willing to make all of this known as a living witness to all.
Antonia García, one of the 11,000 women
incarcerated in Ventas Prison, Madrid

Introduction

Mass imprisonment or the severe deprivation of physical liberty, is considered a crime against humanity under International Law, especially in the Rome Statute that postulates the jurisdiction of the International Criminal Court, under 'Article 7(e) as 'imprisonment or other severe deprivation of physical liberty in violation of fundamental rules of international law'.

The phenomenon of mass imprisonments under Franco was not a logistic aberration, nor an unwanted excess of zeal, nor a simple question of public order. First of all, the phenomenon of widespread imprisonments was an expression of the State's intemperate violence.[i] Secondly, it was a criminal project for secluding its opponents and for providing a means of repressing, punishing, eliminating and re-educating the disaffected and for exploiting them in the form of slave labour. Thirdly, this project had much in common with the III Reich's criminal dogma that is, with the European Totalitarian and Fascist Project.[ii] Beginning in 1933, the mass Nazi imprisonments had the same goals as Francoism: to exclude, select, punish and terrorize political opponents. The Nazi and Francoist penal complexes (concentration camps and prisons) are symbols of the darkest and most shameful period in Humanity. This continues to be a serious concern, especially in Spain where there is such a tendency to sugar-coat, trivialize, ridicule and deny, these events.

There still has been no serious investigation into what occurred in the Francoist jails in 1941 That year, as in Mauthausen, those who were deprived of their liberty under Franco fell by the thousands, like flies. No one has yet offered a detailed explanation for why and how this period became known as the Francoist Auschwitz of 1941. As we shall see in Cordoba, most individuals died in prison rather than executed by firing squads during the post-war period. A huge mortality rate in all Francoist prisons that dropped like a stone in 1942, as it did in Mauthausen, when Spain's 'brother regimes' exchanged their policy of exterminating their opponents for a slave labour strategy.

The massive overcrowding in the prisons was, in itself, an exterminating factor. The official number of imprisoned at the end of 1939 was given as 270,719 in a December 1945 report entitled Brief Summary of the Work of the Ministry of Justice for the Spiritual Pacification of Spain. The validity of this report which has been and continues to be frequently quoted is extremely dubious. First, because the report was published soon after the end of WWII when Franco was not interested in having scandalous data regarding the number of political prisoners made public on the international scene. Second, how was the number of imprisoned in all the overcrowded provincial jails calculated at the end of 1939, considering the numerous large prisons and improvised smaller jails all over the country?

Furthermore, there was a totally chaotic penal situation in which thousands of prisoners were shuffled back and forth from one prison to another. What about the more than 23,000 women imprisoned in 1940, many of them with their children? The aforementioned total number can only be taken as a starting point for determining the total.

Gutmaro Gómez Bravo agrees that there is a marked difference between the official number of imprisoned and the reality, particularly as the official figures only take into account prisoners who were tried and sentenced. Not only were there hundreds of imprisoned who were not sentenced, but there also were thousands of prisoners of war, the imprisoned by the State Security Service, preventive imprisonments and so forth, not forgetting the extra-legal arrests after the continuous raids against clandestine reorganizers and those presumed to be assisting fugitives and guerrillas. He estimated that if one adds the data for the actual victims of the severe deprivation of their physical liberty held in the multitude of Francoist penal complexes – prisons and concentration camps – the true number of imprisoned in 1939 comes close to one million.[iii]

Added to the validity of the official figure at the en of 1939/ beginning of 1940, is that the Francoist Ministry of Justice's report does not include 90,000 who had been sentenced to Workers Battalions, 47,000 who were 'conscripted' into the Disciplinary Battalions of Worker Soldiers and the hundreds in concentration camps. It also appears that the report left out those imprisoned in countless improvised jails (convents, castles, manor houses of all kinds), as well as prisoners of the State whose dossiers continue to be kept under lock and key.[iv]

Ángel B. Sanz cited the official figures in 1945 and they were repeated by Halliday Sutherland in London in 1948: 170,719 for 1939; 233,373 for 1940; and 159,329 for 1941.[v]

What suffering there must have resulted from Franco's mass imprisonments. According to a report by Tomás de Boada, Count of Marsal and President of the National Council of Saint Paul for Prisoners and Convicts, from 1944 to 1948, MORE THAN ONE MILLION families had to contend with having one or more of its members in prison or out on prole and with having to care for young children and/or disabled relatives.

MASS DEPRIVATION OF PHYSICAL LIBERTY

Conditional freedom, provincial jails, mass transfer to Cordoba prisons
Franco's prisons, the epicentre of the repression

Most authors agree that the Francoist prison world was the linchpin and backbone of Francoist repression, dedicated to exterminating recalcitrant (disaffected of political significance) and re-educating the reclaimable (disaffected with little or no political significance), that is, the majority of the population who would be the object of the most ferocious ideological repression with a view to forcing National Catholic thought upon them. Raphael Lemkin[1] describes the transplantation of beliefs as a core feature of genocide, as an attempt to rip out the defeated's personal beliefs and impose the ideology of the victors.

Many authors consider that the Francoist repression was also a type of class justice, an extreme subjectivity that pervaded and disrupted the Francoist prison world, a product of the ambiguity of Francoist legislation. As Ángel Suárez stated in his White Paper on Francoist prisons: "The laws, in addition to being terrible, were sufficiently ambiguous to ensure that they would be applied with all conceivable subjectivity. Some imprisoned were sentenced to longer prison terms than others, only because their father or mother, or uncle, held the same beliefs."[vi] Suárez also pointed to the idea of 'class' as a means of creating a domesticated labour force from the retrievable, until not a single trace of their past activism or traditional class awareness remained.

The question of the effect of the lack of definition on the strict application of the legislation to penal punishment is equally important. It is known that everything depended on the so-called theory of 'reverse justice' used to eliminate or imprison all those who remained loyal to the legal Government. The umbrella crime of 'military rebellion' that Francoists cited with monotonous regularity when punishing half of the Spanish population, was defined and regulated under the terms of the 1890 Code of Military Justice (not amended until July 17 1945). In their rage to persecute the defeated, the military oppressors used little imagination when it came to defining appropriate charges for an indictment. There could be nothing more absurd than this lack of legislative definition.

Franco began by rescinding the Republican penal legislation with Decree of November 22 1936 and replacing it with the pre-Republican Penitentiary Regulation of November 13 1930. New

---

[1] Polish jurist who coined the word "genocide" that was not considered a legal crime, even during the Nuremberg Trials, until December 9, 1948, when the United Nations approved the Convention on the Prevention and Punishment of Genocide. "Genocide, not necessarily meaning the immediate destruction of a nation, is intended rather to mean a coordinated plan of different actions aiming at the destruction of national groups, directed against each group as an entity and the actions involved are directed against individuals, not in their individual capacity, but as members of the national group." *Vide*: United States Holocaust Memorial Museum, Washington, D.C. at https://encyclopedia.ushmm.org .

prison service regulations were not introduced until 1948, by which time the greatest part of the penitentiary disaster was over. During the war, military commanders published a whole series of all kinds of rules and regulations, the *bandos de guerra*, or wartime decrees.

Looking forward, the government created the System for Redeeming Sentences Through Work. October 7 1938, a hypocritical and exploitive invention by the Jesuit José Augusto Pérez del Pulgar. Although Victoria Kent did away with prison chaplains August 1931, Franco brought them back by Decree October 5, 1938.

Despite the apparent legislative ambiguity when it came to suppressing the Regime's opponents, Franco approved a number of complementary laws in support of Francoist authoritarian aspirations: Law of Political Responsibility, February 9 1939; Law Against Freemasonry and Communism, March 1 1940; Law for the Security of the State, March 29 1941, as well as other regulations that dealt more generically with the penitentiaries such as the Decree Creating the Probation Service, May 22 1943 and the 1946 Regulations for Prison Work, introduced long after the 'terrible years'.

Although the study of the regulatory framework for the Francoist prisons has become fashionable – and there is no doubt that this is a positive subject of research – why should this research be done to the exclusion of other essential studies such as those aimed at retrieving the facts of the Francoist prison world at that time. Curiously, in Spain, when something becomes historiographically 'fashionable', it is always accompanied by an anathema for and the proscription of everything that has been done before, including that which still remains to be done or is pending. In this case, all that occurred inside the prisons, how badly the prisoners were treated and how they died. This subject is so studiously ignored that there is still only an extremely limited amount of information regarding the monstrous mortality rate in Franco's penal establishments. The Francoist Auschwitz of 1941 remains largely ignored by historians and nobody has thought to sink his teeth into this task. Ángela Cenarro, Professor of Modern History at the University of Zaragoza, made a remarkable comment about this:

"If we stick to simply listing the infrahuman conditions in which thousands of anti-Francoist prisoners spent their days among such misery and desolation, we run the risk of losing our resolve, of failing to contribute anything to the knowledge of the legal and institutional framework on which the operation of the prisons was based."

Republican-political-prisoners-in-gaol-in-Spain.jpg[vii]

How can one possibly 'lose one's resolve' in studying the inhumane conditions when almost nobody has studied this matter in depth, even less so members of Academia, when almost the only information that has come to the public's attention is Tomasa Cuevas' monumental compendium of eyewitness accounts.[viii] Why cannot the study of the regulations and the study of the inhumane conditions complement each other? There is no doubt that the exclusionary temperament of the Spanish does its damage For example, in 1987 Moreno Gómez published an extremely detailed study of the Cordoba city prisons, the first time that anyone spoke openly of the terrible mortality rate in 1941, when 756 inmates from a prison population of fewer than 4,000 died.[ix]

At the end of her book, Ángela Cenarro recognizes that the Prison Regulations had no more effect on prisoners than soggy bits of paper they could have been written on and that the daily routine inside the jails had nothing to do either with rehabilitation or re-education. In other words, the regulations went along their merry way along one path whilst the terrible daily routines of the prisoners went along another. Prison directors allowed torture and corruption. Inspectors turned a bind eye. Guards facilitated communications in exchange for gifts. Prison chaplains reigned as lords and masters of the men That was the real picture, according to her. Moreno Gómez agrees in that the regulations do need to be studied, of course, but we must not waste the large amount of field work done so far by those who dug deep into the Francoist sewers, by flushing it down the drain.

When working on the draft for his first book on victims of the civil war, Moreno Gómez outlined some notable features of the appalling world of Francoist penitentiaries:

1. the overcrowding and its consequences were absolutely never unwanted or ignored by the new totalitarian State;
2. this destructive situation was neither an isolated or occasional occurrence, but universal and systemic;
3. the defeated never foresaw nor expected to be punished through the deprivation of their personal liberty; nor could they come to grips with the logic behind such repressive furore, so much blood and so many tears;
4. the magnitude of the imprisonments set the stage for the project to exclude half of Spain from the New State, the total absence of any desire to integrate the defeated and the absolute priority given to the plans for punishment, revenge, 'cleansing' and the ideological repression of all the progressive-reformist-leftist ideals that were the soul of the II Republic; and
5. the mass deprivation of personal liberty and punishment was characterized by the victors' unpredictability and randomness, all of which was stoked by the lack of legislative and normative definitions.

The aforementioned White Paper on Francoist prisons described three stages in the Blue Hell of the totalitarian New State's penal complexes, all three of which are in agreement with the Nazi model and practices. The first and most basic, employed all the typical forms of corporal punishment (torture, forced labour, shortage of food, etc.) in prisons, police headquarters, concentration camps and Falangist checas. The victorious Nationalist Army and Political Police had carte blanche to act as they liked in the prisons, to arrest or release prisoners at whim, to arrange for paseos and sacas, interrogations and torture, both in the prisons or in purpose-built interrogation centres.

The second stage was directed at the prisoners' inner self, with methods that were even closer to those practised by the Nazis, as Francoists used human behaviour and psychological pressure techniques in order to obtain an individual's total dependence on his tormentors. More than 're-education', as today they say that it was, it was ideological repression pure and simple. There is no re-education in crushing an individual's self-esteem to the limit of his endurance. When the tormented individual weakens until, in many cases, he becomes incapable of reacting to the continuous harassment of religion, non-stop patriotic ceremonies, the omnipresent cult of the Caudillo, praise of 'redeeming' labour and the impact of the propaganda against the so-called 'Godless', or those who were 'anti-Spain'.

Lastly, the third stage consisted in the psychological effects of the entire process: convincing the tormented man of his guilt and managing to get him to accept the truth of the accusations levied against him. Many victims ended up accepting their guilt, as did Juan Simeón Vidarte, the well-known Socialist, when he wrote *Todos fuimos culpables* – We all were guilty. Because of the persistent lack of information today, the twisted version of the History of this period continues to have a destructive effect on certain sectors of society. Another example of the victors' use of language as a weapon of mass destruction.

Regarding the aforementioned first stage, that of the typical forms of corporal punishment, this is in agreement with a 1949 British report cited by Gutmaro Gómez, from which the following is an extract:[x]

> "Its archives are based on the Nazi model and ensure a systematic watch over all suspicious enemies of the State... As they are paid extra, the poorly paid police tend to apply increasingly violent methods and to prolong the isolation of prisoners to obtain confessions..."

The same author recalls a statement from the Ministry of the Government, regarding that hyperbolic repressive macro-process, according to which THREE MILLION individuals were already on file in 1944. This leads Moreno Gómez to believe that Gutmaro was mistaken when he said that in March 1948, when the state of war was finally declared over, the repressive system was already fully operational and suffered no substantive changes until the end of the Regime. This was in contradiction with everything that was going on in Spain on those very same dates, that period that since 1987 Moreno Gómez has called the Triennium of Terror (1947-1949), and all its precursor and succeeding actions, the thousands of victims of paseos and the Law of Fugitives in all of Spain, from Asturias to Málaga. It appears that nobody is interested in looking at the repressive barbarity of the totalitarian State at the end of the 1940s.

The first period of mass deprivation of physical liberty. The provincial jails (1939-1940)

*"One cannot say what justice consists of if we ignore
the injustices, which is what occurs today.
The memory of the injustice invalidates all those
theories of justice that are currently taught in the
Faculties of Law and Philosophy."*
Manuel Reyes Mate, *La piedra desechada*, 2013

One of the first problems facing the victorious horde at the end of the war in 1939, was how to imagine its imprisonment of half of the Spanish population. The Army of the Republic, half a million men, captured and disarmed, was imprisoned in the almost two hundred concentration camps that existed all over Spain. The Regime also had to imprison the entire Republican governing elite and a considerable number of civilian Republican leftists. Never before in the history of this unhappy country, had anything like this happened, not even remotely so.

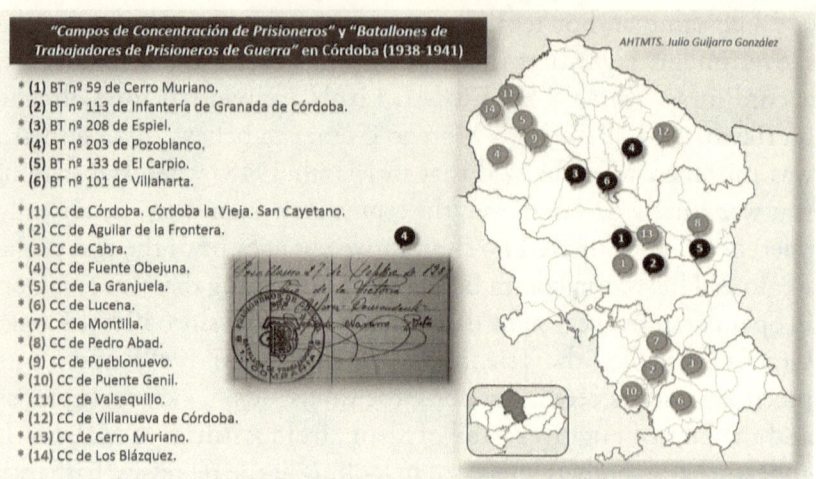

"Campos de Concentración de Prisioneros" y "Batallones de Trabajadores de Prisioneros de Guerra" en Córdoba (1938-1941)

AHTMTS. Julio Guijarro González

* **(1)** BT nº 59 de Cerro Muriano.
* **(2)** BT nº 113 de Infantería de Granada de Córdoba.
* **(3)** BT nº 208 de Espiel.
* **(4)** BT nº 203 de Pozoblanco.
* **(5)** BT nº 133 de El Carpio.
* **(6)** BT nº 101 de Villaharta.

* **(1)** CC de Córdoba. Córdoba la Vieja. San Cayetano.
* **(2)** CC de Aguilar de la Frontera.
* **(3)** CC de Cabra.
* **(4)** CC de Fuente Obejuna.
* **(5)** CC de La Granjuela.
* **(6)** CC de Lucena.
* **(7)** CC de Montilla.
* **(8)** CC de Pedro Abad.
* **(9)** CC de Pueblonuevo.
* **(10)** CC de Puente Genil.
* **(11)** CC de Valsequillo.
* **(12)** CC de Villanueva de Córdoba.
* **(13)** CC de Cerro Muriano.
* **(14)** CC de Los Blázquez.

**Prisoner Concentration Camps [CC] and
Prisoner of War Workers Battalions [BT] in Cordoba 1938-1941**

If in the concentration camps the inmates were predominantly ex-military; in the multitude of improvised jails in the majority of the towns and villages in the countryside, the inmates were a mixed batch: military and civil, because ex-soldiers were straying all over the place and, having abandoned their weapons, were making their way back to their home towns where they fell into the clutches of local Fascists, the worst of the lot.

The victors, drunk with victory; just did not know where to put so many people. Undoubtedly, this was one of the most absurd and clumsy mismanagements of victory and defeat ever seen on the face of this Earth. Only an idiot would order putting half of the Spanish population behind bars simply because they were 'disaffected' or imprison an entire Army of half a million men. Such was the mental power of the Caudillo, of the 'consummate Caesar' as Alberto Reig described him; 'the great manipulator' according to Paul Preston; the 'Criminalissimo' as Rafael Sánchez Guerra called him. April 1939 Spain was, quite simply, catastrophe and gaucherie elevated to the nth degree.

Other European post-war scenarios were not managed with the schizophrenia that affected post-war Spain. Post-war Germany did not dump all the defeated in the same sack, but it divided them into three groups: major offenders (of which there were only 1,600), Nazi sympathizers and minor offenders. Except for the war criminals who were tried in Nuremburg, no death penalties were handed down, only prison sentences and dismissal. The government soon pardoned the youngest offenders, issued another amnesty Christmas 1947 and in 1948 downgraded major offenders to sympathizers.[xi]

All this, three years after the end of the Great War, whilst Spain, eight to ten years after the end of its war, was still suffering the effects of the Triennium of Terror (1947-1949) and the Law of Fugitives was still being applied without restraint in every ditch and gutter in the Spanish countryside. Franco's terror was far from analogous to anything that had happened in Europe. Post-war Italy began to introduce sanctions to Fascism in 1944 and June 1946, and one year after the end of the Great War, the government issued an amnesty for all sentences under 5 years and commuted longer sentences. Soon afterwards, only leading Fascists and individuals condemned for major crimes remained behind bars. Whereas in Spain, the Francoist administrative purging of all public services were directed at half of the Spanish population, in Rome, for example, of 394,041 public servants investigated, only 1,590 were fired. If in Europe the tormentors were the new democracies, in Spain, Fascism was the oppressor.

Back in Spain, the Regime began packing 'disaffected' Spaniards, both civil and military defeated, into the most surprising, improvised jails, where triumphant fascist thugs lay about them with their fists, clubs and whatever they could get their hands on. Historians who have investigated these jails have written very little regarding this first stage of Francoist mass deprivation of physical liberty. Attracted today by the fashion of primarily studying the penal rules and regulations, researchers have chosen to ignore this terrible first post-war year and a half (1939-1940), when there still were no legislative rules or regulations, only systemic, widespread torture. This was the terrible stage of improvised country jails that lasted until Fall 1940, when the prison population was removed from the countryside and concentrated in the prisons in the provincial capitals. There is apparently no present-day detailed account that today's generations can read, that faithfully describes the prisoners' martyrdom in the country jails.

At this first stage, Francoism's social base furiously devoted itself heart and soul, to the New State's task of purging the country from all the 'fiendish Rojos'. A pack of blood bound the Regime's base with its hierarchy to form a carefully fashioned operational triangle that would ensure that all the victors would unite to further their Crusade: to punish and exterminate all the defeated, without the minimum thought for any humanity, clemency or fraternity between Spaniards. No one in the new totalitarian Regime conceived, not even by chance, a single reconstructive project that would assimilate all Spaniards. The New State looked to the most terrible reprisals for its base, the so-called Christian attitude in the midst of the greatest schizophrenia ever seen in Spain, a tripartite association led by the barracks, the Casino and the church vestry.

The provincial jails (1939-1940) in Cordoba province and the rest of Spain.

Here follows a brief revision of the details of the first stage of the mass deprivation of liberty, primarily that of the condition of the individual local jails, most of which were improvised, the lack of local conditions overcrowding and mistreatment of the inmates, with the attending carnage.

In Cordoba province

In the village of Castro del Río, in the heart of the Cordoba countryside, a great many local workers – most of them Anarchists – fled to the Republican zone to fight for the government, and at the end returned to their village looking forward to a time of peace, only to be arrested. When the Municipal Depot became rapidly packed to the rafters with inmates, the authorities improvised a jail in the Santa María de Scala Caeli Convent, a large building with many cells, patios and thick stone walls, where hundreds would end their days. Prisoners from Baena, Albendín, Valenzuela, Luque and Espejo were transferred here in February 1940. A total 1,500 individuals were held behind bars in this village.

In Belalcázar, in the north of the province, prisoners were crammed into the Divina Pastora School, another improvised jail. As the number of death sentences began to rain upon the inmates, a large number of desperate men organized a great escape August 4 1939. 15 of those condemned to death escaped, 3 died in the attempt and the remainder fled to the hills where they joined the bands of freedom fighters and guerrillas who later played an important role in the post-war history of the district.

In Hinojosa del Duque, an improvised jail was created in Concepcionita Convent, where prisoners from Hinojosa, Belalcázar, El Viso, Santa Eufemia and other villages, were kept. Another building, an old manor house named Casa del Condesito, was also turned into an improvised jail in the first few days after the end of the war. (Today, it is an inn.) Here, too, desperate prisoners organized a great escape the night of August 31 1940. 20 inmates escaped through a hole in a wall, 5 or 6 failed to get away and their bodies were left to rot on the town streets. The remainder also fled to the hills where many became famous guerrilla leaders.

In Villanueva de Cordoba, the capital of Republican Cordoba Province during the war, the usual penitentiary chaos began in the early days of April the year of the Francoist Victory. The town was the headquarters of the Nationalist 63rd Division and it also housed a great number of refugees. Prisoners were packed into the Municipal Depot, the anti-aircraft shelter, Romo's House on Calle Herradores, Pepe Barrón's factory warehouses on Calle Industrial, schools and other improvised jails such as La Preturilla, the provisional Guardia Civil barracks, and there still were enough inmates left over to justify setting up an improvised concentration camp in Ángel Díaz's yard, next to the Ramirez Factory. The authorities soon began to send convoy after convoy of prisoners out of town, to Valsequillo and La Granjuela and especially, to Castuera, concentration camps. The bells of victory were still pealing when the notorious SIPM Lieutenant Leopoldo Mena began 'working' with his Information Committee.

Soon after the end of the war was declared, a large number of prisoners in Villanueva were rounded up in Juan Herrero's home, a large manor house on Calle Conquista: men on the top floor and women on the ground floor. Prisoners from Torrecampo, Pedroche and Adamuz were also sent to Villanueva. At the beginning of June 1939, many of these prisoners were transferred to the Fuente Vieja Schools improvised jail, the site of much torture and executions by firing squads until September 16 1940 when all the imprisoned in the town were marched to the Villanueva train station, on their way to Cordoba capital. There, they joined prisoners from all over the province in inaugurating the New Prison, a kind of Auschwitz without gas chambers in which 502 died from starvation and deprivations in 1941 alone.

Appendix VI at the end of the book, records a sample of the multiple accounts of the suffering in the town jails and of the inmate's desperate attempts to escape. One such tory I that of José García Coleto *El Perica*, who managed to escape from the line of prisoners who were being transferred and who vowed to get his revenge on his tormentor – which he finally did in 1940, only to be caught and killed six months later near Villaviciosa. There is also the tale of a planned escape that came to nothing.

In Bujalance, prisoners were packed in the local jail. In Baena, although an even greater number of inmates filled the so-called Tercia, or Posada, jail next to the barracks and the town hall, the authorities still had to improvise a jail in the Plaza Vieja. Prisoners from Albedín, Luque and Valenzuela were also housed in Baena until February 1940 when all the inmates were transferred to the Castro del Río Nunnery in early Fall, and from there, those who had survived the multiple executions, were sent to Cordoba Provincial Prison in the capital. Likewise in Puente Genil, the large number of prisoners far exceeded the capacity of the local jails which is why one was improvised in the La Alianza Factory warehouses in Molino del Marqués until they, too, went sent to Cordoba.

The drama of the Fuenteobejuna prison world was played out on the ground floor of the City Hall building from 1939 until the beginning of 1940 when the prisoners were transferred to the Peñarroya-Pueblonuevo jail, also located on the ground floor of the City Hall building, together with others in improvised jails such as the Miners' Union building and the Trade School.

As the Municipal Depot in Fernán Núñez soon filled up, a movie theatre in the town centre became an improvised jail, with an average 200 inmates, some of whom came from Montemayor. The number of imprisoned rose rapidly at the end of the war: 440 women, 271 civilian men, 688 ex-soldiers and a great many young people under the age of 21.[xii] October 1 1940, 146 inmates were transferred to Montilla jail and soon afterwards, to the Provincial Prison in Cordoba.

The San Luís Convent School in Montilla became an improvised jail, as the Municipal Depot quickly reached maximum capacity.[xiii]

In 1939, Montilla contributed 646 inmates, 239 from Montilla and the remainder from neighbouring towns, to the total imprisoned in all of Spain. July 1940, 27 prisoners arrived from Lucena – not many, because earlier there had been mass executions in that town and besides, there was a small concentration camp with 305 inmates in the town. In La Rambla, there also were mass imprisonments when at least 583 men of the more than a thousand townspeople who had fled earlier to the Republican zone, returned.[xiv]

The fact of the matter was that half of the Spanish population was packed into every conceivable prison and jail throughout the country. According to Casimiro Jabonero's testimony, more than 2,000 inmates filled the Cuenca seminary.

In Guadalajara, in addition to the Polígono concentration camp, the provincial prison was filled to the rafters, there was a military prison and the Convent of the French Nuns women's prison. In Alicante, the penitentiary chaos was total as the authorities strove to find room for all the people they were arresting as they attempted to leave the country. Prisoners filled the Campo de Albatera concentration camp, the provincial prison, movie theatres, the House for Spiritual Retreats, and more. The Alicante Reform School for Adults, designed to house 300, held 3,000. The provincial prison in Albacete filled quickly and so did the improvised San Vicente jail. Juana Doña wrote that 1,000 women were imprisoned in one of those prisons.[xv] Other notorious prisons and jails in the province were in Chinchilla, Yeste castle and Hellin jail. The worst of all was the Chinchilla general complex that had been closed down in 1870 because of the terrible conditions and only reopened by the Francoists.

In Rute, the Municipal Depot was extended to the premises of the Las Palomas anis factory, to house 80 captive members of a Workers Battalion who had been sent to repair the Carcabuey road. To speak of Rute is to remind us of the suffering in Francoist jails during this first stage of the repression, of the constant and rabid harassment of the defeated, of the lines of prisoners taken daily from the jail to the Guardia Civil barracks where they were forced to sing the Francoist hymn *Cara al Sol*, forced to attend Mass under guard every Sunday, or as in Posadas, of group after group of prisoners marched to the town square and forced to kneel before a cross to the Francoist fallen and pray for the victims of the Left.

In Cordoba capital, the Provincial Prison was traditionally located in the medieval dungeons of the Alcázar de los Reyes Cristianos. These extremely unhealthy installations were the venue for the great genocide of 1936, when some 4,000 were exterminated. The New Prison was built in 1940 near the Asland Factory, in the north of the city, on the road to Pedroches. When the first waves of prisoners were transferred to Cordoba from the province Fall 1940, the building work was not complete, there were no panes in the windows and the sanitary installations were insufficient for the needs of the first 3,000, soon to total 4,000, inmates. It is therefore not surprising that the genocide in the Cordoba prisons began within a few weeks and that no less than 502 would be exterminated in 1941 alone.

In the rest of the country

The penitentiary world in Madrid was in total chaos, as every conceivable space served as a prison: Santa Engracia (1,000 inmates); Porlier (4,000); Torrijos on Calle Conde de Peñalver (3,000); Duque de Sesto (8,000); San Antón, a religious school on Calle Hortaleza (2,000); Yeserías, an asylum for beggars in Deicias (3,000-5,000); Convento de Comendadoras (1,000); Atocha (2,000); Santa Rita (4,000); Conde de Torreno (700), in addition to Barco, Príncipe (near Carabancheles), Jaime Vera (another ancient workhouse for beggars), and Cisne on Calle Martínez Campos, an old convent of Trinitarian nuns that had served as a convalescent hospital during the war. All the military courts in Madrid were located on the ground floor and the first floor housed more than 300 prisoners, almost all Republican military There were more prisons (two of them military) for women, namely Vents (12,000 women), Claudio Coello (1,000), Malsaña and San Isidro (women who were breast-feeding children).

Also, according to Rafaela Sánchez Guerra, imprisoned at the time, if to these numbers we add those who were interned in neighbouring concentration camps, in the Security Services dungeons in the ten police headquarters in Madrid and the several Falangista checas, we can easily reach a total of 50,000 individuals deprived of their freedom in the Spanish capital.[xvi]

July 1939, the Puerto de Santa María, Cádiz prison housed 5,4000 inmates, 2,000 of whom were Basque. According to a report from the

National Basque Archives, the situation on that date was deplorable.[xvii] In the tragic San Simón, Pontevedra, penal complex, 2,176 inmates were packed in inhuman conditions during the immediate post-war period, most of whom were over 60 years old (up to 666 inmates were executed here).[xviii]

In Amorebieta, 800 women were interned in each room in a monk's school. Women's' prisons in the North were particularly terrible: Santander, Saturrarán, Durango (housing 2,000 women, many with their children), Segovia, and more. The Cuéllar jail in Segovia was especially notorious, as this was an ancient castle dating from the end of the Middle Ages, dungeons and gloomy installations, housing 1,500 inmates. Survivors of the Torrero, Zaragoza jail continue to talk of their sufferings.

There were three jails in Toledo capital: the provincial prison, the improvised jail and San Bernardo prison. There were many jails in the countryside, but he largest and most overcrowded were Talavera de la Reina and the tragically infamous Ocaña jail (housing 5,000 men and 2,000 women), where 9 inmates were packed into individual cells built for one. In 1940, there were 8,200 prison inmates in the entire province.[xix]

Overcrowding was a feature of the Model prison in Valencia. Built to house 528 in 1907, up to 15,000 inmates were housed at one time during the post-war. In Jaén, there were 6,000 prisoners in the capital in 1941, when 4,000 were housed in the provincial prison built for 80 inmates in 1930.[xx]

The first year and a half of the Francoist penitentiary world (1939-1940) in the local jails, was characterized by an astonishing, extremely humiliating, cruel and inhuman overcrowding. The degree of harassment of the unfortunate defeated reached levels that are difficult to conceive of today. The main terrible feature of that situation was the 'proximity' of the victors and the defeated. In the countryside, the tormenters lived near the victims who were no strangers to them, whom they met regularly with their henchmen when it was time for the defeated to make their daily declarations to the authorities at the barracks or the prisons. A second, very important feature of the local jails was the facility it gave the victors to retaliate for presumed slights during the Republic, both personal and in the form of complaints against employers, labour conflicts, strikes, everything they saw as the 'revolution', as in their minds, all labour unrest was 'revolution'.

The local setting was perfect for anyone who had an ancient quarrel that rankled. Matías Romero Badía, for example, describes the terrible beating that his brother, Crisóstomo received at the hands of a group of Falangists. Among these, Pedro *El Barbero* was revenging himself for the fact that during a strike, a group of men in which Crisóstomo accompanied Alfonso Ibáñez Tamaral, had forced him to close his barbershop. They tossed his brother headfirst, like a ping-pong ball, against the blood-splattered walls of the office [*sic.* in the Preturilla barracks], until he lost consciousness.

Accordingly, during the first year and a half after the war, the New State began its avowed intention to cleanse all of Spain of even the most insignificant syndicalist ideals or activism, both past and present, and this could only be successfully achieved in those places where the defeated were well known, that is, in their hometowns. Although the constant, violent attacks on the defeated was slightly more controlled when they occurred in the major provincial prisons, in the country, the absence of any legislation, rules or regulations of any sort meant that the rule of daily beatings and uncontrolled violence was the norm. When the returnees, returned to their hometowns, they ran the constant risk of being lynched by unruly victorious mobs.

Although there are several academic studies of the situation in the large city prisons, most authors have paid little or no attention to the horror of the country jails in this first post-war period. In the countryside, more so than in the provincial capitals, the inmates suffered what Moreno Gómez calls *retaliatory torture*. Local tormenters' favourite 'sport' was the visitors they brought to the jails, day and night: Falangists, pimps for the New State, spoiled brat Señoritos who amused themselves by pistol-whipping, lashing, throwing chairs and other objects and otherwise humiliating the inmates. Added to this, there was the *judicial torture* that was applied when prisoners went to make their declarations to the authorities, and the *law enforcement torture,* bloody affairs that took place in police headquarters in the cities such as Madrid, and in the Guardia Civil barracks in the countryside and local Falangist headquarters, where so many left their lives.

In an appendix at the end of the book, Matías Romero provides some additional personal details regarding the retaliatory torture that the victors applied at random in the country jails, out of pure and simple repressive fury, day after day, night after night. The cruelty of the repression in the provincial capitals was nothing compared to that which occurred in the countryside where the victors revelled in thrashing the defeated.

Miguel Hernández's prison diary and letters[xxi] disclose how it was in Orihueka, his hometown where he was imprisoned for three months, from September to November 1939, that he received the worst treatment. He expresses his despair for the mistreatment he received from the intolerant, clerical and hateful townspeople whom he compares to those that Gabriel Miró depicted in his painting 'Our Father Saint Daniel and the Leprous Bishop'.

The degree of sadism that local tormentors applied on a regular basis was never matched in the central provincial prisons. Already described in some detail, is the barbarous treatment meted in towns and villages such as Montilla, Fernán Núñez, Villanueva de Cordoba and elsewhere in the Cordoba countryside, methods that were applied throughout Spain. Nightly visits of drunken Falangists are documented in towns of Toledo province, such as in Aldeanueva de San Bartolomé, where one of those who suffered, Jesús Gómez Recio-Quincoces, the Republican Mayor, successfully escaped from the jail and became a famous guerrilla fighter. Also, in Quintanar de la Orden, where they crushed the prisoners' feet to prevent them from running away, as they did to José Manzanero whose wounds took three months to heal.

At this first stage, the Francoist repression was intentionally directed at the visibility of the punishment, to further punish and terrorize the general population. It was the almost daily show of Rojos, of the Godless, the anti-Spain prisoners, expunging their guilt, of lines of prisoners paraded through the towns, men and women with shaven heads, humiliated, their clothes in tatters, signs of the constant starvation and torture they were subject to, etched on their faces.

At this stage, the State spent the very bare minimum share of its budget on feeding its prisoners, whose sustenance was left to their families. Every day, relatives went to the jails to take them their breakfasts in milk cans and jugs and their meals in lunch boxes, baskets and woven cloth bags. Despite their social exclusion and abject poverty, these families wrought miracles to take some food to their imprisoned kin which is how they survived the first year and a half of their incarceration, but all this ceased with the mass transfer of prisoners to the central prisons in Cordoba capital. This circumstance alone may explain but only in a very minor way, the 1941 Francoist Auschwitz in every prison in Spain. Such an astonishing disaster that nobody has yet studied it in any detail.

Mass transfers to Cordoba capital prisons and to Burgos – Summer-Fall 1940

This section begins with the great march of prisoners from the rural jails in the entire province to the central prisons in Cordoba capital at the end of Summer and beginning of Fall 1940. One might presume that the Machiavellian Regime was satisfied that it had done enough with its first stage of penitentiary repression to ensure that it had attained the desired effect on the population as a whole – that is, an uncontested awareness of the terror, punishment and executions that awaited the disaffected. It could now proceed with the second stage, a more systemic prison hell in the New State's large provincial prisons, those that could now be called the Regime's great penal complexes (Burgos, Puerta de Santa María, Alcalá de Henares, Ocaña, Carabanchel, and more).

Enormous convoys and never-ending lines of prisoners made their way to Cordoba capital in a spectacle of humiliation and punishment with which the Regime bombarded the defeated. Great troupes of prisoners were marched daily along the streets of Cordoba capital, from the railway station to the prison, a multitude of men tied to each other by wires, knapsacks over their shoulders, the mark of torture on their faces, gazing blankly over the panorama of a world that was totally hostile to them. Although it is true that many people commiserated with them in silence, the shamelessness of the excited victors expressed itself as insults of all kinds Hysterical society ladies, flippant young Señoritos, jocular wealthy members of the Circulo de la Amistad and frenzied Falangistas, lined the route taken by the prisoners and transformed it into an authentic *via crucis*.

As described earlier regarding Cordoba province, the greater number of trials were carried out in the countryside in the immediate post-war. When at the beginning of Fall 1940, the majority of prisoners were transferred to the provincial capital they were weighed down by the fact that almost all had already been sentenced to death. (Only a minority would be tried and sentenced in the capital in 1941 and 1942.) Behind them, in the hometowns, they left many companions who still faced the firing squads: 167 in Castro del Río, 209 in Pozoblanco, 102 in Villanueva de Cordoba, 88 in Peñarroya, 67 in Hinojosa, etc., etc., although the last legal executions had been carried out in many of the towns and villages during the Summer. The last shots were fired in June in Belmez, Puente Genil, Peñarroya and Pedro Abad. In this last village, they left a pregnant woman who had been sentenced to death, 37-year-old Josefa Ortega Egea whom they would execute October 3, a few days after giving birth.

September 1940 witnessed the last legal executions in Bujalance, Castro del Río, Hinojosa, Montilla, Montoro, Pozoblanco and Villanueva de Cordoba. All these prisoners would now be sent to the capital, where they would be dispatched during the great wave of executions in Cordoba capital in 1941.

Transfer to Burgos

There follows some description of the manner by which the prisoners were transferred from Villanueva de Cordoba to the capital. September 26 was the date set for this transfer and it would be done with considerable military pomp and circumstance (soldiers, Guardia Civiles, municipal policemen, Falangistas with revolvers tucked under their belts, and so forth) down the city streets, effectively cutting off any access roads. A huge chain of prisoners tied to each other, their meagre belongings over their shoulders, was brought from the Fuente Vieja square where the Schools served as an improvised jail. Relatives of the prisoners, especially children, pushed against the crowds on the street corners, trying to make themselves seen, despite continuous insults and slaps from the policemen. There were more than 400 men, not only natives of Villanueva, but also from Pedroche, Torrecampo and Adamuz. Another cordon of 50 women was taken from Juan Herrero's House Calle Conquista.

As they walked, every prisoner tried to see if he could spot his relatives and especially his children and call out words of farewell. Within the city limits, it was difficult because of the large number of soldiers who lined the streets, but as soon as they moved towards and into the avenue that led to the railway station where they were packed into a passenger train, it was all a frantic waving of hands, tragic calls of farewell to their relatives and from their relatives to the prisoners. Witnesses to the events of that day tell how, as the train started off and for quite some time afterwards, relatives, women and children, ran alongside the train, through the fields, hoping for a last glimpse of their husband, father, brother, son, as long as the carriages remained in sight.

The next day, September 27, after a terrible night in improvised shelters in Peñarroya-Pueblonuevo, they arrived in Cordoba capital where they inaugurated the unfinished building of the New Provincial Prison. Of those who arrived from Villanueva de Cordoba, 35 soon faced the firing squads in the capital and at least another 23 would starve to death. Around the same dates, another large convoy of more than 600 men arrived from Pozoblanco and neighbouring towns (there had been 990 in 1939 but 209 had meanwhile been executed); in the capital they shot another 10 and 8 died from hunger. There were two more large convoys to Cordoba capital, one from Hinojosa (with prisoners from Belalcázar and Santa Eufemia) and one from Peñarroya (with prisoners from Fuenteobejuna, Belmez, and other towns). From Hinojosa, 20 were shot in the capital and another 23 starved to death; from Belalcázar, 18 were shot; from Belmez, the most notable is that 16 died from hunger; from Fuenteobejuna, 19 also starved to death in prison; from Peñarroya, 24 more.

As to the prisoners from Montoro, in addition to the 61 who were executed earlier in the town, 40 were executed soon after they arrived in Cordoba and another 25 died from hunger. From Adamuz, in addition to those executed in Villanueva, 20 were executed in Cordoba and 22 died from hunger.

In the countryside – Puente Genil and all the towns and villages in the so-called Red Belt around the capital – the imprisoned in the local ails were transferred to Cordoba city on the same dates in September and October. Fernán Núñez had already sent a group of 146 inmates by truck, tied to each other with wires, to the improvised Montilla jail where they were packed in San Luis Convent School. When the time came to send the prisoners to Cordoba, those who had not been previously sentenced to death were sent to Puerto de Santa María. Those sentenced to death and the prisoners from Montilla, with Manel *El Perla* Sánchez Ruíz, the Republican Mayor, at the head of the group, were sent to Cordoba between October 17 and 19.

A mixed group of prisoners formed the large convoy that left Castro del Río and also included approximately 1,500 prisoners from Baena who had been incarcerated since February 1940 in the Scala Caeli Convent improvised jail. 167 prisoners were executed in Castro and the remainder sent off, tied to each other, to the Provincial Prison in several groups of trucks. 18 were executed in Cordoba and 7 starved to death in prison. Another large convoy left Bujalance. 55 prisoners were executed before it left; 13 more in the capital and 13 later starved to death.

All this was a feature of the Cordoba genocide. Three or four days after the arrival of the first wave of prisoners from the countryside, the authorities began to send part of those who had already been sentenced to death to prison Burgos, were they were to spend the winter. As Eutimio Martín describes it,[xxii] the prisoners were transferred by train, crowded into closed cattle wagons with a bit of straw on the floor, similar to those used by the Nazis to take Jews to the concentration camps. The trip to Burgos, with one stop in Cáceres, took three days, during which they gave the prisoners nothing to eat or drink. Unlike Hitler who charged the Jews for their trip, under Franco, the ride was free. However, as soon as the prisoners arrived in Burgos, they were stripped of all their belongings, including tobacco. The similarities between Franco's New State and Nazi Germany were much more factual than people believe.

Once in Burgos, the bitterly cold winter weather made life intolerable for the inmates; they died not just from hunger but also from the cold. As Diógenes Cabrera wrote, there are only two seasons in Burgos: Winter and the season for travelling by train. His and other personal testimonies agree with Eutimio Martín's comment regarding the role of the weather as a means of extermination, that Franco created a unique Auschwitz of his own where the weather, because of the particular physical weakness of the imprisoned, played the same sinister role as the gas chambers.

The great return trip to Cordoba capital began in 1941, in the so-called 'trains of death'; as soon as a sufficient number of approved death sentences were obtained, a trip was arranged. Diógenes Díaz Cabrera, Freemason and ex-Consul for Venezuela in La Palma, an excellent witness of what happened in the Canary Islands, published an impressive description of his transfer from Madrid to Burgos a couple of years later:

"At the end of October 1942 [from Porlier, Madrid], we were rousted from our cells at 3 a.m. and taken outside where we were forced to stand in formation for more than three hours until a Guardia Civil captain arrived with a large number of Guardias carrying machine guns... Marching to order, backpacks slung over our shoulders, surrounded by Guardia Civiles, soldiers, *requetés* and Falangistas armed to the teeth, we were led to the Madrid railway station On the street corners, we could see prisoners' wives and mothers weeping and waving handkerchiefs from a distance; they were not allowed to approach us.

The enormous convoy is which we were packed, our feet tied together with wire, left at the sound of a bugle This was the most painful trip of all those I was taken on... We were forbidden to speak. The train travelled maddingly slowly... We arrived in Burgos at 3 a.m. November 3 – the trip took 4 days. All our feet were horribly swollen, our whole body ached, we were extremely drowsy and numb from the freezing cold.

They tossed us like bales of hay onto trucks and took us from the railway station to the prison. Our backpacks remained at the station, and we were placed in damp, cold cells like iceboxes. Totally exhausted, we lay on the floor and tried to sleep, but it was impossible. The floor was just too cold... The second night we had not yet been given our backpacks and we were desperate from the lack of sleep because of the cold.... I said [to Antonio Cepero]: 'I cannot stand this any longer. I am from a warm place, and I have to ask the Director of this damned prison to help me if he wants to keep me from freezing to death.' Our backpacks were finally delivered on the 5th, but we were not allowed to get them until the 6th.

Thus, three nights in a row in Burgos Prison without backpacks, without blankets, without anything, plus the 4 days of the trip – 8 days of absolute exhaustion. Such were the transfers the worse being the one to Burgos, symbol of Eternal Spain, and these were the shame of the 'penitentiary tourism'."[xxiii]

Moreno Gómez received reports of another great convoy of Cordoban prisoners who were not sentenced to death but were sent to the Labacolla Penal Camp near Santiago de Compostela. One of those who suffered that misfortune, Antonio Ruíz-Fernández, native of Espiel, wrote that at the end of the war he was a sergeant in a machine gun unit, taken prisoner and sent to Labacolla airfield and the 28th Disciplinary Battalion, to work with a pick and shovel. Of the 5,000 disaffected with Francoism who went there from Cordoba capital and province, few of them returned with their lives. It was a pure Inquisition.[xxiv] Casimiro Jabonero, a native of Cuenca, mentioned earlier, arrived at this camp via Madrid-Ponferrada-Santiago April 28 1939, travelling in the usual cattle cars. Another testimony from Cordoba province, Pedro Gómez González, native of Villaralto, worked out his sentence in the 28th Disciplinary Worker Soldiers Battalion,[xxv] details of which will be discussed later wen we address the matter of Franco's slave labour.

Descent into the circle of the Blue Hell

"Whoever denies Auschwitz is precisely the person who is willing to repeat it." (Primo Levi). There are two deaths in every crime: the physical and the hermeneutic. The assassin does not only kill, but he also strives to conceal it and to do so, nothing is more effective than depriving it of all meaning, that is, to make it appear insignificant. This concealment of the moral and political meaning of the crime is proof that the enemy is still on the loose. The historian trained in the school of Walter Benjamin, must know it and be prepared to face this."
Manuel Reyes Mate, La piedra desechada
2013

José Saramago said: "We are the memory that we retain and the responsibility that we shoulder. Without memory, we do not exist and without responsibility, perhaps we do not deserve to do so." Our scourge today is the erroneous belief in denying History. Those who deny it today and those who persist in sweetening the facts of the terrible Francoist repression, who today pollute the environment like toadstools, are causing irreparable damage to the truth of what happened.

Negationism, as defined by Stanton, is the last stage of genocide with which it is complicit and, as such, gives rise to major problems. The main one being how to determine just who are those responsible for the negationist phenomenon and what are their deep-rooted intentions as it does not appear amongst those who are nostalgic for the dictatorship but amongst influential members of society and a great many academics. It would be disastrous to find the ulterior motive, not in scientific revelations, but in the underlying political ideologies of today and in the terrible effects of postmodern light thought.

Without a doubt, the "Spanish Case" has not enjoyed any historical luck or justice. All the attention has been directed at the "Jewish Case". Neither in Europe, nor elsewhere (except for worthy Hispanic scholars) will you find the most elementary knowledge of exactly what Francoism did within Spanish borders, against cemetery walls, within prisons and jails, inside barracks and police headquarters, in rural and urban areas. The criminality of the Spanish Case is buried amidst one or another of these, consigned to the most deceitful oblivion. Given the above and fully aware that the thematic labyrinth is extremely complex, we are only able to glimpse the reality of Francoist prisons, that is, just what happened inside them, through the evidence given by survivors and provided in the diaries of those who experienced that world.

Cordoba Provincial Prison in 1941
Franco's Auschwitz

Cordoba Provincial Prison usually held an average 4,000 inmates (1,000 in the old prison, the Alcázar, and the remainder in the New Prison). In July 1939, when only the old prison was operational, it was packed with 1,500 inmates. In the countryside, the provincial jails were equally overcrowded. At the end of 1940, prisoners began to be brought to the as yet unfinished building of the New Prison in the north of the city. This improvised jail was not officially inaugurated until July 14 1944, by which time the great manslaughter of 1941 was over. 400 inmates were employed in finishing the building work, under the Regime's original invention of the program that allowed prisoners to redeem their sentences through work. The old jail, the one located in the Alcázar, was emptied of its five hundred inmates and closed Winter 1942.

Table VII
Inmate Mortality from hunger and deprivation in the
Provincial Prison in the 1940s for a prison population of 3,000-4,000
Source: Cordoba capital Civil Registry and Provincial Prison Chapel Record Book only.

Year	
1939	15
1940	30
1941	502
1943	22
1944	5
1945	4
1946	13
1947	7
1948	10
1949	33
1950	30
Total	**756**

Cordoba Provincial Prison in 1941 was an authentic Auschwitz, as were all of Franco's large prisons that year. It was the venue for an extraordinary extermination, no matter how many still deny that it ever occurred despite the early information that Moreno Gómez first published in 1987 regarding the situation in Cordoba and the fats that are contained herein. What is certain, is that what occurred in these prisons was neither ignored nor unwanted by the leaders of the Regime who freely consented to this crime against humanity in 1941, the same year that an equal extermination was occurring in the Nazi concentration camps.[xxvi] In Cordoba capital Provincial Prison the mortality rate among inmates during the 1940s for a penal population of 3,00-4,000, was 756 inmates, 502 of whom died in 1941 alone.

The reasons for such a scandalous mortality are difficult to determine. For one thing, this was not limited to Cordoba province and was widespread all over Spain, yet little has been written about the events of 1941 in Cordoba capital. Also, although this was the year of the great penal manslaughter countrywide, it was not quite the same in every province. For example, in Coto Gijón prison, the great manslaughter was 1938 after Asturias fell to the Nationalists October 1937; in the case of Catalonia, the manslaughter was widespread over time, from 1939 to 1942, inclusive, and in Navarra, the greater manslaughter occurred in 1942 in San Cristóbal Fort.[xxvii]

Fascist Administration of the Provincial Prison

The Prison Staff

In 1939, the *Director* of the Provincial Prison in Cordoba in the immediate post-war period was Miguel Villarrubia y Garcia-Chico, who died soon afterwards in Málaga at the beginning of 1940. From 1940 onwards, during the days of the Cordoban genocide, the tragically notorious Director was Don Enrique Díaz de Lemaire who lost his position March 1941 in the wake of arguments between corrupt officials. The new Director, Juan José Escobar Sánchez, an equally fierce exterminator and every much as corrupt as his predecessor took up his post Aril 1941. The manslaughter of inmates continued throughout the year, only dwindling at the beginning of 1942. Escobar remained in this position until February 1943 when, after he was transferred to Barcelona, he was replaced by Fernando García González.

During those tragic days, the *Sub-Director* was Ramón García Lavella. Rafael Herreros was *Director of the Women's Section*, assisted by the infamous and greatly feared Doña Dolores. Conchita Costa was another such assistant mentioned by witnesses. The *Chaplain General* was Rev. P. García, S.J., frequently accompanied by the parish priest from El Salvador Church, José Torres Molina. *Health Care* was entrusted to a quaint physician, Don Celso Ortíz Megias, under the Provincial Director of Health, Dr. César Sebastián. Dr. Palanca was the national Director General for Health. The *Prison Manager*, who ended up being prosecuted, was Manuel Hernández Vox. The *Purveyor of Supplies* was Rafael Bejarano, who owned a butcher shop on Calleja Marqués de Boíl. The *Caterer* was a so-called Don Francisco.

All these three stole as much as they could, whilst the inmates were served a lethal diet of nothing more than a dirty broth of rotten turnips, according to the magnificent testimony of Dr. Joaquín

Sama Naharro,[xxviii] an inmate at the time, one of the most articulate individuals it has been Moreno Gómez' good fortune to assist him in his research.[2] Dr. Sama strongly condemned the cohort of merciless *Prison Officers* who specialized in beating that multitude of living skeletons. Those who today still can testify to this, refer particularly to Enrique de la Cerda, Antonio Justo, Manuel *Y Pico*, Andrés *El Boxeador*, a so-called Don Ángel, a temporary 2nd Lieutenant, and others such as *El Teleras* and *El Negro Desperdicios*. These were guards who stood in the halls and doorways, throwing out a beating or two as the lines of inmates walked past. One such guard who was especially feared was Ángel *El Dientudo* Baena and his assistant Segundo Rojas, who tortured the inmates in the punishment cells and kept them for days on ed on a diet of bread and water.

The Regime's governing authorities in Cordoba

During those days of genocidal extermination, Manuel Saraz's Murcia, *Mayor* since November 27 1939, was succeeded by Antonio Torres Trigueros November 5 1941. Rafael Jiménez Ruíz was appointed Mayor the following year. The Presidency of the Municipal Assembly was held for a long time by Enrique Salinas Anchelenga[3], whose brother Ángel was murdered by Falangistas in 1936 and who took over from the great Nationalist conspirator, Eduardo Quero.

Joaquim Cárdenas Llavaneras, an Artillery Commander, was *Civil Governor*. He was replaced October 10 1941 by a long-standing Falangista and retired Army officer, Rogelio Vignote, another of the

---

[2] Joaquim Sama Naharro, M.D., interviewed by Moreno Gómez during a day-long Q&A session when he eloquently answered MG's questions and expressed his deep-rooted belief in the Republic and profound anti-Francoist convictions. This was totally contradictory to an interview Sama's son later published in a Cordoba newspaper, in which he said his father belonged to neither the left nor the right. Obviously, like many father-son relationships, the son had a minimal understanding of his father and his ideals.

[3] Enrique Salinas Anchelenga, a member of the high Cordoban bourgeoisie, had a son who was captured by the maquis in 1945 and held for a month in the Villaviciosa mountains, until his father paid a 75,000-peseta ransom. A relative, Jose Miguel Salinas, was a PSOE Member of Parliament around 1980. When Moreno Gómez approached him with some questions regarding the civil war, he shouted, in a threatening manner: "Don't you every again dare to talk to me about this subject." The *Córdoba* newspaper society column ironically published a report of his luxurious jet-set wedding in the famous Círculo de la Amsitad, known in the past as the 'Casino of the Rich' and since then, José Miguel Sanlinas' name has disappeared from public notice.

great 1936 conspirators who died on the job September 16 1942, a few days before his factotum, Eduardo Quero. Vignor was succeeded by Ramón Risueño Catalán[4], from Granada.

Military Governor General Francisco Formoso Blanco served from November 1939 to 1941 when he was replaced by Colonel Antonio Pérez Torrealba who in turn was replaced February 26 1941 by General Saturnino González Badía.

In 1941, the *Director of the Provincial Falange* was Jesús Aguilar, a well-known Falangista from Puente Genil. Fernando Fernández continued as Falange Secretary for Cordoba until he was replaced in mid-September 1941 by Manuel González Ruíz-Ripoli.

Descent into the depths of the corrupt penitentiary world

Having described the administrative world of Fascist Cordoba in 1941, we shall now travel to the depths of the corrupt penitentiary world in Cordoba capital, as seen through the eyes of Dr. Sama Naharro, the Republican physician from Madrid whose journey through the Francoist prison would bring him to Cordoba, where he settled down after his release from jail. Dr. Sama was first transferred from the Scala Caeli Convent improvised jail in Madrid, to Castro del Río in October 1940, after the majority of the provisional prison inmates had already been transferred from the rural jails to Cordoba capital.

> "It took us two weeks to get to Castro del Río from Madrid. Only seven of our party were common criminals and I would entertain them with stories. After we arrived, we spent five days without any food. The Director of the jail, a nattily dressed Army

---

4   Ramón Rissueño was well thought of by the public who spoke of him as the 'good man of the Regime' in the midst of this all this tragedy and outrage. The *Cordobés* newspaper January 16 1943 published an article describing his efforts to ensure that rationed goods would be fairly distributed to inmates and how he would sometimes show up at a queue for goods to make sure they were being properly distributed. He collected beggars from the street and had them clothed and fed. He did his best to keep a lid on prices and prosecuted several tradesmen without scruples. Sadly, this period of Christian charity soon ended as the Cordoba right-wing took over and removed him from office. It was a time of fascism, not of charitable acts.

officer, called us and told us that he had run out of funds. 'If there were more than 25 of you,' he said, 'I could get some outside catering for you, but as you are fewer... you're going to find it pretty difficult.' At long last, they brought in a group of individuals they had arrested on a farm, a whole family. Together, we added up to the minimum 25 inmates the Director needed to have food brought in from outside."

The following November, Dr. Sama was transferred to the Provincial Prison in Cordoba:

"When they took me to Cordoba, I weighed 45 kilos [99 lbs]. When I entered the patio in the old prison, it looked like something out of a novel by Cervantes: misery, lice, filth. Everyone was packed tight. There was a single urinal in the patio and a long lines of inmates waiting to use it. I was dumbfounded. Everyone was mixed together – political prisoners and common criminals. Nevertheless, we were still fed somewhat. I had arrived with a sleeping mat on my back and a suitcase, tied to my companions in line. When I saw that scene, the inmates de-lousing themselves, I asked myself: 'How low have I sunk?' It was so overcrowded, that each one of us was allotted two and a half floor tiles; we had to act together when we wanted to turn around. There were dramatic anecdotes. One day, I was told: 'Don Joaquim, here is one who is very ill.' I came closer and the man was ice cold. 'Yes, we knew that, but we did not want to wake you.' We had been sleeping next to a dead man.
The prison doctor was a Don Celso Ortíz, whom they called Don Ciezo; the famous Director was Don Enrique. We soon were fed turnips only. People's faces would swell a little, under their eyelids, and Don Celso would diagnose 'albuminia'. He prescribed a milk diet, but the milk was watered down, They didn't last three days. I told Don Celso that he should find out what

they were suffering from, and I asked him if he ordered blood and urine tests. "No. I don't do those things. Forget it." In effect, the oedema was a sign of starvation. The person would get diarrhoea and the end result came quickly. There was no such thing as medicine. For Don Celso, everything was a case of 'albuminia' for which he prescribed milk but the orderly who came in from outside, a so-called Paco or Francisco watered down the milk and took the sugar home.

The young prison officers were very disrespectful towards the older ones whose jobs they wanted so that they could share in the hoarding. The prison caterer, Rafael Bejarano, owned a butcher's shop on Calleja Marqués de Boil. According to Bejarano, the prison Director stole everything he could. Both these men were later prosecuted. The prison Governor turned a blind eye to what was going on and he also took his share. The Director of the women's prison which was in a next-door pavilion, the infamous Doña Dolores, was also implicated n the thieving.

During the 'battle of the turnips' as the inmates called our diet, diarrhoea was the last stage of a colitis caused by the lack of food. The meal lists were falsified. They apparently described varied menus when the reality was that the only thing, we were given was a watery turnip soup. Don Celso was fully aware of this as he signed the falsified menus. Still, he sent separate, secret lists to the Governor, informing him that the prisoners were not being fed the regulation meals. It was his signature on these secret lists that later saved Don Celso from prosecution. He would tell me: "Sama, keep out of this. What happens is that the officers are divided: the young ones want to steal and so do the old ones, and they don't want anybody else involved".

I tried to bring some dignity to the removal of the dead. At first, they were just dragged out. I made a stretcher and placed the dead man on one of the benches in the patio from where he was removed with a bit of dignity, as some of them had relatives present. I also organized what they called 'the potato cellar', where those who were very far gone and the highly contagious were kept aside. Everyone knew that anyone who was taken there would not return. It was located at the end of a corridor so that they would not die in front of the rest.

Epidemic Typhus is transmitted by the body louse when it stings the host. The doctor who replaced Dr. Celso – who was fired for negligence – was a so-called Pedrajas who stubbornly insisted that there was no typhus in the prison. So much so, that one prisoner who had all the symptoms of this disease was released and allowed to spread it to others in the town.

Our meals improved with the addition of beans and black carrots (the Nazarene broth) and the number of cases of typhus dropped. The prison Governor, the Director, the Head of the Women's Prison and Rafael Bejarano were all fired, but not the milkman nor the nurse. A special military judge was appointed for their trial. The Governor was sentenced to 30 years and the Director to 20 years, and both were imprisoned in Puerto de Santa María. However, they appealed their sentences, which were reduce, and the following year they were free and back in the same jobs, with back-pay and all kinds of other benefits."

There are several assumptions that one might make after reading Dr. Sama's written, eye-witness testimonial. One, that because it was a physician who diagnosed the cause of the inmates' sufferings, his conclusions should be taken as sound. Second, that the diagnosis of the extermination cannot be exclusively based on the prison officers' corruption but instead of on the perverse nature of the entire system, as this same extermination was occurring all over Spain in 1941. Third, that even if one blames the corruption in the prison, punishment was purely a pro-forma affair as the sentences were quickly overturned. Fourth, that during the entire decade of the 1940s, the majority of the inmates died of starvation, because almost all those who died in 1941, did so from hunger (499 of the 502 total deaths) that year.

Another opinion regarding this dark matter of corruption in Cordoba Provincial Prison is that of a Cordoban attorney, Francisco Poyatos López, whom Moreno Gómez visited in his office at the beginning of the 1980s and who kindly gave him a copy of his book of memoires. According to Poyatos, the Governor, the caterer and two prison officers were prosecuted. The court martial, held in Cordoba with great expectations, was presided by Colonel Aguilar Galinda, Commander of an Artillery Regiment, and notorious genocide of Fernán Núñez. The Prosecutor asked for two death penalties, but the sentences were reduced to thirty years in jail, which in those days meant almost nothing.

Poyatos, in defence of the Governor, pled his innocence on the grounds that the food ration approved by the Directorate General of Prisons, was less than 800 calories a day, whilst a minimum 1,200 calories is necessary for a totally immobile person. Poyatos also argued that the deaths were not the result of a fraudulent decrease in the rations fed the inmates, but on the fact that the Directorate General of Prisons recklessly approved insufficient rations. He concluded his defence by deriding an allusion to a tale from The Arabian Nights, in that when, for example, if a high-ranking official could not be hanged, a lower-level officer would do. The most important being that no crime should go unpunished.[xxix]

This Cordoban attorney's input is extremely clear: the responsibility for the deaths was not so much that of the corrupt officials, but of the Directorate General of Prisons. In other words, it was the fault of the system because the latter approved food rations that would clearly lead to the extermination of those who were expected to survive on them. (Additional testimonials of the post-war conditions in the prisons are given at the end of the book.)

Hunger remains almost the sole topic of discussion amongst all those who survived the Francoist prisons. Hunger was a special repressive instrument of Francoism and of Fascism in general, a tool for the genocide in the prisons, even more than a means of oppressing those who were free. It was a basic principle of European Fascism in the endeavour to bring about the total, material and moral, dispossession of its political opponents, as Ricard Vinyes wrote. Everything was a ploy to bring about the ideological and moral destruction of the individual. It was institutionalized poverty, hunger and destitution. According to Vinyes, this element of the State's dispossession or moral and material plundering, represents the universal point of contact with the great Fascist systems of political reclusion, There is no way that one can deny this institutionalization of hunger in Franco's prisons, exactly as practiced in Hitler's prison camps, although the Francoist did not reach the same level of perfection as the Germans in the 'art' of extermination.[xxx]

The typhus epidemic

It seemed that the Four Horsemen of the Apocalypse – war, hunger, disease and death – were riding roughshod over the entire country as, in addition to the hunger, there was the typhus epidemic. In effect, the 1941 typhus epidemic was not limited to the prisons as it had spread to the general population as a result of shortages and misery of all kinds. The Regime always insisted on minimizing the problem and in covering up the data as much as possible, However, in the Francoist prisons especially, overcrowding, unhealthy conditions and the ever-present parasites provided the idea breeding ground for the spread of the epidemic.

The authorities in Cordoba waited far too long before admitting that epidemic typhus was free in the city. May 25 1941, the *Azul* newspaper first published a report of 16 cases of epidemic typhus. In the meantime, nothing was said about the epidemic that, together with hunger and anaemia, was decimating the prison inmates. The new Prison Director, Juan José Escobar, made some half-hearted attempts to control the epidemic but without much success and it was only in January 1942 that we hear of a campaign against typhus in the city. Civil Governor Reglio Vignote presided over a public meeting of the Provincial Board of Health January 9, when the Provincial Director of Health, César Sebastián, informed the meeting that a new mobile disinfection unit with showers and a gas chamber with hydrocyanic acid had been installed in the prison. Inmates' communication with anyone from the outside was suspended, as was sending clothes out to be washed, all of which had a negative impact on inmates' survival. Measures for quarantining the ill were increased and court martials were stopped for the time being, the only positive effect of the situation.

The *ABC* newspaper in Seville published a rather euphemistic article November 19 1941, regarding the typhus epidemic, in which it stated that during the 1940-41 Winter, 1,097 victims died from typhus nationwide and 8,000 were affected. The article was accompanied by a map, in which we note that the provinces that were most affected were Madrid, Seville and Malaga, followed in second place by all of Andalusia and all Southern Spain. The provincial prisons were pointed to as the centres of the infection.

Table VIII. Number of inmates at death in Cordoba capital Prisons by age group during the period under study								
Infants	-20 years	20-29 years	30-39 years	+40 years	+50 years	+70 years	1941 total	Period total
		17%	17%	26%	26,8%			
		112	113	151	176		**502**	
Also died during Spring 1941 but not included in -20 and +70 age groups above								
5	20					12		**539**

In the article, the newspaper hinted that there were practically no cases of typhus in the North of the country other than a small focus near Victoria, in the Nanclares de Oca prison camp. Cordoba is not mentioned in the newspaper article that does, however, say that those infected ranged from between twenty and fifty years of age, not so coincidentally the age range of the prison inmates. The life expectancy index for Spain, during those ill-fated days, fell like a stone due to the lack of food and health care and all sorts of suffering.

Large periods of malnutrition explain to a great extent, the high incidence of typhus after March 1941, when it was at its height in the Cordoba Prison, This so alarmed the Regime that it attempted to cover it up by instructing the registry offices to issue death certificates for other reasons such as avitaminosis, starvation, cachexia and so forth, anything by typhus, even though many more dying from hunger than from this disease.

The first signs that the typhus epidemic was coming under control in the Cordoba prison were noted in February 1942, when the official death rate from the disease began to fall from an average 40 a month to about a dozen, even though the majority were still starving to death. Whereas in 1941, the death rate rose to 502 cases, in 1942 it dropped to 85; in 1946, the 'year of hunger', 13 inmates died, only to rise again to an average of 30 victims per year after 1948.

Undoubtedly, the great prison disaster in Cordoba occurred in 1941, the worst months being March and April with an average of almost 100 deaths per month. There was no typhus then; starvation and basic privations were the cause. An examination of the towns and villages of origin of the greatest number of dead were those that were at a distance from the capital, therefore making it even more difficult for relatives of the inmates to bring them food: Peñarroya 24 victims; Montoro 25; Villanueva de Cordoba 23; Hinojosa 23; Villaviciosa 22; Adamuz 19; Fuenteobejuna 19; Belmez 16; Villanueva del Rey 14; Dos Torres 13; Pedroche 11; El Viso 10; Baena 13 Bujalance 13; Puente Genil 10; and so forth Of special note, the fact that the majority of dead inmates (150) came from outside the province and were even less likely to receive any outside help.

Also noteworthy, is the highest mortality rate for inmates who were natives of mountain towns and villages, who felt the full weight of the post-war repression and who had been arrested in greater numbers than in the countryside. The fact is that even though the oppressor had already implemented a considerable part of its genocide program in the countryside in 1936, in 1941 it was the proximity or distance of an inmate's relatives that was determinant to the mortality rate in the prison.

The age of the inmates was another key factor. For the period under study, inmates older than 50 years led the number of those who died (26.8%, a total of 176). They were followed by inmates older than 40 years (26%, a total of 151). The remaining two groups, 20-29 years and 30-39 years were equal, each with 17% of the whole (112 and 113 victims, respectively). Although younger inmates appeared more resistant to the effects of the privations and hunger, it was not always so. During the acute stage of the extermination – Spring 1941 – victims aged 20-29 years died in greater numbers than those older than 30 years. There were 20 victims under 20 years and 5 infants who had accompanied their mothers in jail. Lastly, there were 12 victims among elderly inmates, more than 70 years old.

Hence, the impressive disaster in Cordoba Provincial Prison that we can truly call the Francoist Auschwitz of 1941 and that Moreno Gómez further shows was repeated in a similar vein in the rest of Spain.

Meanwhile, we note an example of the great exercises in hypocrisy and cynicism that a Francoism without scruples delighted in displaying, a grotesque poster that the 'magnanimous Caudillo' ordered displayed in his 'beneficent' prisons, to his shame forever:

> Should you visit penal complexes in other countries, and should you compare their systems to ours, I can assure you without fearing of being mistaken, that you will not find as fair, as Catholic and as humane a system as the one created by our Movement for our inmates.
>
> *Francisco Franco*
> *Madrid, 17 July 1944*

This section on the Cordoban genocide closes with the exceptional testimony of an illustrious Cordoban Republican, Rafael Sánchez Guerra, who was interned in several Francoist prison complexed in Spain (Madrid, Cuéllar, Algeciras, Cordoba, Puerto de Santa María, and more) until his 'penitentiary tourism' brought him to Cordoba Provincial Prison in 1941:[xxxi]

> "... The next morning, Don Ramón Carreras Pons, ex-Member of Parliament for Cordoba, the most eminent inmate in all the prison population, having heard of my arrival, invited me to have a bit of coffee and to accompany him on a visit to all the sections of the prison.
>
> The penal regime in Cordoba was an 'open air' regime, by which all the inmates were cruelly forced to spend at least nine hours a day, in a row, in the patio, unless they were sent on some other business. It is there that they went after they were first released from their cells in the morning, there that they were served a meal at half past noon, there they remained after they were called to prayers.

There they had to remain sitting on the ground or walking about to keep from dying from the cold. The patio, occupied by about one thousand five hundred inmates (in the old prison), was really impressive: it was an authentic mass reduced to the most iron discipline.

Don Ramón Carreras Pons, after having been detained for several months, had been absolved by the Courts of Justice but the Governor of the Province, a man who was 'more Papist than the Pope', refused to accept the judge's decision and arbitrarily kept him interned, so that his new period of captivity might be indefinitely prolonged,

As this case of Carreras, I knew of others. In Falangist Spain, sometimes, tyranny was exercised at the margin of the Law, and at other times, without the minimal appearance of justice."

ENDNOTES FOR CHAPTER VII

[i] Michael Ricards. Un tiempo de silencio. La guerra civil y la cultura de la represión en la España de Franco. 1936-1945. (A time of silence. The civil war and the culture of repression in Franco's Spain. 1936-1945). Crítica, Brcelona, 1999, p. 30.

[ii] Renzo Stroscio. Hacia una tipologia de los campos de concentración y exterminio nacionalsocialisto. Congress on the concentration camps and the penitentiary world in Spain during the civil war and Francoism. Crítica, Barcelona, 2000, pp 77 et al.

[iii] Gutmaro Gómez Bravo, El desarollo penitenciario en el primer franquismo (1939-1945) (The development of the penitentiaries during the first stage of Francoism), Hispania Nova, number 6, 2006.

[iv] Ibid., Ist Congress on the Victims of Francoism, Rivas Vaciamadrid, Madrid, April 20-22 2012.

[v] Ángel B. Sanz, De Re Penitenciaría, Talleres Penitenciarios de Alcalá de Henares, Madrid 1945, p 181. Figures reproduced by Halliday Sutherland, Spanish Journey, Hollis and Carter, London, 1948, p. 47.

[vi] Ángel Suárez /Collective 36. Libro blanco sobre las cárceles franquistas (White paper on Francoist prisons). Planeta, Barcelona, 2012, p. 14. (1st edition, Ruedo Ibérico, 1976).

[vii] http://www.asisbiz.com/Battles/Spanish-Civil-War/images/

[viii] Tomasa Cuevas Gutiérrez. Testimonios de mujeres en las cárceles franquistas (Witness accounts of women in Francoist prisons). Instituto de Estudios Altoaragoneses, Diputación, Huesca, 2004.

[ix] Francisco Moreno Gómez. Córdoba en la posguerra. La represión y la guerrilla, 1939-1950) (Post-war Cordoba – Repression and guerrilla. 1939-1950). Buena, Córdoba, 1987.

[x] Gutmaro Gómez Bravo. "El desarollo penitenciario en el primer franquismo (1939-1945) (The penitentiary development during the first stage of Francoism - 1939-1945), Hispania Nova, number 6, 2006.)

[xi] Margalida Capellá i Roig. Represión política y derecho internacional. Una perspectiva comparada (1936-12006). (Political repression and international law. A comparative approach) in La memoria histórica en perspectiva jurídica (Historic memoroy from the legal viewpoint – 1936-12006), Margalida Capellá and David Ginard, Documenta Balear, Palma de Mallorca, 2009, pp. 161 et al.

[xii] Arcángel Bedmar. La campiña roja. La represión franquista en Fernán Núñez (1936-1945) (The red countryside. The Francoist repression in Fernán Núñez). Juan de Mairena, Lucena Córdoba, 2003, p. 71.

[xiii] Ibid. Los puños y las pistolas. La represión en Monitlla. (Fists and pistols. The repression in Montilla). Ayuntamiento, Montilla, 2009, p. 106

[xiv] Jesús María Romero Ruíz. Que el 20 de febrero de 1936, cuando los sucesos del jardín. (It happened in the garden 20 February 1936), Ayuntamiento, La Rambla, 2010.

[xv] Juana Doña. Desde la noche y la niebla (mujeres en las cárceles franquistas) (From the night and the fog. Women in Franchoist jails)., La Torre, Madrid, 1978.

[xvi] Rafael Sánchez Guerra, Mis prisiones (My prisons). Claridad, Buenos Aires, 1946, p.106. Also, Mirta Núñez Díaz Balarte and Antonio Rojas Friend, Consejo de Guerra. Los fusilamientos en el Madrid de la posguerra (1939-1945) (Council of War. Post-war Firing squads in Madrid 1939-1945), Compañia Literaria, Madrid, 1998.

[xvii] National Basque Archives and Sabino Arana Foundation. Informe sobre presos vascos en el Penal del Puerto de Santa María. (Report on Basque prisoners in the Puerto Santa Maria penal complex), Manuel Martínez Cuadrado, in El Penal de El Puerto de Santa María. 1886-1981) (Puerto de Santa María Penitentiaries 1886-1981), Cádiz, 2004, pp. 150

[xviii] Gozalo Amoedo López and Roberto Gil Moure. Episodios del terror durante la guerra civil na provincia de Pontevedra. A illa de San Simón. (Episodes of the terror during the civil war in Pontevedra province. San Simón island.) Serais, Vigo, 2007, p. 66.

[xix] José Manuel Sabín. Prisión y muerte en la España de posguerra (Prison and death in postwar Spain), Anaya, Madrid, 1990.

[xx] L. M. Sánchez Tostado. Historia de las prisiones en la provincia de Jaén. 500 años de confinamiento, presidios, cárceles y mazmorras. (History of the penitentiaries in the Jaén province. 500 years of confinements, imprisonments, jails and dungeons). Jaén, 1997.

[xxi] Miguel Hernández. III Prosas. Correspondencia (III. Prose. Correspondence), Espasa-Calpe, Madrid, 1991, p. 256; and Cartas escritas en la prisión de Oribuela, entre septiembre y octubre de 1939 (Letters written in Orihuela jail between September and October 1939).

[xxii] Eutimio Martín Garcia. "El turismo penitenciario franquista (Franco penitentiary tourism), in Historia 16, number 239, 1996, pp, 19-25.

[xxiii] Diógnes Cabrera. Once cárceles y un destierro (Eleven jails and one exile), Santa Cruz de Tenerife, 1980, p. 102.

[xxiv] Antonio Ruíz-Fernández. Typed letter ated June 15 1985, sent to Moreno Gómez by his daughter, Isabel Ruíz, April 19 2013.

[xxv] Pedro Gómez Gonzálvez. Interviewed by Moreno Gómez in Villaralto, Córdoba, December 16 1986.

[xxvi] Ángel del Río et al. Andaluces en los campos de Mauthausen (Andalusians in the Mauthausen camps). Junta de Andalucia, 2006.

[xxvii] José Manuel Sabín. Prisión y muerte en la España de posguerra (Prison and death in postwar Spain). Anaya-Mario Muchnick, Madrid, 1996, pp, 229 et al.

[xxviii] Joaquín Sama Naharro, M.D. Interviewed by Moreno Gómez in Córdoba, July 8 1983.

[xxix] Francisco Poyato López. Recuerdos de un hombre de toga (Memoires of a man with a robe), op cit., pp. 146 et al.

[xxx] Ricard Vinyes. "El universo penitenciario durante el franquismo" (The prison universe during Francoism). In Una inmensa prisión (An immense prison), Crítica, Barcelona, 2003.

[xxxi] Rafael Sánchez Guerra. Mis prisiones (My prisons), Claridad, Buenos Aires, 1946, p.187.

VIII

THE 1941 FRANCOIST AUSCHWITZ IN THE REST OF SPAIN. PRISONERS AS SLAVE LABOUR. DISCIPLINARY BATTALIONS

"The unalterable characteristics of the great oppressive political systems do not rest on the number of imprisoned or killed, neither in the extermination procedures, but in the denial of the crime against humanity; whether it is to rub it out or to cover it up by means of an appropriately mystifying language that contributed to trivialize the vulnerability of human rights which, under Francoism, was the language of the Christian Church. There can be no doubt that the feature common to the great punitive Fascist systems is essentially the removal of all trace of the crime.
Ricard Vinyes, in *Una inmensa prisión,* 2003

Extermination through hunger and deprivation in Francoist prisons, especially in 1941, was not by any means limited to Cordoba. It was practised throughout the rest of Spain, in prisons to which tens of thousands of natives of Andalusia had been sent. No discussion of the Cordoba genocide would be complete without a description of Franco's Nazi-like methods of extermination of the disaffected throughout the country. This topic, however, is barely addressed by historians who today appear solely dedicated to studying the rules and regulations governing the prison world.

Moreno Gómez begins by providing a description of the sadly infamous Puerto de Santa María prison in Cádiz in 1941 to which Rafael Sánchez Guerra was sent straight from Cordoba:

> "The terrible spectre of hunger had already cast its sinister shadow on Puerto Prison. The inmates, despite the personal efforts of the Prison Director to find food, were starving to death. Avitaminosis, a disease that was unknown in our country until Franco

decided to create a great Spain, caused enormous harm to the poorer classes of society in Andalusia and was responsible for daily deaths in the Central prison, in which I was incarcerated.

I was stunned by the appalling problem of providing food for the inmates in Alcázar de San Juan and in Cordoba prisons. Whereas in the first of these, I once saw two oranges and half a dozen chestnuts shared out for the noonday meal, in Puerto prison, as there was a total lack of bread, the situation was more acute and more tragic.

The several prison infirmaries were packed with young men whose swollen faces were a clear symptom of the disease, impoverished, skeletal, presenting no organic disease but physically unable to stand up. It was a rare day that we did not line up, sad and silent, in the large patio, to witness them carry out the remains of two, three or four cellmates in rough wooden boxes, those who had found such a painful means of 'recovering their freedom'.

Seventy-eight inmates starved to death in March 1941, a huge number that amazed us all. Such was the desperate desire for food by some of the inmates, that each patio had to appoint an inmate to keep watch over the garbage cans to prevent others from poisoning themselves with the waste and filth that were thrown in them. Orange peels, often trodden upon and dirty, were absolutely devoured by the starving men."

Manuel Martínez Cordero discovered a 'confidential' report dated July 1939, in the Archives of Basque Nationalism and the Sabino Arana Foundation, entitled: Report on the Basque prisoners confined in the Puerto de Santa María penal complex.[i] At the end of the civil war, some 4,000 Basque prisoners held in Euskadi and Santoña prisons, were dispersed amongst other Francoist prisons in the country. Of these, 2,000 arrived in Puerto de Santa María, which brought the total number of incarcerated to 5,400 in a space designed for 800 internees. The information contained in the report, referring to 1939, provides important, terrifying information, of the overwhelming situation of those thousands of men. According to the report, presumably submitted by Basque authors:

> "The general impression of the data that we are presenting cannot be more deplorable and once again, it confirms the truth of the matter that has already been repeatedly presented regarding the current penal regime in Francoist Spain, which because of its inhumane characteristics exceeds in its cruel reality, anything that the most partial and unfavourable opinion can imagine. For Francoism, the prisoner has lost his condition of being human and he is treated like an animal. We do not have to try to stress the cruelty, the uncivil and bloody sense that presides over the treatment to which our prisoners in Puerto de Santa María are subjected."

The authors of the report explain that the prison complex is an installation that is open to the Atlantic Ocean and, therefore, subject to blustering winds year-round, especially the east wind during the Summer. The sanitary installations were also disastrous, few and badly placed as there were only two shower rooms, with 57 keys, shared by 5,400 inmates.

Inmates were packed body to body, mouth to mouth, so that there was a profusion of diseases of the respiratory tract, in addition to tuberculosis and pre-tuberculosis. Because of the lack of space in the infirmary, inmates suffering from any of these lived cheek-by-jowl with all the others. There also was a special kind of influenza that attacked many inmates and left them as weak as if they were recovering from an attack of typhoid fever. This type of influenza, sometimes pulmonary, sometimes intestinal, often caused body temperatures above 40°C (104°F) that left the patient totally prostrate.

Water was scarce and poorly distributed. Add to this an unbelievable profusion of bed bugs, lice, fleas and flies, especially on extraordinarily filthy blankets and mattresses, none of which could be washed because of the lack of water, somewhere to wash, boil or even dry these, it is not surprising that nobody escaped suffering some kind of dermatitis, at least once. All these weaknesses were mainly due to the lack of individual physical reserve and to bad food. Added to this, hunger and privations easily explain the resulting mortality.

The prison infirmary consisted of three main wards one for patients with tuberculosis, one for general medicine and surgical treatment, and a special ward for the chronically ill. The most serious problem was that of patients with tuberculosis who lived with the rest of the inmates. It was Basque physicians, also prisoners, who managed to keep them separate. During Sumer 1939, the only quarantined tubercular patients were the 44 who fit in the ward, that is, the most serious cases. There was a very long waiting list for a space in this ward.

There were 17 beds for patients in the ward for the chronically ill. There was no room for anymore and remember, there were approximately 5,400 inmates at the time. According to the aforementioned report, the beds on the wards were little more than tabletops about 50cm from the ground. There were no sheets, blankets, pillows or anything else. If a patient wanted any of these 'luxuries', someone from outside the prison had to buy them in the town and bring them in for him.

Based on the Basque report and the problem of the unsanitary conditions in the prison, there can be only one conclusion: *as the care that the official prison doctors gave the inmates was either negligible or non-existent*, the health care problem was left to any inmates who were physicians to do what they could for their fellow prisoners, at least at the beginning. Eventually, an authoritarian and despotic prison doctor who not only did not give a damn about the inmates' health arrived and began to make arbitrary decisions such as refusing to sign vouchers for the purchase of medicines. Consequently, if the problem of feeding the inmates was serious, no less so was the problem of medication for the ill. Aspirin, streptomycin, luminal, etc., bandages and all kinds of sterile and sometimes unsterilized equipment, were only available in limited quantities and for very short periods of time.

The absolute truth of this is confirmed in Miguel Hernández' diary. Terminally ill with tuberculosis, he was only able to receive treatment on the days that his wife visited him in prison and brought bandages and compresses. In a February 1942 letter he wrote her from Alicante prison, [ii] he begs her to send him three or four kilos of cotton and gauze, with the utmost urgency, without which he will not receive any treatment that day. He tells her that the infirmary has run out of everything and that the day before, they had to resort to rags to treat him as best they could. Obviously, the New State had made no provision for the care of the ill. Was this intentional or by omission?

Also, according to the Basque report, more than 2,000 inmates a month were seen in the infirmary for general consultations. Most of the complaints were related to the digestive tract and the skin, as well as a whole series of afflictions related to the hardships. The few surgical interventions were for abdominal hernias, appendicectomies, epidermitis dermica, etc. No less than 1,943 cases of pulmonary or heart complaints were treated in 1939. Of those admitted to the tuberculosis ward, only three were released. Of all the complaints seen to in the general consultation in the first year, most (40%) were bronchial, mainly involving tuberculosis.

As regards the mortality rate, the report refers not to the entire prison population of 5,400, but only to the 2,000 Basque inmates, of whom 43 died during the first ten months of their stay in Puerto Santa María, that is, 1939 and early 1940. There is no available information regarding the 1941 genocide, nor what happened to the entire prison population.

The Basque report deplores the thoughtless and disrespective behaviour of the prison chaplains and it names two Jesuits, Rev. P. Gutiérrez Silva and Rev. Arjona. The clergymen's negligence was such that the dead were buried in plain, unlined wooden caskets, without the presence of a priest and no religious ceremony, something that rang foul of the Basque people's strong Catholic beliefs. An example of this belief was a whip-round amongst prisoners who did not have any food to eat to purchase a statue of Our Lady of Mercy. As to the chaplains' behaviour, in accordance with the regulations of the Regime, the amazed report states:

> "We have been reliably informed of the existence of secret instructions from Prison Headquarters to prison chaplains, instructing them that, as they are engaged in carrying out their sacred mission, should they find an inmate who is innocent of what he was wrongly condemned by the Courts Martial, instead of taking note of his complaints and transmitting them to the Prison Director, they should attempt to discourage the inmate from complaining and encourage him to

accept his situation as atonement for all the sins he may have ever committed against God and abstain, once and for all, from bothering Prison Headquarters."

There are still very few studies of the other great Francoist penitentiary establishments. Two teachers from the Almendralejo School, Badajoz, Manuel Rubio Díaz and Silvestre Gómez Zafra, who visited Moreno Gómez in Villanueva de Cordoba in 1987, later published a very interesting study of their village where there were two prisons: La Colonia and the Almacén de la Hiz, which they very kindly dedicated to him, August 24 1987, in thanks for his support and collaboration.

One of the eyewitnesses they interviewed told them that in La Colonia, where he had been interned, the inmates were divided into twelve prison gangs. Hundreds of them were crammed into the prison and they slept on the floor on blankets their family brought then when they could; in winter, it was much, much worse. Anyone who was unable to receive some food from his family did not last long. With only a ladle of white beans, what did they expect? Their witness was transferred to a Seville prison in May 1941 but his brother José, who remained in La Colonia, was tried and executed a few days later.[iii]

In their study, Díaz and Zafra include an enormous list of 333 inmates executed following numerous sacas, mostly in 1941. In addition to these, no less than 144 inmates starved to death in the sixteen months between July 1940 and November 1941.

It is not easy to explain how it was possible for 144 individuals from a single village to die from hunger and hardships during 16 months' imprisonment under Franco. It is extremely galling that, still today, when such a tragedy is brought to the attention of individuals who consider themselves members of the intelligentsia, despise as much as they ignore the facts and insist that there was no extermination, nor eradication, just re-education.

There were many holocausts in Franco's jails where men fell like flies, but one that can be truly called an Auschwitz was the Penal Colony on the island of San Simón, off the Pontevedra, Galicia, coast. This prison was set aside as a kind of scrap yard for extremely elderly men from the South of Spain.[iv] Diego San José de la Torre, in his memoires, describes the drama of prisoners from the South of Spain who were transferred to prisons in the North of the country.

He writes that the majority of the individuals 'invoiced' to the most humid and coldest regions, were elderly peasants from Andalusia who were so poor that they only had the clothes they were wearing and who were leaving the land where they were born, never to return.[v]

In May 1941, when autocratic Miguel Cuadrillero was appointed Director, the hunger in San Simón reached such a level that the prison became a slaughterhouse. More than 300 inmates died in less than three months of that terrible year. As Diego San José reports, the Director horse-traded with contractors and suppliers and those who were responsible for the canteen. He would buy the cheapest cereals and vegetables and rotten fish. The mortality was so great during that Summer, that the stench from the cadavers that were stacked in a warehouse on the seafront waiting for caskets that were not being supplied by the local carpenters because they were not being paid, alarmed the villagers of neighbouring Redondela. The 1941 mortality rate did not spare a neighbouring penitentiary, the Camposantos concentration camp, where 70 inmates starved to death and another 100 were executed.

Amoedo-Gil, in their study of San Simón prison[vi] calculated that there were 666 prisoners in the prison, of whom 161 (24.2%) of a total 2,176 inmates, were from Andalusia. Given the specialized nature of this penal complex, it is not surprising that over half of the men who died were aged over 60. In conclusion, these authors stress that the excessive mortality of the period cannot be considered as exceptional, typical of only this or that prison establishment, as many of those who blame the Prison Director claim. On the contrary; 1941 must be seen as a tragic year during which this was the norm in prisons all over Spain. Basically, they say, Spanish Fascism was no more than a crude, would-be farcical imitation of Italian Fascism or German Nazism, more appropriate to a satanic music hall. There lies the difference. All that happened during the Spanish civil war was nothing more than a try-out, or a rehearsal if you prefer, of the wave of terror which the Nazis would later unleash over all of Europe.

Eutimio Martín García came to almost identical conclusions as Amoedo-Gil regarding the extermination program and the Hispano-Nazi similarity, with which Moreno Gómez agrees. After establishing a resemblance between the Francoist repressive methods and those of the Nazis, *muatatis mutandis*, knowing that the German barbarity shall never be surpassed in all the world and that any future attempts would be more or less bungling, albeit equally deadly, copies. Eutimio says the following regarding the San Simón penal colony:[vii]

> "On the island of San Simón, Francoism devised an Auschwitz of its own creation, where the harsh climate and to the particular physiological weakness of the inmates, would play the same sinister role as the gas chambers. It rains day and night, for months at a time. The rainstorms that whip the island are frequently accompanied by gusts of wind so powerful that they can bring down the strongest trees by the dozen."

Eutimio Martín points to similarities with Nazism, not only when it came to the extermination (it was never a case of total extermination in Spain, not even in Mauthausen) but also when it came to the method of transferring prisoners in cattle wagons on trains, day after day, without any food or water. In Spain, that is also how prisoners were transferred in closed carriages, in never-ending exhausting journeys, from one prison to another.

There is data from the Military Prison Hospital in Guernica, established between 1938 and 1940 in the Augustine college building as a penitentiary for prisoners and also as headquarters of a Battalion of Workers from Devastated Zones. Designed as a 650-bed hospital, 265 sick prisoners (almost half) from all over Spain (8 from Cordoba) died from disease from June 1938 to May 1940, mostly from tuberculosis, typhus and typhoid.[viii]

Cordobans who died in Guernica Military Prison Hospital

Andrés Blanco Castro (Villafranca), Francisco Cantarero Castillo (Bujalance), Ramón Ferrer López (Peñarroya), Manuel López Castillejos (Alcaracejos), Jesús Molina Villaga (Villanueva), Eusebio Murillo Ortiz (Fuenteobejuna), Juan Muro Acedo (Pueblonuevo) and Antonio Rojas Ruíz (Cordoba capital). Buried in the extension of Zallo Cemetery. As to other prison establishments in Spain, Eutimio Martín reports that in El Dueso prison in Santoña, Cantabria, with a capacity of 3,000, 53 inmates died from starvation on a single day – 9 January 1941.

A study by José María García Márquez provides reliable data on the mortality rate in Seville prisons of 786 victims, most of them during the 1940-1942 period of extermination, of which 500 in the Provincial Prison and the remainder in several concentration camps. The most scandalous of these was the Las Arnas concentration camp where 144 of its 300 inmates died from starvation during the tragic 1940-1942 Triennium. Even more shocking is the fact that this camp was especially created for the imprisonment of 'beggars', in those days, anyone from the thousands and thousands who were starving because they had been reduced to extreme poverty.[ix]

For the first time, in all the studies of the civil war, we are told that a concentration camp was destined for 'beggars', half of whom would be 'exterminated'. This brings us to reflect on other aspects of the Francoist repression that are similar to the III Reich and add to the description of the genocide in Franco's Spain. We are continuously told that there was no 'technical genocide', but this is clearly doubtful because the totalitarian New State did not tolerate, in any form whatsoever, the poor, beggars, the impoverished rabble nor shirtless individuals who wore *alpargatas*[1]. The truth is, we still have much more to learn about Francoism.

García Márquez provides significant evidentiary insight regarding the degree of neglect and total lack of interest with which Falangistas impassively observed the great national scandal of prison mortality. Actually, it was the Regime, wishing to turn a blind eye to that which was happening in the prisons, that abolished the *compulsory inspection* required under Articles 684 and 685 of the Code of Military Justice.

When June 22 1937, the Military Governor of Cádiz informed the Inspector of Prisons that there had been no inspection of the prisons as required, the Inspector replied June 30 that "in accordance with National Defence Junta Decree 88 of September 18 1936, all regulation prison inspections were deferred *sine die.*" Right from the beginning, the Regime was well aware of exactly what was happening in the prisons, but it preferred not to hear of it.

As to the hypocaloric diet mentioned earlier, Moreno Gómez agrees entirely with García Márquez when he says that the great mortality in

---

[1] Rope sandals, the typical footwear of country people and the height of fashion worldwide today.

Francoist prisons could be directly attributed to the Regime that knew full well that its policy for feeding inmates was causing the death of thousands of prisoners all over the country. To deny the evidence of this is an exercise in unscientific and unethical frivolity. Obviously, victorious Francoists did not give two hoots for what was going on in the prisons; the Rojos simply did not deserve to live. If they died in prisons, the Regime would save bullets. Admitting that the death of three or four inmates in a prison during one year might be the norm, there can be only one name to describe the scandalous mortality rate death from starvation in post-war Francoist penitentiaries: *a crime against humanity and extermination*, regardless of what negationists want to call it.

The following table gives some figures for the extremely little amount of data currently available regarding Franco's prison holocaust. More than 6,000 dead in a dozen prisons, all but no data for major prisons such as Puerto de Santa María, Málaga, Hellin, Chinchilla, Cuéllar, Segovia, Madrid, Ventas, Alcalá de Henares, Burgos, Palencia, Santurrarán, Amorebieta, Santander, El Dueso, Santoña, Zaragoza and so many more. How many more thousands died?

Table IX
Sample prison mortality from starvation across Spain in the 1940s

Location of penal establishment	Died from starvation
Cordoba capital	756
Seville	786
Almendralejo (Badajoz)	144
4 Pueblos de La Serena	90
San Simón (Pontevedra)	666
Oviedo	251
Gijón	84
Guernica	265
San Cristobal (Navarra)	328
Catalonia	648
Valencia	813
Alicante	240
Toledo	680
Cáceres	150
TOTAL	**6,013**

In conclusion, Franco's crafty, underhand extermination policy, negligent by action and omission, was a colossal, scandalous phenomenon in the prison world that we might call the Four Horsemen of the Penal Apocalypse: 1) Mass imprisonment and large-scale overcrowding, with lethal effects of their own; 2) Official starvation diet, that is, planned hunger, excused as the greed of corrupt prison employees; 3) Almost total lack of health care and medical assistance, where not even aspirins or bandages for wounds were available; 5) Climate as a weapon of mass destruction, where prisoners from warm Southern Spain were sent to the cold, damp jails of the North, where the inmates dropped like flies, especially the weak and the elderly as in San Simón, or in the Artic conditions of Burgos prison in winter. Few historians have drawn attention to this particular feature of the Regime's perversity.

Other Aspects of the Multi-Repression
Conditional release from Prison

"The Historic Memory still continues to be dominated by the victors; it remains the memory of the victors. In Spain, nobody comes out to protest because the Church has initiated the beatification of members of the clergy... Nobody protests in favour of the Basque priests who were executed by Francoism, nor because there is no 'beatification' of the other dead. People appear to think that this is normal they accept it... This is, however, an indication, an aftertaste of the absolute domination that the victors have had over the memory of the people, they who have imposed their memory and their interpretation of the civil war. Today, there are books that tell that it was not always so, but in the collective memory of the population, it is still the victors' memory that prevails."
Jorge Semprún, *Los caminos de la memoria*

Those who have little knowledge of this topic believe that Franco's reprieves and pardons at the beginning of the 1940s were a philanthropic gesture by a magnanimous, honourable and benevolent Caudillo. Nothing could be further from the truth. It was not a question of magnanimity but of getting rid of, as it were, the chaos of half a million prisoners-of-war interned in concentration camps and prisons of all kinds, all over Spain. A chaos for which there were no infrastructures, no penitentiaries, no budget. After the first post-war years, when all kinds of persons faced the firing squads, regardless of their political importance, or lack of it, Francoism had no choice but to begin releasing prisoners, from the bottom to the top, timidly beginning with the lesser sentences in 1940.

This marked the first application of the Francoist theory for classifying prisoners that it divided into two categories: the confirmed disaffected (politically significant) and the recoverable (of no political significance). Nonetheless, this was not an exact classification as prisoners were often executed or starving to death in prison. There is another factor to bear in mind also, that until the end of 1941, as in the III Reich, extermination as a measure predominated and overcrowding was intentionally used as a method for eliminating the unwanted. This is why, both in Nazi camps and in Franco's prisons, 1941 was the year with the highest mortality, when little attention was paid to the lives of the imprisoned.

Beginning in 1941, both in Spain and in the Nazi concentration camps, the regimes started to implement their 'magnificent plan' for using prisoners for forced labour, which led to some improvement in their living conditions. Furthermore, the slow decrease in the number of inmates gradually contributed to prisoners' survival. Thus, the appearance of all kinds of so-called humanitarian and charitable programs that had nothing whatsoever to do with the earlier punishment, extermination, submission and educational measures. No matter how hypocritical or cynical its propaganda, the victorious New State intended to resort to the new programs to squash all ideas of trade unionism and Republicanism, once and forever. Reprieves (and the commutation of sentences) were mechanisms to 'rid itself of some penal ballast',[x] never benevolent measures, no matter what the Regime said in its propaganda.

Thus, we see the onset of a wave of reprieves (until 1945 they were not called 'reprieves' but instead 'extraordinary measures of conditional freedom') as a means of alleviating the penal chaos and its insupportable budget. Drop by drop, beginning with inmates condemned to lesser sentences, prisoners who had been 'cleansed of their dissolutionary ideas' and old grievances as evidenced by their 'good behaviour' and who had demonstrated their submission to the New State and acceptance of its principles, were gradually released on probation.

They were reminded that this was 'conditional freedom' and as such, they and their families, were subject to the supervision of the local authorities and their movements restricted. The released prisoner was obliged to make regular visits to Guardia Civil headquarters to sign in, which frequently implied a beating. As far as benevolent measures were concerned, not a single one. To make matters worse, as the released inmates were always subject to the hostility of their Francoist neighbours, they and their families lived in a social vacuum, despised, marginalized and excluded from all kinds of activities. Sometimes it was better to remain in prison than to have been released and forced to live amid the pack of fanatic victors, the constant singing of *Cara al sol* and the fascist salute.

April 5 1940 marked the first release on probation, of prisoners older than 60 who had been condemned on the grounds of Marxist rebellion, had served at least a fourth of their sentence and exhibited impeccable behaviour.[xi], [xii]

June 4 1940, probation was granted to those 'condemned by Military Courts to sentences from six years and one day to twelve years and who have served at least one half of their sentence, have exhibited exemplary behaviour and obtained a favourable report from the local authorities and the Falange'. The latter requirement made it almost impossible for an inmate to be released in this manner given the great amount of local hatred of the defeated in those days. In order to lessen any opposition from the local Falange, the Regime provided an alternative in the form of 'probation in exile'. Whichever way you look at it, this was an additional punishment that was applied in Spain after the 1919 repression of the Bolshevik triennium in Andalusia. Besides, all totalitarian regimes exile individuals as a standard repressive measure.

October 1 1940, release on probation was granted to those sentenced by a Military Court to serve twelve years and one day. Considering that the majority of the inmates had been condemned to thirty years in prison or given the death penalty, these lesser sentences had very little impact from a numbers viewpoint.

October 16 1942, release on probation was granted to those 'sentenced on the grounds of Marxist rebellion to up to 14 years and eight months'.

March 13 1943, the 1942 law was amended to apply to those 'sentenced to no more than twenty years in prison'. The insignificant impact of all the laws is shown by the fact that, in this case, only 1,087 inmates were released (311 in exile and 776 on parole).

September 29 1943, release on probation was granted to those sentenced on the grounds of Marxist rebellion and were at least 70 years old, regardless of the length of their sentences. Their dossiers were amended when an appeal had been denied because of an unfavourable report from the local authorities, with the indication 'Approved by the recently created *Servicio de Libertad Vigilada*.[2]

November 17 1943, release on probation was granted to those sentenced to twenty years and one day and in the case of those sentenced to longer periods, 'on health grounds, extraordinary behaviour and other outstanding accomplishments'. This covered all those considered 'recoverable' in which the ideological repression was successful, who abjured their principles and accepted, without question, the National Catholicism platform. Again, these were not 'benevolent' measures, even though they were promoted as such (especially in view of the adverse scenario that was emerging as it appeared that the Axis might not win World War II). They were the result of the unavoidable pressure of the chaotic and costly situation of the Francoist prisons.

These release measures were, to a great extent, also motivated by the sluggishness and lack of effective release operations of the Redemption of Sentences through Work Board, which is why the Regime had to resort to additional parole measures. Even so, these measures did little to resolve the Regime's problems.

From the beginning of 1940 to mid-1941, only 28,787[3] of the 300,000 inmates of Franco's prisons at the beginning of 1940, were released under probation.[xiii], [xiv] In 1942, 29,353 were released.[xv] By then, this 'benevolent measure' was available to anyone who managed to survive death from starvation during the great prison famine of the previous year.

This brings us to the mythical 'total pardon' decreed by the Regime October 9 1945, which had nothing of the 'total' to it. There

---

[2] Probation Service, i.e., National Parole Board.
[3] Domingos Rodríguez in El Régimen de Franco points out that Stanley G. Payne exaggerated when he spoke of 40,000 prison inmates released on probation during this period.

is so much misinformation regarding this matter that it has led to considerable confusion. The October 9 1945 Decree for Total Pardon does not refer to the total number of the condemned but, as indicated in the Introduction, to a total or full pardon for all sentences for crimes of military rebellion against the internal security of the State or public order, committed *before* April 1 1939. In other words, war crimes, *not* to the multitude of arrests for post-war 'crimes', crimes relating to clandestine activities and association, to the persecution of fugitives and their supporters, and so forth...

The Introduction to the Decree contains a great piece of propaganda directed at World War II's victorious allies, in that it affirms that "90% of all those who were condemned for their activities during the Communist Revolution have been arrested". Continuing in a conciliatory manner, the Regime speaks of the normalization of life in Spain, with reference to those Spaniards who fled the country.

Still, it will escape nobody's attention that the Regime classifies the prisoners as either black or white sheep, because when it speaks of 'normalization' it distinguishes between "those who fought because they were conned by the political fervour and those who led and incited the masses". This is an important distinction as it separate the "mass of criminals" into two groups: those who were tricked into fighting (the deceived) and those who led and incited them (the deceivers).[xvi]

Item 1 of the Decree stresses that a pardon will be granted "in all cases where it is shown that the delinquent never participated in acts of cruelty, death, rape, desecration larceny and other deeds that because of their nature are abhorrent to all honest men". Other individuals were disqualified from benefitting, such as Masons, Franco's nightmare, for whom he would never contemplate any kind of reprieve. Clearly, the door remained open to all kinds of repressive pretexts.

Item 2 states that the "reprieve will be granted upon the request of the condemned" whenever the information is favourable.

Item 3 proposes a curious extension of the reprieve to "those who are still insurgent on the condition that they turn themselves in within one month" (in other words, fugitives who fled to the hills or are in hiding elsewhere). This was a trick to capture them. There also is a message for those who had gone into exile: "individuals who are living outside Spain and who return within a maximum of six months".

Item 6 qualifies the terms of the reprieve that "shall not be granted to accessory sentences and shall be considered null and void in the event of a repeat offense".

Anyone who still believes that the Total Pardon of 1945 opened the prison doors wide and freed all the imprisoned, is living in cloud cuckoo land.

José Manuel Sabín has reported on the consequences of the 1945 reprieve in Toledo. There were no such reprieves in the jails of the capital as these were mainly inmates with short-term sentences. Some inmates in the major prisons of Ocaña and Talavera de la Reina did, however, benefit from the new regimes, as 35 inmates had been reprieved by the end of 1945 and considerably more, 523 in 1946. Sabín says that one had to wait until 1964 when, as part of the celebration of twenty years of peace, all prison inmates jailed under the terms of the 1945 Decree who benefitted from the extinction of the accessory sentences exclusion, were reprieved and their criminal records destroyed.[xvii]

Classification of Prisoners

The Committees for the Classification of Prisoners, created by Decree January 9 1940, were another of the Regime's useful inventions for keeping inmates wrapped up, and well wrapped up, at a time when no less than 300,000 individuals were imprisoned all over Spain. Again, this was no new mechanism for releasing inmates, nor a charitable measure as some claim, but a means of organizing and classifying prisoners in an attempt to bring penitentiary chaos under control. Under this Decree, inmates were divided into four large groups:

A. For whom the reason for and the authority who ordered the prisoner's arrest are unknown.
B. Arrested by the Government.
C. Tried by emergency summary courts.
D. Under 16 years of age.

The classification of young men of military age who had been in the Republican zone during the war and whom Franco was now conscripting, was much simpler. The lists of young men eligible for the draft in each township were examined by the local army enlistment boards, one by one and year by year, to determine their status:

- *Desafectos:* having been involved in politics, supporters of the Left or the Republic (more than 90% of all). Sent to the Disciplinary Battalions of Worker Soldiers.
- *Encartados*: already tried and either on their way to jail or already sentenced. If the sentence was a light one, also released to a Workers Battalion.
- *Indiferentes*: with no leftist tendencies, drafted into the Nationalist Army.

The processing of all prisoners was somewhat straight-forward, depending on which group classification they belonged to.

- *Class A:* Information was taken out on these, somewhat similar to the earlier requests for certified good conduct declarations, and if the resulting report was favourable, they were released from prison.
- *Class B*: These were detained for a further maximum 30 days and unless otherwise retained by the authorities, released.
- *Class C*: These prisoners were the meat of the Francoist repression; they were the cannon fodder for the court martials and the firing squads In these cases, the Examining Magistrates were told to speed up their deliberations.
- *Class D:* Surprisingly eight months after the Nationalist victory there still were minors in the prisons – both teenagers and children. The latter were infants who remained with their mothers in the women's prisons. This group of inmates was put under the supervision of the Child Protection Services.

The National Parole Board

The Servicio de Libertad Vigilada, or National Parole Board, created by Decree May 22 1943, was another New State invention with which to control the movements and activities of prisoners who were released on probation. Accordingly, those who were released remained under the constant watch of the victors, never truly free. Tagged as criminals, they were excluded and marginalized within their communities.

Established under the aegis of the Directorate General for Prisons, the National Parole Board was mandated to 'maintain an effective supervision of both the released prisoner's activities and channel them along safe paths of conduct, particularly their politico-social behaviour'. It was as if, once the prisoner was released, the seed of National Catholicism that had been sown in prison were to fall on fallow ground or fail to germinate.

The fundamental principles of the Movement and of National Catholicism were also directed at ensuring the submissive and converted change in any prisoner on prole who exchanged permanent incarceration for a life of submission and silence. The Regime's official line was that the Release on Parole program had the social duty to 'find work (for the ex-convicts) and to ensure that they were accepted by society without any misgivings'. What blatant lie! The reality was totally different: it was the pure and simple strict control of the released inmates, within and without the penitentiary. Neither the Nazis nor the Italian Fascists wielded such a hard and long-standing control over their political opponents.

Each town set up its Parole Board, consisting of three members: the Mayor, the local Commander of the Guardia Civil and a representative of the Falange. For the first time, the clergy was not represented on such a body as it would have been in the past. These Parole Boards could oppose the conditional release of a prisoner, something that often occurred given the enduring local vindictive fanaticism. In such a case, the prisoner might still be released on probation, but he would be exiled, or deported, to a town at least 250 kms distant from his place of residence. Sabín's research has shown that of all those released on probation, 25% were exiled locally. Francoists presumed that the internal exile or deportation of local disaffected was an effective weapon against local trade unions whose organizations were thus dismantled forever. (This measure was also applied to the families of individuals who were in the maquis during the persecution of the guerrillas.)

Most importantly, we must keep in mind the difference between the longstanding penitentiary regulations as compared to the real penal hell that the disaffected had to suffer. The following is an explicit example of how vindictive a Parole Board could actually be and often was.

Manuela de la Cruz Cabrera, from Almadén, Ciudad Real, was a young 16-year-old when war began. She studied typing and was an active member of the JSU and the UGT but not otherwise particularly involved in politics. May 2 1939, like so many others, Manuela was arrested and taken to Almadén prison, but she was not tried until October 1940 when she was charged that '...during the Marxist rule she belonged to the group of antifascist women, she applauded the Red cause and insulted the National Army, making propaganda in the press and in meetings... and she generated arrests....', all of which was highly unlikely for a young person of her age. She was condemned to six years and one day for 'promoting rebellion'. Had her activities been truly punishable, she would have been guaranteed a 30year sentence. Six-year sentences were those given to people who were totally neutral.

April 18 1941, Manuela was transferred to the Ciudad Real Prison for Women. Soon afterwards, procedures began for her Conditional Release on Probation for which she applied for 'permission' from the Almadén Parole Board. May 1 1941, Justo Sánchez Aparicio, the Mayor and one of the members of the Parole Board, sent the following report to the Directorate General of Prisons:

> "It is this Municipality's firm opinion that this person must not be set free and let alone, agree to her returning to live here as she will resume her obsession with slandering the National Cause as she did before."

Meanwhile, the Directorate General of Prisons who wanted to get rid of those who had been condemned to lesser sentences in order to alleviate the chaos, resubmitted her application, regarding which the Mayor redoubled his opinion regarding the young women, whom he said was:

> "...totally incorrigible as it seems that she was born a Roja and she decided to continue to be one; such imbecile behaviour on her part could be excused on its own, if it were not to cause such harm."

Young Manuela was finally released on parole, but she was exiled to Madrid, where she went to live with an aunt. In June 1942 she obtained permission to live in Almadén, but she was so ostracized that she moved to Puertollano. She finally obtained a full release in May 1945. A few months later, it so happened that two of the guerrillas in the hills killed an Army Sergeant and in the following encounter, two guerrillas were killed. It was the Regime's custom, whenever something happened in the mountains, to cast a web in the region, not especially to catch any accomplices, but in most cases, as an excuse to terrorize members of the Left and ex-convicts on parole.

Manuela was arrested by the Guardia Civil in the middle of July 1946 and charged with complicity with the rebels in the mountains, which was untrue. She was sent to the Ciudad Real provincial prison, charged under Military Justice by the authorities who dispatched her to Madrid to be tried by the Special Court Against Crimes of Espionage and Communism.

The last we hear of Manuela de la Cruz Cabrera, is that she had been confined in Ventas Women's Prison since August 1946. So much for her 'conditional freedom'.[xviii]

The Regime continued its demagogic and cynical propaganda campaign in favour of the National Parole Board, full of lies about counselling and protecting those who were released on parole. The reality was that this was no more than supervision and control, repression, marginalization and exclusion. Theoretically, the parolee had to sign in at police headquarters every fifteen days, but not always. Frequently, it was not just the parolee but all the members of his family – wife and adolescent children – who had to sign in. In the evening, all the parolees were required to gather at the gates of the barracks, sign in again, sing *Cara al Sol* and suffer further humiliations; several were also regularly chosen at random to receive a beating.

Although there clearly was nothing paternalistic nor protective about the parole regime, there are still quite a few scholars who, in their enthusiasm for the new prison regulations, still believe that these New State regulations had anything to do with the reality of the situation. As we saw in Ernest Caballero's testimonial in Appendix IX, his description of how as a 9-year-old child, he had to accompany his mother who had been released on probation and was required to appear at the Villanueva de Cordoba barracks every evening (and because his father was still fighting the guerrillas in the mountains).

There can be no lingering doubts as to the false and cynical root of the 'conditional freedom' system in Franco's Spain.

The Exploitation of Prisoners as slave labour

"Without memory, following generations will not, obviously, have any idea of what happened. Furthermore, without memory it is as if there never was any injustice, and the world could proceed as if no barbarities had ever been committed... We have to erase the traces of the crime, not with a crude Negationism, but by depriving the crime of any meaning. Western culture has been a masterful creator of the invisible crime."
Manuel Reyes Mate, *La piedra desechada*, 2013

Three can be considered the number of Francoism's great barbaric acts during the post-war period: the mass deprivation of personal liberty of more than half a million Spaniards; the new wave of executions of 40,000 victims of reprisals (added to the 100,000 already executed the previous three years); and lastly, the monumental phenomenon of widespread forced labour applied to hundreds of thousands of prisoners, forced to work as a punishment and exploited as a means of economic management by both the State and by corrupt private companies.

The forced labour phenomenon began with the Regime's usual subtle rhetoric, when June 1 1937, the official Government Bulletin published Decree 281 of May 28 1937, by which the imprisoned are granted the 'right to work'. In reality, this was no such 'right' but a 'requirement'.

Furthermore, Article 3 decrees that the imprisoned shall be considered 'military personnel, thereby subject to the Code of Military Justice and the June 27 1929 Geneva Convention'. This reference to the Geneva Convention is as surprising as it is cynical, considering that the Francoists never ceased resorting to the summary executions of prisoners whenever they felt like it. What is clear, is that these references to International Law were cheap rhetoric for the consumption of the democratic governments of the day. The aforementioned decree also established the amount of the measly daily wage paid to each slave:

2 pesetas a day (the standard wage was 10 pesetas), from which 1.5 pesetas would be withheld for food and lodging, which means that each 'worker' actually received a daily wage of 0.5 pesetas, paid at the end of each week.

The exploitation was scandalously evident and, as one would expect, immediately triggered the greed of the great private enterprises, as well as the public ones. Of course, the Church also wished to have a slice of this cake. Thousands and thousands of men who should have been at home taking care of their homes and families, enjoying a modest, happy and productive life in their hometowns as part of society, lost their future and all personal aspirations, no matter how humble, including sharing the love of their young lives. All of this was destroyed by the barbaric acts of an upper class that was stupefied by the repression, by the influence of European Fascism and by the greed of their class, as they partook in the frenetic management of the Nationalist victory that had been left in the hands of the Barracks, the Casino and the Church Vestry.

Slave labour camps or Disciplinary Worker Battalions

Whilst the Regime's great idea of exploiting the prisoners was taking shape in the middle of the war, and as Rev. Pérez del Pulgar was preparing the Holy Water, the Workers Battalions were being set up. January 1 1939 there already were 119 such battalions with 87,589 worker prisoners, engaged in many tasks behind the lines and housed in campsites under deplorable conditions.

Thousands and thousands of prisoners, some who had not been tried and others who were already sentenced to 'light' sentences of a maximum 12 years were required to enrol in a Workers Battalion if they wished to be conditionally released after they were considered to be duly repentant and willing to accept a two-fold ransom for their freedom: 'physical redemption through work as a convict', and a 'spiritual redemption through positive acts' (in other words, the full acceptance of the Catholic doctrine) as certified after examination by the chaplain and the prison Parole Board.

This was nothing else other than applying the ancient Inquisitional mechanism of forcing prisoners to abjure their beliefs as a means of instilling National Catholicism. For three centuries, the Catholic Church in Spain looked to the Inquisition as the means to punish dissidents (torture and pyre), and during post-war National Catholic Spain, under the *purgandus est populus* motto, to eradicate all heretic ideas, namely Republicanism, syndicalism, democracy and secularity.

During the war, after Francoists had began blowing off steam at will, by eliminating *in situ*, political commissars or skilled individuals, the imprisoned were sorted by the so-called Classification Committee that would class them according to one of the four groups we already know:

A. Supporters or favourable to the Movement
B. Insignificant disaffected with no responsibilities
C. Disaffected with political or military connotation
D. Suspicious individuals or common criminals

Based on the above, prisoners classed C and D (and some B), were handed over to the Military Examining Magistrates to be tried by court martial, the outcome of which was extremely gloomy. The remainder were sent to concentration camps or to Workers Battalions to wait for their release documents to arrive.

The Workers Battalions (later, Worker Soldiers Battalions) were comprised of prisoners or young men of military age held in concentration camps and classed as disaffected. They were sent directly from the concentration or prison camps to the Battalions, reaching the enormous number of 100,000 slave workerss in 1940.

These were men who had no taste of freedom after the war; and who in the majority of cases, had never been found guilty of anything because they were never tried. They were punished on the basis of political criteria, simply because they were opposed to the victorious fascists.

Hence, the double purpose of the Battalions: 1) to provide cheap or free labour; 2) as an instrument of physical and ideological repression. Punishment, humiliation and submission, by means of miserable living conditions and an iron discipline, to ensure they learnt the lesson of exactly what their marginal role in the New Spain was.

This 'appetizing' scenario of a mass of cheap, slave labour was now influenced by another divine Francoist invention. The <u>Redemption</u>

of Sentences through Work Regime, Ministerial Order of October 7 1938, the brainchild of Jesuit Reverend Pérez del Pulgar (and some say, the magnanimous brain of the Caudillo), as a continuum of the May 28 1937 Decree's philosophy regarding prisoners' right to work. The Council for Redemption of Sentences through Work was appointed December 15 1938, effective January 1 1939. The number of prisoners who turned to this Regime was insignificant: 4.5% in 1939 and 6.6% in December 1940 (18,781 prisoners of a total 380,000 imprisoned nationwide). In Cordoba, only 600 of a total 4,000 applied.

The Workers Battalions were the first forced labour modality during the war and the immediate post-war period. The highest number of these slave workers was recorded in 1939-1940, total at the end of the first year: 90,000 (34,143 in 1937; 40,690 in 1938). A gradual change to the new designation appeared during 1940 (new name, same thing) when they were renamed Disciplinary Battalions of Worker Soldiers (BDST), focused not so much on prisoners of war or concentration camp inmates, but more especially on the young men of military age that the local Army Recruitment Centres had classified as C's - disaffected, with some leftist tendencies. If a worker had already been sentenced his file was marked BDST1.

Table X
Number of slave workers in the Disciplinary Battalions[xix]
[Both Workers Battalions & Disciplinary Worker Soldiers Battalions]

1937........................ 34,143
1938........................ 40,600
1939 & 1940 90,000
1941 47,000
1942 46,380
1943 4,800

Thus, in addition to being a source of slave labour, the battalions were now also responsible for punishments, purges and repressive discipline.

The two names for these Battalions were used interchangeably and mostly the usual reference was just to Workers Battalions. All of Spain was full of these centres of slave labour. Still, when mention is made of great public works carried out by this means, one notes that Franco clearly preferred the centres in the Centre-North of Spain, especially the North.

Andalusia appeared to be somewhat forgotten in Franco and his henchmen's plans. There, the most famous public work slave labour construction was the *Canal of the Prisoners* in Seville, on the lower reaches of the River Guadalquivir, at Los Merinales. Also built with slave labour, the Torre del Águila Seville reservoir[4] and the Algeciras-Bobadilla railway between Cádiz and Málaga.

Cordoba Province was assigned Workers Battalion No. 130. One company was billeted near Espiel and another in Peñarroya-Pueblonuevo, where the battalion Commander lived. In all these battalions, where the workers were housed in very poor housing and feeding conditions, almost like concentration camps, special emphasis was given to the spiritual education of the Rojos. The task of morally disinfecting and re-educating the workers was entrusted mainly to the Jesuits who spearheaded National Catholicism. Workers Battalion 130 was placed in the hands of José Luís Díez, S.J.

Pedro Gómez González, from Villaralto in Cordoba,[5] was sent to Workers Battalion No 28, Company 4, in La Bacolla, near Santiago de Compostela, where they were put to work building an airfield. His description of the situation there is reproduced in Appendix IX, at the end of this book.[xx]

It was not just the New State that took advantage of this colossal exploitation of labour, but also and surprisingly so, numerous private enterprises all over the country. Great public works of today were built by those who preyed on the slave labour. The Cordoba Civil Registry Office contains records of quite a few deaths in Workers Battalions, undoubtedly victims of the inhuman living conditions, as well as some who were shot on the spot as they attempted to escape. The Official Bulletin for Cordoba published Wanted Ads for deserters from these Battalions and, when they were caught, details of the courts martial which tried them, stressing the grave consequences of their acts. This happened to Miguel Caballero who, having been sent to the Los Barrios Workers Battalion in Cádiz, escaped but was turned in by

---

[4]   Magdalena Gorrell Jaen's cousin, Antonio Jaén Romero, the young Commander of the Cordoba Volunteer Militia founded by his uncle, Antonio Jaén Morente, spent many years as a slave worker on both these public works before he obtained his freedom in 1952.

[5]   The first mention, in this section, of Cordoban town or village as the hometown of a slave labour prisoner is underlined to highlight the effect of this program in this particular province.

a fellow countryman. He was sent before a court martial which he managed to survive by miracle.[xxi]

During 1939, the Year of Victory, thousands and thousands of Republican prisoners were distributed among all the concentration camps and Workers Battalions, the length and breadth of the country. Cordobans, for example, could be found everywhere. José María Romero from La Rambla, published a book containing more than 500 statements from those who returned to their hometowns, with details of the Workers Battalions to which they had been sent:[xxii]

Workers Battalions Nos. 41 and 116 – Zaragoza
Workers Battalion No. 51 – Teruel
Workers Battalion No. 179 – Ceuta
Workers Battalion No. 152 – Palencia
Workers Battalion No. 26 – Medina del Campo
Workers Battalion No. 91 – Oviedo
Workers Battalion No. – Melilla
Workers Battalions No. 159 – Salamanca
.... and more, to name but a few.

During the first months of the victory, prisoners were not only continuously shunted from one concentration camp to another, but also between camps and Workers Battalions. As Casimiro Jabonero told Moreno Gómez, in April 1939 the government began sending convoys of inmates from the more than 2,000 prisoners packed in the Seminary and La Serrería prisons in Cuenca, to concentration camps and Workers Battalions. April 25, a shipment of 900 left La Bacolla, La Coruña, followed by another convoy to a concentration camp in Madrid that also housed a Workers Battalion.[xxiii] In Cordoba May 21 1940, 12 inmates from Carcabuey were sent to the Rota Workers Battalion in Cádiz.

Prisoners were treated like so much cattle as they went from one concentration camp or Workers Battalion to another and back, before they were released. In his study of Montilla, a fellow historian, Arcángel Bedmar, records several such cases. Antonio Arroyo was first sent to Number 93 Disciplinary Battalion of Worker Soldiers, then from one to another battalion, going through Bilbao, Palencia and North Africa until he was able to return to Montilla in 1943. José Gómez

Márquez, aged 19, was sent from Padul, Granada, concentration camp to Cádiz Workers Battalion No. 6 in December 1941, then to Workers Battalions in Cherta, Tarragona, and Logroño. He only returned to Montilla in 1945 after serving in the Francoist armed forces. Antonio Alcaíde went from the La Bacolla Workers Battalion where he remained for a year and a half, then to a roadworks in Algeciras, until, after doing his military service in the Canary Islands, he returned to Montilla in 1944.

Living conditions for the prisoners in the Workers Battalions were inhuman, particularly when it came to the manner by which the guards, most professional soldiers many of whom had served in the African Legion, disciplined the inmates. Cristobal Carrier Díaz, from the village of Santa Cruz, Montillo, was sent to the Cherta, Tarragona, Disciplinary Battalion of Worker Soldiers where Sergeant Aurelio Azcona Zabalza forced the men to run with sacks of sand tied to their backs, after which he would lash them with a whip in a hut. Another Montillano, Miguel García Ruíz spoke of the hunger he suffered when he was interned in Rentería Disciplinary Battalion No. 51, where they were only fed boiled cabbage. When he complained, he was punished by having to run with sacks of earth tied to his back, a typical Foreign Legion punishment. Another form of torture was to make the inmates walk past a companion who was being punished, slapping him, as they went by. Francisco Carmona Priego, interned in the Cerro Muriano, Cordoba, Disciplinary Battalion of Worker Soldiers, saw a sergeant break a prisoner's back with blows from a pickaxe handle. This particularly brutal punishment is also mentioned by another historian, Antonio D. López, when he relates the testimony of Manuel Esperilla, from Castuera, Badajoz, that he saw just how many beatings with sticks a person could endure, as the handle of the pickaxes broke a man's ribs.[xxiv]

Young men from every town in Spain, ex-soldiers of the Republican Army, were swallowed up by the mass imprisonment whirlwind: camps, prisons, workers battalions, exploitation of slave labour, a gigantic oppressive scene that almost nobody today can comprehend. Every town's history reflected a Dantesque picture, inside and outside Spain. Arcángel Bedmar studied this in depth in the village of Rute: Brothers Francisco and Pedro Caballero Tirado died as members of the French Resistance. Pablo Baena served his sentence in the Cherta,

Tarragona, Workers Battalion, as did Manuel Jiménez and Gabriel Porras. Bernabé Montes, Gregorio Puerto and Francisco Viso worked as slave workers on the building of the *Canal of the Prisoners*, Dos Hermanas, Seville. Siméon Rojas worked out his sentence in a Workers Battalions in Tetuan, Morocco.

And so on, in every Spanish town and village. Next, a sample from <u>Baena</u> of 172 natives of this town who were exploited as slave labour.[xxv]

Workers Battalion No. 3 – Melilla	36
Workers Battalion No. 8 – Los Pastores, Algeciras	34
Workers Battalion No. 26 - Labacolla	19
Workers Battalion No. 29 – Labacolla	12
Workers Battalion No. 33 – Tetuan, Morocco	6

<u>A few more Workers Battalions</u>

Workers Battalions & & 58, Los Barrios, Cádiz	Workers Battalions 17 & 116, Zaragoza
Workers Battalion 42, Oyartzun, Guipúzcoa	Workers Battalions 12 & 140, Barcelona
Workers Battalions 9, 5 & 11, San Roque	Workers Battalion 32, Tarifa, Cádiz
Workers Battalion 145, Tetuan, Morocco	Workers Battalion 152, Palencia
Workers Battalion 33, Ceuta	Workers Battalion 125, Manresa
Workers Battalions 54 & 124, Algeciras	Workers Battalion 21, Astorga
Workers Battalions 53 & 155, Madrid	Workers Battalion 51, Teruel
Workers Battalions 130, Peñarroya	Workers Battalions 212, Morocco
Workers Battalion 151, Alsasua	Workers Battalion 128, Navarra
Workers Battalion 63, Camprodón	Workers Battalion 23, Arañones, Huesca
Workers Battalion 178, Serós, Lerida	Workers Battalion 166, Sam Blás, Teruel
Workers Battalion 64, Maya, Navarra	Worker Battalion 15, Tortorella de Mont

J. F. Luque Moreno's catalogue of 55 forced labour inmates from <u>Montemayor</u> is interesting in that the name of the camp varies as either BDST (Disciplinary Battalion of Worker Soldiers) or BT (Workers Battalions):

BDST No. 8 Los Pastors, Algeciras	17
BDST No. 29, Labacolla, Santiago de Compostela	14
BDST No. 51, Oyarzun, Guipúzcoa	12
BDST No. 3, Melilla	8
BT, Cherta, Tarragona	4

There were more, dispersed one at a time in other camps, in 20 other Workers Battalions all over Spain, several in Morocco, as well as Ceuta and Melilla. The men from Montemayor, classified as disaffected without any other crime, lost several years of freedom as slave workers in the Workers Battalions and/or doing military service, only regaining their freedom in 1944 and 1945.

One cannot help but be impressed with the way that the Regime dispersed so many thousands upon thousands of men from every town and village in Spain, among Workers Battalions and Disciplinary Worker Soldiers Battalions all over the country. Just take the example of a small village in Cordoba province, <u>Fuente Palmera</u>, for which Alberto González Sojo reports on 48 men and how many were dispersed, as follows:

BDST No. 6, Facinas, Tarifa Cádiz	4
BDST No. 16, Tarifa, Cádiz	5
Belchite Penal Camp	2
BDST No. 211, San Roque	2
2nd Penal Colony of Montijo	2
BDST No. 28, Santiago de Compostela	3
BDST No. 51, Oyarzum	3

Others from that town were the only prisoners from in other Workers Battalions such as BDST No. 214, Tifasor, Melilla; BDST No. 108, Zaragoza; BDST No. 162, Gerona; BDST No. 55, Tarifa; BDST No. 50 Cordoba; BDST No. 212, Bab-Tazza, Morocco; BDST Nos. 5 & 213, Melilla; BDST NO. 13, Ceuta; BDST No. No. 2 Santiago de Compostela; BDST No. 35, Palma de Mallorca; Penal Colony Dos Hermanas, Seville; Alberche Prison, etc.

Mortality in the Workers Battalions was high and frequent. Luque mentions five deaths for Montemayor: Manuel López Sánchez, in the Algeciras Battalion, died 'from eating poisonous roots'; Miguel Moral Nadales, Battalion No. 29, Labacolla, from where he was released already moribund, only to die as soon as he got home; Juan Moreno Gómez, same as the previous one, from Battalion No. 8, Los Pastores, Algeciras; Antonio Galán Sillero, Battalion No. 13, Larache; and María Navarro Bernal, a Montemayor matron, who died after she was released from a disease she contracted in jail.

After considering all the above, one's attention is drawn to a very interesting feature: the great majority of all the prisoners from Cordoban towns regarding whom the above data refers to, were doing forced labour in 1942.

There are several conclusions we can draw from this: 1) After 1942, Francoism preferred to increase its income from slave labour, rather than let prisoners die in prison. 2) That the drop in the number of prison inmates noted from 1941 onwards, is an ambiguous bit of data because many inmates, especially those condemned to minor and intermediate sentences, were being released from jail, not to freedom on the streets, but to servitude in Workers Battalions, Disciplinary Battalions of Worker Soldiers, Penal Colonies, etc. 3) The 'official' number of prison inmates that are published and the only ones that are taken into account, are totally useless, considering that a great many inmates were only 'released' to become slave workers.

The migratory/deportation movement in Franco's Spain was awesome. It is impossible, today, to understand how it was possible for the victors to turn Spain upside down in this manner. How was it possible for the Regime to manage the victory in such an absurd and irrational manner? Could it be that the only possible answer is that only the Regime's barracks mentality can help explain such bestiality and savagery, all raised to the maximum degree by the influence of European Fascism and its cult of radical violence.

Furthermore, it is worth pointing out that young men of military age, classified as such by their local army enlistment boards, were not only first sent to Disciplinary Battalions of Workers Soldiers, then also to Africa to a Regular Army Regiment, where they would be subjected to an extremely harsh disciplinary and punitive treatment. This is something that Moreno Gómez has not seen mentioned in any study, but that was confirmed to him by several eyewitnesses, such as Manuel Bustos, one of those young men.[xxvi]

> "In 1942, I was one of 16 young men of military age from <u>Villanueva de Cordoba</u> who was examined by the local enlistment board and declared to be disaffected. All 16 of us were sent to the IVth Regiment, 4th Company, whose barracks were in Alcázarquivir. The senior officers, all veterans of the civil war, boasted of the abuses and excesses they had committed in Spain. Some regular soldiers had already been promoted as officers. They had as confidents, affected or Falangist Spaniards.
>
> The discipline was so strict that one day, when a recruit arrived late for the roll call, he was spread out

on the ground and given 'half a beating', or 50 lashed, 25 on one side and 25 on the other.

Eugenio *El Ramo* also came with us from Villanueva, despite the fact that he was disabled, for the only reason that his family were leftists; they beat him to death one day. I also remember a bugler from Ciudad Real who when on leave, stole some sheets to sell and make some money Unfortunately, he was caught and beaten; he died the next day.

The disciplinary terror declined when the Americans landed in Casablanca. They lined us up along the River Luco, to defend our position from attack, but there was a general desertion: more than 200 of us went over to the Americans."

If the Francoist prison nether world, inhabited by hundreds of thousands of men and women, many accompanied by their children, is a clear case of genocide, no less so is the nightmarish scenario of the thousands of slave workers interned in Disciplinary Workers Battalions all over Spain. The general situation in the prisons and Battalions was so astounding, it appears impossible that so many atrocities could have been committed in a country like Spain, a nation that was celebrated for its enlightened culture.

Tragically, the above was the product of a Regime created by the union of a hawkish, loud-mouthed and pedantic Spanish military class whose mindset remained bound by its experience in Africa, and a no less manic, Inquisitorial National Catholic clergy whose minds were liberally seasoned with the savagery exported by Nazi Germany. There could be no other consequence than a human catastrophe.

No wonder that there are those who still today would like to conceal, cover-up and silence the events of the post-war period, because crimes against humanity are so beyond belief that any acknowledgement of these is capable of shaming the most callous scoundrel.

Historians' field work on the Workers Battalions in the Basque province of Navarre, has provided some references and testimonials regarding the 'slaves of Francoism' in Valle del Roncal, Erronari, and Zaritzu, where some 2,000 slaves, interned in Workers Battalions Nos. 106 and 127, later transformed in 1940 into Disciplinary Battalions of Worker Soldiers Nos. 6 and 38, built the Igari-Bidangoze highway in the Navarre Pyrenees. This slave labour came from many places in Spain, especially Andalusia, Asturias and Vizcaya. Of these, 245 came from Vizcaya, 211 from Granada, 165 from Jaén, 126 from Asturias and 78 from Cordoba and elsewhere.

The Franco Regime designated Navarre as a zone of preferred action as part of what it called the 'Plan for the Defence of the Pyrenees'. Under this program, more than 10,000 slaves were employed on a great many public works throughout the province: roads, railways and more. Improvised housing in terrible conditions was found in schools (from which they removed the children) and ramshackle huts. In the Vale del Roncal battalions, there was 'nothing but hunger, cold and beatings', according to Fernando Mendiola and Edurne Beaumont's magnificent documentary.[xxvii]

As one of those interviewed in the documentary said: "As I worked, I did nothing else but think: am I going to spend the rest of my life here? I would sit down and think... I spent 6 years in Roncal". To these years, the younger survivors had to do an additional 3 years military service. In all, the lost youth of an entire generation.

As elsewhere, the living conditions in Roncal were horrendous, particularly in terms of the hunger everyone suffered. In addition to the meagre rations, the prison officers stole as much as they could and sell it. The inmates felt sorry for those whose families lived far away and could not send them anything. "Those poor Andalusians", they would say. "As they did not receive anything, there was nothing for them to eat..." One day, there was no dinner. "We were called to attention and the sergeant told us that considering that there was no dinner that night, we could stand easy". As to food packages from home, "It took fifteen days to one month for a package to reach us; when it arrived, the contents were rotten".

The working conditions were primeval. "There were no machines of any kind. Everything had to be done by hand, with a pick and shovel.... We were practically barefoot, and our feet were cut by the stones... They later gave us some new boots, but one of us who had received a new pair from home, decided to sell the new ones. He was caught doing so and they beat him to death." A frequent punishment was to strip the slaves naked and make them stand in the cold all night. "They tied a large stone with wires to the backs of others and made them run around the parade ground."

The documentary also refers to the Regiment's chaplains. "The priest was a bad one. He was a Francoist. He was really awful! He would tell us that we were evil and that we were redeeming our sins through work. We had to listen to all this garbage and keep quiet… We were the bad ones."

The worst treatment was reserved for any inmates who tried to scape, a crime for which the punishment was death. One of the eyewitnesses in the documentary said: "There was a fellow from Catalonia who spoke to nobody; one day he escaped but he was turned in by a companion who had planned to escape with him. The snow was a metre deep. The sergeant ordered his men to kill him when they found him. They discovered him in a mountain refuge and there he died. Later, they brought his body in on a stretcher.

We worked in groups and if one of us tried to escape, they shot all the others and their families had to suffer the consequences."

Another tragic incident, the subject of another Mendiola and Beaumont documentary,[xxviii] was the case of Cecilio Gallego García, a 24-year-old who belonged to a family from Dom Benito, Badajoz, all of whom had been very active members of the PCE. July 25 1939, a convoy of forced labour workers for the Roncal works arrived at the same Workers Battalion we have been speaking of, Number 127, 3rd Company.

October 23 1939, Cecilio decided to escape but he was caught and shot on the spot. In 2009, one of his brothers went through hell and high water to obtain permission to dig up his grave in Roncal public cemetery and take Cecilio's remains back to his hometown to be buried. The family was denied a formal wake in the *Casa de la Cultura* because of the opposition of the PP Mayor. "Had he been a Falangist, they would have welcomed his remains with open arms", lamented his ninety-year-old brother.

Escapes from the Workers Battalions (or Labour Camps as the people also called them), were quite frequent and a number of successful escapees joined groups of other fugitives in the mountains. A number of famous native Cordoban guerrilla leaders came from these groups, such as Francisco *El Gafas* Expósito who turned himself in in Andújar, Jaén, at the beginning of 1944, together with *Aragonés*, both of whom had escaped from the Anguiano, Logroño, Labour Camp and were leaders of part of the II Guerrilla Group in

Ciudad Real. The famous Sebastián *Chichango* Moya from Albacete, escaped from a Disciplinary Battalion of Worker Soldiers in Santoña, Santander. Emílio *Escobero* Pérez of the Albacete maquis, escaped from a Labour Camp in Belchite, Zaragoza. Felipe *El Castaño* Moya, from Pozoblanco, fled to the hills from a Disciplinary Battalion in Galicia. Another famous guerrilla, Francisco *Veneno* Blancas, from Adamuz, escaped from the Toledo Penal Colony in Fall 1944. He was known as 'the last guerrilla from La Mancha'.

One of the more remarkable escapes, which became famous after the publication his memoires, was Albino Garrido's[xxix] from Castuera concentration camp, March 4 1940, when he was about to be executed. He was one of a group of 5 who, except for one who was shot by the Guardia Civil, managed to make it safely to France March 22.

Re-education Camps

The Re-education Camps, directly managed by the Council for the Redemption of Sentences, were another form of forced labour, one that was most directly associated with private enterprises. These consisted of larger or smaller groups of individuals condemned to less than twelve years and one day (and some whose higher sentences had already been commuted), who were allocated by the Council to public companies, to private enterprises or to Church-owned companies. To be sent to one of these was considered somewhat of a privilege as the labourers worked under a semi-free regime, although they had to return to prison for the night. Communists and Masons, for which there was no possible redemption, we banned from these camps.

Theoretically, a prisoner was paid a daily wage of 14 pesetas but in actual fact, he only received half that amount. His family, if he had one, received 3 pesetas; 1.40 pesetas went to pay for food and housing and the remaining 9.60 pesetas were withheld for taxes. This was a pot of gold business for the New State, as it also was for the Regime's emblematic construction companies, similar to the cheap labour arrangement that enriched the great private companies during the III Reich.

In 1940, there already were 70 re-education camps with 5,155 slaves working on several projects, most of them in the Centre-North of Spain. At the end of 1943, there were 92 re-education camps in

Spain with a total 11,554 slave workers.[xxx] There were practically none in Cordoba province.

The Hato Blanco Re-education Camp was located in Ciudad Real, in the Port of Niefla, almost on the border with Cordoba. From there, 50 slave workers allocated to the A Carretero company organized a mass escape June 29 1943, most of whom went to join the groups of fugitives in the mountains.

Other well-known construction work carried out by workers from these camps were: Hermanos Nicolás Gómez, Madrid; Vich Cathedral, Bishopric of Vich; Valle de Los Caídos, Banús, San Román and Huarte & Cia; Pozo de Fondón, Asturias, Duro-Felguera; Alberche reservoir, Talavera de la Reina; New Prison, Cordoba, with 120 workers; Valdemanco, Madrid, Sociedad Marcor (from which the guerrilla stole a large quantity of dynamite in 1947, which in turn led to a court martial that ended tragically with torture and executions).

Devastated Regions

Work in regions devastated by the war was another way of putting convicts to work, very similar to that of the Disciplinary Detachments. Also controlled by the Council for the Redemption of Sentences, the inmates were individuals condemned to lesser sentences, or at the prisoner's request. Again, this was also forbidden to Communists and Masons. The work that was to be done were public works or work that the State requested from the Council, with a view to rebuilding installations or centres of population that had been damaged during the war. "Let them rebuild that which they destroyed" was the motto, as if the defeated were responsible for the destruction, not those who caused the war and fought it with the assistance of the destructive machinery provided by Rome and Berlin.

A Ministry of the Interior March 25 1938 Decree assigned responsibility for this reconstruction to the Directorate General for Devastated Regions, for which it would use convict labour, although the work would not start until after the war was over. By the end of 1940, 2,034 slave workers were working in Teruel, Belchite, Potes and Oviedo, According to José Manuel Sabín, Ministerial Notice of 1943 reported that 4,075 convict workers were deployed in Figueres, Guernica, and other locations.

Paramilitary Labour Camps

The Regime's most important post-war 'redemption and exploitation' invention was the Paramilitary Labour Camps, under Law of September 8 1939 with a view to providing a systemic organization of the slave labour as part of the New State's growing programme of public work. Actually, this was just another version of the Regime's program for punishing and cleansing the defeated whose penance was work, although in practice it was the State and the great private construction companies who were especially engaged in hydraulic works, who would benefit.

As those in power always use rhetoric to cover their shameful intents, in this case the explanation was that the Government had to 'compensate the increased amounts that it was spending on prisons'. (Just who told the Government to put half a million Spaniards behind bars...?) The militarization of the service (organized by the Battalions and the companies, under the command of the military), was justified on the grounds that as these convicts were working in areas far from any jails, they required strict discipline.

Eight such camps were created between 1939 and 1957 (at the end, common criminals were also sent to these labour camps). Slave labour from Camps 1 and 6 worked on the famous canal on the Lower Guadalquivir River, so-called *Canal of the Prisoners*, at Dos Hermanas, Seville; Camp 2 was assigned to the Montijo Canal, Badajoz. Many Catalan prisoners worked here, some of whom had been condemned to 30 years in prison (157), others to 20 years (22), and the majority, the remainder, who were serving 6 years until 1946.[xxxi]

From August 1940 to September 1946, 2,826 slave workers from Camp 3 worked on the Bajo Alberche canal, Talavera de la Reina, Toledo. This camp was divided into three sections: San Román, Real de San Vicente and La Sal. Camp 4 was assigned to the construction of the Rosarito canal in Añovar de Tajo, Toledo, on the right bank of the River Tiétar. Camp 5 was sent to the work on the new Infantry Academy. Camp 7 was engaged in several works for the National Colonization Institute in Aragon and Catalonia. Camp 8 worked on the mining railway from Samper de Calanda to Teruel.

There was another such camp on Formentera Island regarding which practically nothing is known other than that the living and working conditions were so bad that in 1941 and 1942, 58 convicts starved to death (avitaminosis, weakness and lack of nutrition are the causes of death given in the records), most of them from the province of Badajoz.

The most notorious of the Paramilitary Labour Camps was put to work on the construction of the Canal of the Prisoners. January 20 1940, the first batch of workers for the canal arrived from the La Corchuela, Dos Hermanas camp. The canal was designed to cross the entire province of Seville, from its border with Cordoba to its border with Cádiz, a distance of 158 kms, and it would take from 1940 to 1962 to complete. Although the initial work was done with slae labour, free market workers were hired to put the finishing touches on this megalomaniac project for the sole benefit of the local landowners. The first workers came from Camps 1 and 6 and after 1946, only from Camp 1.

As the work progressed further from the La Corchuela base, three nearer camps were created in 1945: La Corchuela, El Arenoso and Los Merinales, the latter of which was the most notorious, The prisoners came from all over Spain and when released, found themselves living conditional freedom in exile at great distances from their hometowns. As their families had to travel far to visit them, many settled in the area. At first, they lived in shacks in a number of peculiar shantytowns on the flood plain and as time went by, they built permanent homes. These are the origin of today's popular neighbourhoods on the outskirts of Dos Hermanas and Seville, such as Torreblanca, Bellavista and Valdezorras. In this respect, the Regime's policy of deporting the disaffected workers as another means of ensuring the dispersal and the deep-rooted destruction of the Spanish labour movement, was successful.

The award-winning *Prisoners of Silence* 2009 Spanish television documentary[xxxii] presents a collection of accounts that clearly illustrate the dark world of the *Canal Prisoners*, extracts of a few of which follow, others in Appendix IX.

- "After waking, we were lined up in rows of 3 persons each, surrounded by soldiers, and marched to the canal works. If you wanted coffee in the morning, you first had to sing *Cara al Sol*; the same at noon and in the evening before you got your soup."

- "During Holy Week, a van would arrive with four or five priests to take confessions and give sermons. Many prisoners did not listen to them, which infuriated the priests who complained to the commander who then ordered beatings. Attendance at mass was compulsory."
- One day when I was at work, two apprentice priests came to see me: "We have come to get the boy and take him to the children's home, because as your father is in prison…" I replied: "My father may be in prison, but I am free, and I have the balls to care for my son; and if I go to look for you, you come to me. My son is going nowhere…"

One is struck by the imposition of religion by force and the daily humiliation of requiring the inmates to sing the victors' hymn. Moreno Gómez already stressed that one of the distinguishing features of genocide, as Lemkin states, is the imposition of the oppressor's ideology and the demolition of the oppressed's beliefs. Equally important and deserving consideration were the inmates' complaints regarding the enormous difficulty they had to communicate with their families, particularly in view of the Regime's underhand attempts to take control of their children.

This description of the desperate situation of Francoist prisoners is best summed up by what two ex-prisoners who were interviewed in the documentary had to say:

- "It was not one day: it was many years without seeing our families, without seeing anybody, which is why some were not quite right in their heads when they were released. Others were forever marked by their experience, because all of this remains engraved forever on our hearts."
- "I have been kicked out of many places because I was a Rojo. One cannot speak of anything. We have to shut up. People are still afraid. Franco died and we were released to walk the streets as free people… If you do not speak of something, you erase it; you rub it out…"

Prison Workshops and Special Assignments

This discussion of the monumental phenomenon of Francoist exploitation of prison labour, ends with the Prison Workshops, a form of free labour within the prisons. The first workshops were created by Order of April 30 1939, in the Alcalá de Henares prison. Some of the first items produced by this Graphic Arts and Carpentry Workshop included furniture for the Social Welfare organization, pews for churches and 15,000 crucifixes for schools. The management and exploration of these workshops was also the responsibility of the Redemption of Sentences Council. Soon, every large prison created its own workshops: Burgos, El Dueso, Gijón, Guadalajara, Ocaña, San Miguel de los Reyes, Yeserías, Barcelona, Cordoba, to name but a few.

Lastly, there was another means by which prisoners could work to reduce their sentences: Special Assignments. These were reserved for inmates who had a record for good conduct. That is, absolutely no Communists, Masons, adulterers, individuals who had not married in the Church, who blasphemed, etc. Special assignments could be worked in the prison kitchen, bakery, storehouse, barber shop, school, infirmary, cleaning, offices, etc. It is believed that these inmates represented a wide range of other professions: craftsmen, doctors, university graduates, construction workers, bricklayers and so forth. Inmates could also use their experience to work, when available, to reduce their sentences, as plumbers, electricians and masons and as 'assistants' to the chaplains (altar boys), to the teacher and in the infirmary. Women were allowed to work as seamstresses, embroiderers or other 'feminine crafts', the product of which was put on exhibit whenever the Director, Mother Superior or Bishop visited the prison.

ENDNOTES FOR CHAPTER VIII

i   Manuel Martínez Cordero. El Penal de El Puerto de Santa María, 1886-1981 (The Puerto de Santa María Prison Complex, 18861981). Cádiz, 2004, pp. 149 et al.

ii   Migiuel Hernández. Undated letter number 309 in Epistolario (Collection of letters). Espasa-Calpe, p. 271.

iii   Manuel Rubio Díeaz and Silverio Gómez Zafra. Alendralejo (1930-1941) Doce años intensos. (Almedralejo (1930-1941). Twelve intense years.) and Los Santos de Maimona, Badajoz, 1987, p. 355.

iv   Gonzalo Amoedo López and Roberto Gil Moure Episodios de terror durante la guerra civil na provincia de Pontevedra A illa de San Simón (Episodes of terror during the civil war in Pontevedra Province. San Simón island). Xerais, Vigo, 2007. In Moreno Gómez' opinion, one of the finest academi studies of Franco's penitentiary world published to date.

v   Diego San José de la Torre. De cárcel en cárcel (From prison to prison). Do Castro, La Coruña, 1988. P. 159.

vi   Gonzalo Amoedo López and Roberto Gil Moure, Op. Cit.

vii   Eutimio Martín García, Op. Cit

viii   José Ángel Etxaniz Oriñez and Vicente del Palacio Sánchez. "Dossier. Morir en Gernica-Lumo" (Dossier. Dying in Gernika-Lumo), Aldaba magazine, issue 122, April-May 2003.

ix   José María García Márquez. Las victimas de la arepresión militar en la provincia de Sevilla (1936-1963) (Victims of the military repression in Seville province. 1936-1963). Aconcagua, Seville, p. 174.

x   Mirta Núñez expression, La gran represión (The Great Repression), Op. Cit., p. 207.

xi   José Manuel Sabín Rodríguez. La dictatura franquista (936-1975). Textos y documentos (The Francoist dictatorship 1936-1975. Texts and documents.) Akal, 1997, pp. 406 et al.

xii   Domingo Rodríguez Teijeiro. "Excarcelación, libertad condicional e instrumentos de control poscarcelario en la inmediata posguerra (1939-1945)" Prison release, release on probation and postpenal control in the immediate post-war period (1939-1945). University of Vigo.

xiii   Rendención, 28 June 1941.

xiv   Domingos Rodríguez Teijeiro, El Regimen de Franco (Franco's Regime) Alianza, Madrid, 1988, p. 240.

xv   Rendención, 16 January 1943.

xvi   Mirta Núñez Díaz-Balart et al., La gran repressión. Los Años de plomo del franquismo (The great repression. The dark years of Francoism). Madrid, Flor del Viento, 2000.

xvii   José Manuel Sabín. Prisión y muerte en la España de posguerra (Prison and death in postwar Spain), Op. Cit., pp. 212-213.

xviii   Ángel Hernández Sobrino. "La joven Manuela" (Young Manuela). Article kindly given to Moreno Gómez by the author, April 2013.

xix   Fernandio Mendiola Gonzalo and Edurne Beaumont Esandi (Associación Memoriaren Bideak). "Batallones Disciplinarios de Soldados Trabajadores. Castigo político, trabajos forzados y cautividad." (Disciplinary Battalions of Soldier Workers. Political punishment, forced labour and captivity). In Revista de Historia Actual, number 2, 2004, and in the documentary Desafectos, Esclavos de Franco en el Prineo. (Disaffected. Slaves of Franco in the Pyrenees), 2007.

xx   Pedro Gómez González. Letter to Moreno Gómez from Cordoba, December 16 1986.

xxi   Testimony of Miguel Caballero Vacas, Madrid, September 1979.

xxii   José María Romero Ruíz, Recuperación the la memoria histórica de La Rambla (Recovery of the Historic Memory of La Rambla).Ayuntamiento, La Rambla, Cordoba, 2010.

xxiii   Casismilo Jabonero. Diário del solado republicano Casimiro Jabonero. Campo de prisioneros de Lavacolla. Prisión de Santiago de Compostela, 1939-1940. (Diary of Casimiro Jabonero, a Republican soldier. Lavacolla concentration camp, Santiago de

Compostela Prison – 1939-1940). Ed. Víctor Manuel Santidrián Arias, Ayuntamiento, Santiago de Compostela, 2004.

xxiv Antonio D. López Rodríguez. Cruz, bandera y Caudillo. El campo de concentración de Castuera. (Cross, flag and Caudillo. Castuera concentration camp). Badajoz, Ceder-La Sernam, 2006.

xxv Arcángel Bedmar, Baena Roja y Negra. Op. Cit., pp. 215 et al.

xxvi Manuel Bustos. Interview with Moreno Gómez, Villanueva de Cordoba, June 23 2002.

xxvii Fernando Mendiola and Edurne Beaumont. Desafectos… Op. Cit.

xxviii Ibid. 827 kms, sin retorno. (827 kms, one-way). Creative Commons, Iruñea, 2010. Documentary.

xxix Albino Garridom Una larga marcha. De la represión franquista a los campos de refugiados en Francia. (A long walk. From the Francoist repression to the refugee camps in France). Milenio, Lérida, 2013, pp. 85 et al.

xxx 1944 Report from the Ministry of Justice.

xxxi Francisco Ruíz Acevedo. Memória Antifranquista del Baix Llobregat. El genocidio franquista en Extremadura (Anti-Francoist Written Report on the Baix Llobregat. The Francoist genocide in Extremaura). Number 12, 2012, p. 109.

xxxii Mariano Agudo & Eduarado Montero. Directors. Presos del Silencio. Documentary produced by La Zanfoña Producciones, Canal Sur T.V., March 2009. Uploaded by Intermedia at: http://www.kaosenlared.net/noticia/presos-silencio-documental-fortaleza-historica.

Women's Prisons in Spain

IX

WOMEN IN FRANCO'S PRISONS

**Eyewitness Descriptions of the Penitentiary.
Ideological Repression in the Hands of the Church.
Subjugation of Childhood. 1947-1949 Triennium of Terror.
Final Comments.**

*"You had to live this to believe it. Even though we say it, even though
we talk of it, those who have never been in jail will ever understand
what that was like. Those who have been outside may have some idea,
but you have to have lived in the prison environment, else you presume
that you made friends and that your life was full and happy. Your
companions were called, you said your goodbyes and the next day you
knew that they were no longer alive. That is something that one only
really knows when one has been in jail."*

Tomasa Cuevas
Eyewitness Accounts of Women in Franco's Prisons
1st edition, Madrid, 1982

It is estimated that between 40,000 and 50,000 women, about
10% of the total post-war prison population, were imprisoned
during the hardest times in the 1940s. In reality, many more women
were punished but an exact tally is difficult because of the mass
imprisonment of both sexes and the constant comings and goings
from the one prison to another. In order to obtain a clear picture
of what was going on in the Francoist penal hell, we need more
than academic studies of the penal regulations. We need the direct
eye-witness accounts of those who were imprisoned such as Tomasa
Cuevas's outstanding account, *Testimonios de mujeres el las cárceles
franquistas,*[i] and four or five other such reports. Shocking as these
accounts are, they let us see what differences we can find between
Mauthausen, Auschwitz and the Francoist prisons.

Tomasa Cuevas' book is a monumental compendium of
declarations from women who were incarcerated under Franco, a
type of Spanish *Diary of Ann Frank* written by many authors. The
Diary of Ann Frank has been thoroughly examined and researched

over the years since it was published: this written testament to what women endured in Francoist prisons deserves no less consideration.

As Moreno Gómez reproduces these women's words with extracts from Tomasa's book, this section goes directly to the heart of the matter. Readers are warned that unless he, or she, is able to approach their descriptions from a more dispassionate academic viewpoint, they may find it so distressing that they might prefer to skip this section and go on to the next one, the Role of the Church.

Toxic overcrowding

Salvadora Luque, imprisoned in Illana, Guadalajara, described how they were crushed together, men on one side, women on the other, like a bunch of grapes. The silk factory had been turned into an improvised jail. When they loaded the women onto the truck to transfer them, the crowd wanted to lynch them, calling them murderers, whores, Rojas, etc. [p. 97]

Flor Cernuda was taken from Lillo jail in Toledo, December 28, 1939, and transferred to Ocaña prison, where they put her with eight other women in a cell built for one. As they got in, they had to stand with their arms on their sides and they could not move. Never one to keep quiet, she called out to the guard and told him that there were too many people in the cell. "Too many?" he asked. He went into the corridor and called to another guard: "We shall see. Bring in two more." Instead of nine, there were now eleven women, without any light, without water, without anything. She seemed to remember that during the three months when she was incarcerated in Ocaña prison, they were only taken out into the patio two or three times. [p. 153]

According to Rosario la Dinamitera, seven hundred inmates were packed in a single room in Durango. [p. 178]

Another witness, [p. 275], spoke of Ventas Prison in Madrid where, added to everything else, there was overcrowding, a lack of living conditions, not a single moment of privacy, the total lack of mental stimulus, an absence of human kindness. There were twelve women in each cell. All vestiges of the previous purpose of the rooms had been removed: the building had been transformed into a gigantic warehouse, a storehouse of women. There was a lack of food and water – equipped with a kitchen designed to prepare meals for a maximum five hundred inmates, it was impossible for the prison to feed the thousands of prisoners twice a day.

Antonia García also described Ventas Prison which housed eleven thousand women in that prison designed for five hundred. Naturally, all the toilets blocked up, pipes broke and there was no water to wash clothes. Furthermore, as there were so many inmates, they were only permitted to receive one food package every fortnight. The women were covered in body lice and despondent. Added to these conditions was the fact that the inmates were impoverished, that they had eaten badly during the war, and their families were too poor to send them any food. [p. 322]

Antonia continues to describe the Dantesque scenario [p. 326]:

> "The first day that I was put in the room, the time came for rollcall. When it was over, everyone threw herself on the floor to grab a bit of space for the night. I fell against a wall, and I remained in that position all night. The next day, my legs were terribly swollen, and I had a terrifying image of the kind of life that was waiting for me. In the morning, it was like a painting of Bedlam. Some women were de-lousing their bodies; those who had scabies, scratched themselves. My first reaction was that they were turning us into animals. You could not shower. You went to the toilet, and it was filled to the brim with faeces.
>
> There was no way that we could live like this. Several of us began to talk to the other women and say that we should take turns with the showers. That we should arrange to sleep in shifts so that each would have more space on the floor and that each group should sleep in three-hour turns. That we somehow had to get better conditions for the children."

Systemic torture

The half of Spain that was imprisoned by the totalitarian New State was frequently subjected to the torment of torture, a form of repression that in Spain dates back to the Inquisition. Now, however, the new and improved techniques introduced by Fascism, produced an improved, equally toxic, prisoner management cocktail.

Tomasa Cuevas, herself a prisoner, continues with the testimonials as she reports that in 1945, she was again arrested together with the man who became her companion and husband, Miguel Núñez. His spine was seriously damaged due to the beatings and mistreatment that he received from Commissar Polo in Barcelona during endless interrogations, a lesion that he still suffers from to this day. [p. 13]

Blasa Roja refers to Guadalajara prison as she remembers the moans and groans from the beatings given the inmates. She states that she will never forget what they did to an inmate called *El Chinés*, a bonny young man with black hair when he was arrested, who was so badly beaten, so martyrized, that eight days later his hair had turned white. [p. 77]

Everyone was deeply affected when Juan Raposo was beaten to death in Guadalajara and she could not forget the way they beat the man, his screams, the way they murdered him.

> "He was killed like Jesus Christ, tied to the iron prison doors and whipped". [p. 78]

Nieves Waldemar, also from Guadalajara, opened her heart to Tomasa Cuevas:

> "What can I say? All six of us, brothers and sisters, were in jail. One of my brothers was sentenced to death, although he never harmed anybody. Another sister spent ten years in jail: first for five years, after which she was released then rearrested and sent to Governación jail in Madrid, for three months' punishment: a beating in the morning and another in the afternoon. She was so badly injured that she lost her mind and for the next ten years, did nothing else but sit listless in a chair, taken care of by her husband." [p. 94]

Another witness in the same jail described how they would often spend hour after hour, with their nerves on edge, hearing the guards beating the men and the men's screams. There were moments that they could not avoid suffering and crying out with rage... [p. 94]

Carmen Machado, from Madrid, told what happened at Police Headquarters at Calle Jorge Juan number 5, under the torturer Aureliano Fontela. The entire time that she was there was a nightmare because of the horrific beatings. One case, specifically that of a Youth Leader from the Cuatro Caminos neighbourhood who was so mistreated that they set a bed up for him under the stairs. During the day, a doctor was sent to treat him and at night, they would take him again for more beatings. He was later executed. [p. 129]

Flor Cernuda from Quintanar de la Orden, a Durango inmate, described a young girl from Ciudad Real, María Fernanda, who slowly became totally disabled and unable to walk after they broke her back during a beating. [p. 155]

Pilar Calvo, from Talavera de la Reina, recalled how her husband, Julián López, from Santa Olalla, suffered after he was arrested in Madrid until he was executed in Ocaña prison where he was terribly badly treated. They attached electric wires to his private parts and did many more terrible things he did not deserve, but he had been an active supporter of the Communist Party and had served in the 5th Cavalry Regiment. [p. 160]

Mari Carmen Cuesta, member of the JSU in Madrid, was a friend and survivor of the famous *Thirteen Roses*. Like those women, she was taken to the Calle Jorge Juan Police Headquarters where she was interrogated every fifteen minutes, night and day, so she could not sleep. As soon as she started to fall asleep, they woke her up. Three or four days later, they first heard terrible, blood-curdling screams as the guards began plunging companions into ice baths and giving them electric shocks. [p. 199]

Agustina Sánchez Sariñena, from Madrid, went through her own private hell. Her mother-in-law was executed July 24 1939, her husband was beaten to death in the Gobernación jail, and she miscarried after being tortured. When they arrived at the Gobernación, they were not allowed to speak.

> "They told me: "You bitch, speak and we will kill you." The interrogations began and were terrible. Beatings and communion hosts came one after the other, without considering that I was pregnant, and if what they were doing to me hurt a great deal, my

suffering was three times worse because they beat my husband and a friend in front of me.

Half conscious, I became aware that they called the doctor, and he told the policemen: "Don't interrogate her any more today, leave it for tomorrow, because her mouth is too dry now". I did not know that we need our saliva to talk and what happened to me was that my mouth had gone dry from what I was seeing. I could not even feel my tongue. The next day they took me to some special cells, and they really began to beat me hard. They had stopped beating my husband, but he was more dead than alive and had begun vomiting blood.

I was left alone in the cell, bleeding copiously, but the foetus was not coming out. I no longer felt the baby move and I was afraid that it was dead, as it was. The doctor came and urged them to transfer me to Ventas Prison as soon as possible. When they came to get me, I said I would not leave until I saw my husband. They refused, but when they saw how much in despair I was, they said: "You can see him, but it will be the worse for you." I saw a bundle in a corner that looked like an old man with grey hair. I heard someone call out weakly: "Agus, Agus, it is me." I turned towards him. He was no longer a human being, just a bloody lump who could not move except to hold out hi arms. My last hug ever." [pp. 236-237]

Returning to <u>Antonia García</u>, she described what she witnessed the day that she was arrested and taken to Calle Núñez de Balboa where she was put in a cell. Sometime afterwards, the door opened, a light was turned on and six or more men brought a woman and a young man in. "They began to beat the young man, so much that he began bleeding from his mouth and his nose. The woman cried: "My son! My son!" and she turned her face to the wall so that she could not see what was happening. Finally, she threw herself against one of the men and scratched his face badly. They shoved her and she fell against the corner of one of the stone benches. As she hit the bench, one of her eyes popped out, I never saw that woman again, not even in the prison I don't know whether they killed her." [p. 324]

This section concludes, continuing with <u>Flor Cernuda</u> from Quintanar, Toledo, who was interned in Lillo prison and whose testimonial is proof that Toledo Falangistas were indeed no more than some kind of animals. It also shows how moral damage resulting from mistreatments can ruin a person's life.

> "In the end, my father was left alone helpless and there I was in prison, suffering and watching my fellow inmates suffer, women who were treated horrendously as were their husbands. Every alarming and terrifying thing you can possibly imagine. Relatives who came to visit us were made to stand in line and were occasionally beaten which a whip.
>
> One day they arrested my mother, a person who would not say two words out loud in order not to offend anyone. According to her death certificate, she died from 'moral debility'. In other words, the doctor recognized that she had not been suffering from any disease; what she had was emotional pain that she could no longer endure. They say that you cannot die from your emotions, but you can. My mother saw so much, and this upset her so much that she only lasted a few days." [p. 152]

Evidently, the great repression targeted not only individuals of some 'political importance' as it did all sorts of inoffensive country town and village residents. It was the great repression against 'all those who do not think like us'. The Regime mistreated all the disaffected and even those who were indifferent to politics, The sole purpose was to ensure that the terror that upheld the basis of the Regime's power, would permeate every humble working man's and every Republican's home.

Every kind of humiliation

<u>Carmen Machado</u> described how everyone in Venta Prison was routinely lined up in the patio to sing hymns. If you could avoid it,

you did, by trying to hide in some room. There was a prison officer, a tall, good-looking man, who whenever he was on guard went from room to room with a thorny stick, looking for inmates who were, or pretended to be, sick. Without saying a word, he would poke the thorns against their legs to force them to get up. He was always very watchful when it came to singing the hymns. If he caught a woman who was pretending to sing or just was not singing, he would take her, together with other he had caught *in flagrante*, to the middle of the patio and make them stand with their arms raised in the Nazi salute and sing the hymns over and over again until recreation was over, when he would again poke against their legs with his thorny stick. [p. 134]

Tomasa Cuevas, wrote of Guadalajara where many women in the patio, particularly young women, had arrived with shaven heads. The Guadalajara inmates had a full head of hair. One day, a group of prisoners arrived from a country town, and they were all as bald as billiard balls. The Falangistas found all this hilarious. Laughing, they said that the law was the law, like it or not: either every woman had a shaven head, or they all had a full head of hair... and as you cannot glue hair back on... They called several of us out and, after setting up two inmates who were barbers by profession in a hallway, put them to work shaving our heads. [p. 106]

Flor Cernuda told how the Falange in Quintanar de la Orden, Toledo, celebrated the first days of the victory.

> "We were taken to the local jail and the first thing they did was shave our heads and insult us – there was no end to the swear words and invectives they threw at us, April 2 or 3 1939. In the morning, once our heads had been shaven, they arrived and told us they were taking us to the town square, to hear Mass. They said they had intended to burn us at the stake in the square after Mass, but the day before, a group from the African Harem had come to town and had stopped them. They threatened us with their pistols and rifles to force us to attend Mass, they insulted us at will – both the local priest and the Fascists." Flor also said that, when she was in Ocaña after the end of 1939, the humiliations continued. There was an old prison guard who suffered

from tuberculosis, called Don Marcelino, who called out "Come up whores", each time the women were taken from the patio to their cells. [p. 153]

Pilar Calvo, from Talavera de la Reina, spoke of humiliations and, especially, the great social vacuum that enveloped the defeated families. She went to Santa Olalla because her husband had been arrested there. His family told her that she had to leave because there had been so many threats against them, they were terrified of what could happen if she stayed. She left with her infant daughter in her arms. On the way out of town, she met a man who was riding a donkey and he called to her and asked where she was coming from. When she replied, "from Santa Olalla", he said: he knew who she was: Bicho's wife, a Communist and an evil Roja. He asked her how many she killed. Pilar replied that she had killed nobody, that she had an ideal that she would believe in all her life. She could not deny it, because it was deep in her. He then told her to hand him her daughter; she could ride on the donkey, but Pilar would have to walk. Better still, he said he would tie her to the donkey's tail and drag her all the way to the railway station. As he ushered her out of town, he swore incessantly at her and said that they would kill her if she returned. [p. 159]

In Ventas Prison, Agustina Sánchez said that one morning they called all the women to the judge's chambers where they were greeted by a group of young Falangistas who shaved the women's eyebrows and their heads. Her eyebrows never grew back, her hair only a bit, and some of them remained bald for the rest of their lives. [p. 228]

Women in Ventas Prison, Madrid[ii]

Hunger as an instrument of submission and extermination

Of all the painful hardships prisoners had to endure in Francoist prisons, hunger was the worst. <u>Pascuala López</u>, from Brihuega, Guadalajara, described what occurred to her when they were attending Mass, the day after her niece Lola was executed. She was too weak to kneel because she was so distressed by what had happened to Lola. One of the nuns, Mother Gertrudis, insisted that she kneel but she could not. Although Pascuala totally lost her appetite after Lola died, she was still so very hungry because they were not given enough to eat – a pot of water with an onion skin floating in it and some days, nothing at all. She might occasionally eat an orange that they sold in the Commissary – first the orange, then the orange peel and, because of that, her stomach was a mess.

<u>Blasa Rojo</u> told Tomasa Cuevas that she saw such terrible suffering for eight years. In Amorebieta, she watched women die of hunger. She remembered that those who arrived from Santander, at the same time as Tomasa's group, called them the yellow people as their skins were yellowish from not being allowed out in the patio and because they were dying of hunger. She remembered a casket with the body of a dead prisoner from Asturias in the patio. Every day, the casket left the prison on a trolley. They removed the casket, took out the body, came back and prepared the casket to receive the next one. [p. 78]

The 2010 Basque television documentary *Prohibido recordar* (Remembering is Forbidden), describes the high mortality of women in Francoist prisons. In Santurarán, there was typhus, scabies, tuberculosis and diphtheria. When the inmates attended Mass, they might hear a thump – a woman had dropped dead. An inmate might be taken to the infirmary and never heard of again. A donkey cart would arrive at the prison, and it would be laden with caskets with the bodies of children and adults.[iii]

Saturrarán, one of the sadly most notorious Francoist women's prisons, is where they imprisoned women classified as 'extremely rebellious and dangerous', which was a lie as these were women who had done nothing more than exhibit leftist sympathies. More than 2,000 women between 16 and 80 years of age were imprisoned there from 1937 to 1944.

Returning to Tomasa Cuevas' Diary, <u>Nieves Waldemar</u> talks of her internment in the Convent of the French Nuns women's prison in Guadalajara. While she was there, she became friendly with several inmates, women who received no help from anyone as they were country people whose families had also been punished and left destitute. The food they fed the inmates was pretty awful: an onion boiled in water, and they called that food... [p. 92] Those who were unable to receive any help from outside suffered even more, because all they got was water; they either fell ill or died.

Nieves remembered that one day that they brought the food, a cauldron that they placed on the steps that led down to the patio and from which the inmates were served. Her sister filled her plate, placed it on the stairs, and in front of the prison officers who were serving the meal, proceeded to use this 'water', because that is all it was, to wash her face! For some unknown reason, she was not punished for that. Since then, Nieves was operated on four times, because the life they lived in prison was no life at all. In the end, she said it is not a matter of talking or not talking about all this, but of shouting loud. [p. 95]

<u>Tomasa Cuevas</u> complained that in Guadalajara, hunger chipped away at their bodies and there were inmates who were so weak, they fainted. The doctors spoke with the prison administrators and told them that the situation could not continue in this way and that they had to allow the inmates to receive food packages, otherwise people would start dying. But the administrators could not care less as this is exactly what they wanted. Eventually, the doctors managed to get them to allow food packages for the neediest. In order to qualify for the packages, inmates had to have a medical examination and not all were allowed the packages. The male prisoners were worse off, so the women, with the help of several belts they tied together and a satchel, sent them some food through a window. [p. 103]

Tomasa continues to describe her transfer to Durango where she joined other prisoners in the Convent of the French Nuns. They had a hard time as the food was terrible. The building was three stories high, and the kitchen was on the ground floor. If they wanted to eat, they had to go down to the patio with their plate to get their food. It was icy cold, and they were served a disgusting mash that stuck to the plate if you turned it upside down; it looked more like carpenter's tar with which to glue their stomachs. The oldest inmates suffered badly. Some broke down, others went five or six days without a bowel movement. Batches of new inmates arrived daily from all over Spain. There just were not enough jails in the country for so many prisoners and many convents were turned into improvised jails.

From Ventas Prison, they sent a batch of youngsters, many of whom were minors. They taught her this song: "People of Spain/we the prisoners are calling you,/this injustice/cannot continue/as hunger, a widespread evil, is doing its damage..." [p. 120][1]

Carmen Machado also had something to say about Durango prison where there was another scourge, avitaminosis: lack of food and little variety. For example, the food in Durango was almost always only rice. Because of this, the women's legs began to be covered with sores full of a watery liquid. They had to set a room aside for the enormous number of women afflicted in this way. There was no food, there was no cleanliness, but of hymns there were plenty and they had to sing them at least twice a day. [p. 134]

In 1941, Antoñita Hernández, the most head-strong of their group when it was time for their meal, would collect the orange peels that the Basque women kept for her, and she would return with her plate piled high with orange peels. Emilia, their communal mother, would divide them into eleven little piles and each of the women would go into her corner and relish the peels. [p. 135]

The famous Rosario la Dinamitera described her experience with hunger in Saturrarán prison. This prison had the inconvenience that the sea air opened their appetites, and the food was awful as it was everywhere and there was even less of it. The women were put to work in the kitchen to peel carrots and a nun would stand on a stool to make sure that no woman even attempted to put a tiny piece in her mouth. What kind of religion was that? What kind of Catholicism did they have in Spain? Rosario said that every woman who passed through Saturrarán could confirm the truth of this. [p. 135] Saturrarán was an icy cold prison... the inmates suffered so much from hunger that many women's' legs were severely damaged... because all that they were given to eat was some kind of mouldy flour full of bugs." [p. 332]

---

[1]   There is a magnificent book and documentary by María González Gorosarri and Eduardo Barinaga, entitled *No lloréis: lo que tenéis que hacer es no olvidarnos* (Do not cry. What you must do is never forget us). Ttarttalo, Donostia, 2008. The documentary is entitled *Izarren Argia*

Augustina Sánchez added some information regarding her stay in Palma de Mallorca prison. The food they gave the inmates was disgusting, nothing more than boiled marrows and, sometimes, a few noodles. [p. 232]

Antonia García also speaks of Palma de Mallorca where she says it was so terribly windy and so dusty that their hair was always covered in a chalky kind of dust. The water for the showers came from a well where the water was so chalky that their bodies were white when they finished washing. They were fed some greens that the nuns cultivated, and they were full of ants. You have no idea how bitter those ants tasted. As their families were usually unable to send any food packages, they began to become terribly weak, so much so, that many of them could barely move. It got so bad, that the vertebrae in their bottoms stuck out so much they could not sit. In that prison, you died standing; during every roll call, some would fall down, dead. [p. 330]

Tomasa Cuevas ends with a comment that there was so much hunger in Segovia prison, that many young women no longer menstruated. The floor was cemented and not tiled, and it was so, so cold that you felt it even more than your hunger. [p. 293]

Transfers of inmates, so-called penitentiary tourism

The continuous transfer of prisoners from one jail to another, similar to the practice in Nazi Germany, was another instrument the oppressors used to punish, when not to kill, the unwanted. As Pascuala López, from Brihuega, relates, they were transferred up to four times during the winter in cattle wagons, because the longer it took for them to complete their journey, the more the guards were paid. The trains would be left standing in the railway stations for as long as they liked. In some villages, where a train was left standing on the line, the locals would throw some food through the netted window, but elsewhere, they treated the prisoners badly. There was a bit of everything: some felt sorry for the prisoners, others did not. [p. 71]

Cecilia Abad, from Guadalajara, describes her penitentiary tourism:

"I was sentenced to twelve years and one day. I spent four years in prison and four years exiled in Zaragoza, then in several jails in Guadalajara, then to Ventas

Prison. From Ventas, I was sent to Durango via Burgos, where we spent the night. We were put in with all kinds of women and given no food or water. From Burgos, we went back to Durango where I spent six months. The prison was closed down and we were sent to Santander; I was there during the hurricane and the fire, for a year and a half, when they closed the prison and we sere sent to Amorebieta. I was finally released from Amorebieta in 1942 and sent home." [p. 83]

Tomasa also describes the first of her trips from prison to prison. Guadalajara, Madrid, Durango. December 28 1939, only four from each cell were taken, one of them the poor old Letona. The inmates were handcuffed two by two. In one hand they held their little bags and very little in the other hand because they were attached to a companion, and then they were sent into the street. Her father, her mother and her brother were outside, in front of the prison, but the Guardia Civiles prevented them from speaking to each other. She was marched to the railway station with the other prisoners, where she saw her sister, her brother-in-aw and her nephews, but not her father. She was unable to ask about him because the guards would not let them get close to each other. She later found out that her father went home as he could not stand seeing her handcuffed and surrounded by Guardia Civiles. She was not told that actually, he had suffered a stroke. He had reached the end of his tether. Her father died a year later, always asking after her.

A carriage had been set aside for these inmates in Guadalajara station. There were girls from Sierra villages who had only left their hometowns to go to prison. They had never seen a train before and had no idea how one worked. When the prisoners arrived in Madrid, at Atocha Station, they were loaded onto a Guardia Civil truck and taken to the North Station. It was terribly cold, there were roadworks all over the place, there were no doors in the waiting rooms, and everyone was frozen. They were not allowed to get onto the train for four or five hours as they waited for a group of three hundred and fifty women to arrive from Ventas. [pp. 117-119]

When the prisoners finally arrived at Zumárraga, they had been travelling for three days. As they had been given food for 24 hours and it was already 72 since they had left, they had gone without food for two days. When the residents of this town heard that there was a train full of women prisoners, they brought them some food. They arrived in Durango, in the Basque country, having taken four days to get to the cold, damp prison in the North, from the warm Centre-South of the country.

The lack of health care

The lack of health care was one of the great shortages in the Francoist penitentiary world and still another lethal tool for decimating what they considered were no more than Rojo rabble. Carmen Machado, from Madrid, who had been travelling from jail to jail since she was 20 years old, as so many other women prisoners whose youth Francoism stole, had an interesting tale to tell:

> "There was a very curious case that illustrated just how little they cared for our health and well-being. Just a few days before the Durango prison was closed, we were vaccinated against typhus and just a few days later, sent to Orúe prison, Imagine that although this prison knew that we had already been vaccinated, they repeated the injection as we arrived. As one would expect, many had very high fevers and were frankly very ill. They really cared little for your health." [p. 134]

Flor Cernuda told Tomasa that she did not remember the exact date when they were told that they were going to be transferred to Durango. Her legs were already bad when they left Ocaña, and they had to wait for a long time at North Station until the trains arrived. They were made to walk all over the place, get on a train, get off it, change tracks, so much so that she could not make it, not even sitting down. They had to push her into the carriage with her legs apart and she developed sores on her legs, leaving scars that are still there today... As she could not walk, she was taken into Durango prison on a stretcher. She was not treated because there was nothing with which to treat her. The infection went up legs, right to her hips; she was terribly ill but there was no medicine chest, no doctor, nobody to treat her sores." [p. 154]

Other situations where there was an absolute lack of hygiene, were the standard fare. <u>Rosario la Dinamitera</u> described Getafe jail, near Madrid:

> "Worst of all was Getafe and that is where I was sent. In that prison... there were no toilets: all your body functions had to be done in cans, in the three or four rooms, without doors, where we were imprisoned. Once a day only, at 7 a.m., they came and removed the shit cans. There was very little food and what there was, was awful, but worse still, was the lack of hygiene. There were no washbasins, nor running water for us to wash in. We got water in a can of condensed milk, that is, a cupful per person and that had to last us 24 hours. We could not ask for more, not for the elderly, not for the children." [p. 177]

<u>Tomasa Cuevas</u> told of Ventas Prison for Women in Madrid where medical assistance was totally absent. It was only in 1943 that they appointed a gynaecologist. Until then, the doctors were, heaven know why, practicing dentists. Dr. Delfin Camorredondo who is best forgotten, was the dentist on call in Ventas in 1939. Despite this, there was one exceptional individual, Dr. Juan del Cañizo, a doctor from Segovia, whose name honours all doctors and who remains unforgettable for his self-denial, interest and kind-heartedness. Once the prison hospital was finally set up in Ventas, Dr. Castrillón, a specialist in treating tuberculosis, was another excellent doctor." [p. 277]

There is the story of Tomasa's companion Rosita who arrived from Malaga, already sick. She was taken to the hospital in Segovia, a fake tuberculosis sanitorium, where she fell into the hands of an unscrupulous individual, Dr. José Luís Canto, the nephew of a high official in the penitentiary system. He never examined a woman, never even took her pulse. He would make his rounds wearing a hat and overcoat, his hands in his pockets, looking at the skeletons who lay in the beds, with revulsion and loathing. The only thing that distinguished this hospital from an ordinary warehouse were the very large, bare, icy cold rooms with beds.

More than anything, Rosita was hungry, as we all were, and this young girl's black eyes were terribly sad. She was a tall brunette, with a face that was so pale that it stood out amongst all the other greyish, pale faces. Because she was still underage, she had been condemned to twenty years. Both her parents had been executed: first her father, then her mother. Back home, just two young brothers. About to come of age, Rosita was totally helpless, yet another of the Roses who had no youth, victims of the oppression of the barracks, the Casino and the vestry. [p. 286]

Tomasa continues with her description of the Segovia prison hospital:

> "There were four large, cold, tacky wards devoid of everything except a few of the patients' meagre belongings. Added to this, the very real illnesses of most of the women, some of whom were considered to be such a nuisance that the staff tacitly hoped that they would hurry up and become even sicker and die once and for all. Of course, this type of patients were nightmares for the hospital administrators[Gusen now comes to mind] who sent them back to where they had come as soon as they could.
>
> The prison conditions, the lack of food, the total lack of medical assistance – the doctor was 'allergic' to these women – and the previous lack of all health care for the patient meant that calling this place a 'sanitorium' was no less than a bloody joke. Furthermore, the theft of the few 'extras' that had been allocated for the patients was highly profitable for the Sisters of Charity community of nuns who ran the wards. This has been proven because we located the procedure for delivering the food stolen from San Sebastián." [p. 290]

Paz Azati, from Madrid, stated that until 1945, the situation in the prisons was beyond belief in all aspects: hygiene, food, illnesses. All of the inmates were swollen, the doctor would not give them any medicine and there was an illness that the doctor said affected all women in jail and that was that they stopped menstruating, which led them to swell like balloons.... They also suffered from avitaminosis. Then there were problems with the water.... The women had to cut their hair because they could not wash it. In spite of all this, it must be said that Spanish women, in the midst of tragedy, surrounded by deaths, executions, beatings, lice, bed bugs, hunger and so much more, nevertheless managed to keep their spirits up and never allowed anyone to debase them, nor has anyone every done so. [p. 357]

Petra Cuevas, from Orgaz near Madrid, gave birth in jail (New Prison for Mothers, next to the Manzanares), to a daughter who was born without difficulty, weighing 4 Kgs. Sadly, she gave birth in a room in which a little girl had died of whooping cough, and it was not surprising that her new-born daughter caught the disease. She had been placed in the same bed from which they had just removed the sick baby's body and had not changed the sheets. Lacking medical assistance, the baby worsened every day; she coughed all night. One day she heard that a doctor was coming to examine a child, La Topete's goddaughter. Petra took advantage of this to go to the infirmary unannounced, but they closed the infirmary door in her face and refused to speak to her, no matter how much she banged on the door insisting that the doctor see her daughter. Finally, the ward staff informed her that there was nothing wrong with her daughter and that was what the prison doctor was for. Four or five days later, as the baby got sicker and sicker, Petra returned to the infirmary and told them that there was nothing more that she could do. "Can't you see that the child is dying?" All the prisoners began calling out in protest. The next day, when Petra took her daughter back to the infirmary, they called four doctors in. By the time they arrived, she was already dead. [p. 369]

The general destruction of the family

Francoism placed special emphasis on de-structing and dismembering the Republican families by breaking them up either in the jails, through exiles, or before the firing squads, all with a view to destroying all the links forged by working men's families and the transmission of their ideals. As we shall see later, this destructive process culminated in snatching the children from their families and bringing them up according to the Regime's philosophy in foster homes, in the hands of the Social Welfare Commission and in religious institutions. First, however, there was the appalling treatment of mothers and their children in prison.

It is said that the Regime had a strategy for the wholesale ideological repression of the defeated, but if the truth be told, it was a program for pure and simple unvarnished oppression. This was no kind of 're-education' of disaffected individuals; just new, refined methods for subjugating and repressing those who supported the Republic until no traces were left of their ideals, their family connections or their links with others of their social class. The Regime and the victors knew perfectly well that the most direct and efficient way of de-structuring these families was to arrest and imprison mothers, without whom families would crumble without remedy.

Pascuala López, from Brihega, remembers a mother who was executed, and her children left uncared-for.

> "Many, many of us were sentenced to death. The number of women they took out with the sacas! They killed Señora Antonia, from Yunquera. Did you know her? She had eight children. I remember that she would say: 'Pascula, killing me, a mother of eight children. They are such criminals! Murderers! They cannot be forgiven. If you sometimes meet one of my little boys, could you wash his shirt?' I never me them, poor little ones. I wonder what happened to them?" [p. 71]

You could write a book about how they destroyed Bala Rojo's life. They destroyed her life, but they did not kill her, unlike Paca and Gregoria, who were from the same town, as she describes. "You ever know", she said, "what the last reaction of a defendant will be like", because she saw how brave Señora Paca, from Auñón, was the night they took her niece, her daughter-in-law and her daughter out to be shot... "She was seventy years old when they executed her. Yes, and there was Gregoria who was married to one of the revolutionaries in Auñón; her husband was quite a man. He had been quite a fighter for many years, but she was not involved. She just cared for her home and her children. They killed them all that night. Poor Gregoria had three young children, the oldest of whom was only eleven, imagine that, and you could see the poor kids walking, getting rides on the roads. The eldest came and he brought the other younger ones to Guadalajara, begging along way until they reached the Convent of the French Nuns that had been turned into a prison. They went up to a window and called out 'Mamá, mamá!' The poor kids were half barefoot, and their noses ran with snot. Gregoria cried that she would kill herself. 'Is there no human being who will take care of my children? Is there nobody who can get them into *La Inclusa*?[2] How are my little children? They are going to kill me before I know what is going to happen to them...' After she was executed, the Municipality arranged for her children to enter the hospice."

Blasa continued, bemoaning her own misfortune.

> "They destroyed the best thing in my life. When Raimundo [Serrado] left, I was twenty-six years old. I was imprisoned and released eight years later, totally devastated, without a home, without clothes, without money... When you have nothing, you bother everybody, even your mother, the poor old lady. I left prison and went to one of my brothers' house, but I saw that I was an inconvenience, so I went to another brother's house. The only thing that kept me going was finding out what had happened to my husband. I soon found out that he was living with another woman with whom he had made his life. I was the first Communist woman in Guadalajara... During a war, everyone is a Communist." [pp. 79-80] Tomasa Cuevas added: "War does not only kill. It also destroys homes without killing."

Another prisoner from Guadalajara, Julia García Pariente told how they arrested her mother, her sister and her husband and they all met up in prison. She had another son who was hiding at his grandfather's, Vicentes Montes, but as he was a Fascist he made the

---

[2] *Hospital de La Inclusa.* Hospice for abandoned and orphaned children.

boy's life impossible, so he left. The poor child asked the nuns in the Inclusa Hospice if he could stay there, and they let him in. He stayed there until his mother was released from jail when she took him home with his brother Tomasito who was born in jail." [p. 85]

Tomasa Cuevas used her Diary to help alleviate the pain of her troubles. She recorded how she found out about the many difficulties that her family was enduring, despite the fact that her brother was living with them. His son had died aged 2 and he also lost his wife. He also told her that their father would spend every dawn in front of the cemetery where they carried out the executions. The prison officers had frequently told my parents that if they were Catholic, they should pray for me because one day they would take me out to be shot. So, the poor old man spent many months in front of the cemetery, watching the firing squad at work, hiding amongst the bushes outside. His nerves must have been destroyed by those daily spectacles. That was his daily routine: prison, home and the cemetery. He went to the prison because it was there that he felt close to Tomasa even though they couldn't see or talk to each other. Still, he gained some solace by going to the back of the prison that gave onto the fields, and sitting down there he remained until nightfall, when he went home. [p. 112]

She continued talking of her father. The day that she was taken from the jail for her court martial, August 27 1939, they were taken out handcuffed, as you would expect. As only Granny Letona and Tomasa were removed from their cell, they were cuffed together. Letona was such a painful sight, a terribly frail old lady who had done nothing they could accuse her of, but they still court-martialled her (she was accused of having fried some eggs for some guerrillas and was condemned to twelve years and one day). When Tomasa walked out of the prison, she saw her father. He was alone and his eyes were shining; he must have cried when he saw her handcuffed and also when he saw the little old lady who was attached to her. Her father walked behind them as far as the building where the court martial was sitting. [pp. 115-118]

When Tomasa spoke of the Saturrarán prison, located in a Salesian Monastery, she recalled Daniela Piazo, a long-standing Communist whose entire family belonged to the Party. Her husband and one of her sons were executed and she had no news of her other son who was listed as missing in action – had he died at the end of the war? She and her husband had previously been arrested in 1917 for participating in some strikes in Madrid and Guadalajara. As the authorities had a detailed file on her, Tomasa never understood why they did not execute her as well. [p. 181]

Isabel Huelgas' was another family destroyed by the Regime whilst she languished in Ventas Prison under sentence of death:

"Very elderly, weak and febrile, she lay on a makeshift bed, more like a gurney, placed in the corner of a hallway, between the schoolroom (packed with women) and the entrance to the third hallway on the right. She was so ill that everyone hoped, somewhat perversely, that with a bit of luck, she would die before they executed her. During the war, she had lost her husband, a doctor, and her two daughters, from tuberculosis. She presumed that both her sons, Antonio and Joaquim, who had fought on the front, had been taken prisoners-of-war as they were imprisoned in Comendadoras. At five o'clock in the morning on a day in Spring, Isabel was called to the office. An ex-prisoner, Pilar Millán Astray, had come to tell her that both her sons had been executed. She was helped back to her corner as she was too weak to make it back on her own and as they lay her on the bed, they noticed that she had died. They removed her body that evening." [p. 281]

Tomasa was released from jail in Segovia, on parole, and exiled to Barcelona, far from her roots and family in Madrid and Guadalajara. She said that when she was released, she was concerned about a friend of hers, a member of the JSU, who had also been exiled to Barcelona. 'Young Bene', as she called her, just 19 years old, had spent 5 of those years in jail and had nowhere to go. Her mother was in Ventas Prison infirmary, sick with tuberculosis, and there she died. Her father and one of her brothers had been executed; another brother had lost a leg during the war and was selling black market tobacco at the steps to the Metro, to make a living. She had another younger sister who was in a children's home. A family of fighters, one more of so many others like it all over Sain, heavily punished by the victors. [p. 268]

After all that, there is little more that one can say other that this is an extremely painful illustration of Francoism's effective destruction of left-wing supporter families. Moreno Gómez has already spoken of the intent, the punishment and the physical repression of the disaffected and of the ideological repression and radical cut in the chain of transmission of class ideals, all of which the above is an example. The Regime killed the parents or reduced them to penury, so that it could take their children and place them in hospices, foster homes or even put them up for adoption where it could more easily brainwash their young minds. Tomasa's Diary continues...

Women and children in Ventas Prison for Women, Madrid.
European Observatory on Memories
University of Barcelona Solidarity Foundation[iv]

Local Basque solidarity with the prisoners

Almost all the reports that we have refer to workers prisons in the Basque Country. Several times, in Tomasa Cuevas' book, we find references to the great disaster of children in women's' prisons (a tragedy that is addressed in great detail further ahead). Although there is no date, the following is a testimonial from the Basque Country.

"The townspeople of Durango were very supportive. They came to speak to the Director of the prison, and they told him that they would take the children in with them until their families could come and get them, and they took all the children over two years of age into their homes. They cared for them and fed and dressed the children very well. On visiting days, they took them to see their mothers, until little by little the children disappeared from the town as relatives or friends of their families were able to come and collect them." [p. 121]

Some time later, when the Convent of French Nuns in Durango prison closed, some of the imprisoned women were sent to Amorebieta, others to Saturrarán, Tomasa Cuevas and others, to Santander. They day that they all left the prison, on their way to the railway station, there was a wave of solidarity of such a kind as few had ever seen.

> "The departure to the station was extremely moving as it seemed that all the townspeople had agreed to be present. Everyone wanted to give us packages, everyone wanted to say goodbye. Of course, they could not come close to us because the guards would not let them, but they were still allowed to give us their packages. We began to sing, and we sang as we got on the train and until it left the station, as the women of Durango cried. The song went like this:
> 'Fare thee well, Durango, Durango of my heart. Fare thee well Durango of my heart. When I am free, I shall see you again. I do not leave because of the townspeople; they are good people, good people. I leave because I am being taken to another prison'."
> [p. 281]

Although the Basques flew the banner of solidarity very high, elsewhere the women were called whores and the men scoundrels and assassins. What happened in Durango was extraordinary indeed, as other inmates such as <u>Carmen Machado</u> said:

> "There is something I wish to make clear, something that has marked me a lot, something that has made me love the Basque people. In Durango, there was a solidarity that I, in Madrid, perhaps because of the astounding repression, I had never before experienced."
> [p. 135]

and also, <u>Pilar Cernuda</u>, from Talavera de la Reina:

> "From Santurarán I left with the family that had taken me in. It was in the Basque Country that we

were received the best. I live there, in Eíbar, and since then I have struggled a great deal." [p. 160]

When Ángela Mora, from Puertollano, was released from Durango, she took a tiny train from Durango to Bilbao. She left the prison wearing a very long coat and realized that fashions had changed. Carrying her bag, she bought a ticket with the money that her companions had collected for her. In the train, everybody looked at her and asked: "Poor dear, are you coming from the prison?" A man and a woman got up and they went to everyone in the carriage asking for some food and money for her. She didn't dare take it, but they insisted. "This is for you, so that you will have something to eat for the first few days". You cannot imagine how moved she was to see how, since she left the prison, she was sheltered by a chain of true solidarity. She cried but could not eat. They bought her ticket onwards at the station. Another group of people gave her some bananas and wrapped in a bit of paper, some money they had collected, saying: "Have a good trip". [p. 165]

Angelina Sánchez Sariñena, from Madrid, recalls that although everywhere they were put through a lot, it was in Palma de Mallorca that she found much more solidarity. The people from Palma helped the imprisoned women so much so that they set them fresh fish every day. Although it was a gift, to supplement our diet, the nuns sold it to us in the canteen. [p. 281]

Tomasa Cuevas describes her arrival at the Bilbao provincial prison, in the pouring rain, carrying their bags and surrounded by Guardia Civiles. They shall never forget, she says, the emotional memory of those groups of people who, right in the town centre and under a violent downpour, pushed at the guards, yelling at them because they were making us walk, soaking wet, carrying the numerous scruffy bags that were all that we owned. [p. 291]

Lastly, Antonia García, declared how "In Saturrarán, almost all the food we got was brought in by the Basques who were wonderful with us. They brought us work, they would sell what we made and spend what they got for it on food for us." [p. 331]

ENDNOTES FOR CHAPTER XI

[i] Tomasa Cuevas. Testimonios de mujeres en las cárceles franquistas. (Eyewitness Accounts of Women in Franco's Prisons). Instituto de Estudios Altaragoneses, Huesca, 2004. (1st Edition, Casa de Campo, Madrid, 1982).

[ii] https://encryptedtbn0.gstatic.com/images?q=tbn:ANd9GcTzKgFX8nnSP1rYI-uleOGwHwV7Vy3jpyqDSg&usqp=CAU. Obtained from a Google search

[iii] Basque TV documentary. Proíbido Recordar (Remembering is forbidden). Tentazioa Rec, Moztu, País Vasco, 2010.

[iv] https://europeanmemories.net/activities/franco-4040-las-ventas-the-history-of-a-womens-prison-1933-1969/

X

GREAT IDEOLOGICAL REPRESSION IN THE HANDS OF THE CHURCH.

Religion as a form of blackmail. False accusations: confess or die. The role of the Church as a manager of the slave labour program. Subjugation of Childhood. 1947-1949 Triennium of Terror. Final comments.

> *"Whose are those men who, after kissing their children goodbye in the morning, would go off to work as model employees, to torture and murder political prisoners? How did the hired assassins feel and what did they think? The authors of this extraordinary piece of research state that these were normal individuals, delinquents when immunity allowed them to break all bounds."*
> **Baltazar Garzón & Vicente Romero**
> *El alma de los verdugos,* January 1 2008
> Book and TVE documentary

One of the features of genocide described by Raphael Lemkin in 1944 refers to the ideological repression of the subjugated that is, the imposition of the oppressor's (national, ideological) identity on the repressed. In *Axis Rule*,[i] he often referred to groups who lose their identity and are forced to take on the identity of their oppressors. National Catholicism did exactly that in Franco's victorious Spain. It put into action a radical ideological repression (ideological genocide) to destroy all of the defeated's Republican or working-class ideas: trade unionism, Republicanism, secularism, free-thinking, progressiveness of all kinds, as well as their faith in the wide range of democratic freedoms that had erupted during the great generation of 1930.

Some have attempted to mellow the impact of this action, calling it re-education, but it was much more than simply a matter of 're-educating'. It was much more radical and all-inclusive. Franco turned to the Catholic Church as his instrument for his monumental repressive operation, especially the prison chaplains but also the clergy in general, both inside and outside the prisons. One must not forget that the clergy put its great ability to mobilize the masses at Franco's service, occasionally assuming the typically fascist role of a great political party of the masses: huge processions, re-Christianizing the people, befitting the New Spain, a "missionary country".

Prison chaplains

The fulcrum of the great ideological repression began in the jails where rabid chaplains saw themselves as 'purveyors of the Divine'. Although prison chaplains were abolished by Victoria Kent in August 1931, Franco reinstituted them on October 5 1938. Priests and nuns became prison wardens, as the Church was given total and absolute power. Chaplains acquired the absolute privilege of reigning as lords and masters in Franco's penitentiary world, free to proceed with their intent to regenerate, re-educate, control and subjugate the prisoners' resolves.[ii] Priests and nuns ruled as much as the Guardia Civil did.

The Church was the cornerstone of National Catholicism's great penitentiary building. It created a penal discourse based on its rules for purging, expiating and degrading 'dissolutionary' i.e., democratic, ideas. This was a new and much worse Inquisition. The Church and the victors considered that they had the absolute moral authority to impose their ideas until they had 'forced a conversion'. They may have not been totally successful in this, but submission and silence still exist, even today.

One of the major instruments that the prison chaplains used was the Regime's great invention of the Redemption of Sentences Through Work Regime of October 7 1938, Jesuit priest, Pérez del Pulgar's brainchild. More than a means for releasing prisoners from jail, which it practically did not, it became a means for manipulating, brainwashing and destroying the inmates' morale. This deceitful system for commuting sentences was overseen by the Council of Our Lady of Mercy, created by Regime April 1939, whose chairman, Maximo Cuervo, the Director General of Prisons, a fanatic Catholic who was assisted by a clerk, an inspector and a priest, appointed by the Cardinal Primate. The ideological pressure directed at the inmates was so great that the release of a prisoner depended on how well he had assimilated the doctrine of the Catholic Church. "There will be absolutely no redemption of sentence for anyone who does not know the principles of our Religion." In other words, not on how much an inmate had worked to redeem his sentence, but as set forth in the

November 23 1940 Decree from the Ministry of Justice, on how well he had learnt his Catechism. Amoedo-Gil give a precise description of the way this worked:

9 a.m.   Morning mass, with a sermon
Noon to 1 p.m. and 3 p.m. to 4 p.m. - chores
6 p.m. catechism
7 p.m. rosary novena

One could say that the medieval Cistercian monks were required to engage in fewer religious activities than Franco's prisoners. The National Catholics sometimes called civilians from outside and seminarians in to assist with this work. Diego San José, a journalist from Madrid who was interned in San Simón prison, described their being "directed by several impulsive and 'pious' young men from Redondela, one or more of whom would later enter the priesthood without a doubt as a means of washing their mouths out with Holy Water in order to get rid of the bad taste of the paseos in which they had been involved..."[iii]

Of course, from the beginning the Church took it upon itself to carry out the ideological repression in the Francoist prisons, in order to cleanse the 'Godless' and 'Anti-Spain' of their dissolutionary ideas and to implement the cerebral transplantation of the new Fascist ideology. According to Rev. Menéndez-Reigada's *Catecismo Patriotico Español*,[1] the enemies of Spain were liberalism, democracy and Judaism.[iv] "It is the Church's ultimate responsibility to totally eliminate Communism, as well as all ideological and behavioural deviations", stated Jesús Riaño Gaoiri, a member of the National Catholic Association of Propagandists and a Judge on the Special Court for the Repression of Freemasonry and Communism.

There can be no doubt that the Church's repressive work was one of the fundamental pillars of Francoism. It is not a question of whether the Church was simply collaborating with Francoism, but that it actually was part of Francoism. To this purpose, the chaplains were given *carte blanche* to act at will in the prisons, not only regarding spiritual and religious matters, but even more importantly concerning

---

[1]   *The Patriotic Spanish Catechism.*

all things ideological and political. In other words, the behaviour of the Church and the prison chaplains was the principal vehicle for the Francoist Regime's propaganda.

Chaplains were forbidden to talk with the inmates about their trials or their sentences and especially forbidden to do anything in favour of an appeal. Faithful to Francoist instructions and to the repressive program, the chaplains limited themselves to stress the importance of resignation, submission and waiting for Divine grace. This absence of any kind of intercession when faced with so much injustice from the courts, was another obvious feature of the Church's complicity with the great Francoist repression.

Religion as a form of blackmail

Attendance at lessons in the catechism or partaking in the sacraments were incentives continuously dangled in front of the inmates if they desired the most elemental favours such as being allowed to receive correspondence and packages from their families, as well as a means of commuting their sentences. Even being buried in the cemetery could depend on whether the inmate had made confession and kissed the crucifix at the last moment.

Religion as a form of blackmail permeated every aspect of an inmate's life. If a prisoner's children wished to attend school, they had to be baptized. For a marriage to be considered legal, the couple had to have been married in the Church. A woman who was married in a registry office was not considered a member of the prisoner's family and was, therefore, prohibited from visiting her husband in prison, Never before had the Church gone so far, not even at the height of the Inquisition.

Before a prisoner could be released on parole, he had to first pass an examination on the catechism (Decree of November 23 1940). This was a major problem because illiterate individuals, or persons with little schooling, were incapable of understanding the twists and turns of a doctrine that was totally foreign to them. In addition to this there was supreme humiliation: men and women who ha been militant anticlerical during their entire lives, were forced to submit to the victor's ideology if they wished to survive.

May 1942, prison chaplains from all over Spain met in Madrid, at Chamartín de la Rosa, to set the guidelines for their activities. These can be summarized as follows: a) always act in agreement with the Prison Director; b) attendance at catechism classes is compulsory for all inmates, at least at the elementary level, if they wish to benefit from 'special assignments'; c) chaplains must find the most qualified inmates and have them organize classes or study groups, and perform Saint Ignatius spiritual exercises; d) any inmate who does not repent his sins before his death shall be denied a Christian burial; e) before an inmate can get 'special assignment' work he must have been married in the Church; if he has not, he will be listed as single or having had a civil marriage.

Again, a Church wedding was pure blackmail because it was *conditio sine qua non* to receive mail, visits or packages and this applied to both men and women inmates. Miguel Hernández, for example, tells how he was forced to have a Church wedding in Alicante prison before his wife and son were allowed to visit him.

Demonization of the defeated – the 'evil' Rojos

The Church and the chaplains' opinion of Republicans in general and prisoners in particular, was that they were no more than animals or evil demons, infra-human beings, the dregs of society. The horror of the Inquisition was insignificant when compared to the New State's demonization of the defeated who had defended the Republican democratic government. The campaign for demonizing the Rojos was massive. The half of Spain whose hopes had been based on a frustrated modernity, found themselves thrashed daily by official doctrine and by huge doses of reactionary Tridentine Catholicism.[2] As Mirta Núñez wrote, the chaos within the Church during those days was terrible, unique and unusual in 20th century Europe.

An inmate's principle contact with the clergy took place during Sunday Mass, in the form of a rousing sermon that all prisoners had to listen to, standing at attention in the open-air prison patio, regardless of the weather. Some testified that they spent the entire time with their arms raised in the Nazi salute. Others stress how those sermons insulted them as the priest would rage against atheists and Marxists, lecturing them with fits of passion.

---

[2] Reference to Roman Catholic doctrine as defined at the Council of Trent in the 16th century.

José Merino Campos,[v] who was captured in the Cordoba mountains in 1947, recalled his experience in Seville jail and the notorious, authoritarian Rev. P. Ibarra, S.J. Standing along before the inmates, he began by ordering them to stand at attention. He then made them stand still while he scolded: *"The punishments you are receiving here are nothing compared to that which awaits you in the hereafter."* The priest was followed by an equally authoritarian female catechist whom they called 'Señorita Cero', who lectured them as if they were already dead:*" Souls, pay attention. Your chances of success in the afterlife are few, because you have done nothing worthwhile in this one."* One Sunday, the Bishop who came to Seville prison to give the sermon told them: *"All of you who are in this patio are criminals. That colour that was on your flag you will have to dye with your blood."* [3]

An example of the demonization of the defeated is seen in the Director of Modelo Prison's, classical statement in Barcelona: *"You must know that a prisoner is worth no more than the ten millionth part of a crap."*

Concentration camp patio.[vi]

The Church's total disdain for the 5th Commandment reached unimaginable heights. Already in 1937, a <u>Collective Letter from the</u>

---

[3] The Spanish Republican flag bears three parallel stripes: red, purple and yellow, The Francoist flag, the current one, only ears two stripes: yellow and red. Oddly, the Republican flag is still legal, but it is only very recently that anti-Franco Spaniards have begun daring to display it in any form.

<u>Bishops meeting in Synod</u>, stated their acceptance that there would be a mass execution of Communists who "*...as they die, according to the sanctions of the Law, the immense majority of our communists will have made their peace with the God of their fathers.*" Killing them did not matter; was mattered was that they should have said confession before they were executed.

Pablo Uriel, in his book of memoires,[vii] reports a dialogue between two prison chaplains in Aragon:

- Father B: "You who on occasion had the chance to come close to the Rojos as their last moments, did you see any signs of repentance among them?"
- Father A: "None. They are not men, they are pigs. The majority of them do nothing else than protest their innocence and the others even insult us. Believe me. I often had to contain myself. If I could do what I wanted, I would shoot them myself with my revolver. I know them well. There are vermin and there is only one way to deal with vermin."

Uriel also tells of an incident involving a nun in the Sanidad Barracks, Zaragoza, in 1936, who exclaimed:

- "Dear God! I have just been into the autopsy room. There must be at least two hundred bodies. It is horrible."

When she realized that everyone was looking at her in astonishment, without speaking, she blushed, dropped her eyes and spoke some words that Pablo would never forget:

- "Dear God. How many evil people there are in this world!"

As far as all the nuns and priests were concerned, the 'evil people' were the Republicans.

False accusations: confess or die.

The Church closed ranks with the victors through another terrible mechanism: false accusations. It was a system of allegations in which the entire social base of the Regime wallowed, a widespread and typical net of complicity with the Fascist victors, a complicity of blood, a rotten environment in which the Church participated in full as one of the three pillars of Francoism.

If we think back to the words of the Basque monk, Rev. Gumersindo de Estrella, he says that many of those innocent men whose execution he witnessed had been falsely accused by priests (a shameful task in which priests in Galicia and Castille-Leon, especially, delighted in).

This was the case of the sadly famous 'certificates of good behaviour' that the priests in the prisoners' hometowns were asked to issue. In the great majority of these, the priests added to the defamatory allegations at will, in order to present as negative a report as possible for inclusion in the prisoner's military dossier. The same thing occurred with the cases tried under the Law of Political Responsibility where negative reports from the clergy did such damage. In Cordoba capital, two clergymen were especially notorious for the number of bad reports they issued: a priest, Rev. Ildefonso Hidalgo and a Capuchin monk, Brother Jacinto de Chucena.

In North Spain, Julián Casanova[viii] cites the case of the priest in the village of La Segarra, who went into great detail with the defamatory allegations he included in the 'revolutionary curriculum' of eighteen inmates in the village jail. At the same time, the priest asked for more townspeople to be punished, on the grounds that they: 'blasphemed, were irascible, anti-religious, encouraged expropriations and the murder of members of the clergy, namely one who in addition to being ugly, gross and a coward, was an active member of the CNT.'

The Church's imposition of religion by force, much more aggressively even than in the distant past when it did so against the Moors, the Jews and the native South Americans, is frequently referred to in the incredible testimonials of those who survived the Blue Hell. As Dr. Sama Naharro told Moreno Gómez July 8 1983 when he was interviewed regarding his experience as a prisoner in Cordoba:

> "When Escobar was the Director, there was a Carmelite chaplain called Don Justo who was not very clean and somewhat clumsy, wore a stained habit and never preached sermons 'because they never did anybody any good', as he would say. He was removed

and replaced by a wily Jesuit, Rev. García, who was about to have me exiled to the Canary Islands because I did not attend Mass. The Director warned me, so I went to Mass, unwillingly, always standing in the front row. Several doctors were also involved in the conspiracy as they wanted to get rid of me.

When they wanted to punish a prisoner, they exiled him or transferred him continuously from one prison to another, for a long time. This is what did to Dr. Sufo who spent two years shunted from one place to another, on Varela's orders."

Amoedo-Gil described another case involving a 'wily Jesuit' in San Simón, Pontevedra, prison who was actively involved in the firing squads and would tell the inmates they had to go to Mass. Taking his revolver out, the assembled prisoners were informed that they had to tell him who did not go to Mass. He did not care whether he killed on or a hundred and the more he killed, the more honoured he would be.

Another survivor, Melquesídez Rodríguez,[ix] also wrote a Diary in which he described another denizen of the Black Museum, Nicasio Martín Nieto, the Alcalá de Henares prison chaplain, whom they nicknamed *Palo Largo* (Blue Stick):

"He frothed at the mouth every time he invoked the need to cut the weeds. Days before a saca, regarding which he had already been informed, he repeated that he had to discover and exterminate the black sheep that were hiding among the white ones. The prison chaplain would also beat those who were sentenced to death when they refused to say confession. He incited the prisoners' hatred."

A uniquely dramatic moment certainly was that when the prisoners were subjected to religious pressure during the ceremonies that preceded their execution. Manuel Espejo[x], a Socialist alderman in Dos Torres imprisoned in Cordoba, tells how in the early evening a Guardia Civil truck would arrive at the prison and call out the names of those who were to be executed.

"They were taken to a room to confess. There was a different priest at each saca, monks or Jesuit priests from Cordoba. The prison chaplain was Rev. García, S.J. and he took charge of confessing the most reluctant. He took them to his room and afterwards, whether they had confessed or not, reported that they had. Because of that, many also refused to go into his room. Approximately half in each saca confessed, many of these convinced by Rev. García to whom they owed favours at hard times or in exchange for some food when they were hungry."

In Pozoblanco, the confessor to those who were about to be executed was the Salesian priest, Antonio Do Muiño, whom the inmates called the 'bird of death' because each time he showed up, it was a sign that an execution was imminent. In Villanueva de Cordoba, the confessor was Marcial Rodríguez to whom those about to die were taken to confession the day before. They say that only two inmates confessed in this jail: Matías Villarreal and Juan Antonio Bustos (the latter did so only to protect his sobs, one of whom is a priest today). Moreno Gómez believes that many more confessed as he knows of another, unidentified priest, who was also called for confessions. Adding insult to injury, Tomás Cantador told Moreno Gómez what one of his prison companions said after he had confessed: "Can you believe what the priest just said to me? 'My son, if you have any property and no heirs, you might as well leave it all to the Church, because you are still going to pay for all you are guilty of."

The prisoners' confessor in Montilla was Luís Fernández Casado, a pompous National Catholic priest who would appear in photographs wearing cassock, wide-brimmed hat and cloak, surrounded by Falangists. One day, Francisco Solano Martínez' mother went to Rev. Luís to complain, sobbing, about the mistreatment to which her son was being subject in a Palma de Mallorca Workers Battalion. The priest replied: "Forget it. He's atoning for his sins."

Again, Tomasa Cuevas' work is an encyclopaedia of knowledge of the clergy's repressive actions in the Francoist prisons, by both priests and nuns, as we see in the several testimonials that are reproduced herein. Although not all priests were quite as nasty as those reported in the testimonials, there was the occasional exception, such as Dom José María, a Basque priest in Santurarán, whose kindness and interest in them all although there was very little he could do to help, they never forgot. [p.277]

It was a very different matter when Tomasa spoke of Amorebieta, where everything was very much worse. "The way that the staff (nuns) treated the women was horrible, blackmailing the starving women to force them to repent their sins by controlling how much they fed them. The prison chaplain was a pretentious offensive devil who considered them as none other than thieves, murderers and prostitutes." [p.292]

Despitei the continuous threats in the event of non-compliance with their order, prison chaplains did enjoy some success in their attempts to morally 'disinfect' the prisons, even organizing choirs to sing at the solemn Mass at Easter, on special Saint's Day, and so on.

As mentioned earlier, the Church's influence continued beyond the prison walls as it had its say whenever a prisoner attempted to be released on parole, before which he first had to pass this catechism examination, as Francisco Herencia reports herein.

Our discussion of the chaplains would not be complete without a word about the role of nuns as prison workers. Under the New State, chaplains were sent to men's prisons and nuns to women's prisons. In the latter case, nuns were much more powerful than chaplains, as in addition to their role as Ministers of the Devine, nuns were also members of the prison staff; chaplains never were.

Following a proposal from the Conference of Bishops, October 2, 2005, the Prince of Asturias de la Concordia prize was awarded to the Saint Vincent of Paul Sisters of Charity, the order of nuns who controlled several Francoist prisons such as the Provincial Prison of Les Corts, Barcelona; Alcalá de Henares prison; Palma de Mallorca prison and others.

Their job was to take care of all the internal governance of the prison, maintain discipline among the inmates, teach classes and oversee the workshops. In practice, they administered the prison with favouritism and a chilling lack of humanity... The inmates were only fed watery soup with cabbage... Some women fainted with hunger when at attention in the prison and the nuns did nothing at all. The nuns were so cold, they were inhuman.[xi]

Although Victoria Kent expelled nuns and chaplains from the prisons in 1931, precisely because of their inhumanity, Franco called them back in 1940. December 1940, 342 nuns belonging to 15 different orders were scattered amongst 40 prisons.

In Ventas Prison in Madrid, Máximo Cuervo, Director General of Prisons, appointed a young nun of the Order of St. Teresa, Carmen Castro, as Director of the prison after the victory. During the war, she had been part of the 5th column that eventually went over to the Francoist zone in 1937. She got her training as a prison director in San Sebastián, Saturrarán and Santander prisons, was appointed Director of Ventas Prison and governed it with an iron fist. When María Sánchez Arbós, a secondary school teacher in Madrid and in Huesca was imprisoned in Ventas, much to her surprise she discovered that the Director of the prison, Carmen Castro, had been one of her students. It appeared that the latter was somewhat embarrassed by the fact as she approved some of her ex-teacher's, now one of her prisoners, requests such as creating a section in the jail for women who were there with their children and a school for young inmates directed by María Sánchez.

María Sánchez was also allowed to collaborate with the prisoners' workshop whose work with crafts was supervised by Matilde Landa, another inmate. The latter, unable to withstand the constant harassment from the Church, finally committed suicide in Palma de Mallorca prison; María Sánchez was finally release at the end of 1939; Carmen Castro remained with the Central Inspectorate of Prisons under her death in 1948.[xii]

In brief, the Church, as one of the three pillars of the New State, enjoyed its share of power as it embarked on its task of achieving the ideological repression of the defeated. To this end, the Church's tactic was to combine methods from the Inquisition with modern psychological methods of European Fascism. As Moreno Gómez pointed out earlier, the ideological repression of stripping oppressed peoples of their identity to replace it with that of the oppressors is one of the features of genocide, as described by Raphael Lemkin.

The Church set the pace of the indoctrination and submission in the prisons, as it also engaged in a dark collaboration with the Regime, providing false denunciations for the courts, and bureaucratically, by issuing 'bad conduct reports' which caused so much damage and ruined the lives of the defeated. Furthermore, the National Catholic Church reinforced the Regime's penitentiary discourse by decreeing that the 'offense' was a sin, the 'sentence', an affliction, and the punishment, a painful process necessary for atonement.

In other words, the physical and mental purge was the prisoner's public repudiation of his heretical doctrine. If that were not enough, the Church contributed to entrenching the spirit of Francoism, with its great capacity for mobilizing the Catholic masses who prostrated themselves before the Dictator. Clearly, these latter-day Inquisitors thoroughly enjoyed punishing and forcing the 'Godless' to atone for their beliefs.

Commuting a sentence through work.
The role of the Church as an overseer and
manager of the slave labour program.

At the beginning of 1937, as Franco was beginning to realize that he had no idea of what he was going to do and where he was going to put the many thousands of prisoners who were piling up in the prisons, the Church appeared on the scene as an advisor, in the form of a Jesuit, Reverend José Agustín Pérez del Pulgar. Scientist and student of social problems, he fled Rojo Madrid at the beginning of 1937. From Valladolid Pérez del Pulgar became Franco's theoretician for the great mass of prisoners in the Francoist zone, beginning with <u>Decree 281 of May 28 1937</u> recognizing the prisoners 'right to work'. The Church-inspired version arrived with the creation of the <u>National Board for the Redemption of Sentences through Work</u>, Ministerial Order of October 7 1938, based on Decree 281, as it applied to prisoners found guilty of acts of military rebellion. The Council was formally established December 15 1938 and the program started January 1 1939.

Do not think, by any stretch of the imagination, that all prisoners were eligible for this program. The system was only open to prisoners who had already been condemned and were accepted as exhibiting good behaviour. Women, Masons and common criminals (accepted as eligible in 1944, especially black marketeers), among others, were excluded.

Furthermore, the Jesuit's brainchild, the *Redemption* program, had nothing to do with any commutation of sentences. The supposed redemption was nothing other than the ideological manipulation and exploitation of labour. All the talk of 'redeeming souls through instruction and persuasion' was rubbish; this was pure and simple brainwashing. The concept of "intellectual work" was added in November 1940, apparently with the same redeeming effects but actually to enable the Church's representative to grade the inmates on the religious and similar courses under the chaplains and nuns' supervision.

The system was in the hands of the National Board for the Redemption of Sentences through Work, which added: 'and of Our Lady of Mercy' in 1942. The council was chaired by the Director General of Prisons, a member of the clergy and some virtuous lady member of the Catholic Action movement. The local branches of the Council (Mayor, parish priest and the usual virtuous lady) were those who, in the towns, exercised the real social control both of prisoners out on parole and their families. It was they who determined how much the families would receive from the miserable salary the prisoners were paid for their 'redeeming work' and they who wrote the indispensable certificates of good conduct for the prisoner's release.

In addition to its role as an instrument of social control, the Council provided important national and international propaganda services for the regime, as they proclaimed *urbi et orbi*, the benefactory actions of the 'magnanimous Caudillo', a means of concealing his true nature as a wolf in a sheep's clothing. Some less astute historians such as Julius Ruíz, have shown a surprising lack of discernment when swallowing this argument, by letting themselves believe that 'the extermination was not so much of an issue' as the Caudillo provided beneficent alternative such as the Redemption system.

There was no official information regarding the Redemption of Sentences before 1943 when the first data regarding the number of prisoners who entered the Redemption system each year during the first five post-war years was published 1944 in an official Government Memorandum: 1939 (12,781); 1940 (18,781); 1941 (18,385); 1942 (23,610) and 1943 (23,884).[xiii] This memorandum also provides information regarding the number of children of prisoners being taken care of in charitable institutions: 9.050 in 1942 and 12,042 in 1943. Later memorandums publish the number of prisoners enrolled in the system in 1944 (26,518) and 1945 (17,162).[4]

Clearly, these figures pale in comparison with the total number of imprisoned. For example, in 1940, 300,000 individuals were held

---

[4]   José Manuel Sabín in *Prisión muerte en la España de posguerra*, Madrid 1996, and Josep María Solé i Sabaté in *História de la presó. Model de Barcelona*, Leida, 2000, are two historians, among others, who have written on this subject.

in Franco's Spain, of which only 18,000 had been accepted by the Redemption system. Through the period under study, although this proportion varied little from 5%, the Regime used these figures for propaganda purposes but inflated them greatly, particularly when referring to the data for prisoners released on parole (the numbers usually included the number of prisoners reprieved or pardoned and even of those who were executed).

The 1945 Memorandum gives the following numbers for prisoners freed on parole: 3,654 (inmates condemned up to 20 years in prison) and 7,791 (inmates condemned from 20 years and 1 day to 30 years in prison).

Regarding the Cordoba prison, according to the 1943 Memorandum, 600 inmates were enrolled in the Redemption system; 400 in Jaén; 800 in Seville; 200 women in Málaga; and 1,300 men in the Dos Hermanas, Seville, Paramilitary Labour Camp. Regarding their professions, of the 6,347 who were enrolled in 1943, almost 4,000 were manual workers, followed by 1,242 bricklayers and stonemasons.

When you come down to it, the reality of the Redemption program was not the release of prisoners but their ideological repression and even worse, their source of slave labour. This is clear in the terminology of the Decree that created the Council when it speaks of work as 'an act of submission and reparation'. Submission, in the lines of the genocidal philosophy of replacing the defeated's identity and ideology with that of the victors'. The ideological moral and class awareness repression. The renunciation of a person's ideals, following which the defeated's lives are forever stained with shame.

Then there is the use of the word 'reparation' that, rather than referring to an atonement of guilt, hides one of Francoism's ugliest intents: 'making the defeat rebuild that which they destroyed'. In other words, it is the defeated who are guilty of creating the damages for which those who instigated the coup, and their allies were responsible.

Furthermore, reparation by way of the slave labour of the prisoners, the most vile means of exploiting that immense mass of cheap labour, to the benefit of the New State and the great opportunistic corporations, just as in Nazi Germany. The Church also took full advantage of this source of slave labour to rebuild some damaged churches and convents.

Rev. Pérez del Pulgar, S.J., never ceased proclaiming that this program cemented the Church's ministry and its mandate to bring about the spiritual and social appeasement of Spain through material reconstructions. The Director General of Prisons, Ángel B. Sanz also publicly rejoiced that this program had put this mass of delinquent unemployed to good use.

The newspaper *Rendención*

The National Board for the Redemption of Sentences Through Work newsletter was the weekly newspaper *Rendención* [Redemption], published by the National Catholic Association of Propagandists (ACNP) that spearheaded National Catholicism in Spain. The first issue was published April 1 1939, the First day of Victory, and it was continuously published until 1978. The first Editor was José María Sánchez de Muniaín, past secretary to Ángel Herrera Oria.[5] General Máximo Cuervo, Head of the National Prison Service, was also a member of the ACNP.

Rendención was the only newspaper allowed inside the prisons. As a barefaced propaganda instrument as well as a tool for indoctrinating the reader, the newspaper recruited its staff among inmates enrolled in the Redemption system and who, after having been 'redeemed' would become champions of the Regime. Each time that an inmate 'was converted' with the First Communion, a Church wedding or his public abjuration of democratic principles, the event was published in *Redención*. Totalitarian and oppressive regimes have always encouraged this type of personal indignity, thereby forever destroying their victims' self-respect.

The purpose of the great physical, ideological and slave labour repression was famously broadcast in a NO-DO[6] government cinema newsreel of the period, that begins with a very suntanned man whose closed fist he slowly opens as he raises his right arm in the Nazi salute. He speaks:

> "It is here that Franco's Spain regenerates these men as they regain the dignity they had lost. If they demolished a bridge, they are now rebuilding it. If they destroyed a house, they are now reconstructing

---

[5] Roman Catholic journalist and politician. Bishop Emeritus of Málaga and later, Cardinal.

[6] NO-DO is colloquial for "Noticias y Documentales", a series of cinema newsreels created n 1943 as a major source of news, public information, censorship and propaganda for Franco and the Fascist State. Absorbed 1981 into RTVE, the government-controlled Spanish TV and radio.

it. Their days as prisoners are devoted to learning some skills that will make useful beings of them, whilst they atone for their existence as pariahs. All that human trash shall owe Spain its regeneration.

From the proletarian mass we have created order and accord, we have dispelled malice, like the cripple whose closed fist slowly opens, these men open their fists as the brotherhood of the open hand and the raised arm receives them with the generosity that the Spanish empire of the past has always had for the defeated. That is our Justice."[xiv]

SUBJUGATION OF CHILDHOOD
Kidnapping. Segregation. Brainwashing.

"The Francoist dictatorship, unlike other dictatorships, lasted forty years and had its origins in a genocide that has not yet been openly recognized. Franco killed or exiled half of the country. Spain, today, is still incapable of seeing itself in the mirror of that concealed, silenced, genocide.
Miguel A. Rodríguez Arias, *Esa Memoria*

Children during the Spanish Civil War
Archives of Ontario
Photographer unknown[xv]

Children behind bars

The repression of childhood, the manipulation of the children, the social segregation as well as their theft and disappearance, is another of the most pestilent and monstrous chapters of Francoism. The military coup caused death and suffering to everyone, especially to children. Infant mortality, just from the inhuman living conditions, boke all records. To that we have to add other calamities, such as the bombings. A particularly painful chapter was the evacuation of children out of the country; the children of the war", especially from the Basque Country and Madrid. No less than 37,487 children were sent out of Spain, of which 20,266 were repatriated over time. The remainder were forever lost to Spain.[xvi]

There follows a brief panorama of the post-war situation regarding children. The New State basically had two general objectives when it came to children: a) the destruction of families by breaking them up and scattering their members, with a view to erasing inherited ideological and class traditions, especially those of the working class; b) secondly, by segregating the children, to subject them to ideological re-education to prevent their being contaminated by their parents' ideas. Again, this is a feature of genocide according to Lemkin: erase the oppressed defeateds' identity and impress the victors' identity on them.

At the same time, the Church took advantage of this great operation to impress National Catholicism on Spain. The operation of imposing National Catholicism also involved considerable widespread abuse such as changing a child's identity, arranging for irregular adoptions and out-and-out kidnapping. The UN Working Group on Enforced or Involuntary Disappearances, following a visit to Madrid in September 2013 when the delegation met with several organizations, associations and relatives of the missing, calculated at 30,960 the number of children who disappeared or were kidnapped.[xvii]

If under Francoism, parents and families lost all their rights under the law, so did their children. A first step towards the chaotic management of children during the post-war period was the March 30 1940 Order regarding the removal of all children over 3 years of age in jail with their mothers. This began the mass removal of children who were seized from their mothers and taken into care by the State.

Nothing better to understand the tragedy of the situation, than to read the direct statements of mothers who were interned with their children and who suffered this humanitarian catastrophe. Beginning with Tomasa Cueva's Diary, one of the women interviewed, Julia García Pariente, from Guadalajara, reports how in a horribly hot August, when there wasn't enough water to drink. Her small child who cried because she couldn't give him any water. Worse, when she asked for some, they would give her an empty bottle. [p. 85]

Tomasa Cuevas' description of the problem of the children in Durango prison and the remarkable support their mothers got from the townspeople, as well as several other testimonials from inmates, are reported herein. The Basque television documentary, *Forbidden to Remember*, also refers to the deaths of 120 women in Saturrarán and of 57 children, 32 of whom died in a single fortnight.[xviii]

A Catalan television documentary records many testimonials from inmates such as María Villanueva and Juana Doña, regarding the lack of everything, especially milk for the children, many of whom fell ill and died. It drew attention to the problem of how could young children survive when they were packed in cattle carts with their mothers, for seven days in a row, from Alicante to Madrid, a horrendous trip during which many children died. As one witness stated there is no way to describe what they did to them and to their children, unless it is to underline the Regime's intention to exterminate everyone to ensure that nobody would be around to take their revenge.[xix]

Ventas Women's Prison in Madrid, the largest worldwide, held 4,000 women in 1945, which gives one an idea of the number of children behind bars of abandoned. Ventas, under the orders of the notorious 'boss ladies' (La Topete, La Chanclitos, La Veneno, La Susana, La Castro, etc.), was part of the same great crusade to turn everyone into a Catholic that was sweeping Spain. The Director ordered that all the children had to be baptized. Their godmothers would be ladies from the Falange. The inmates rebelled, stating that if they had no choice but to allow their children to be taken to the baptismal font, at least let their own mothers be their godmothers. The question of religion stuck in many inmates' throats because half a century of working people's lay culture could not be erased in one fell sweep.

La Topete (María Topete Fernández, sister of General Topete, Director General of the Guardia Civil), actively implemented Vallejo-Nájera's[7] theories. As she wore a military uniform, she was considered to be a 'Guardia Civil in woman's clothing'. María Topete was influential in opening a new prison, San Isidro Prison for Nursing Mothers, next to Segovia bridge in Madrid. No beneficial measure, it was a means to segregate and manipulate the inmates.

According to Trinidad Gallego, interviewed by M. Armengou in the *Los Niños Perdidos* documentary, in the San Isidro prison. "La Topete applied Vallejo-Nájara's thesis of the need, from the very beginning, to separate the prisoners' children from their parents' 'evil influence'. She took over the children and enjoyed making their mothers suffer.

The fascist way of thinking, both within and without the prison, was that the children had to be separated from their parents' ideas and re-educated. One thing was certain. Their mothers knew that if they did not lose their children physically, they would lose hem morally and that they could become future enemies of everything that their parents' life represented."

San Isidro prison was inaugurated September 17 1940, with La Topete as Director.[8] Segregation was immediate: mothers in the upstairs gallery; the children, in the patio, were not allowed to be with their mothers who might nurse them with Communist milk. Their mothers were only allowed one hour a day to be with them. The segregation was such that even their toys were given to them by the prison Director. Their mothers rebelled and began making toys themselves.

Disappearance and theft of children

Francoism wanted to emulate the Nazi experiments with those it had detained. Apparently lacking the technical perversity of the Germans, Franco fortunately only managed to outline a grotesque

---

[7]   Antonio Valllejo-Nájera, a high-ranking Army doctor who, after a trip to Nazi Germany, decided to practise psychiatry along Nazi theories which he brought back to Spain. He believed in a close relationship between Marxism and mental inferiority and because of the association of antisocial Marxist psychopaths and social unrest, offspring must be segregated from their parents, from childhood, in order to protect society from such a terrible plague.

[8]   La Topete died in 2000 at the age of 100.

scenario based on the theory of a 'Rojo gene' that he also considered was contagious. His right-hand promoter of such nonsense was the aforementioned Dr. Vallejo-Nájera, on whose advice Franco created the <u>Army Bureau of Psychological Research</u> (August 23 1938), an imitation of a similar organism Himmler created in 1935 in Nazi Germany. The purpose of this Bureau was to research the bio-psychiatric roots of Marxism, in an attempt to find the 'Rojo gene'. Coronel Dr.Vallejo-Nájera set to work studying 297 members of the international brigades captured in Burgos and 50 women imprisoned in Málaga. His study purported to demonstrate the 'mental inferiority' of the Marxists.

Vallejo-Nájera concluded that Marxists were not only mentally inferior, but they were also dangerous and consequently, unqualified to bring up their children. This then led to the proposal of 'segregating thee subjects from childhood, so as to free society from such a terrible plague'. Recently, Juan Sánchez Vallejo has shown how Francoism also had recourse to insane asylum inmates to justify its repression of Republicans on the grounds of their presumed madness.[xx]

Vallejos-Nájera's theory was applied during the 1940s and 1950s as in Spain, as in Argentina and elsewhere, the theft of children was institutionalized under law. <u>Law of December 4 1941</u> endorsed the kidnapping of children as it permitted the government to change the children's name and surnames, which made it almost impossible for their families to find them and get them back. As A. Rodríguez Arias stated: "With this theft of children, Franco committed the whole gamut of crimes against humanity, the entire range of crimes that are continuously being presented to the International Criminal Court."[xxi]

As families were destroyed, scattered or destructured, and any that managed to survive were subjected to social exclusion and starvation, particularly as may fathers had been executed and their mothers imprisoned, the Regime developed its arguments in favour of its theft of children, decreeing that:

> "All families who cannot show that they have the means to educate their children according to the principles of the Glorious National Revolution, must give the State custody of their children."

Understandably, few defeated families could comply with that requirement. As parents were unable to obtain certificates of economic, ideological and religious compliance with the Regime, etc., the State took their children, just as they did with children who had attained 3 years of age and were with their mothers in prison. A great many mothers whose children were thus kidnapped from their arms, never saw them again.

The stew that fed the wave of the theft and disappearance of children was, most especially, misery and poverty. The extreme state of misery broke the traditional domestic universe of the woman, and this triggered an entire chain of acts against childhood: children taken into care by the Welfare Authorities, placed in foster homes and in hospices, ending with the dark slippery slope of irregular adoptions.

There was, however, another recourse for children of poor families whose parents were unable to feed them, which was to place them with foster parents. This was a common custom in Andalusia during the 1940s and 1950s, but Moreno Gómez has not seen any reference to this anywhere. Only those who came from country towns such as he, knew of this. The method was to send minor children to live with small and medium-sized landowners where they would help with the field work or in taking care of the livestock, in exchange only for food and lodging, and some clothes.

Rojo children, regardless of how charitably they were treated in a community, were always looked down upon by the victors, scorned, shunned and despised, regardless of the official paternalistic rhetoric. One who was interviewed by Günter Schwaigaer in his documentary, describes his experience as a 9-year-old in 1936: [xxii]

> "I had to put up with some very disagreeable moments, many insults. We could not leave the house, we were always insulted with calls of 'Communists and thieves', almost always so. When eleven of us made our First Communion [in Santa Cruz de la Salceda, Burgos] and were taken for a meal I had to go home because some said: "Communist, what are you doing here?" Other times, when I went to get a piece of bread, a lady would call out from the balcony of her house saying "Rojo, go away, you are going to dirty my door", and I would go home in tears."

A woman also interviewed in the same documentary told of how, a young girl at the time, she was called Roja and told to go home to Russia. She did not want anything to do with that town and vowed that as soon as her father was released, she would take him away from there. She still had not forgotten. Similar testimonials in the *Prisoners of Silence* documentary: "They kicked me out of many places because we were Rojos. You could not speak of anything; you had to keep quiet. I am still afraid to speak out. When one does not speak of something, that something slowly disappears, slowly disappears."

Another misfortune that fell on children during those dark times, was the problem of the "children of war", those who had been sent outside the country during the war. Francoism created a pirate scheme, the Falange Foreign Service, whose sole purpose was to hunt down and kidnap the "children of war" who lived outside Spain. Many were captured in Leningrad and Hitler sent them back to Franco where custody of these children was usually given to the Welfare Authorities.

In 1943, 21 children were sent back to Spain without, as Montse Armengou said, telling their parents. The parents of a child from Santander found out that their son had been returned but when they went to get him, the authorities refused to hand him back. He was sent to the Welfare Authorities where he was forbidden any contact with his family 'because they were unsuitable to educate him'.[xxiii]

On another occasion, 180 children arrived in Madrid. Some parents heard of this, but many others never found out because the Regime had ordered that they should not be told. It was all carefully organized to prevent the children from having any contact with Republican families and to encourage their adoption by right-wing families Kidnapping these children abroad was very profitable.

As mentioned earlier, the UN Working Group on Forced Disappearances calculated that 30,960 children were stolen by Francoism, based on the Report of Penal Court Number 5 (Diligencias Pr. 399/08 V, October 16 2008), the report of the Audiencia Nacional (a type of Supreme Court), on Montse Armengou's book *The Lost Children of Francoism*, and on data from the Brotherhood of St. Paul (1944-1954) which was given custody, by the State, of that number of 'disappeared' children, distributed over 258 centres, who had lost all contact with, the education and loving care of their parents; i.e., 'disappeared' both physically and emotionally. All of this was part of the Regime's program for the children of the Rojos, as declared in another NO-DO newsreel:

> "The sacred mission of recovering thousands of children of Spain and of saving them from misery, to hand them over, healthy and regenerated, to the Fatherland."

One of the children sent to the Welfare Authority, Francisca Aguirre, is quoted in Montse's book as saying:

> "The nuns would assemble us and tell us clearly that we were rubbish, children of criminals... We knew that we were guilty, but we did not understand of what."

Eugenio Álvarez, from Asturias, is quoted as saying:

> "I know all the hymns of the Falange and Church backwards. We were punished by sending us to bed without dinner, night after night, we were so hungry... They stole everything from me: my childhood, society, my family, my parents' ideals."

It was a truly Machiavellian program destined to erase their identity: Their names (name and surname), their identification with their family, their social class, their ideology. It is inconceivable today how Francoism managed to destroy the backbone of the Republican family. One cannot imagine, nor does one want to know, nor is one told about it, not at school, not in the community, not in the media.

This was a monumental fraud against historical justice. One of the great irreparable damages that the Allies committed against a progressive, innovative, hopeful and noble Spain was to allow the country to remain in Franco and the National Catholic Church's hands for 40 years, allowing them to act at will with all the time in the world to continue their terrible Nazi-inspired programs with impunity. The European democracies are to be blamed for a large part of this catastrophe whose effects endure to this day and will continue to do so.

The 'official' theft of Rojo children, a massive New State operation, had unimaginable dimensions and was accompanied by all sorts of irregularities and abuse of power, against which the children and their parents were totally impotent. This was a clear crime of lèse humanité.

Returning to Tomasa Cuevas Diary, she tells of the tragedy of her fellow inmate, Anastasia, mother of six, the youngest of which was eleven months old at the end of the war. Both parents were arrested, and the children were left to shift for themselves in a maddened and terrified Madrid, where everyone was suspect, everyone was considered Rojo unless he could show proof that he, or she, had been a Falangist for years. Both parents were sentenced to death and her husband was shot. Anastasia, saw her sentence commuted to thirty years in jail and she was sent to Santurrarán prison. She had tried to find her children when she was in Ventas. All that her neighbours new was that someone, they believed from the Welfare Authorities, had taken them away. Anastasia continued to search from Santurrarán. Nothing. Her children were not to be found in any hospice or children's home. Rev. José Maria, a kind Basque priest and the Saturrarán chaplain, did his utmost to find them. It took him some time and trouble. The children, who had been given a different surname, had been interned in a provincial hospice, apparently in Ciudad Real. [p. 286] This was the road taken by thousands of children who 'disappeared' under Franco.

Tomasa tells of another companion, Clara, from Andalusia, who was arrested with her husband at the end of the war, both charged with being Communists. He was executed and she was condemned to death, but her sentence was commuted to thirty years in prison, and she was imprisoned with her younger daughter who was only a few months old. Clara was allowed to say goodbye to her husband as he was taken to the firing squad. Many years later, her mother-in-law and her mother were able to take her daughters to Segovia to visit her in prison. The eldest was working as a servant in Madrid; the younger, very little ones, were able to visit her. She did not know them, nor did they know her. The girls looked at her with wonder. For many years they had been told that she was an evil woman which is why she was in prison. The por creatures looked and listened to this woman who was looking so fondly at them. Clara managed to keep from crying until the gate of the jail closed behind them when they left. [p. 289]

When they took <u>Elena Tortajada</u>, from Ciudad Real, out to be executed, she handed her infant son to her fellow inmates, in front of the guards and the soldiers, saying: "I give you my son and ask that you educate him and teach him to respect my beliefs and to never forget why his mother died." [p. 165]

It was very difficult for these destroyed families to retain their beliefs amongst themselves. It was because of them that the Regime beat them. Yet the State and the Church who saw themselves as the traditional defenders of the family were responsible for destroying more families than in all Spanish History.

Today, sanctimonious clergy speak out in defence of the family, but they do not deceive those who know their history and what they did against thousands of hundreds of families in Spain.

"There are at least 30,000 persons in Spain today who still do not know who their parents were They were given to Catholic families, and their true identity wiped from the records."

That is what the Historic Memory is all about – to fully understand the present and correct the errors of the past. The problem is that in Spain, the State did its best to erase 40 years of the Historic Memory of its people. Outside Spain, some 80 years after the beginning of the civil war, almost nobody knows the history of the genocide. A tragic history that is gradually being unveiled despite lingering fears among the descendants of the supporters of the legal government, with every electoral victory of right-wing political parties in the country.

The Social Welfare Authority

In pre-war Spain, charity was always brought to public attention more frequently than justice, especially at times of economic depression, unemployment, hunger or calamity, when the ruling class in Andalusia opened their purses to set up soup kitchens for the hungry, although helping people find work or get better wages would have been more philanthropic.

In 1936, faced with the great wartime humanitarian catastrophe created by the military coup and the war, the better-off classes' inclination towards charitable actions was elevated to the nth degree by the great propaganda umbrella of National Catholicism kindness, a social welfare apparatus promoted as the "Falange's work of love to provide 'mortal and material' assistance to hundreds of poor, abandoned or orphaned children".

In the beginning, this was primarily a volunteer organization run by Falangist men and women, namely Mercedes Sanz Bachiller (widow of Onésimo Redondo) and Javier Martínez de Bedoya, from Valladolid, together with several clergymen, of course, to make sure that bread was accompanied by a good dose of religious instruction. Rev. Pedro Cantero Cuadrado, one of the best-known of its chaplains, made of this work his lifelong career.

Order of December 29 1936 establishing the Assistance to Children and the Elderly program, under the Ministry of Governance, published January 3 1937 in the Official Government Bulletin, was accompanied by public appeals for charity to help fund the program. The volunteer Winter Welfare organization obtained increasing autonomy following the publication of a series of decrees such as General Orders of February 4 and March 10 1937.

The first Social Welfare Congress was held in Valladolid (September 13-18 1937, the second, in the same town, one year later (October 16-23 1938), and the third, now under the bugle calls of Victory, at the end of December 1939. The latter was inaugurated by a euphoric Mercedes Sanz Bachiller, the first to coin the designation Social Welfare. It was closed by Serrano Súñer who led a delegation from the Falange and caused turmoil by removing Social Welfare from the aegis of the Ministry of Governance and assigning it to the Falange. This destroyed any relative independence the welfare program to provide assistance to orphaned children had so far enjoyed. Mercedes Sanz was forced to resign and was replaced by Pilar Primo de Rivera.

New Social Welfare regulations were formalized by Decree May 12 1940. The institution was now established an official organization, a member of the FET and the JONS, dependant on and under the aegis of the State. The official nature of the Social Welfare Program was thus recognized for the first time, which enabled it to have assets of its own, the product of donations. Its budget was set by the State.[xxiv] The new body, clearly modelled on the Nazi *Winterhilfe*, was yet another example of Franco's attraction to European fascism.

Decree of November 23 1940, in addition to reaffirming that the orphans or children of prisoners were subject to the guardianship of the State, snubbed the numerous existing welfare organizations, emphasizing its duty to entrust the education of orphans to persons 'of impeccable moral character who would ensure their education from a religious, ethical and national viewpoint'.

As time went by, as the victors began to tire of so many appeals for charity and of giving to so many alms to Rojo children, the program began to find itself in serious financial difficulties. Order of May 23 1942 declared that all Spaniards had to buy and wear a Social Welfare badge, and that the so-called 'Blue sticker' had to be displayed in all public places (cinemas, bars, restaurants, etc.).

October 7 1937, the Regime created the Social Service for the New State's virtuous women. This was a kind of 6-month military service intended to provide staff assistance to the Welfare Authority, either on a daily basis or as resident wardens in homes for women in religious establishments or the so-called Recovery Colonies-Homes. The 'Welfare' ladies and the members of the Women's 'Militia' rarely got on with each other.

The key to all this Francoist charity work was not just its material assistance, but most especially, religious instruction. The Social Welfare's evangelization work was praised, particularly the mass christening campaigns, first communions and church marriages of parents, as well as its multiple actions as part of the troika of civil, military and clerical authorities. In 1939-1940, the Madrid branch of the Social Welfare Authority celebrated 9,872 christenings, 6,642 first communions, 1,116 church weddings, several religious vocations (9 in Oviedo; 22 in all of Spain in these two years alone). There were 24,513 christening in Spain in 1940.[xxv]

The daily Social Welfare routine included saying the rosary, praying before and after meals, days of spiritual retreat and sermons from clergy expert in attracting the masses. A declaration on the occasion of the 1944 general meeting of Social Welfare Advisors, clearly showed how over and above welfare itself, the majority of their tasks focused on Francoist and religious manipulation, re-education and ideological repression.

- "Children who eat in the refectories shall attend religious instruction at the parish church and at school.
- Adults who are receiving assistance shall be given religious instruction and written propaganda.

- Social Welfare activities must be energized by the spirit of the Christian and Spanish family.
- Thanks are given to the Catholic organizations for their collaboration.
- Cordial gratitude and pledges of unbreakable support of Franco."

No matter how charitable the founders of the *Winter Welfare* wanted the organization to be, the Social Welfare Authority was corrupted from the inside by the manipulation of Francoist religious-patriotic fanatics. It in a recent (2002) heated declaration, Mercedes Sanz Bachiller insisted that she was not trying to turn the children into Francoists. She just wanted to make them become anti-Communists. Ricard Vinyes, speaking in Montse Armengou's documentary, affirmed that the Social Welfare Program was no more than a set of concentration camps for children, not with a view to exterminating them but to totally make them over. It is interesting to note, when you look at old photographs, that the Welfare children are all wearing the Falange uniform: dark blue shirts, light-coloured trousers and badges.

The welfare organization was divided into several different sections that became increasingly more complex with time. The first sections were **Children's Soup Kitchens** for abandoned orphans from Valladolid, whose parents had been assassinated. They were cared for by volunteers, with prayers and patriotic songs. Next were **Brotherhood Kitchens**. These began in 1937 as assistance for the 'freed populations' – adults and children over 12 years of age. By the end of 1938, they had already fed 71,430 adults to whom they served 4.4 million meals. There were **Refugee Soup Kitchens** for right-wing individuals who had returned to the Francoist zone, especially via Portugal and France. Destitute women in general were cared for in **Homes for Pregnant Women and Sick Mothers**, in **Recovery Colonies.**

Not all of the above existed everywhere. More widespread were the Soup Kitchens and the Brotherhood Kitchens, and Refectories because of the general misery and great number of orphans. A great many parents had been executed, other died in battle and others had disappeared.

The **Infants Feeding Centres** were for poor and orphaned children up to the age of two and they were nothing special – provision of some milk and medical care. Do not forget that any child suffering from a contagious disease, and there were a great many, were prohibited from going to these centres. **Nurseries** took care of and housed poor and orphaned children from the age of one month to three years, and even these had an educational touch. **Day Care Centres** took in children of the same age but only in capital cities and large towns as they were for children with working mothers and fathers. **Children's Homes** boarded children from the ages of three to seven who were segregated according to their sex. **Residential Schools** boarded children from seven to twelve years. In 1942, there was one such school in Valladolid whose rules of admission were the following: 50% places for orphans of the revolution and the war; 25% for children who are disabled because they are orphaned or poor; and 25% for beggar children. **Summer Camps for Children**. Originally created for weak or convalescing children, they also had to attend intensive indoctrination events in the form of politico-social and religious talks.

The Welfare Authority was in total chaos during the first days of victory, as it tried to meet the needs of numerous 'liberated towns' in the Centre-South of Spain and Madrid. During the first days of April 1939, it distributed more than 750,000 meals. With victory, charitable fascists began at last to act in Madrid in an attempt to compensate for their failure in November 1936, when they appeared with trucks full of melons that rotted as the trucks waited unloaded on the side of the road.

No sooner was victory announced in the North of Cordoba, at the end of March 1939, the Social Welfare organized a caravan to take help to the towns and villages in the Los Pedroches district. The caravan arrived in Alcaracejos March 28, but they found the town deserted and partially destroyed. They drove around several other towns with their trucks of food, until they reached Almadén. 1940 reports, presumably referring to the end of 1939, speak of 50 Social Welfare Food Centres for Children providing assistance to 50,000 children. This number was obviously insufficient as of this number, 28,000 were children fed in 16 centres in Madrid alone.[xxvi]

In 1941, the 5th Anniversary of Social Welfare was celebrated throughout Spain with great pomp and circumstance and multiple newspaper articles praising the 'magnanimous Caudillo's' great welfare work. A great fuss was made of the 61 welfare centres in Spain (Soup Kitchens for Children and Children's Homes) that had assisted 48,186 children, including 11,869 orphans in residential care. Again, it is presumed that these figures refer to 1939.

Figures reported for 1941 were 2,254 Children's Soup Kitchens, feeding 288,548 children and 1,355 Brotherhood Kitchens feeding 333,396 persons.[xxvii] There is additional information. In Cordoba, Public Soup Kitchens were created in 1937, supplying daily meals. In 1940, long lines of starving Cordobans waited for food at the Kitchens on Calle Santa Marta. There was a Nursery on the Paseo de la Victoria. In 1942, the Cordoban press published numerous photographs of huge lines of hundreds of children on many streets in central Cordoba, waiting for their meal. All are wearing uniforms: girls with white skirts and boys, also in white, with a Sam Browne type belt across their chests, in the Falangist manner.[xxviii]

1941 was a terrible year all over Spain. This was the year of the great slaughter in the Francoist prisons. Countless executions by firing squads in the capitals and an upsurge in the application of the Law of Escapes. Inmates were being decimated by the increasing lack of food. Earlier, in this book, we spoke of the Pro-Inmate Association that had been formed in Cordoba to which the regime reacted violently by arresting some hundred individuals, Hunger in 1941 went well beyond Social Welfare expectations and in January, City Hall was forced to create a **Municipal Winter Soup Kitchen** in the Mother of God and Saint Raphael Asylum. A month later, the Municipal Soup Kitchen was providing more than 2,000 meals to beggars.[xxix] So much for the 'Great, Free Spain'.

There was a decrease in Welfare services in October 1942, that at the time were providing help to between 12,000 and 15,000 persons in Cordoba capital and province.[xxx]

In Cordoba, the Welfare Authority headquarters were on Emilio Luque square. The provincial Delegate was Demetrio Carvajal Arrieta. The Woman's Section was on Calle Sevilla and the provincial Delegate was Mercedes Ordóñez Oria. The National Delegate, Pilar Primo de Rivera, visited Cordoba March 1942.[xxxi]

Undoubtedly, this was social chaos with a hyperbolic dimension, totally beyond belief today, even though today we speak of another uncontrolled catastrophe during the years that followed. In reality, country people always lived from one catastrophe to another.

Table XI

Social Welfare Assistance – Cordoba end of 1941		
Organization	*Type*	*Number Assisted*
Province		
77 Soup kitchens	Children	9,000
54 Brotherhood kitchens	Adults	4,500
Palma del Rio Nursery	Poor/orphaned children	429
Capital		
2 Soup kitchens	Children	1,500
2 Brotherhood kitchens	Adults	1,500
San Rafael nursery	Children	841
San Gonzalo residential school	Children	85
San Acisclo & Santa Victoria home	Children	73
Calle Manríquez Orphanage	Children of all ages	Countless
Soup kitchen	Beggars	2,000

Table XII

Welfare Assistance Cordoba at the end of 1942 as per *Cordoba* newspaper		
Organization	*Type*	*Number Assisted*
Province		
63 Soup kitchens	Children	7,334
22 Brotherhood kitchens	Adults	1,990
Palma del Rio Nursery	Poor/orphaned children	429
Capital		
2 Soup kitchens	Adults	1,260
4 Children's soup kitchens	Children	431
Food Centre	Children	500
Children's Homes December	Resident children Day-care children	183 602

The Social Welfare Authority had a budget of half a million pesetas in 1941, which was not much. That sum included the 80,000 pesetas it received from the sale of the required Blue Stickers countrywide. The Pozoblanco Municipality contributed with 225 pesetas a month in 1940. Not all municipalities contributed in this manner. Already in February 1941, the Villanueva de Cordoba Municipality had expressed its alarm at the great number of people benefitting from welfare (children, elderly, widows, relatives of inmates). Nothing else was to be expected because on those dates, half of the townspeople, both men and women, were imprisoned in Francoist jails. Villanueva de Cordoba did not officially support the Social Welfare's 'work of charity' until 1947, when it paid a monthly subscription of 100 pesetas for Blue Stickers but only after it was pressured by the provincial Welfare delegate.

In Villanueva de Cordoba, the Welfare Soup Kitchen was located on Calle Real (today the *Hogar del Pensionista*[9]). It is here that one saw long lines of hungry people, many of whom came from the Los Bretes Group of Schools where many people who had lost everything with the end of the war, were housed. A legion of children roamed the streets begging and a number of these were taken in by local families. In January 1941, the Villanueva Municipality increased the number of food rations served daily at the Soup Kitchen by 100.

The Municipality also took advantage of public holidays and religious festivals during the year, to organize charitable works in favour of the hungry. September 29 1939, the Feast of San Miguel (Saint Michael), the patron saint of the town, it served an 'extraordinary repast for the imprisoned and poor people of the town'. The Falange would also participate in these charitable actions for the children of the tortured, executed, imprisoned and generally excluded, during the 1940s. The Juvenile Section organized a *Reyes*[10] campaign for used clothing for poor children in 1941. At the end of that year, the *Acción Católica, Obreras Parroquiales* and *Conferencias de San Vicente de Paúlo* (religious sister/brotherhoods to which the flower and cream of the upper class belonged) organized a Christmas party for the poor.[xxxii]

As mentioned earlier, the Social Welfare Soup Kitchen in Villanueva de Cordoba was on Calle Real, whereas the children in residential care

---

[9] Home for Pensioners.
[10] Twelfth Night, the Spanish Christmas festivity in January in honour of the Three Wise Men.

were taken in by the nuns at Cristo Rey Convent. There were around 80 girls in residential care. They were primarily taught how to sew and how to pray. In November 2012, Moreno Gómez spoke with <u>Virtude</u>, from Fuentes de Andalusia, Seville, who told him her story. In 1936, they assassinated her father, her mother (who was eight months pregnant) and her grandmother. Virtudes, 14 months old at the time, was taken in by some aunts but the nuns showed up and took her to their convent by force, where they kept her until she was 13 years old.

"They taught me how to sew, how to pray and to beg for the convent, but they did not teach me my sums." Today, Virtudes is a woman of unlimited energy. "Before, one could not speak, but today I speak out loud and nobody pays attention."[xxxiii]

As to the boys, at the beginning of the 1940s, about 50 children from destitute Rojo families (parents executed, imprisoned, guerrillas, etc.) were cared for under a day-care regime at Cristo Rey Convent. The boys would arrive in the morning when they were taught how to say the rosary, then some ABCs and numbers, and finally given military training. Curiously, it was the nuns who taught them to march to the following rhyme: *Jay, jap, jaro! Los fusilados de los rojos son de palo.* After parading around the patio, they were given lunch and then sent back.[xxxiv]

In Montemayor, the children under care of the Welfare Authorities and the elderly from the Brotherhood Soup Kitchens, were treated with a special meal December 8 1941, in celebration of the Duchess of Frias' saint's day. The tale of this event is a perfect example of what was common practice on such occasions:

> "Once desert was over, both guests and the servers raised their right arms high and sang *Cara al Sol*, ending with cries of Viva! Duke of Frias! Their spokesperson added: 'We thank their Excellencies the Duke and Duchess of Frias, for this example of Christian charity, and we hope that it will be extended in favour of the humble and the needy, and for the good of the New Spain to which we all aspire."[xxxv]

The work of the Social Welfare Authority (later renamed Ministry of Social Affairs under the democracy) in Cordoba was overseen by the National Delegation under Mercedes Sanz Bachiller until the end of 1939, with headquarters at Calle Abascal number 39. Some of the heads of the Cordoba provincial delegations were Amparo Bahamonde, appointed September 15 1937; Luís Nicasio Garrido Lama, February 1 1938; Manuel León Adorno, February 1 1940; Salvador Maro Martí, May 1 1941; Demetrio Carvajal Arrieta, August 1 1941. Thus, we conclude this overview of a work of charity that is so rarely spoken of today, following what could be considered a Regime catchphrase: Execute the parents, then be charitable with their children.

Saint Paul National Board of Trustees for Prisoners and Convicts

In addition to the Social Welfare Authority, additional assistance was provided by the Saint Paul National Board of Trustees for Prisoners and Convicts, created by Ministry of Justice Decree July 26 1943 to administer the State's guardianship of the children of prisoners enrolled in the redemption program.[xxxvi]

This new organization was nothing else than a branch of the Board for the Redemption of Sentences through Work that, in October 1941, was already overseeing the care of 3,000 children of prison inmates, for whom it paid for their education and other needs. This was a dramatic increase in assistance as earlier that year, in February, the Board only reported 202 such children under its care. At the same time, reports in the local press referred to more than 6,000 children of inmates who were still not getting any help. Much of this could be explained by the confusion in the manner by which their parents were being shifted from prison to prison.

In theory, the St. Paul Board of Trustees' uplifting mandate was to take in prisoners' children and care for their upkeep and instruction. In practice, however, 'instruction' was a euphemism for brainwashing, ideological repression, preventing them from being contaminated with their family's Rojo ideology, ensuring that they are cleansed of all of the oppresseds' beliefs and impressing the oppressors' ideas on them. All this for the Rojo children whom they sanctimoniously described as "That divine mass regarding which we have to rectify their parents' mistakes".[xxxvii]

An important feature of the St. Paul Board of Trustees' actions was its avowed purpose to take advantage of the separation of parents and children – parents behind bars and children out on the street and in religious institutions – to control the families of all the disaffected. This in pursuit of the so-called magnanimous Caudillo's great task of 'disinfecting' Spain and abolishing all dissolutionary, i.e.., democratic, ideas. As Ricard Vinyes stated, this was solely another name for Franco's massive re-education project, whereby the control of the children had already been programmed by Order March 30 1940 regulating how long children were allowed to remain with their mothers behind bars.

The Francoist penitentiary system did not only target the prisoners with physical and ideological repression, but it also kidnapped their children in order to 'disinfect' and re-educate them and erase all memories that the children might have of their parents and the ideals they defended.

Ángela Cenarro Laguna described the manner by which the prison universe institutionalized these practices.[xxxviii] Children's visits to their parents in prison could be stopped if the wardens felt that there was a risk that the children might be 'scandalized by their parents' advice'. During the holiday months, the Board only allowed parents to see their children after it had received a favourable report on the 'living and moral situation' of the family. Worst of all, with its brutal National Catholic campaign against children, its loathing of their parents and the brainwashing of heir offspring, the Regime actually was successful in getting many children to disown and hate their families.

There is a famous October 5 1945 letter to *The Times* from Tomás de Boada, President of the St. Paul National Board of Trustees for Prisoners and Convicts, containing some facts, some lies and especially, several scandalous "declarations". A fragment of this letter is reproduced below.[11]

> "During my current visit to England, I noticed that there is a campaign of lies regarding how Justice is administered in Spain... Fewer than 23,000 inmates

---

[11] The full text of this letter can be consulted on the Internet but in Spanish only. The original letter could not be found after a search of *The Times* archives. The above text is translated from Spanish.

are currently housed in our penitentiaries. The prison population immediately after the end of the civil war was some 250,000 criminals convicted of common crimes punishable under the laws of all civilized nations. I have personally examined many sentences, without having found a single one that was based on a political motive. The Council cares for a large family of about 1,500,000 persons, among inmates, convicts and families."

Boada's only correct statement was that Francoist penitentiary repression affected 1,500,000 persons; prisoners and their immediate families. The number he gives for prisoners under Franco soon after the end of the civil war is a lie as the total exceeded 300,000. Another blatant lie was that those sentenced under Franco were tried only for common, not political crimes. This was pure propaganda, a typical example of Franco's usual mendacious policy when addressing an international audience as he played lip service to International agreements. Foreign governments, either ingenuously or hypocritically, believed him. At the end of 1947, only 2% of all prison inmates in Spain were officially classified as common criminals. In other words, 98% of the remainder of those imprisoned in Francoist prisons had been tried and sentenced by court martial law for purely political reasons.

FINAL COMMENTS REGARDING A PERIOD OF REPRESSION NEVER BEFORE SEEN IN SPAIN

The 1947-1949 Triennium of Terror

Located at the final stages of this book, having reached the limits of permissible space, it is still important to give the reader a minimum indication of the massive repression with which Franco terrorized the Spanish countryside during the period that Moreno Gómez has called the Triennium of Terror (1947-1949), a period regarding which there were serious precedents in 1946 and a bloody epilogue in 1950.

Since 1987 Moreno Gómez has consistently drawn attention to this new Francoist despotic turning of the screw in its repression of the civilian population (fathers, mothers, sisters, wives, and other relatives of those who had fled to the mountains, including many with presumed links to them, which in many cases simply did not exist). Today, however, many historians of the period have not yet reached this last great chapter of the Francoist repression, the one that sowed the greatest amount of terror in the rural world, with some assistance from the city police forces.

In guerrilla vocabulary, this is often described as 'repression in the plains', as opposed to 'repression in the mountains'. In the case of the repression in the plains, more than looking for fugitives or those opposed to the Regime, the real purpose was to sow such terror in the countryside that no guerrillas would be able to survive in the maquis. It was a scorched earth policy whereby the liberal application of the Law of Fugitives[12] or simple summary executions, were responsible for 1,500 deaths throughout Spain during those three years.

Pay attention, all you half-baked negationists, literati of the Regime, members of the 'third Spain', and liars and prevaricators of all kinds. Also take note, you, who like Julius Ruíz, without knowledge of the cause, fatuously nourish the theory that the Francoist repression decreased significantly after 1941, totally disregarding the fact that there still were many more deaths to come after that dark date in the history of Francoism.

Moreno Gómez regrets that the scope of this book is not sufficient for him to go into great detail regarding the tragedy of the Triennium of Terror, especially in the Cordoba countryside, but in other Spanish provinces as well.

He does, however, refer the reader to two of his previous books, both in Spanish, that deal with this subject and all the horror of Francois that came into play, unrestrained, over all the Spanish countryside:

> *Cordoba en la posguerra (La represión y la guerrilla, 1939-1950)* (Post-war Cordoba. Repression and Guerrilla, 1939-1950), Cordoba, 1987. In which he introduces his preliminary study of this dark period in Spanish history.

---

[12] The term employed here as *Law of Fugitives* and previously as *Law of Escapes*, represents in both cases, the same legislation. The application of the law, however, requires a different translation of the Spanish "Fugir" which can be translated as 'to run away' or as 'to escape'. In the first case, the 'Fugitives' refers to individuals who flee prisons, camps, to the mountains and elsewhere and who when captured were summarily executed, as opposed to 'Escapes' where these are individuals already captured and on their way to prison but allowed to 'escape' so that their guards could shoot them in the back as they ran away from their captors.

and his primary work on this subject:

> *La resistencia armada contra Franco. Tragedia del maquis y la guerrilla.* (Armed Resistance Against Franco. The Tragedy of the Maquis and the Guerrilla.), Crítica, Barcelona, 2001. In which he presents an exhaustive study of that which was the serious criminality of the Triennium of Terror.

In addition to several articles, Moreno Gómez co-authored another book, also in Spanish: *Morir, matar, sobrevivir. La violencia en la dictadura de Franco* (Die, Kill, Survive. Violence under Franco's Dictatorship), Crítica, Barcelona, 2002, which provides data regarding the Triennium of Terror in the rest of Spain, as he talks of those who fled, who became guerrillas and freedom-fighter members of the resistance.

Regarding documentaries regarding the 1948 terror, Moreno Gómez refers the reader to two in particular, that he considers especially awesome.

The first of these, *Muerte en El Valle*[13] (Death in El Valle), by Cristina M. Hardt, CM Pictures for Channel Four Television, 2005, is about a crime in a small town in León province in 1948, when the Guardia Civil applied the Law of Fugitives to two farmers, one of whom, Francisco Redondo was Cristina's grandfather.

Many years later, his granddaughter who was living in New York, decided to tell the world what Franco's Regime had done to him. She went to El Valle and asked questions, searched and researched the facts, no matter how scared the townspeople, and her relatives still were of being seen speaking to her.

Francisco and his wife Florentina sheltered guerrillas in their little old farmhouse in the country. One day, an informer turned them into the Guardia Civil who raided their farm and burnt the house down. The guerrillas who were there escaped but Francisco and his wife were arrested and taken to Bembibre prison where they were horribly tortured. Six days later, they took Francisco and Florentina into a field where they shot them both.

---

[13] The documentary, *Muerte en el valle [DVD]*, dubbed in Spanish, is available from amazon.es

Back in El Valle, Cristina had the guts to discover which Guardia Civile had shot them, she went to his house and recorded her interview with him. Only an American is capable of such bravery. In Spain, nobody would dare to do such a thing.

The second awesome documentary that Moreno Gómez mentions is practically unknown in Spain: *La isla de Chelo* (Chelo's Island), Film IBCinema, 2008, by Odette Martínez, daughter of the ex-guerrilla *Quico*, the story of another tragedy in El Barco de Valdeorras, Galicia.

The film recounts the sufferings of the Rodriguez Montes family because two of their sons had gone over to the guerrillas. The Guardia Civil went to their farmhouse, ordered the parents to come out and shut their daughter Consuelo in the sheep pen. After a while, they called her to come out and say goodbye to her parents. Her mother held her hand tight, begging them to leave her alone, but they put Consuelo back with the sheep. They took her parents down a footpath, where the Guardia brought them down with a burst of gunfire that the young girl heard from the corral. Hours later, neighbours came and took her to the place where her parents had been killed and where they buried them. Some time later, the Rodriguez brothers came down from the hills and took their sister back with them. She spent three years in the hills where she was known as *Chelo*.

The relentless Francoist repression of those it considered its opponents, both before and after this bloody period, lasted until 1950. Fifteen years killing people so that "Thirty years from now, none of this scum will be alive" the Seville executioner, Captain Díaz Criado liked to say. In 1950, the Law of Fugitives was still being applied in some places in Spain, as in Nerja, Malaga.

Beginning July 18 1936, Spain suffered a never-ending ordeal that was so much more than simply the slaughter of so many people. It was the concentration camps, mass arrests, killing through starvation in the prisons (just as in the Nazi camps), slave labour, forced labour battalions, deporting prisoners from their hometowns after their release the lack of food and widespread hunger, social exclusion, blacklists of defeated prevented from working, purges, rationing, black-markets, courts martial, mass torture, firing squads, summary executions or Law of Escapes, until 1950. Especially as regards the defeated or disaffected, it was shaving women's heads, the castor oil treatment, stripping women naked in public and other gross humiliations, the disappearance and kidnapping of children, imposition of religion by force brainwashing children, official theft and seizure of the defeateds' property, the purposeful destruction of families, and so much more.

All of this Moreno Gómez describes in great detail in this work as he provides solid proof, based on his in-depth research and interviews with survivors and eyewitnesses of Fascism's crimes against humanity, many of which are in agreement with Lemkin's descriptions of acts of genocide. Crimes that the reactionary sectors in Spain (the Military, the Falange and the Church, in the barracks, the casinos and church vestries) committed as they attempted to emulate the terrible form of Fascism practised by the Rome-Berlin Axis. The July 18 1936 military coup led to a terrible civil war whose end in 1939 brought neither peace nor reconciliation, nor forgiveness, nor amnesty, none of that by which all Spaniards, both victors and defeated, could come together.

On the contrary, far from encouraging a reconciliation of parties, the victors embarked on a cruelty in which they wallowed without restraint, with an indescribable fury and an unmeasurable desire for revenge that is unique in the history of Spain, as they forced the defeated to bite the dust for almost four decades. The result was a colossal humanitarian catastrophe as Francoists and National Catholicism rode roughshod over every form of human rights, past and present, added to which, the departure into exile of half a million eminent intellectual Spaniards, was an irreplaceable brain drain forever lost.

For all the wealth of detail, this book only pretends to disclose the tip of the iceberg of that which really occurred. He who wants to know, knows; he who does not, let him continue to hide his head in the sand or to contribute to the great lie, another popular national sport.

Moreno Gómez concludes, for the reader's reflection of this tale of a terrible conflict, by quoting the Greek Herodotus', father of History, introduction to his work *Histories*:

> "Herodotus of Halicarnassus hereby publishes the results of his inquiries, hoping to do two things: to preserve the memory of the past by putting on record the astonishing achievements both of the Greek and the non-Greek peoples; and more particularly, to show how the two races came into conflict."

ENDNOTES FOR CHAPTER X

i   Raphael Lemkin. Axis Rule in Occupied Europe. Carnegie Endowment for International Peace, Division of International Law, Washington D.C., 1944.

ii   Gonzalo Amoedo López & Roberto Gil Moure. Op. Cit. p. 85.

iii   Diego de San José. De cárcel en cárcel (From jail to jail). Do Castro, La Coruña, 1988.

iv   Julián Casanova. La Iglesia de Franco (Franco's Church). Crítica-Bolsillo, Barcelona, 2005, p. 299.

v   José Merino Campos. Interviewed by Moreno Gómez in Cordoba, February 22 1981.

vi   Google search: https://encrypted-tbn0.gstatic.com/images?q=tbn:ANd9GcRTjFZr-R4vycJF1-U8k9Ybw7FG32Ji3u4JUaw&usqp=CAU

vii   Pablo Uriel, Mi guerra civil (My Civil War). Introduction by Ian Gibson. Self-published, Valencia, 1988.

viii   Julián Casanova, Op. Cit.

ix   Melquecídez Rodriguéz Chaos. 24 años en la cáracel. (24 years in jail). Forma, Madrid, 1977, p. 162.

x   Manuel Espejo. Interviewed by Moreno Gómez in Madrid, January 1984.

xi   María Salvo i Iborra. Interviú maazine. October 17 2005. Also, testimonials from previously referenced work by Ricard Vinyes, Thoasa Cuevasm Fernado Hernández Holgado and an email from the Andalucia CGT October 29 2005.

xii   Carlos Fonseca. Trece Rosas Rojas (Thirteen Roja Roses). Temas de Hoy, Madrid, 2005.

xiii   Memoria que eleva al Caudilli de España y a su Gobierno el Patronato Central para la Redención de Penas por el Trabajo, de 1943. (Memorandum from the Caudillo of Spain and his Government regarding the Central Board for the Redemption of Sentences Through Work. 1943). Madrid, 1944.

xiv   Text transcribed from Presos del Silencio (Prisoners of Silence), documentary by Mariano Agudo & Eduardo Montero, Intermedia Produciones, Seville, 2005.

xv   Children during the Spanish Civil War, [between 1936 and 1939] Alexander Albert MacLeod fonds Photographer Unknown Reference Code: F 126-4-0-7 Archives of Ontario, I0013272.

xvi   Francisco Moreno Gómez. "La represión en la posguerra" (Postwar repression) in Victimas de la guerra civil (Victims of the civil war). Temas de Hoy, Madrid, 1999, p. 286.

xvii   Preliminary report of the UN Working Group on Enforced or Involuntary Disappearances after a visit to Madrid, Spain, 30 September 2013.

xviii   Basque television documentary Prohibido recordar, Op. Cit. 2010.

xix   Catalan television documentary Los niños perdidos del franquismo (The lost children of Francoism), Montse Armengou & Ricard Belis, 2002.

xx   Juan Vallejo. La locura y su memoria histórica. (Madness and its Historic Memory). Ediciones Atlantis, Madrid, 2013.

xxi   A. Rodríguez Arias. Interviewed in the documentary Esa Memoria (That Memory) by Dominique Gautier & Jean Ortíz, Crea. Atlantique, France, April 2011.

xxii   Interview in Günter Schwaigaer's documentary Santa Cruz, por ejemplo (Santa Cruz, for example), Austria, 2005.

xxiii   Montse Armengou, Op. Cit.

xxiv   Mónica Orduño Prada. El Auxílio Social. (1936-1940). La etapa fundacional. (The Social Welfare Program. Foundation). Escuela Libre Editorial, Madrid, 1996.

xxv   Ángela Cenarro Lagunas. "Historia y memoria del Auxilio Social de la Falange" (History and memory of the Falange's Social Welfare Program), Pliegos de Yuste, issues 11-12, 2010.

xxvi   Azul newspaper, Cordoba, October 18 1940.

xxvii   Córdoba newspaper, October 30 1941. Headline: "Social Welfare obeys. Intense so-
        cial, moral and religious reconstruction. 6,000 christened, 150,000 first communions,
        2,000 marriages legalized".
xxviii Córdoba newpaper, November 3 1942.
xxix   Azul newspaper, Cordoba, January 24 1941. "Campaign against begging in Cordoba.
        City Hall creates a Municipal Winter Soup Kitchen with a sum of 50,000 pesetas. The
        Falange will collaborate with all its means."
xxx   Cordoba newspaper of April 23 1942 published a very complete list of all those assist-
        ed in the Soup Kitchens.
xxxi   Azul newspaper, Cordoba. March 17, 1942.
xxxii   Minutes of the Meetings of the Villanueva de Cordoba Municipality dated 28-8-39;
        31-12-40; 28-1-41; 25-2-41; 16-12-41; and 18-3-1947 [day-month-year].
xxxiii Interviewed by Moreno Gómez on the occasion of a Historic Memory Conference,
        Seville, November 24, 2012.
xxxiv   Ernesto Caballero Castillo, testimonial from one of those boys.
xxxv   Azul newspaper, Cordoba. December 13 1941.
xxxvi   There is an ancient publication on this matter with several authors: Crónica del Patro-
        nato Nacional de San Pablo (Chronicles of the National Board of Saint Paul), Ministry
        of Justice, Madrid, 1951, 379 pp.
xxxvii Mírta Núñez, Op. Cit.
xxxviii Ángela Cenarro Laguna. "La institucionalización del universo penitenciario franquista
        (The institutionalization of the Franco prison universe), in Uma immensa prisión (An
        Immense Prison), Crítica, Barcelona, 2003.

APPENDIX I

COURT MARTIAL PROSECUTOR JOSÉ RAMÓN DE LA LASTRA Y DE HOCES' COMMENTS IN A COURT MARTIAL.

Other rebuttals, appeals, statements of fact.

<u>**Baena Court Martial May 20 1939**</u>, excerpts of which are reported in the Cordoba *Azul* newspaper, May 23.

- "**Prosecutor José Ramón de La Lastra y de Hoces** began by stating that the most serious feature of the tragedy was that a criminal fury was raging through every town and across every field in the province. He resisted the desire to analyse the horrifying events described.

Referring to one of the accused, he began: "Cristóbal Díaz Borrego, aged 6, accompanied his father to the Casa del Pueblo. At the age of 12, he had become a Socialist Republican. That father, instead of giving his son a Christian education, instead of guiding him along the paths of culture, allowed him to go to places where hate and resentment came together. He opened the road to where his son would later go to commit crimes of this nature, crimes that would bring him close to the gallows. That father taught him how to hate God and not to love his Fatherland. The responsibility of his having done so should weigh like a slab of stone on his conscience. This is a painful and terrible lesson for those who do not supervise their sons with true love, who do not instil the virtues of men in their souls as opposed to the instincts of wild beasts.

To think that one can throw three fellow men alive, into a bonfire, and that one can stuff another person's mouth with a burning rag when he called for his mother in the hour of his anguish and torture, is horrible. And that this is done in the name of liberty, equality and fraternity is doubly monstruous. The dastardly Republic has made us witness these inhuman vandalisms. Those responsible for these dramas do not deserve to live, neither inside nor outside of Spain.

I stand before the Court to ask for justice. We must remember Pradera the martyr's words: 'The life of a criminal is worth nothing, when compared to the life of his victim.' Our Caudillo, the man who providentially has saved the Fatherland with his magnanimous heart, has offered his pardon to those whose hands are not besmirched with blood. But that promise is also a promise for justice to be applied to those who have blood on their hands. Given these facts, I have only one word for the accused: Since you have not known how to live as Christians, know how to die like those who do not behave like men.

On behalf of His Excellency, the Head of the Spanish Government, I ask for..."

Rafael Bedmar Guerrero. Testimony in rebuttal to **Prosecutor José Ramón de La Lastra y de Hoces'** allegations at Rafael's trial in Cordoba Court October 25 1939, as he spoke in his own defence. Presiding Judge: Evaristo Peñalver.

■ "Don José de La Lastra began his speech as follows: «President, gentlemen of the Court! What you have here is the scum of society. This is the Marxist rabble that we must remove from every town in Spain. Everyone here will say he is innocent, but who killed our priests? Who burned our churches? Who killed our lawmakers? How may of ours have fallen for God and for Spain? The blood of our finest asks for Marxism to be exterminated from our society. »

He concluded his lengthy discourse, one directed at the emotions of the Francoist cause, with these last words for the presiding Judge: «President, gentlemen of the Court: all the accused are guilty of high treason against the Fatherland, as their hands are sullied with blood, all prefer to make a first rather than extend an open hand, all are Marxists and for all of them without exception, I ask for the sentence of death. I have spoken. »

The officer who was acting as Public Defender limited himself to asking the Court for clemency for the accused. Turning to us, the Presiding Judge declared: «As we know beforehand that you will all say the same thing, let one of the accused stand and speak for all. » Not even one did so. So, he turned to me and said: «You, the youngest, stand up and speak on behalf of your companions. » The guard who was standing behind me indicated that I had to stand up.

I began: Sir, I was only 16 years old on July 18. I did not think that by putting myself at the service of the legal Government of the Republic I had committed an act of high treason. I have served the Spain that you all one day swore to defend. If I had burned churches, as I am accused of doing, which is not true, I would have destroyed buildings of stone, but you are destroying human buildings.

Mr. President: Have we been permitted to see an exact information of everything we have been accused of? Have we been allowed to appoint our own defence? How is it possible to condemn innocent men who were not permitted to prove their innocence? How, seated in front of a cross of Christ, can you apply the same methods that were used by those who crucified him?

The Presiding Judge slammed his gavel on the table and shouted out in anger! «Enough! This session of the court is adjourned. »"

Other rebuttals, appeals, statements of fact.
[*The sentences this Judge always asked for were death, by firing squad or by garrotte, or by both.*]

<u>Eugenio Jurado Pozuelo</u>. From Villanueva de Córdoba. Appeal against death sentence filed May 9 1940.

▪ "Dear Sir,
EUGENIO JURADO POZUELO, aged 33 years, married, carpenter by trade, born and resident in this town, tried March 13 in this town by court martial that condemned him to the 'ultimate sentence', respectfully presents the following comments regarding certain statements recorded in his file, to Your Excellency, for your consideration:

1. As I was in the fields on July 18, the date on which the Movement began, I had no knowledge of it until the 19th at 3 p.m. which is when I returned to the town. I remained at home, or nearby, without anybody asking me to do anything until the 22nd, when I was ordered to leave the town by the Evacuation Committee. I duly set off with my wife and a niece to a farm some 15kms distant from the town, having spent the night on the way in a shed belonging to

road workers in the place called Custea del Jaro, together Alfonso Tintorero, Francisco Serrano and his wife.

The next morning, as we attempted to continue our move, we were stopped by a truck driven by Arturo Díaz López and occupied by several individuals armed with rifles who forced me to abandon the women who accompanied me and return to a place near the town called La Zorrera, where there was a camp of armed Rojo men and where several members of the right-wing were being held under arrest. I protested vigorously against the mistreatment of some of them and I was listened to, as Juan Pulido Díaz who was one of them, can testify.

I remained in that location and near the camp from the 23rd until the morning of the 25th, during which there was an attack against the Guardia Civil barracks and the town was retaken [*sic.* by Republicans], events in which I took no part, precisely because I was elsewhere, as I have stated.

On the morning of the 25th, once the fighting had ceased, I entered the town by the street on which my parents and I lived, accompanied by Pedro González Valle and a brother and a nephew of his. We separated at the door to right-wing leader Diego Romero Rodríguez's house, without bothering or asking anybody for anything, even though we noted that he was hiding. I returned immediately afterwards to my parents' house.

2.   When I went to the Totana Communists, which I did, not for the reasons given at my trial that I wished to aggravate the situation of the right-wingers who were arrested there, but just because I wanted them to do what they could for one of my brothers who had been arrested and who, despite everything I tried to do for him, was still sentenced to 30 years in prison.

3.   I was later appointed to the War Committee but after all the bloody events that occurred in this town, I only remained as member for a fortnight. I was appointed Director of the Electrica in this town and was replaced by somebody else on the said Committee.

4.   I joined the Communist Party in 1931, the year in which the party became legal, and I had no position of any responsibility until well before July 18.

5.   I always did what I could to assist right-wingers to avoid as much unpleasantness and trouble as I could, to Crístobal Arellano, Diego Romero, Diego Higuera Díaz, Andrés Cabrera Valero,

Francisco Ochoa Ortega and José Benítez Caballero. It is totally false that my declarations in Jaén harmed the accuse, as Basilio Villareal and Sánchez Gómez said. Proof of this can be obtained from Blas Carbonero and Bartolomé Torres and others who will speak for me.

Given all the above, the truth of which I swear to by all that I hold dear, that this represents all my actions. It causes me great pain to think that others who were much more extremist than I, have been sentenced to 30 years in jail, whilst I have been condemned to the maximum sentence under the Law.

I BEG YOUR EXCELLENCY to take all the above into consideration and ask for confirmation of what I said, if you should consider it appropriate, so that my sentence might be lessened.

A favour that I hope I have deserved from Your Excellency, on whose life God may shine for many years to come, for the good of Justice.

Eugenio Jurado Pozuelo
Villanueva de Córdoba
May 9 1940."

Pedro *Cuadrado* Torralbo Gómez. Villanueva de Córdoba. Appeal filed June 29 1940. Ex-Militia Captain, executed one year later, June 3 1941 in Cordoba.

ADDRESSED TO THE CORDOBA JUDGE ADVOCATE

- "Dear Sir,

I, PEDRO TORRALBO GÓMEZ, aged 40 years, married, weaver by trade, resident in this town at Calle Egidio number 23, interned in the prison of this town and submitted to an emergency summary court martial by this Party's Military Tribunal, respectfully presents the following to Your Excellency.

WHEREBY May 1 1940 I was tried by the Military Tribunal and as requested by the Prosecution, was condemned to the maximum sentence on the grounds that, under the terms of Article 238 of the Military Code, I was guilty of having adhered to the military rebellion.

The charges laid by the representative of the Law and the subsequent request for the maximum sentence were motivated, without any doubt and I say so with all due respect, to a defect in the summary information that was presented to the Court, given that there are missing elements of information which had they been included in the charges, would have meant that the Prosecutor's arguments and conclusions would have been different and always more beneficial to the undersigned petitioner. The following information was missing from my case file:

FACTS

1. It is not true that I took part in the taking of the towns of Valle de los Pedroches as a Militia Commander. I never held that rank and nobody can say that I participated in that battle in any of those towns because I was not there nor was I a member of the Rojo Militia who went there, as I can prove with witnesses (Miguel Camacho Illescas, ex-resident at Calle del Plazarejo).

2. It is also false that I took part in the attack on the Guardia Civil barracks, as I was elsewhere and not in this town when that occurred, as Juan Vacas Capitán, prisoner in this prison, and Diego Camo Rico and Miguel Camacho Illescas (both members of the right-wing), and others who were detained can attest to, and furthermore, I attempted to stay with them to ensure that nothing would happen to them.

3. When I arrived at Fuente Vieja square where two people died and several were injured the events had already taken place, as Juan Gómez Calero (Calle Conquista), Juan José Fernández Moreno (Calle de Herradores) and others who were wounded there can attest to. They can also confirm that I was threatened by death by someone from out of town called *El Trapero* who pointed his gun at me because when I arrived at the place where these events had occurred, I said that those responsible were murderers and strongly condemned what they had done.

4. Julián Caballero, the Mayor at the time, ordered me to La Charquita, where the Guardia Civil and several other countrymen surrendered, to prevent, if I could, any bloodshed In the first place, I did not participate in the fighting there, because that was not my mission, and in the second place, the fighting was over when I arrived there.

Soon after the Mayor and other members of the Frente Popular took charge of the detained and took them back to town along the

Pedroches road to Villanueva da Cordoba. I returned along the road to Torrecampo to search for my family which was in Dehesa de Navauenga, and I did not return to town until the next day when I discovered that almost all those who had been detained at La Charquita had been shot.

5. I strongly deny all moral and material responsibility in those murders. I was, at all times, against such occurrences, as Juan Camacho Castillo, imprisoned in this jail, and others can attest to. In fact, I wrote to the Frente Popular condemning that which was going on. Seeing that this was not working, I suggested that Juan Camacho Castillo should call a meeting of the Communist Party, which he did, and I repeated my condemnation of the crimes that were being committed, describing them as murder. Faced with my opposition and that of others, such as Juan Camacho, we were able to get many to react and more than a few took our side until we all agreed that nobody should be killed. I firmly believe that if there had not been the intervention of individuals who were strangers to the town, we would have managed to enforce the agreement we had just reached.

6. My actions and behaviour towards the members of the right-wing was to protect them as much as I could, as can be proved by José Liñán (Calle del Torno) and by Pedro Cano Rico (Calle de la Preturilla), in whose case, when I discovered that they were looking to arrest him, told him to remain hidden in the house in which he was until I could come and get him out without danger. Nobody can say that I arrested anybody, that I pillaged any house, that I molested anyone with words or deeds, as I never was an organizer or member of the War Committee. Likewise, nobody can say that I was ever seen with that Committee, not even at a distance because I found its actions totally repulsive.

I BEG YOUR EXCELLENCY to read the above, to reflect upon my account and after considering everything that I have said, nullify the sentence to which I was condemned and order the revision of the indictment. This is the Justice that I hope to obtain from the high principles advocated by Your Excellency, on whose life God may shine for many years to come.

Pedro Torralbo
Villanueva de Córdoba Military Prison
June 29 1940

<u>Names of witnesses, written in the margin:</u>

Casildo Cabezas, Juan Cantado Díaz, *Vizco* Zurita, José Liñán (Calle del Torno), Pedro Gómez Calero (Calle Contreras).

Also written in the margin:

Manuela, have somebody type this for you and send it to me so that I can sign it. Kisses to the children.

With my love, your husband, Pedro Torralbo."

<u>Adriano Romero</u>. Villanueva de Córdoba. Important pre-war labour leader in Cordoba. Belonged to the Central Committee and the Political Bureau of the PCE during the Republic, Frente Popular Member of Parliament for Pontevedra, fought as a Major in the Militia. Arrested March 11 1940 in Ciudad Real, onlyt tried by court martial August 1941 in Seville, when he was condemned to death. Retried September 1943 in Granada by a second court martial, was sentenced to death a second time. Thanks to the efforts of several influential individuals, his sentence was later reduced to 30 years in jail.

- "José Espina told me that my indictment was not yet finalized and that it contained some very serious accusations and if the case was quickly taken to trial there was a risk that I would be speedily liquidated, but if I let it rest, which was the best thing to do, it might one day be taken down a rung or two. We agreed that he would gradually work on this."

A whole new year passed during which he was able to gain some time, until there was a change in Judges.

- "My dossier was closed, just as it was, and given to a Plenary Judge so that I could be informed of the charges against me. I was alleging that the case had not yet been prosecuted and that the serious accusations that were made were false, and I asked that a series of witnesses be interrogated to corroborate my affirmations.

The Judge told me that what I wanted would need six months to process and he could not agree to my request, which led me to understand that he had been ordered to proceed... Consequently, I was also not interrogated. Finding myself in the worst conditions, without any witnesses nor any documents in my defence, I was taken to trial by court martial in August 1941...

This first court martial was also remarkable, because as I was very well known in Andalusia for my political activities, they wanted to give the people of Cordoba the idea that I was being tried in a Court of Justice. Contrary to the custom by which defendants were tried in groups, I appeared alone before the Judge. The Prosecutor, who had long worked for that Provincial Court and who knew of my activities before the war, presented a very detailed description of my political activities, that I never denied.

As I knew that I would not be allowed to speak at the end of the trial, I adopted the tactic f defending myself during the interrogation stage in order to destroy all the non-political accusations against me, to wit: of having served as Secretary of a People's Court that tried and sentenced to death the Guardia Civiles who rose against the Government in Villanueva de Córdoba, my birthplace, and of having organized a Column Unit in Almería with which I disembarked in the Port of Morril. Both these charges, under Article 37 and 28 of the Code of Military Justice, were penalized by death on the grounds of military rebellion and leader of a rebellion.

As those charges were so obviously false, I had no trouble rejecting them. I could not have participated in the said Court because on that date I was on the front line in Granada. The Column that I was accused of organizing was organized by the Almería Military Headquarters and I only took command of it after it had disembarked.

The Prosecutor, after making a long and detailed report on this, could no longer justify the charge of military rebellion and asked the Court to adjourn so that he could organize his notes. When he returned, he presented another report in which he modified his conclusions, stating that he could only accuse me of 'supporting the rebellion', which required a sentence of life imprisonment.

When the Court returned, I was condemned to death, on the grounds that during the court martial I had shown that I was the same enemy as always."

Letter from Colonel Joaquín Pérez Salas to Bartolomé Fernández.
Letter written June 16 1939 from Garay Barracks, Murcia prison, in reply to Bartolomé Fernández' request for assistance with his own appeal. Pérez Salas was executed August 4 in Murcia.

- "Murcia
To Bartolomé Fernández, Cartagena
June 16 1939

My dear friend,

From the first days of April, I have tried by every means available to me, which naturally were not very many, to find out how you were doing. I had earlier tried to get you out of Murcia, which I was unable to do because your arrest almost coincided with mine. In mid-April I thought I heard your name called on a list in the San Julián patio, where people replied to their second surname, but I could do nothing – I was incommunicado – please confirm whether you were there.

On the 17th, I was transferred to this place and again I tried to find out about you, but I could not find anyone who could give me any news. Today, I finally got a letter from Cifuentes, whose whereabouts I also did not know, where it appears that you are there, to where I am writing.

In addition to my natural interest in getting news from you, the reason I decided to bother you was to offer to do anything that I can to help you. If you need, or for any reason are interested, in a declaration from me, always in agreement with the truth that I believe in and that you deserve, you must tell me as soon as you can, with details of what you need and to where I should send my statement.

It may appear more logical for such declarations to be submitted indirectly, rather than at the request of the interested party, most especially in cases such as yours, whereby telling the whole truth the declarant is doing the greatest favour. Even so, I did not wish to do so without first telling you, because it would be simpler and more effective if you could tell me exactly which points could be of greatest interest in my declaration.

Of course, as I said earlier, everything that one can declare regarding you, in respect of the truth, must be in your favour. But I must again strongly insist that it is far better for me to do so, you must at least in part tell me what it is you wish me to declare and to whom I should address my declaration.

I believe that you did very well, as I did, even though you had, as I did, all kinds of means at your disposal to avoid these small bothers, precursors of other greater ones. At least, one way out is to confess to the accusation of a crime that neither you nor I committed, regardless of whether others believe us or not, and to accept the punishment, no matter how severe it is, as in my case with the full knowledge that we did our duty.

I presume that you have heard of Castro's march, regarding which I have hard no more since March 29, which is not surprising. If you have any news about any of our friends from Andalusia, I would like to hear it.

I await your reply so that I can work in the way that you ask of me and, meanwhile, please receive a warm hug from your friend,

J Pérez Salas
Garray Barracks

The construction of a wall at Mauthausen at the start of 1941. In the foreground, two Spanish inmates.
MUSEU D'HISTÒRIA DE CATALUNYA[1]

---

[1] https://images.english.elpais.com/resizer/DOV9o7tWcLLAmNxQJXy20D53Hs= /980x0/cloudfronteucentral1.images.arcpublishing.com/prisa/WOTAA2FCAPKAZV ZHKU634Z7UYM.jpg

APPENDIX II

THE FRANCOIST DEATH RITUAL.
FOUR EYEWITNESS DESCRIPTIONS.

Ángel Horillo. OJUELOS ALTOS, FUENTEOBEJUNA.
Description of the moments leading up to the execution, by a survivor.
Letter to Moreno Gómez from Peñarroya-Pueblonuevo, July 1 1983.

■ "When I returned to my village in May, even before I entered it, I was surrounded by armed men and taken to the Falange. I could easily deny the first allegations, because when they met in their committee where they used to determine the charge, they produced such a number of false accusations, with the sole purpose of eliminating me, I was tried by a summary court martial, with another nine men, and there were ten death sentences.

There were some 80 of us in jail during 1939 and the beginning of 1940. From the patio, we could see the arrival of those who came to sign the list of those who would be executed that night. In this terrible situation, we found ourselves at 8 or 9 at night and, arranged in circles, the jailor - *Don Manuel* he was called – made us sing *Cara al Sol*, once or twice. He then gave us five seconds for all to go through the 70-centimetre door, but before doing so, hit us, left, right and centre.

A little later, speaking very slowly he would read out the list and sarcastically say "To the chop" after each name. Four of my group were shot on four different nights: Antonio Ruíz, Luís Romero, Antonio Muñez and Juan Pedro Hidalgo, the last of whom was my uncle. He and his wife were kept in a separate room, but of the same lot of prisoners, and when his name was called, she let out a terrible cry. I found it remarkable that I could not control either my heart or my nerves."

Eva Ruíz. SEVILLE. Granddaughter of Antonio Ruíz Quiles from Alcalá del Río, Seville, executed following a saca from Seville prison October 22 1936. Report presented by Cecílio Gordillo at the Meeting of the Historic Memory Association September 2006.

- "The night my grandfather was shot, a cousin of his or of my grandmother's, I am not sure which, and who worked with him at the power plant, was nearby. It was dawn and he heard voices next to one of the cemetery walls, and he recognized my grandfather's voice. My father tells me that, according to this relative, he recognized his voice shouting and begging not to be killed because he was the father of four."

Antonio Chaparro. VILLANUEVA DE CÓRDOBA. Eyewitness account of the trip to the cemetery and the execution of three men November 7 1940.

- "We were at Fuente Vieja when we heard Pedro Juan *El Chunga* Martínez shouting *Let's go to the bullfight!* We followed the truck and stood behind some walls close to the cemetery. There were a great many soldiers and Guardia Civiles. The condemned men got off the truck like sleepwalkers, white as sheets, and without a word, walked towards the wall. They stood with their back to the firing squad. Corporal *Pepinillo* Moreno Sevilleno gave the signal with a white handkerchief, without saying anything. The shots rang out and suddenly, the head of one of the men with white hair and a ruddy complexion, turned red, as if it had exploded. He and another man fell to the ground, but Pedro Juan jerked upwards, fell on his back, one leg bent at the knee and moving. Corporal Moreno approached him administered the shot of mercy."

José Moreno Salazar. BUJALANCE. Eyewitness to the execution of three young men from Bujalance, without due process of law, November 22 1941. Reported from a position near the cemetery.

- "I witnessed the execution of those four[2] companions, friends of my brother's. They were shot at 9 a.m. against the cemetery wall, where the executions were carried out daily, at dawn. In the case of my brother's friends, they wanted to advertise the event and issue a word of terror, which is why they chose daylight.

My father and I stood behind a nearby wall, and we watched it all. The two brothers, Nievas and Joseillo, stood with heads bent and said nothing before they died. However, one of them was a true anarchist,

---

[2]   Three, according to the Civil Registry records.

19-year-old Alfonso Alharilla Morales, a close friend of mine, whose father and older brother had already been executed in the same place. He often went out with me to look for food for his mother and young brothers. He was a courageous person...

It was a very clear day, and you could see and hear everything perfectly. From my look-out point, I heard the fatal 'Ready, Aim...!' The two brothers, Nievas and Joseillo, dropped their heads but Alharilla, unbuttoning his shirt and presenting his chest to the firing squad, shouted: "Shoot, cowards! We anarchists do not fear death! Viva the FAI!" I couldn't see anything else because my eyes filled with tears."

APPENDIX III

FAREWELL LETTERS FROM PRISONERS TO THEIR FAMILIES, WRITTEN IN THE PRISON "CHAPEL" AS THEY AWAITED EXECUTION

Joaquim Moreno Muñoz. BAENA. Letter to his family written from jail February 1942. A returnee from Baena, aged 43, he was imprisoned in Córdoba. One month later, 10 March, he was executed in San Rafael cemetery, Córdoba.

- "My dear parents and brothers… Mother, I received everything that you say you sent me, as you do not know how much good it did to make me better from the weakness I was suffering. If you can continue to come or send me clothes and what food you can, I will be better off; if you cannot, it will be impossible for me to go on… You have no idea how grateful a prisoner is to know that he is remembered, more so in the circumstances in which I find myself; nobody knows that better than one who has to endure these situations and how they devour you. Tomás of my heart: do whatever you can, both for the sake of my health and to relieve me of this unbearable burden…"

Francisco Copado Sánchez. VILLANUEVA DE CÓRDOBA. Executed in Paterna, Valencia, 1 November 1939. First cousin to Fr. Bernabé Copado, S.J., the famous Jesuit chaplain to the Nationalist Redondo Unit conscripts. Arrested on a trumped-up false allegation. His wife wrote letters and tried to visit his cousin in Málaga to enlist his help in obtaining his release, but he ignored her. A few days before his execution, Francisco wrote Maria, praising her attempts in his favour but clearly expressed his resignation to his fate which he decided was predetermined.

- "Yesterday you wrote of your determination and plans to go and see cousin Bernabé in Malaga and have him come to our hometown. Dear girl, I do not want you to undertake another difficult trip, but when I consider what is and has been done, go ahead. Because, even if you can get Bernabé to go to Villanueva and despite that, he is not successful in getting the reversal of that denunciation whilst he is there, it is because their 'secret committee' met, and they did not agree to do so. I imagine, Maria, that you will arrive in Málaga and you will be nicely and well received by Bernabé, but even though you are going there with good intentions and firmly decided to prove what you can, and try to convince him to accompany you to Burgos, you won't manage it, because these men have their agenda for every moment, and in Villanueva, the slanderers will continue with their own plans. It was a great triumph on your part that you were able to get a pre-trial hearing, the results of which are already in my case file, and if I were not already a 'caged cat', this would be enough for my sentence to be commuted… I am thinking, although I have not yet decided to do so, of writing to that stupid guy, influenced as he is by his wife and other relatives, the slanderer Rodríguez."

- "Valencia, Modelo Prison
  31 October 1939.

My dear parents and brothers,
Today is the last day of my life and I am facing it as calmly as any other day because I am innocent, as you know, of that which I am accused. I wish you good fortune and that you get some pleasure in this world. Receive my heartfelt gratitude for all that you did, to the very end, to save me, but those men did not want that.
Farewell my parents and brothers. Many hugs and kisses.
Francisco Copado."

Francisco "Curro Beatas" Sánchez Muñoz. VILLANUEVA DE CÓRDOBA. Member of the Izquierda Republicana and alderman for the Frente Popular. Executed 17 My 1940in Villanueva. Wrote several letters in chapel, two of which, both undated, are reproduced here.
- "Please deliver to:
  Agustina Cerezo Garcia
  Plaza del Carmen 11, Villanueva de Córdoba

To my dear wife and children,
When you receive these four letters, if they arrive, I will be no more, but as I have no other way to bid you farewell, I do so in this

way, also because I believe it is more suitable as if it were any other way, none of us would have the courage to bear it.

Agustina, I trust that you will be strong enough to bear this new trial that you must face, even though for no other reason than for the sake of our children who, given their ages, more than ever need your care and love. I am very aware, dear Agustina, of the extremely difficult situation in which you are left, but trust in the future; this situation will not last forever and as they grow older, our children will soon be able to help you. So, be strong, accept matters and try to give them a good education. Make them respect you as you deserve and as is due to all honourable and hardworking people. Keep them from suffering from the lack of a father.

To you, my dear children, I charge you to be respectful and obedient to your mother, who deserves it all, for no matter how much you do for her, you will never be able to repay her for all that she deserves and has done for all of us. You, dear Rosalía, as the eldest, must care for and rebuke your younger brothers whenever necessary and help Mother in every way you can; avoid giving her any trouble and do not allow anyone to do so.

Also, no matter how much you hear others say about me, you have nothing to be ashamed of, as your father died without ever having killed or robbed anybody. On the contrary, I always did my utmost to help those who needed it, never failing in my duty, even though some say not, some because they are afraid, others because of ill will, as to the best of my knowledge nobody has done anything to help my situation.

Dear Agustina: I would also like you to show these letters to my brothers, so that they may become aware of the situation in which you find yourselves and, as I would expect, help in any way they can, as I always did when they needed it. The only person I do not want you to even bother to contact is that shameless nephew of mine, whose veins I very much doubt run with my brother's blood since he has behaved so badly towards you and me after I did so many favours for him. I die convinced that sooner or later he will reap his just reward.

I will not tire you any longer, my dear wife. Please give my regards to your parents and your brothers, to my parents and the rest of my family and friends. To you, my dear children and wife, receive this last heartfelt embrace from your father and husband who will keep you in his thoughts until his very last breath.

Francisco Sánchez"

A second letter which he most certainly had delivered by a different means, although also addressed to his family and also written in chapel is not personal, as the above one, as in it he denounces those who tortured him in Villanueva, adding the wish that some day they shall as they deserve.

- "My dear wife and children:

If I am trying to make sure that you receive these lines, I do so because I want you to know what happened to me during all the time that I was in prison and who are the principle persons guilty of the monstrous accusations with which I was charged, as I believe that, in the not too distant future, you can denounce them to the powers that be, not with lies, as they did with me, but with the truth.

When I went to the barracks where I was beaten with rods until I signed a confession. I was forced to sign everything they put before me, and whenever they wished to add something, they would beat me again. My principal tormentors were "Berenguer", Matias Pedraza, "El Tiraor", Miguel Higuera's son and Manolito el Panadero's son, all who beat me viciously, as you would not do even to the wildest animals.

Gregorio Pedraza, Miguel Higuera, Pedro Cano, Pablito and two or three more whose names I don't remember, made an endless number of false accusations, something that none of them had a reason to do so in the way that they did. They had warned me, when I defended Don Dionisio's sons, that if ever I continued to do so they would repay me in this way.

As to the witnesses for the accusations, they may have said nothing more than what was read to me: Pedro "Cascanueces" said that I had taken statements from those who were arrested and Alfonso [Fernández] that I, as a member of the War Committee, had prevented his execution, contrary to the other twenty-one, implying that it was I who was responsible for ordering the executions; this, so you know what it is that you have to thank them for.

For now, I do not ask you for anything more, but if you can avenge me, do so. I also beg my sons not to be ashamed of me because

I never stole from, nor killed anybody. I am guilty of nothing else other than supporting a legally constituted Government; also, I never exceeded my powers when I duly carried out my duties as a member of that Government.

I have nothing more to say, except to you, my dear children: I charge you to be always respectful to your mother and that you love her as she deserves, for no matter how much you do for her, you will never be able to repay her for all that she deserves and has done for you.

To you, my dear companion, be strong and care for our beloved children as if I were with you and wait for the better times that shall soon come, for the good of Spain. With a heartfelt embrace from your father and husband who will keep you in his thoughts until his very last breath."

Eugenio "Palmera" Jurado Pozuelo. VILLANUEVA DE CÓRDOBA. Executed 26 May 1940 in Villanueva.
- "My dear wife:

Convinced as I am that, despite all the efforts of our family to save me, it will all be to no avail because those who hold my fate in their hands, although insignificant in number, today are the people's arbitrators. I am writing you these lines so that you know who are the principal authors of our misfortune and the cruel manner by which they persecuted me during the year and half of my Calvary.

I am convinced that the future of the world is in play today and that it lies in the fight to death between tyranny and the freedom of the people, which has only just begun, and I die knowing that those who are responsible for my death will not wait long before they are made to pay for the blood lust that led them to commit so many crimes. When the time for justice comes, I beg you do your utmost to avenge my death.

I assure you, as if you ever had any doubts, that I have never taken part directly or indirectly in any of the deaths that occurred in our town and that I did as much as I could to avoid problems for many who today may have turned their backs on me. Of these, I can truthfully mention, among others, Diego Higuera, Paco Ochoa, Pepe "El Florista", Andrés Cabrera and Cristóbal Arellano. As regards my political behaviour, I did no more than do my duty, not as a member of a certain political party, but as a simple citizen, in support of a

Government whose legality even its enemies recognize. Circumstances obliged me to accuse (not with lies as they have done with me) before a Court, those who to their shame are the material authors of my death and of the suffering that I have been subjected to.

You need to know that the first to arrest me was Andrés Pontes (a carpenter) from Daimiel, who undoubtedly wishing to curry favour, presented me to the authorities of that town as a terrible criminal, which led them to shut me up in an attic with other unfortunates. Later they brought me here the day that we met in the jail, but there is no point in listing the suffering and aggravation that I have endured since then.

My tormentors have not missed a chance nor a pretext, no matter how base, to mistreat me with word and deed, resorting to shocking and shameful things. Those who most persecuted and were angry with me were primarily: Matías Pedraza and Pepe Delgado, who beat me two or three times, and Manuel Delgado (Pepe's father), and Blas el Sillonero, who were guilty of many beatings that I received in the barracks, as well as the brothers Valero and Vicente "Salado" Muñoz.

Others who did their best to make sure that I would not be saved, by making declarations in my dossier against me: Diego López, Gregorio Pedraza Cámara and Ángel Díaz, son of Manolo el Panadero, many of whom I never bothered at all, and some of them, such as don Gregorio, who at least in part owes me his life. These are those who ostensibly prepared my death, but I have no doubt that as regards those who collaborated with them, you shall one day reward them in the way they deserve.

As regards my brother Zacarías, I want you to remind him to whom he owes his existence, as nobody ignores that he is my brother, if for no other reason because when he was arrested and tried [*sic.* in Jaén by the Republican People's Court, at the beginning of the war], I went and spoke to the prosecutor and to Nemesio on his behalf, to try and influence the jury, which is why he was not sentenced to death. Behave with him with as much consideration as he has behaved with us, as I do not know with any certainty how he will react to my situation. Try to do for my father, if he survives so much disgrace, all that you possibly can and comfort each other.

I know that your affection is sufficiently sincere and deep to understand the causes of our misfortune and I am sorry that because of me you will also suffer what you shouldn't. Be strong in your adversity and trust that the time shall soon come when my death shall be avenged.

My affection to all, as I hold you all in my heart. First you and our daughter, with whom you shall soon share your misfortune and who will help you with her love and company; my poor father, who has struggled so much; my brother José, who has always been so kind to me; my sister and all my nephews and nieces; your parents who have done so much for us; my friends who interested themselves in my fate. To all, a warm embrace from your husband who loves you with all his heart.

Eugenio."

Pedro "Cuadrado" Torralbo Rico. VILLANUEVA DE CÓRDOBA. Executed 3 June 1941 in Córdoba. Copy of letter given to Moreno Gómez by Pedro Torralbo's son, José, from Sallent, Barcelona.

- "Dear wife and children:

I dedicate this to you at the last moments of my life. Dear wife: You know how I have acted and that I have nothing to be ashamed of. Continue to be good with our children, don't impress anything bad on them so that they do not absorb any of the hatreds that poison Humanity. A very deeply felt remembrance to all my brothers and to your parents. Embrace our children, as I would, and you, my dearest wife, receive my last embrace with love.

Darling daughters Petra and Paquita: It is at a very sad time that I write these lines, a remembrance for you. As I think of you, my emotions prevent me from telling you everything I wanted, but I will try to recommend two things: be honest and good with your mother and brothers. Think of your father every so often, but you need be ashamed of nothing, as I always was an honourable man. When you are grown up, you will understand why your father died.

Dearest sisters and brothers-in-law: At these last moments of my life, I write these words of farewell. My spirit is calm and my conscience even more so because I never did any harm to anyone as I always followed my conscience and am ashamed of nothing. At these last hours of my life, I tell you with all sincerity, that I never participated in any crime of blood, either directly or indirectly, therefore, neither you nor my children have to bow your heads before anyone. I only ask you to look after my children, as I die for my ideals. I don't mind dying; I only feel for them. Regarding the rest, I am content.

Please do not shed any useless tears for me and help my children understand why their father died. I ask you to look after them, as you will do my wife, and care for her as you would a sister.

Many hugs and kisses to the nephews and nieces whom I never met. Hugs to Santiago. Say farewell from me to all the family. And to you both, forgive me if I ever offended you in any way.

With a warm embrace from your brother.
Pedro Torralbo."

Juan José "El Conejero" Mohedano Sánchez. VILLANUEVA DE CÓRDOBA. Executed 26 May 1940 in Villanueva de Córdoba.

- At this very tragic moment in my life, when seventeen of my companions have just been taken out to be shot, I pick up my pencil, dearest companion and dear children, to ask you to forgive me if I have ever, at any time, offended you and, at the same time and if you receive this letter, to let you know my last Will and what you shall have to do if there ever is some justice...

I only wish that you are never ashamed of your father, who committed no other crime than to guard the prisoners at Los Grupos, from 23 July to 23 September 1936, when I thwarted outside attempts to free the prisoners from the prison, but as they could find no crime to accuse me of, they introduced four false accusations that I was prevented from defending with true facts, with which they ensured that I was given the maximum sentence."

Juan Luna Enríquez. VILLANUEVA DE CÓRDOBA. Communist member of the Villanueva Committee. Fragment of a letter he wrote to his wife while in the chapel.

- "...and to my María Josefa, tell her to be good and, as the eldest, that she helps you with your hardships, and that she nailed a thorn in my heart the day she came to see me and said: "Father, do not worry." Also, my Magdalena, mi Teodora and my boy, tell them to never forget their father, not to be ashamed of him: their father was neither a criminal nor a thief; nobody could prove that. Also, help your mother. With no further ado, I, your dear father, bid you farewell, and you, my Luna, as this is the last time I shall speak your name..."

Letters from residents of Córdoba Province who were executed elsewhere in Spain

Miguel Ranchal Plazuelo. VILLANUEVA DEL DUQUE. Socialist leader and Mayor of Villanueva del Duque during the Republic. Executed 13 June 1940 in Modelo Prison, Barcelona.

- "To: Maria Josefa Luna
5th Gallery, cell 448, Barcelona
Barriada de la Estacion, Villanueva del Duque, Córdoba
13 June 1940

My dear wife and children,
As I have always told you, I saw the coming of the last hour of my life and it arrived this morning, 13 June. I go in peace and in the knowledge that you shall never be ashamed of my actions, as I neither robbed nor killed anybody. All the people know that. I shall only tell you one more thing: teach our children as much as you can so that tomorrow they will be men of good will. For my sister, for Bartolomé, for your father and all the family many hugs, in the knowledge that I always remembered them. Last, for you and our dear children, thousands of hugs and kisses, until eternity.
Your Miguel."

José Cantador Huertos. VILLANUEVA DE CÓRDOBA. Banesto bank employee resident in Játiva where he was an alderman for the Frente Popular. Executed 29 August 1940 in Valencia.

- "To: Isabel Sánchez de Cantador
Calle Reina 12, Játiva
Valencia
29 August 1940.

My dear wife,
When you receive this by such an extraordinary channel, I who was your husband, will have ceased to exist. There is nothing we can do about that. I suggest that you clothe yourself with calm and patience, and that you try to organize your life for the future in the best possible way for yourself and for our children. As last favours, I ask that you do your utmost to give our children a good education

so that they can learn a profession that will bring them a good living and that you don't spend any money on my remains.

Try to move back to our hometown as soon as the means you have at your disposal make it possible. There, surrounded by family, you may possibly feel better. Please thank all those who, in my absence, did something for me and for you, and tell them that I am only sorry that I cannot repay them.

I believe that you are keeping the receipts from the Montepío de Banca. Do not ever let them go and if the situation shall ever change, and I wish it does, if somebody advises you, you may perhaps be able to recover some of our savings. This is the only capital that I, this man who worked his entire life, can leave you. Regarding our household goods, if there is anything left, they are all yours during your lifetime and afterwards, our children's.

Bid farewell to the children for me and tell them that it is their father's wish that they do their best to be good, honest, and hardworking and that they give you all the love that you deserve; that they learn a lot from their father's example and, most especially, that they only make friends with honourable people. Tell them that their father dies because of those who owed him the most favours, as they could not accuse him of having done them any harm.

The war, in my opinion, was a waste of energy, albeit dedicated to good and to the defence of my class – bank employees – yet it is this family that abandoned me at the worst moment and even let me die. Do they have documents? I don't know. I saved their lives and for them I went to as many places as I had to. Except for one you know; the remainder deserve to be reviled. Especially Ricardo Diego Ruiz who is a prototype of perversity.

It does not pain me to die in such circumstances, not more than not being able to repay you for the many sleepless nights you suffered on my account; but if it brings you some solace, I swear that if it is true that there is a life beyond this one, I will watch over you all… Never think of revenge. Revenge dishonours the person who exercises it! Think of justice and use it to defend me and defend yourself.

Many hugs for the children and for our family; for you, receive the heart shot through with pain and bullets of he who was your husband,
José Cantador.

(Postscript)

Dear wife,

At this moment (5 p.m.), I am being taken to fulfil the sentence dictated by the court martial. I repeat, go to Villanueva, care for our children, and do your utmost to educate them in the best way possible. Many hugs and kisses for everyone, and for you, the love of he who shall forever watch over you. José Cantador. Regards to my brother."

Rafael Porras Caballero. POZOBLANCO. Son of Antonio Porras the Republican poet and politician. Executed 19 May 1943 in Madrid. Wrote two letters in chapel, one to his sister Carmen and one to his parents and brothers which is reproduced below.

- "My dearest and never forgotten parents and brothers,

In a few minutes, I will deliver my soul to God. I have said confession and I die as a Christian in that I forgive my enemies, with all my heart. Therefore, my last request is that you do the same as your son is doing in the last moments of his life.

Carmen shall arrange my funeral and she will always offer you the solace of being able to pray at my grave and every so often, to bring me some beautiful flowers. I die as a Christian, with resignation and a clean and tranquil conscience, as you and I know that I never hurt anybody. If I have a great regret, it is that I cannot embrace all of you for a last time and tell you that I always, always have held you in my heart and in my thoughts.

Carmen will explain how I invested the money in the house; if I spent too much for some reason, please forgive me. I do not want to remain quiet regarding Carmen and her husband's behaviour towards me: they never abandoned me for a single moment. They did more for me than they could imagine, which is why I trust that you will know how to acknowledge this as they deserve.

I may be rambling a bit, but I cannot concentrate on writing it as I should. I just want to tell you all, dear parents and brothers, that I die thinking of God and you. Farewell. I do not want to go on any longer as you will suffer more. Be resigned, be good and live in peace, you certainly deserve it. My last embrace holds all of you together, to make it stronger and more heartfelt.

Rafael.

José María San Ildefonso dies with me. He leaves a widow and children. Please do as much as possible for them. They live at Carretera de Aragón number 15, Ventas."

Eduardo Bujalance López. HORNACHUELOS. Member of Parliament for the Frente Popular. Executed by the Nationalists on 30 July 1936 in Córdoba. Letter sent to Moreno Gómez by his sister, Salud Bujalance, Hornachuelos, April 1981.

■ "Dear father, wife, brothers, nephews, and other relatives.

Today, the last day of my existence, I am writing you these last lines, that you will remember for a long time, never forgetting, above all, that when I was tried for the first time, everything that I said to my sister and wife.

As you know, because of what I have told you before, I never took a direct or indirect part in blood crimes. There are charges against me, but you know who is accusing me, so that my conscience has nothing to regret and your pride as father and honourable family, as always, remains unblemished.

There is only one certainty that I have had from the very first day (I noticed this when I became aware of the way they were focusing on my declarations): They do not want me to live so that tomorrow I cannot press charges against those who were responsible for shooting my brother. It is not by chance that the military judge told me that the worst that could be said of me was that I was the brother of Antonio Bujalance, the Member of Parliament; that is why they are interested in that I no longer continue to exist.

It does not hurt me to die because of death, no; the only thing that pains me is that they are eliminating me without just cause, that is, for false reasons.

The irony of Destiny planned dubious things for us; for you, my father, it allowed the loss of your three sons, in the flower of their lives. Sister of mine – I beg you to not to abase yourselves for anything in this life, fight for the future of your children.

Carmen, my dear wife, what can I say to you? When I try to speak to you, I get a knot in my throat, not because of the pain in my heart, but because I feel responsible for destroying a young life such as yours, although you will never have to bear the burden of having loved a man who loved you and in which you deposited all your trust... Today, destiny separates us. You are alive, you are young; do not waste the chance of living your life well, because I want you to be as happy on this earth as when you were at my side. That is my wish.

Lastly, to all of you, do not waste time or energy on useless matters and on shedding tears. Be as strong as I am in these last moments. Get along well with each other, as true sisters who help each other. Look after Father, who is already very old, and remember your brother, son, husband, and uncle, who sends you his last kiss.

Eduardo Bujalance.

(Written in the chapel, hours before my execution.)"

Children, unaware of the horror of war, play in front of the shrapnel-hit facade of No 10 Peironcely Street, Madrid 1936. Photograph: Robert Capa © International Center of Photography/Magnum Photos[3]

³ *https://i.guim.co.uk/img/media/e40ef1b2e152aacdc275fff4bd3ab7f7f2ee363b/0 _295_4600_2761/master/4600.jpg?width=620&quality=45&dpr=2&s=none*

APPENDIX IV

WRITINGS AND LETTERS FROM MEMBERS OF THE CLERGY OR CHURCH-RELATED EXCERPTS FROM FATHER GUMERSINDO DE ESTELLA'S DIARY

Reverand José M. Gallegos Rocafull, Professor at the San Pelagio Seminary and of the Universid Central de Madrid. <u>Extracted</u> from a manifest entitled *Palabras cristianas* (Christian teachings) by Rev. José M. Gallegos Rocafull and Rev. *Leocadio Lugo.*

- "The Church will never stop teaching the respect and obedience due to the established power..." Collective declaration by Spanish bishops 20 December 1931.
- "...failure to obey and to encourage sedition is a crime of lèse majesté..." Pope Leon XIII, Inmoratale Dei.
- "...the Church has always condemned the doctrines and the men who rebel against the legitimate authorities." Pope Leon XIII, Au milieu.
- "Fight, Catholic men, in defence of the rights of the Church, with perseverance and energy, but without ever using sedition and violence." Pope Pius XI, Gravissimo.
- "The truth is that a few men have burdened the shoulders of the innumerable multitude of proletarians, a yolk that is very little different to the one of the slaves." Pope Leon XIII, Rerum Novarum.
- "The economic organization violates the true order when capital enslaves the workers." Pope Pius XI, Quadragesimo Anno.
- "In the Catholic world (albeit more in Europe than in America), at the first there was a spontaneous movement of surprise, of protests, of fear of the intervention of the clergy in the war, even more so because this was a civil war between some insurgents and a legally constituted government."

Ernesto Caballero Castillo. Son of Julián Caballero, ex-Communist Mayor of Villanueva de Córdoba, who at the time of these events was a guerrilla commander in the North of Córdoba province. He was taken from home by the Welfare authorities and placed in a convent school in Villanueva de Córdoba.

- "Those days of profound beliefs were horrendous times for me. I was permanently terrified of the punishments that God might hurl against me. Everything was a sin and God saw everything.

In summary, my bad memories of the Cristo Rey Convent are of old and musty chickpeas, lentils, and dried beans, ridden with bugs, served with only a bit of fat, inedible... I hated lentils for years afterwards, broad beans forever. I had nightmares of the scary stories Mother Maria Josefa told us; of the loud bells; of Mother Rucio's wicker cane and the way she beat us on our heads with the pot in which the food was served, to shut the children up... of the terrifying religious lessons and, quite obviously, the day of my confirmation...

When I was 7 years old, the nuns sent me to work as an apprentice blacksmith at Domingo Torres' forge on Calle Cañada Baja. Domingo Torres was, unwittingly, the person who most influenced me to stop believing in God, as much as the nuns tried to make me believe in Him. I had to go to daily mass before I began work in the forge, early in the morning.

Domingo swore constantly, blaspheming as the nuns would call it, slanging Christ, every saint, virgin, blessing, and consecration of the communion host in the Book. The first time that I heard such things, I looked at him in fear and amazement. I expected God to strike him dead as the nuns had taught me that this is what happened to all who blasphemed."

"The missionaries appeared in 1943 or 1944. These were young Jesuit priests intent on Christianizing every left-wing individual who was still alive and not yet imprisoned, as well as his relatives and the relatives of those who had already been executed. The imprisoned were Christianized in the jails.

For quite some time, the missionaries taught a whole series of classes, giving those who attended some economic compensation in exchange, a few bags of food, clothes. As the authorities were again compelling the rich to accept that the workers had to be believers before they could hire them, anyone who did not convert found it difficult to find any kind of work.

It saddens me greatly when I remember those poor people, crowding around those priests like so many sheep, hoping for some clemency, a bag of food, some clothes to cover their bodies and a reference that they were good Christians."

ADDITIONAL EXCERPTS FROM
FATHER GUMERSINDO DE ESTELLA'S DIARY

22 June 1937. Terse dialogue in the chapel with Don Tregídio, Socialist, Secretary of the Escatrón Municipality near Casp. He was a friend of Father Gumersindo's. Condemned to 30 years in prison by the court martial, his sentence was overruled in Burgos, and he was condemned to death.

- "He was a tall gentleman, about 50 years old... He was nervous as he entered... When he saw the altar, he stood straight, raised his arms, and exclaimed:
- Why have I been brought here? Let them kill me quickly, with four shots; do not keep me here suffering.

He sat down and again asked:
- Am I permitted to know why I have been brought here?

This was a good time to speak to him, so I said:
- You can imagine my deepest sympathy at seeing you suffer. I would like to help with lessening something of the pain you must be suffering. This is why I have come to offer you some religious comfort.

He looked me straight in the eye and replied:
- What are you saying? What religion are you talking about? If you are referring to the religion that I learnt in my mother's arms, that one is very good for consoling someone... But the religion that now has set your lot to killing a million Spaniards, that one comforts no one; don't even mention that one; that one is a fascist religion..."

TORRERO, ZARAGOZA, CEMETERY. Execution of Don Tregídio and another prisoner.

- "...We had arrived at the cemetery. We drove along the wall to get to that part of the wall at the front, facing the city and neighbourhood of Torrero, from where we had just come. And … we found a detachment of some one hundred soldiers. They were formed in rows facing the wall, but about fifty metres distant. Sixteen of them were closest to the wall…

As we arrived, our truck stopped. We got off and began to walk towards the soldiers. They looked at us with curiosity. I walked next to Don Tregidio. The other prisoner was accompanied by Fr. Victor, who could not contain his roughness. We walked past a Red Cross van, almost touching it… Next to the van, two gurneys that had been prepared to receive the bodies of both men. And they could see it all! What a terribly sad walk! Sixty or seventy bitterly difficult steps for the condemned men and for anyone who had been born with a bit of heart…

Nobody asked the prisoners whether they wanted a blindfold. I still did not abandon my friend. I stood at his side, stroking his right arm and neck with my hand, and I repeated the prayer: 'Merciful Jesus, save my soul'. He repeated it and he kissed the crucifix. I offered it to the other prisoner for him to kiss, but he shook his head. The silence was deafening. I realized that the officer who had to give the sign to fire was waiting for me to leave. I walked away and stood behind the squad of soldiers.

The officer shouted: 'Aim!' Don Tregidio shouted: 'Long live God and Socialism!'. The officer shouted again: 'Fire!'. The fatal shots rang out. Each body was riddled by eight bullets. They fell backwards, onto the ground… Some Guardia Civiles approached to remove the metal handcuffs that bound their hands.

I approached to administer extreme unction to one and absolution and a prayer. Both bodies were lying in a large pool of blood that had run down their legs and was mixing with the dew… A lieutenant shot them each twice in the head. The doctor approached to confirm their death. The members of the Brotherhood of the Blood of Christ picked them up and placed them on the gurneys…."

21 September. After consoling six Nationalist soldiers who had been condemned to death: three from the quartermaster's, three health workers and one from the artillery.

- "The six confessed, attended Holy Mass and took Communion with great devotion. Some cried. They were so very wretched! One of them sobbed: "What is my poor mother going to think that I did, when she hears that I have been shot? How horrible! All for nothing…!"

22 September. Report of three women who entered the chapel because they had attempted to go over to the Republican zone. Two of them were carrying young children. The execution the followed was horrific and so upset the Capuchin monk that he left alone, walking like a robot.

■ "I had entered the chapel, in which everything was prepared for Mass, when I heard the heartrending cries of women outside… Sobbing appeals…

- My daughter! Don't take her from me! Have pity, don't steal her from me. Kill her with me! I want to take her to the next world with me!

- I don't want to leave my daughter with these executioners…! Daughter of my heart, what will become of you?

…Meanwhile, a fierce fight broke out between the guards who were trying, with all their strength, to tear the children from their mothers' breasts and arms and the poor mothers who held onto their treasures as tightly as they could…

When I heard those poor little creatures cry, as they did not want to be taken from their mothers, terrified at the sight of the guards, as I heard the heartrending cries of those unfortunate women, I felt my heart breaking… I never thought that I would ever have to see such a sight in a civilized country. I never believed that there could be a king, or leader, or caudillo on this earth who could decree such a thing. To anyone who intended to do such a thing I would say: Either you pardon those poor mothers, or you admit that you are the shell of a human being without charitable feelings…"

At the place of execution, there was an entire company and a squad of 24 soldiers who formed the firing squad – six for each of the four prisoners (three women and one man):

"…We began the slow walk towards the place of execution. It was the most horrible walk of my life. The three women wobbled as they walked, their hands had been tied, their clothes were in disarray, their hair was a mess (the babies of two of the women had been torn from their breasts as they entered the chapel) … One of the women shouted: 'So many men just to kill three women…!'"

11 October. Execution of a miner from Asturias. As the prisoner entered the chapel, seeing the Dictator's picture on the altar, he protested and refused the comforts of the Church: "Can't you remove Franco's photo from the altar?"

3 February 1938. Execution of two prisoners-of-war, the first who was taken prisoner in Celadas, on the Teruel front, and the second, taken in Santoña when Vizcaya was lost to Franco:

- "When I spoke to him of confession and other religious practices, he flatly refused them. Relatively calmly, he said that religion had been falsified by those who called themselves supporters of the right, priests included, those, he added, who were responsible for everything that we are suffering…"

- "When Mass was over, he became notably more dejected, and he began to panic at the thought of the execution. He asked that he be given some chloroform to deaden his awareness. But they refused, as was expected. When he saw the soldiers, he faltered a moment and refused to walk. Finally, with wobbly steps, he arrived at the site of the execution. He had nothing to say. You only heard him crying: *Ay, Dios mío! Ay, madre mía…!* As I had suggested, he turned his back to the rifles."

17 March. Following the execution of two high-ranking Republican prisoners and prisoners' state of mind as they arrived in the chapel:

- "I did not ask these condemned men to make a full confession, because their state of mind was comparable to a person who was seriously ill. Some prisoners were truly moribund; in some cases, their nerves were so on edge that they shook violently, and their arms and legs jerked. Others vomited. Some fainted."

12 May. Description of a saca of nine inmates, possibly all prisoners-of-war from Alcañiz, one of whom was a woman. When Maria de Asis Figueras entered the chapel:

- "…She began to cry again and shouting at the other prisoners, especially the one called Tomás: 'Look at our misfortune, Tomás, Tomás. This is horrible. Why are they killing us?' That is not all she shouted and then all the others began shouting at the same time, protesting their innocence. They accused the courts of cruelty.

Their yells formed a strident and extremely tragic concert. I confessed another two during all that shouting… They all took Communion, except for the oldest, Miguel Andrell, who was 61 years old. When several Guardia Civiles entered the chapel and began to bind their wrists, Maria stood up and shouted: 'Don't take me, don't tie me up, shoot me right here! Tomás, why don't they just kill us right here…?"

At that moment, Father Gumersindo noticed that the woman appeared to be pregnant, so he approached the Judge Executioner:

■   "Take a look at the young woman, María Figueras. Look for yourself, if you have any doubts as to what the female prison guard said, that she could be pregnant.

-   'If for every woman who had to be tried, we had to wait for seven months' the judge replied, 'you will understand that it is not possible…' and Maria was tied up, like all the others.

In the truck in which we all went, and which was filthy with bits of earth inside, we continued to hear the mournful concert of heartrending *ayes*, of anguished cries and weeping. I attempted to console them, holding my crucifix in my hand, speaking unending words of comfort.

They sat on wide benches on either side of the truck. I knelt on my knees in the middle of the truck, as there was nowhere else for me to sit. Old man Andrell, who was the calmest of all, looked at me with anger and finally said: 'You really know how to play your role…"

When we arrived at the place of the execution, we placed ourselves between the wall and the firing squad. I suggested that they turn themselves to the wall so that they would not suffer from looking at the soldiers. Old man Miguel Andrell, seeing the soldiers arrive, spoke to them: "Men! You are about to kill sons of the people…!"

I stood between the wall and the condemned, unceasingly begging them, one by one, to have faith in God… Suddenly, I heard the prison director's voice shouting: 'Father Gumersindo! Get out of the way!' He wanted to give the order to fire, and I hadn't realized that. In fact, the officer hadn't noticed that I was still talking to the prisoners. There still was very little daylight.

Young Maria was twisting and turning in place. She continued to sob and cry: *Ay! padre mío! Ay, padre!* Why are they killing me…?' This excited the soldiers and I saw their indifference and disgust mirrored in their faces.

No sooner did I get out of the way than the shots rang out. Four rifles for each prisoner. None of them died instantly. They were not mortally wounded. One of the wounded, rolling about on the ground, cried out: 'They have finished us; we only have a minute left to live!'

They all cried out with pain, and some begged for the mercy shot. I went up to each one and absolved them. An officer administered the mercy shot, sometimes repeatedly shooting the dying three times in their heads."

14 July. Description of the execution of eight prisoners from Alcañiz. The firing squad forgets their ammunition.

- "As we approached the soldiers, the truck stopped but we were ordered not to get off. A quarter of an hour later and we still had not been ordered to get off. I jumped down and asked a soldier what was happening and what was the reason for the delay. He told me that the soldiers had not brought their ammunition. Fifteen minutes later, some soldiers arrived in a car...

That delay was very harmful to the condemned. Some lost patience and began complaining: 'Why are they keeping us suffering here? How pleased they are to make us suffer! Then they dare say that it is the Rojos who are cruel!... Hurry up and kill us! Please, kill us and be done...!'

Finally, we got there. The soldiers made another mistake that added to the suffering. After lining up the eight condemned in a single row, they decided that they would execute four first, then the other four. Right there, in full view of the last four, the officer gave the order to fire against their companions.

They saw the soldiers take aim and fire and they heard the shots. They then watched their companions lying on the ground, rolling in the pools of their blood, and crying with pain. Then the same soldiers took their positions in front of the surviving prisoners, who fell next to the first.

Looking at the eight wounded lying on the ground, you got the impression that this was a battlefield."

26 July. Description of the execution against the wall of Torrero cemetery, of 7 victims, almost all prisoners of war, several from Gelsa and one from Belchite.

- The victims of the 26th suffered a great deal. The soldiers shot badly. They were apathetic. One of the executed fell to the ground shouting 'Hurry up and kill me!' All cried and screamed with pain. A deplorable and sorry spectacle…!"

18 October. Description of the execution of another 4 prisoners.

- "The soldiers who on that day had the misfortune of being the executioners or part of the firing squad, shot badly. The unfortunate prisoners rolled around on the ground, with pitiful cries and screams that tore at my soul. The unfortunate Martín raised his feet and legs, whilst harsh, deep *Ayes*! from his chest, sounding like a death rattle.

The two who refused confession, exclaimed just before they were shot: "Viva the Republic! You will soon suffer the same fate!".

Regarding Isidoro Franquesa, a prisoner-of-war from Vich, on the Teruel front who was denied all spiritual comfort:

- "No sir don't ask me to get involved in religious practices. The right-wing is killing in the name of religion and it has declared war in the name of religion. I want nothing to do with a religion that inspires so much cruelty."

It was 11 June 1938, and he was still in the chapel when he stood up and started walking towards the door. The matter was that the Director and several military officers, as well as the Judge Executioner, had entered the chapel and were standing next to the altar. On the altar, above the crucifix, hung a portrait of Franco. In his opinion, it was a typical political and military farce.

I accompanied the unfortunate prisoner to the identification room. Two prison officers followed us. When he saw those officer, he regretted that as a prisoner of war, he was going to be killed, contrary to the rights of Man. He added that in the Republican zone they did not kill their prisoners, they respected their right to live; as he himself had done."

APPENDIX V

SELECTION OF CAUSA GENERAL FILES FOR CORDOBA PROVINCE

Causa General archives can be consulted on the
Spanish Archives website at www.pares.mcu.org.
Not all the CG file numbers are available for the following.

Bartolomé Fernández Sánchez. [Case number 5,753/39. Pre-trial hearing in Cordoba, March 16 1943] POZOBLANCO. Sentenced to death but the sentence was commuted.

- Bartolomé Fernández Sánchez was one of the most famous defenders of the Republic in the Cordoba mountains. His case file states that he was a Socialist, 'an extremely active propagandist in the February 1936 elections,' and that as a result of these elections, 'the accused and other lowlife of the same kind, attacked City Hall and appointed themselves town councillors.' This was totally false, in addition to being absolute nonsense.

Although there are entries to the effect that he organized the Pedroches Militia Battalion August 31 1936 and served as its first Commander, which was correct, the authorities did not discover that he was also Commander of the 73rd Battalion of the Mixed Brigade. It is noted that he was promoted Major in the Militia September 1 1938 but not that he commanded F Column during the Cordoba-Extremadura battle of January 1939. In addition to all these military-related charges, in themselves punishable by death, his case file includes a typical, presumably false, allegation from a private individual: 'According to a reliable witness, José Plazuelo, he commanded firing squads against noted individuals and himself administered the *coup de grâce*,' to ensure that the accused would be executed.

Moreno Gómez was unable to discover how Bartolomé Fernández managed to get the latter, totally false, accusation erased from his file. It may have been that as he was tried in 1943, when the Regime's deadly arrows were already soaked in blood there was a certain lessening of the repressive fury. Besides, this type of accusation could be easily proven false because the firing squads and the coup de grace always were a senior officer's responsibility, never the commanding officer's. Nevertheless, in this case Fernández Sánchez Bartolomé was still sentenced to death, but miraculously, his sentence was commuted.

Antonio Baena Moreno. POZOBLANCO. Executed in Cordoba November 17 1941.

- Antonio Baena Moreno, a 37-year-old Socialist schoolteacher, was in a worse predicament than Bartolomé Fernández. He had served as President of the local War Committee, although at heart he was a kind and reasonable individual who tried to impose law and order but did not know how to go about it. Despite his efforts, many of the right-wing prisoners in Pozoblanco in 1936 who were evacuated to Valencia after Pozoblanco surrendered August 15 1936 to the insurgent troops, were executed by Republicans in Valencia. To add to the seriousness of his situation, he also served as a high-ranking political commissar: no less than Commissar for the VIII Army Corps and finally, for the entire Army of Extremadura. There was no escape for him. Although he had been sentenced to death by strangulation [garrotte], he was shot by a firing squad in Cordoba November 17 1941.

José Madueño Serrano. [Case number 26,298/39]. POZOBLANCO. Originally sentenced o death but later commuted to 30 years in prison.

- José Madueño Serrano, an attorney, was tried by court martial September 30 1939. His case fille states that he had once been a member of the right-wing Acción Católica, the Juventud Liberal and the Patriotic Union. However, when the Socialist party arrived on the scene, he ran unsuccessfully for office as Socialist Member of Parliament for Seville in 1936. Served as town councillor in Pozoblanco and later as Provincial Deputy for Cordoba. His file acknowledges that he remained in his house during the insurrection, but one informer declared that 'he remained in contact with the Rojos'. The file also stated that although he was a mine owner and had access to dynamite, he did not give any to the Rojos. A particularly negative allegation states that at the beginning of the war, Madueño Serrano travelled to Elche, Santa Pola and Madrid 'and that during his journey he did not avoid the challenges and crimes committed by the Marxists'. Lastly, the absurd charge: 'considered to be conceited'. Although the case file clearly showed that the accused was never implicated in any kind of disorder, he was sentenced to death, but this was later commuted to 30 years in prison.

Antonio Varo Granados. [Case Number 26,454/39. Pre-trial hearing in Pozoblanco Aril 22 1940.] AGUILAR/POZOBLANCO. Sentenced to death by strangulation but commuted.

- The pre-trial hearing case file for Antonio Varo Granados, from Aguilar, a public prosecutor in Pozoblanco by profession, was an anthology of false allegations and other irregularities. The notation _educated_ on his dossier is underlined. All the charges against Antonio Varo focussed on the supposed nefarious influence of his being an educated person. He was accused of:

- great leftist importance, being an effective propagandist of leftist ideas and using the press to disseminate them, propaganda that was extremely dissolutional, and in view of the personality of the accused and the lack of culture on the part of the worker element, of having considerable influence over the workers, the constant target of his propaganda;

- the beginning of the Glorious Uprising finding him in Madrid and having in his hands the destiny of the people, far from presenting himself to the authorities [_sic._ in his town of residence] to prevent the commission of outrages and excesses, he continued to remain in Madrid;

- having employed Masonic tactics, as there is a strong presumption that he belonged to that sect; _(Unbelievable! He must have been a Mason because he had all the characteristics of one! FMG)_

- his house in Madrid is said to have been the regular meeting place for all the significant Marxist tycoons;

- a supposed allegation by relatives of a right-winger who had been murdered in Pozoblanco, Moisés Moreno, stating that Varo was the 'instigator' of that death and that someone heard somebody else say: 'Antonio Varo murdered your father', although Varo was in Madrid at the time.

Based on that mountain of garbage and insidious comments, the special Court convened in Pozoblanco presided by Pedro Luengo Benítez, one of the major fascist magistrates in Cordoba, sentenced Varo to death by strangulation, after declaring that all the information given agreed with the statements that the defendant was the leading individual responsible for all the crimes committed in that town and that he was an extremely dangerous person. Antonio Varo did not stop appealing. January 10 1940 he appealed to Military Court No. 9 in Pozoblanco, in which he:

> "Requests that you receive the sworn declaration of Don José Sánchez, Captain of Infantry Regiment 3, of the Badajoz garrison", stating that the defendant sheltered him in his house in Madrid, despite his belonging to an army that was fighting the Republic. Varo adds: "Many right-wing persons from Pozoblanco came to my house to ask for advice and assistance and I received them all". He concludes: "My daily routine basically involved going to work in my office and not going out when I returned."

Without a doubt, the testimony of the Badajoz soldier bore some weight as Antonio Varo's terrible sentence was commuted. He was transferred back and forth from one Francoist prison to another, among them Burgos prison.

Cesário Romero. [Case Number 11,113/39. Pre-trial hearing Pozoblanco, June 15 1939]. TORRECAMPO. Executed.
- Cesário Romero's case file was another pack of lies. His problem was that he had served as Mayor before the coup and the right-wing in his town were baying for his head. His family told Moreno Gómez that Cesáreo was in Espiel when the military coup broke out and he remained there until October 1936, totally removed from any kind of disturbance that might have occurred in Torrecampo. He returned to his town and was not reappointed Mayor until February 1938, many months later. Although he could not be accused of anything, unless it was his political beliefs, his file contains the following allegations and charges for which he received the death penalty:

- bad behaviour
- was a Socialist
- supporter of the class war
- instigator and leader of the masses

- serving as Mayor of the town until four months before the Movement and later re-elected to the same position
- giving unfavourable information against brothers Alfonso and Esteban Márques, who were murdered, and
- every witness considers him to be an extremely radical individual and a danger to the National Cause.

Juan Pulido Cantador. [Ministry of Agriculture proceedings against Juan Pulido Cantador and multiple interviews with Moreno Gómez who was provided with many details of those tragic times.] VILLANUEVA DE CÓRDOBA. Sentenced to death but commuted.

▪ Juan Pulido Cantador was sentenced to death, not for blood crimes, but on such as flimsy charges as 'the Movement found him in Cuesta del Hornillo, in the county of Montoro where, influenced by his ideals and the aims of the Rojo revolution, he supported the latter by walking to a farm located one kilometre from the place denominated El Niño Herruzo, where Marxists were staying. The accusers alleged that he got into contact with those Marxists and participated with them in the fight to recapture Villanueva de Córdoba, where they committed murders, then served their cause by first collecting arms from the farms and later, grain. The reality of the matter is that the Falangistas wanted to get rid of this individual because he had been a leader of several agricultural cooperatives, but he survived, his sentence commuted.

A sample of the many Causa General files for Villanueva de Córdoba, with backdated allegations

José Romero Cachinero. [Case Number 26,709. Pre-trial hearing, Villanueva de Córdoba, April 24 1940]. Sentenced to 30 years in prison.

▪ José Romero Cachinero was accused by Sebastián Cepas Díaz (brother of the notorious *Berenguer*) of having participated in the attack on the town in 1936 and as a member of the guerrilla, having destroyed bridges in Monterrubio, Castuera and Cabeza del Buey. He was sentenced to 30 years in prison. He died soon after leaving jail in extremely poor health from the mistreatment he had suffered.

Luís Sánchez Torralbo. Executed May 17 1940.

• Martina Castro Díaz accused Luís Sánchez Torralbo of having arrested her husband, Antonio Ruíz Justos, who was later shot in Fuente Vieja. Although he denied this, Luíz was sentenced to death and executed May 17 1940 in Villanueva de Córdoba.

Juan José Serrano Cepas [Case Number 27,324/39. Pre-trial hearing, Villanueva de Córdoba, January 31 1940]. Executed December 27 1940 in Cordoba.

• The charges against Juan José Serrano Cepas were listed on a form signed by a group of Falangistas who dedicated themselves to collecting allegations. Miguel Higuera Días, Alfonso Fernández and Matías Pedraza appeared as witnesses to the charges that Serrano Cepas 'participated in the occupation of the town', 'had been a guard in local prisons', 'was a member of the Confiscation of Property Committee', and 'had volunteered' to serve in the militia. No crime was listed, but Serrano Cepas still was sentenced to death and executed in Cordoba December 17 1940.

Juan Miguel Amor García. [Case Number 1,200/39. Pre-trial hearing, Cordoba, February 5 1941]. Sentenced to 30 years in prison.

• Juan Miguel Amor García, a teacher at the Centro Obrero, was accused by Pedro Serrano and Roque Díaz. Pedro Jesús Torres and Juan Ocaña were witnesses for the defence. Juan Amor García had been an active member of several political parties and was accused of being a member of the local War Committee, which he denied. He was also charged with witnessing accusations against Francoists in the Republican People's Court of Jaén. Sentenced to 30 years in prison.

Diego Fernández de Haro. [Caser number 27,453/39. Pre-trial hearing, Villanueva de Córdoba, March 14, 1940]. Sentenced to 30 years in prison.

• Diego Fernández de Haro (whose brother, Fernando Fernández de Haro, a private schoolmaster, was executed May 25 1940). Diego faced the flimsiest of accusations by Matías Pedraza and Arcadio Herrara who alleged that he beat the priest Rev. Rafael García and was a member of the Supply Committee. In this case, there were two witnesses for the defence, the priest himself who said that when he was a prisoner of the Republicans, he was only hit once because he was walking too slowly in line, and Ángel Chaparro who said that he owed him his life. Despite the inconsistency in the accusations, he was condemned to 30 years in prison. By the time Diego Fernández was released, he had lost his wits as a result of the mistreatment he had received.

Pedro *Cuadrado* Torralbo Gómes. [Case Number 27,404. Pre-trial hearing, Villanueva de Córdoba, May 1 1940]. Executed June 3 1941 in Cordoba.

- One of the most upstanding Republicans in Villanueva de Córdoba, Pedro *Cuadrado* Torralbo Gómes was one of the founders of the PCE in 1921, city councillor in 1931, Provincial Member of Parliament in 1936, active organizer of the rebuilding of the town after the military coup and a Captain in the Militia. He was not an extremist, but a prudent and well-organized individual. He was beaten to a pulp after his arrest in 1939. They could not charge him with blood crimes, just of being a leader. His accuser, the landowner Bartolomé Torrico, declared: "I know that Pedro Torralbo himself did not kill anybody, but he did nothing to prevent others from being killed". Pedro was executed June 3 1941 in Cordoba.

Selection of Causa General files for women in Villanueva de Córdoba

The case files for the numerous women who were charged in Villanueva de Córdoba exhibit the same bizarre accusations and allegations.

Carolina *La Mojina* Buenestado Herrero. [Case Number 26,708. Pre-trial hearing, Villanueva de Córdoba, December 21 1939]. Sentenced to 20 years in prison.

- Catalina *La Mojina* Buenestado Herrero was accused by Mayor Gregorio Pedraza Cámara of belonging to the JSU and to *Mujeres Antifascistas* [Antifascist Women], of taking part in registrations and seizures and of supporting the International Red Aid. Nothing of any importance. Sentenced to 20 years in prison.

<u>Ana María Gómez Ruíz</u>. [Case Number 37,131/39. Pre-trial hearing, Cordoba, July 4 1943]. Sentenced to 30 years in prison.

▪ Ana María Gómez Ruíz was accused by Francisca Coleto Gutiérrez (widow of Miguel Gutiérrez), of being a Communist and in the presence of right-wing prisoners, having called out loudly for their death. Sentenced to 30 years in prison.

<u>Juana *La Flora* Pozuelo Expósito</u>. [Case Number 27,416/39. Pre-trial hearing, Villanueva de Córdoba, April 15 1940]. Sentenced to 20 years in prison.

▪ Juana *La Flora* Pozuelo Expósito, sister to the famous leader Nemesio Pozuelo, was accused of the following absurd charges: being a Communist and great propagandist, and of having lent her house as a prison for Guardia Civil widows and orphans, which was a monstrous lie. What had happened was that when the barracks were closed July 24 1936 after the Republicans re-took the town from the insurgents, she offered her house as a shelter for the families of the Guardia Civiles. An altruistic action that the victors turned into an accusation, also alleging that she went into the fields and countryside looking for weapons. Sentenced to 20 years in prison. Obviously, both the allegations and charges were far-fetched nonsense, simply used to pad the file for a case whose resolution had been decided beforehand.

**Women whose heads were shaved in Toledo
because they were related to known Republicans.
(Photo by: Photo 12/ Universal Images Group)**

APPENDIX VI

EXTERMINATED IN THE PRISONS IN CORDOBA CAPITAL DUE TO HUNGER, DISEASE AND HARDSHIPS

PP = Provincial Prison, Plaza del Alcázar 5
NP = New Prison, on the road to Pedroches
[for emaciation, read starvation]

1939
May 8 – PP
  Francisco González Heredia, 61, bricklayer, acute nephritis, Almería
May 23 – PP
  Antonio Gómez Leal, 29, farmer, suffocation, Hinojosa del Duque
  Alfonso Martín Rojas. 25, suffocation, Villa del Río
June 13– PP
  Juan Moreno Tejada, blacksmith, suffocation, Badajoz
  Germán Ramirez Madrid, 25, farmer, suffocation, Espiel
  Francisco Juárez Zapata, 21, suffocation, Granada
  Rafael Luque Serrano, 32, farm worker, suffocation, Carcabuey
August 23– PP
  Jacoba Centeno Tena, 59, housewife, perforation of the stomach, Badajoz
September 17– PP
  Juan Boquizo Delgado, 79, manual labourer, collapse, Lopera
October 9 – PP
  Rafael González Roldán, 22, manual labourer, tuberculosis, Nueva Carteya
Octbober 21 – PP
  Luís Pérez Escués, 74, manual labourer, gastroenteritis, Lopera
November 1 – PP
  Joaquín Palma Delgado, 62, salesman, asystole, La Victoria
November 5– PP
  Antonio Ramírez Mesa, 54, road worker, myocarditis, Cabra
December 1– PP
  Pedro Muñoz Burgos, 59, manual labourer, myocarditis, Lucena
December 28 – PP
  Féliz Pulgarín Agenjo, 44, manual labourer, acute nephritis, Fuenteoejuna

1940
  January 3 - PP
  Francisco Guerrero Cáceres, 43, tar worker, acute nephritis, Puente Genil
January 11 – PP
  Francisco Sáez de la Torre, 50, soldier, tuberculosis, Madrid
February 10 – PP
  Rafael Cruz Fernández, 40, manual labourer, heart attack, Monturque
February 11 - PP
  Cristóbal Jiménez Sevillano, 44, manual labourer, bronchitis, Aguilar
  Francisco Rodríguez Palos, 52, manual labourer, chronic nephritis, Puente Genil
February 15 – NP
  Francisco López Hierro, 52, manual labourer, bronchitis, Peñaflor
February 20 – PP
  Pablo Barja Peláez, 47, tool sharpener, chronic nephritis, Porcuna
  Juan Fructuoso Quesada, 74, manual labourer, laryngeal epithelioma, Posadas

March 7 – PP
    José Blázquez Milara, 53, miner, chronic bronchitis, Belmez
    Miguel Garcia Madrid, 61, blacksmith, cerebral haemorrhage, Montoro
March 10 – PP
    Antonio Vázquez Pérez, 57, manual labourer, chronic nephritis, Cerro Muriano
March 13 – PP
    Luís García Orihuela, 58, manual labourer, valvular heart disease, Villaralto
March 14 – NP
    José Moreno Hernández, 47, manual labourer, bronchial asthma, Posadas
April 21 – NP
    Francisco Talballido Blanco, 53, manual labourer, pneumonia, Navas Concepción
June 3 – PP
    Tomás Martins Barber, 20, entrepreneur, Villaviciosa
June 16 - PP
    Manuel Montserrat Leonard, 25, manual labourer, heart attack, La Carlota
July 4 – NP
    Francisco Cebrián Amil, 32, manual labourer, tuberculosis, Adamuz
July 18 – PP
    Antonio Luna Hidalgo, 28, manual labourer, tuberculosis, Morón de Frontera
    José Siles Jiménez, 29, bricklayer, epilepsy, Almodóvar del Río
August 2 – PP
    Isabel Cañete Molina, 40, housewife, diabetes, Baena
    Miguel Álvarez Tena, 37, miner, suffocation, Pueblonuevo
October 13 – PP
    Domingo Valenzuela Salamanca, 27, market gardener, bronchitis, Albendin
Nvembaer 3 – NP
    Miguel García Gómez, 28, manual labourer tuberculosis, Villaviciosa
November 15 – NP
    Antonio Pozo Gañán, 68, manual labourer, myocarditis, Adamuz
December 7
    NP - Eusebio Corrales Trujillo, 54, miner, heart attack, Belmez
    PP – Antonio Ruíz García, 58, manual labourer, uremic coma, Priego
December 15 – PP
    Manuel Hurtado Torres, 57, manual labourer, enterocolitis, Cabra
December 24 – NP
    Alfredo Ramos Vázquez, 58, farmer, nephritis, Fuenteobejuna
December 29 – NP
    Manuel Cárdenas Delgrima, 42, farm worker, rheumatic endocarditis, Granada
December 30 – NP
    Ramón Girado Gonzáez, 48, farm worker, mitral insufficiency, Villanueva Rey

1941
January 3 – NP
    Juan Miguel Lao Gordillo, 58, manual labourer, arteriosclerosis, Valenzuela
January 6
    NP – Julián Prieto Carmona, 30, manual labourer, apoplexy, Puenblonuevo
    PP – Vicente Martín López, 22, manual labourer, anaemia, Alicante
January 8 – NP
    Juan J. González Hidalgo, 42, miner, enterocolitis, Pueblonuevo
January 12 – PP
    Juan Peña Delgado, 50, manual labourer, nephritis, Villanueva del Rey
    Juan Sierra Rabanera, 49, manual labourer, nephritis, Granada
January 14 – NP
    Aquilino Estévez Risquez, 65, farm worker, myocarditis, El Viso
    Francisco Alberca Albarca, 66, barber, gastric ulcer, Posadas
January 23 – NP
    Manuel Torresilla Ferré, 43, butcher, endocarditis is, Constantina
January 27 – PP
    Alfonso Moreno Bajo, 63, miner, angina, Villanueva del Duque

January 29
    PP – Antonio Benitez Frutos, 63, manual labourer, heart attack, Zalmea Serena
    NP – Julián Lama Maiz, 51, manual labourer, anaemia, Cañete de las Torres
January 31 – PP
    Antonio López Bravo, 60, cattle breeder, heart attack, Trasierra

February 2 – NP
    José Cano Rodríguez, 53, manual labourer, cirrhosis of the liver, Baena
February 4 – PP
    Agustín Roiza Morales, 50, manual labourer, heart attack, Calzada Calatrava
February 6 – NP
    Luís Hemica Gómez, 51, manual labourer, anaemia, Azuaga
    Dímas Marta Flores, 60, carpenter, enterocolitis, Hinojosa
    Ramón García Martínez, 37, manual labourer, nephritis, Bujalance
    Ernesto Díaz Naranjo, 57, manual labourer, bronchopneumonia, Villanueva Rey
Februry 8 – NP
    Alfonso Valverde Shepherd, 62, manual labourer, gastric epithelioma, Adamuz
February 9 – NP
    Juan B. Sánchez Pizarro, 50, manual labourer, mitral insufficiency, Villanueva Rey
February 15 – PP
    Argimiro Sánchez García, 24, heart attack, Villanueva de Córdoba
    Antonio López Sánchez, 65, heart attack, Villanueva de Córdoba
    Ramón Molina Gutiérrez, 65, policeman, heart attack, Cordoba
February 16
    NP – Pablo Sánchez Moreno, 78, farm worker, rheumatism, Castro del Río
    PP - Juan Miguel Jiménez Santiago, 61, painter, heart attack, Benamargosa
        José Castro Murillo, 51, manual labourer, heart attack, Granada
February 17 – PP
    Luís Dobao Martínez, 42, manual labourer, emaciation, Palma del Rio
February 18 – PP
    Vicente Riañoz Martín, 33, manual labourer, emaciation, Peñarroya
February 20 – NP
    Francisco Horcas Castro, 57, bricklayer, anaemia, Valenzuela
Ferbruary 22 – NP
    Francisco Pareja Valdivia, 54, manual labourer, myocarditis, Priego
February 23 – PP
    Enrique Martín Junca, 57, Municipal Secretary, emaciation, Cordoba
    Jesús Chacón Rodríguez, 40, manual labourer, emaciation, Alzázar de San Juan
February 24 – NP
    Melquiades Ruíz Porras, 50, pneumonia, Peñarroya
    Estanislao Tirado Roldán, 45, nurse, heart attack, Madrid
February 25 – NP
    Juan A. Rodríguez Roig, 49, miner, pericarditis, Peñarroya
    Bartolomé Regalón Sabariego, 56, manual labourer, anaemia, Cordoba
February 26 – NP
    Pedro Guerrero Ruíz, 52 manual labourer, myocarditis, Murcia
    Antonio Sánchrz Villaseca, 51, manual labourer, myocarditis, Hinojosa
February 27
    NP – Diego Redondo Rodríguez, 32, manual labourer, pneumonia, Pozblanco
    PP – Hilario Celaya Pozo, 46, manual labourer, emaciation, Zalamea Serena
February 28
    NP – Antonio Pérez Caballero, 65, manual labourer, cerebral haemorrhage,
        Doña Mencia
        Alejandro Alamillos Muñoz, 42, manual labourer, anaemia, Vn. De Córdoba
    PP – José Olmo Martínez, 57, manual labourer, emaciation, Espiel
        Florentino Figueroa Expósito, 47, manual labourer, emaciation, Puente
        Genil

March 1 – NP
   Juan Perea Agarcía, 56, miner, arteriosclerosis, Hinojosa
March 2
   NP – Juan Rabadán Lucena, 42, manual labourer, endocarditis, Epejo
   PP – Rafael Guerrero Malaver, 27, manual labourer, emaciation, Malaga
March 3 – NP
   Pedro Caballero Cardito, 44, coal merchant, anaemia, Palma del Río
   Antonio Barrios Lozano, 32, shoemaker, gastric ulcer, Peñarroya
March 4
   NP – Blas Acosta Méndez, 57, manual labourer, pneumonia, Montoro
        Martín García Jurado, 29, tradesman, acute anaemia, Belmez
        Justo Monge Ramos, 52, manual labourer, endocarditis, Hinojosa
        Pablo Alcalde Plazuelo, 49, miner, myocarditis, Espiel
   PP – Alfonso Mansilla Calvo, 59, manual labourer, emaciation, Baena
        Manuel Expósito Márquez, 50, manual labourer, hepatic epithelioma,
        Cordoba
March 5 – PP
   Manuel Reyes Guerra, 68, manual labourer, emaciation, Palma del Río
March 6
   NP – Vicente Mesa Polo, 62, manual labourer, mitral insufficiency, Doña Mencia
        Antonio Lorenzo Torremocha, 52, shoemaker, nephritis, Constantina
        Manuel Moreno Ramírez, 46, shoemaker, cerebral congestion, El Viso
   PP – Félix Migallón Román, 38, manual labourer, emaciation, Añora
        Rafael Onieva Huete, 61, mailman, emaciation, Cordoba
March 7 – NP
   José Tarrajo Ochando, 44, manual labourer, nephritis, Villaviciosa
   José Baena Lozano 66, manual labourer, myocarditis, Baena
March 8
   NP – Pedro Moreno Jurado, 53, bricklayer, nephritis, Pozoblanco
        Demetrio Parra, farm worker, 54, anaemia, Hinojosa
        Bautista Ventura Jurado, 66, manual labourer, heart attack Villanueva Rey
        Juan Montilla Gallardo, 43, manual labourer, arteriosclerosis, Valenzuela
        Antonio Cosano Mansilla, 20, manual labourer, pneumonia, Puente Genil
        Rufo Madrid Sánchez, 35, shoemaker, heart attack, Valdepeñas
   PP – Felipe López Serrano, 27, manual labourer, heart attack, Villafranca
March 10
   NP – Antonio Pérez Muñoz, 57, manual labourer, myocarditis, Pueblonuevo
        José Blázquez Pérez, 74, miller, heart attack, Fuenteobejuna
        Juan Horcas Castro, 66, manual labourer, arteriosclerosis, Valenzuela
   PP – Rafael Caracuel López, 18, manual labourer, emaciation, Montilla
March 11
   NP – José Elices Pérez, 42, railway worker, acute anaemia, Rota
        Acisclo Jurado Villarejo, 56, manual labourer, anaemia, Pozoblanco
        Pablo Morillo Castro, 49, farmer, endocarditis, Fuenteobejuna
   PP - José García Sánchez, 39, manual labourer, emaciation, Espiel
        Críspulo Daza Peña, 67, manual labourer, emaciation, Santa Eufemia
March 12
   NP – Manuel Delgado Buenorostro, 31, manual labourer, acute anaemia, Puente
        Genil
        Francisco Colorado Fernández, 25, baker, anaemia, Marchena
   PP – Jacinto Vega Luque, 40, manual labourer, emaciation, Castro del Río
March 13
   NP – Francisco Sánchez Mediavilla, 58, manual labourer, anaemia, Almodóvar
        José Barroso Gómez, 51, mechanic, anaemia, Azuaga
   PP – Juan López Vallejo, 20, manual labourer, emaciation, Bujalance
        Guillermo Muñoz Neira, 30, miner, emaciation, Huelva
        Rafael Córdoba Ariza, 53, manual labourer, emaciation, Cordoba
March 14 – PP
   Francisco Escobar García, 34, manual labourer, emaciation, Cordoba

José García Peralta, 18, carpenter, emaciation, Cordoba
Marco Majuelos Torres, 39, manual labourer, emaciation, Villanueva de Córdoba
March 15
  NP – Antonio Vega Amaya, 28, stonemason, emaciation, Morón de la Frontera
       Francisco González Serrano, 47, manual labourer, pneumonia, Montoro
       Francisco Bonilla Cubiles, 60, manual labourer, anaemia, Cádiz
  PP - Luís Ruíz Mora, 20, manual labourer, emaciation, Cordoba
March 16
  NP – Juan Ortega Fuentes, 54, manual labourer, anaemia, La Carlota
  PP - Antonio Ortega Baena, 27, manual labourer, emaciation, Rute
March 17
  NP – Ángel de la Fuente Ramírez, railway worker, acetonemia, Malaga
       Ramón Garcia Guerrero, 53, mule driver, pneumonia, Montoro
  PP - Leopoldo Pernil Huertas, 21, shoemaker, emaciation, Valverde del Campo
March 18
  NP – Antonio Ruíz Lozano, 39, manual labourer, acute anaemia, Villaviciosa
       Antonio Calabria Molero, 56, manual labourer, epithelioma, Posadas
  PP - Juan Lavado López, 39, manual labourer, emaciation, Castro del Río
       Estanislao Trujillo Garía, 30, cattle breeder, emaciation, Las Quemadas
       Antonio Ruíz Navajón, 50, manual labourer, emaciation, Santa Eufemia
March 19 – PP
  Antonio Martínez González, 26, manual labourer, emaciation, Cazorla (Jaén)
  Francisco Pérez Sillero, 34, manual labourer, emaciation, Loja (Malaga)
March 20 – PP
  Antonio Sánchez Núñez, 28, carpenter, emaciation, Cordoba
March 21
  NP – José Vargas Gutiérrez, 49, miner, emaciation, Belmez
       Fernando Martínez Magaña, 51, manual labourer, anaemia, Almeria
  PP - Antonio Gamazo Díaz, 49, farm worker, emaciation, Arcos de la Frontera
March 22
  NP – Narciso Ruíz Ceballos, 57, manual labourer, anaemia, Villanueva Cordoba
  PP - Miguel Romero Cabezas, 46, manual labourer, emaciation, Villanueva de
       Cordoba
March 23 – NP
  José Gómez Márquez, 71, baker, anaemia, Dos Torres
March 25
  NP – Manuel Calvo Manzano, 36, manual labourer, anaemia, Cabeza Buey
       Pedro Gallardo Pérez, 63, manual labourer, anaemia, Badajoz
  PP - Juan Fernández Díaz, 46, manual labourer, emaciation, Villarreal
       José Murillo Risquez, 49, manual labourer, emaciation, Villanueva Rey
March 26 – NP
  Valeriano Rodríguez García, 64, manual labourer, anaemia, Belmez
  Basílio Horrillo Sereno, 56, miner, anaemia, La Granjuela
March 27 – PP
  Julián Ruíz Gómez, 45, entrepreneur, asystole, Madrid
March 28 – NP
  Gonzalo Delgado Sánchez, 49, railway worker, bronchitis, Toledo
March 29 – NP
  Juan Serrano Ruíz, 46, manual labourer, anaemia, Montoro
March 30 – NP
  Juan López Romero, 47, manual labourer, emaciation, Villanueva de Córdoba
  Santiago Ambrosio Arroyo, 54, miner, acute anaemia, Belmez
  Juan Ibáñez González, 34, soldier, emaciation, Valencia
  Antonio José Peña Agenjo, 55, miner, emaciation, Villanueva del Rey
March 31
  NP – Ernesto Medina Aceitunom 25, manual labourer, acuate anaemia, Alcalá
       Antonio Ruíz Muñoz, 19, metalworker, anaemia, Cordoba
  PP – Francisco Sanz Moreno, 38, manual labourer, coma, Villa del Río

April 1
NP – Francisco Salinas Toledano, 53, manual labourer, emaciation, Bujalance
Pedro Sánchez González, 57, manual labourer, emaciation, Villanueva Rey
PP - Francisco Céspedes Vázquez, 55, mailman, tuberculosis, Almeria

April 2
NP – Miguel Abril Molina, 28, manual labourer, staphyloma, Espiel
Eustaquio Lozano Pedraja, 29, bricklayer, emaciation, Peñarroya
Ángel Bejarano Blanco, 49, miner, anaemia, Peñarroya
Zacarías Pérez Trujillo, 37, manual labourer, emaciation, Iznájar
PP - Miguel Jiménez, 40, plasterer, emaciation, Puente Genil

April 3
NP – Juan Gómez Jiménez, 45, manual labourer, anaemia, Adamuz
Juan García Guerrero, 56, manual labourer, anaemia, Montoro
Manuel Estévez Barragán, 55, bricklayer, anaemia, Fuenteobejuna
PP - César Buenahora Pascual, 49, road worker, heart attack, Madrid
Roque Herrera Bejarano, 45, coal merchant, tuberculosis, Villanueva de Cordoba

April 4
NP – Manuel Tanay Dávila, 48, manual labourer, anaemia, Zalmea de la Serena
Julio Caballero Madueño, 30, salesman, nephritis, Villaralto
José Aguilar Espada, 57, manual labourer, emaciation Palenciana
PP - Antonio Ortiz Pérez, 25, employee, emaciation, Seville

April 5
NP – Rafael Pons Luque, 58m, market gardener, emaciation, Posadas
Francisco González Cobos, 59, manual labourer, emaciation, Granada
Felipe Caballero Polonio, 47, foundry hand, emaciation, Belmez
PP - Antonio Soto Arte, 25, bricklayer, tuberculosis, Priego

April 6 – NP
José Moya Díaz, 54, manual labourer, anaemia, Peroche
Juan A. Rodríguez Ruíz, 50, manual labourer, anaemia, Belalcázar
Joaquim Espinosa Gómez, 40, manual labourer, anaemia, Cabra

April 7
NP – Antonio la Torre Nevada, 30, manual labourer, tuberculosis, Villaviciosa
José Rodríguez Calvo, 28, manual labourer, pneumonia, Belalcázar
Rafael González Berengena, 41, manual labourer, anaemia Villanueva Rey
Juan Ruíz Criado, 40, shoemaker, acute anaemia, Montoro
Luís Serrano Yepes, 46, plasterer, emaciation, Espejo
PP - Roque Arenas Garrido, 28, manual labourer, enterocolitis, Arjonilla

April 8 – NP
Luís Bagre Maestre, 52, manual labourer, anaemia, Fuente Palmera
José Ticiba Montoro, 61, mechanic, anaemia, Priego
Rafael Sánchez Galán, 52, miner, anaemia, Villanueva del Duque

April 9 – NP
Antonio Buenosvinos Rebaño, 47, manual labourer, emaciation, Cañete
Antonio de la Fuente Arribas, 52, manual labourer, anaemia, Villaviciosa

April 10 – NP
José Olmo Rísquez, 46, market gardener, anaemia, Cordoba
Toribio Valderrábanos Agredano 64, manual labourer, emaciation Fuenteobjuna
Salvador Molina Mora, 48, manual labourer, emaciation, Luque
Pedro Giménez Berenguer, 24, manual labourer, anaemia, Posadas

April 11
NP – Juan Franco Navarro, 50, manual labourer, mitral insufficiency, Palma Río
Juan López Gisado, 62, manual labourer, anaemia, Fuente Palmera
Andrés Márquez Moncayo, 66, manual labourer, anaemia, Fuenteobejuna
Juan Fernández Romero, 46, shepherd, anaemia, Torrecampos
Antonio Arroyo León, 64, manual labourer, anaemia, Baena
Miguel Prieto López, 39, manual labourer, nephritis, Granada
Celestino Calvo Calvom 48, anaemia, Villaviciosa
PP - Antonio Cuadradro Cuadrado, 55, carter, emaciation, Adamuz

April 12 – NP
Antonio Expósito Caracuel, 53, manual labourer, heart attack, Montilla
Benito Márquez Muñoz, 38, manual labourer, anaemia, Cardeña
Juan A Margarín Millán, 30, farm worker, anaemia, Peñarroya
April 13 – NP
José Padilla Rojano, 60, manual labourer, emaciation, Baena
Francisco Calzadilla Barbado, 64, manual labourer, anaemia, Montilla
Antonio Gordillo Luque, 28, manual labourer, anaemia, Valenzuela
Juan Fernández Barrigón, 32, manual labourer, anaemia, Hornachuelos
José Marín Palacios, 24, mechanic, anaemia, Jaén
April 14 – NP
Martín Vallejo Aljarilla, 48, manual labourer, emaciation, Valenzuela
Antonio Flores Serrano, 51, manual labourer, anaemia, Belalcázar
Antonio Viso Peña, 56, manual labourer, emaciation, Villanueva del Rey
April 15 – NP
Alfonso Calles Peinazo, 62, manual labourer, anaemia, Montoro
April 16 – NP
Rafael Romero Benavente, 20 months, enterocolitis
Prudencio Bravo Martín, 48m hat maker, heart failure, Badajoz
Luciano Capilla López, 37, shoemaker, cerebral vascular attack, Hinojosa
April 18
NP – Maximiliano Bravo García, 46, mailman, gastric ulcer, Belmez
PP - José García García, 26, barber, emaciation, Pueblonuevo
Pedro Ceballos González, 58, basket weaver emaciation, Fuente Tójar
April 19
NP – Eugenio Galán Cepas, 44, manual labourer, heart failure, Villaharta
PP - Eduardo Guerrero Aguilar, 65, bricklayer, Cordoba
April 21 – NP
Manuel Castro Orellana, 32, manual labourer, tuberculosis, Seville
April 22 – NP
Francisco Murillo Sánchez, 62, blacksmith, enterocolitis, Hinojosa
April 23
NP – Amador Jiménez Roldán, 43, manual labourer, enterocolitis, Pedro Abad
José Luna Rivas, 41, manual labourer, heart failure, Puente Genil
Joaquim Monzonis Alemán, 25, manual labourer, anaemia, Peñarroya
PP - Antonio Vázquez Barrera, 18m manual labourer, emaciation, Peñarroya
Manuel Rodríguez Hidalgo, 52, manual labourer, pellagra, Dos Torres
April 24 – NP
Alfonso Alcalde Palacios, 27, blacksmith, tuberculosis, Belmez
April 25 – NP
Juan Caseos Gallardo, 46, manual labourer, anaemia, Belalcázar
Miguel Amor Arroyo, 28, baker, heart attack, Cabra
April 26
NP – Esteban Manuel Rubio Vinagre, 35, manual labourer, pneumonia, Fuenteobejuna
Antonio Obrero Hernández, 37, bricklayer, cerebral haemorrhage, Posadas
PP - Rafael Sedo el Águçila, 17, cook, heart attack, Almeria
April 27 – NP
José Filigrana Silva, 61, dealer, myocarditis, Llerena
Liderio Revuelto Pedregosa, 22, manual labourer, tuberculosis, Villaviciosa
Antonio Solis Romero, 45, manual labourer, anaemia, El Carpio
Tomás Cachinero Olivros, 52, manual labourer, enterocolitis, Montoro
Antonio López Blanco, 40, miner, anaemia, Alcaracejos
April 28 – NP
Antonio Baza Granados, 47, farmer, anaemia, Los Blázquez
Manuel Moreno Bedmar, 56, manual labourer, anaemia, Villa del Río
April 29 – NP
Antonio Luque Alcántara, 40, manual labourer, enterocolitis, Castro del Río

April 30 – NP
   Julián Páez Nieto, 38, manual labourer, septicaemia, Almedinilla
   Pedro Moreno Pulgarín, 58, manual labourer, anaemia, Cordoba
   Manuel Cuevas Serrano, 53, septicaemia, Cordoba

May 1 – NP
   Antonio Monterroso Guadalupe, 31, manual labourer, anaemia, Navas de la C.
May 2 – NP
   Antonio Torralbo Caballero, 47, manual labourer, anaemia, Villanueva Cordoba
May 3 – NP
   Gabriel González López, 60, manual labourer, pneumonia, Pedroche
May 4 NP
   José Tena Paredes, 65, manual labourer, anaemia, Badajoz
May 5 – NP
   Rafael Leiva Manrique, 40, electrician, tuberculosis, Puente Genil
May 6
   NP – Juan Bailén Orega, 45, metalworker, anaemia, Torrecampos
       Antonio Torres González, 45, manual labourer, anaemia, Almeria
   PP - Pedro Cabrera Gálvez, 30, manual labourer, pellagra, Montoro
       Francisco Cabrera García, 24, manual labourer, tuberculosis Villanueva
       de Córdoba
May 7 – NP
   Antonio Pedregosa Blanco, 33, manual labourer, anaemia, Villaviciosa
   Ildefonso Hidalgo Escribano, 57, entrepreneur, arteriosclerosis, Montoro
May 8 – NP
   Andrés Salido González, 43, waiter, tuberculosis, Jaén
   Antonio López García, 59, railway worker, pneumonia, Cabeza del Buey
May 10 – PP
   Alfonso Pérez Lorencio, 26, tradesman, tuberculosis, Madrid
   Leopoldo Galán Cepas, 48, manual labourer, emaciation, Villaharta
May 11 – NP
   Jesús Fernández Moreno, 24, anaemia, Dos Torres
   Juan Lara Muñoz, 52, manual laourer, pneumonia, Villa del Río
May 13 – NP
   Pablo Aranda Prado, 27, manual labourer, enterocolitis, Hinojosa del Duque
   Francisco Garrido Molina, 56, manual labourer, enterocolitis, Pedroche
   Manuel Prieto Navas 42, tree feller, tuberculosis, Puente Genil
May 21
   NP – Francisco Muñoz Pozo, 60, manual labourer, enterocolitis, Cardeña
   PP - Críspulo Pontes Lópe, 24, manual labourer, enterocolitis, Belalcázar
       Juan Santofimia Cantdor, 51, manual labourer, embolism, Villanueva de Cordoba
May 22 – NP
   Mariano Ginesta Santos, 31, mechanic, pneumonia, Valencia
May 23 – PP
   Antonio Moral Rider, 34, manual labourer, heart attack, Posadas
May 25
   NP - Plácido Campaña Gutiérrez, 21, manual labourer, enterocolitis, Castro del Río
       Cristóbal Vílchez Rodríguez, 53, manual labourer, pneumonia, Jaén
   PP - Antonio Pérez López, 19 manual labourer, emaciation, Guadalajara
May 26 - NP
   Juan Rojas Holanda, 61, manual labourer, bronchopneumonia, Pedro Abad
   Alfonso Rojas García, 3, bricklayer, enterocolitis, Pedro Abad
   Manuel Fernández Paz, 34, manual labourer, enterocolitis, Villanueva del Rey
May 27 – NP
   Juan Cuadrado Salnia, 54, manual labourer, anaemia, Adamuz
May 28 – NP
   José Cruz Ortíz, 36, manual labourer, enterocolitis, Baena
May 29 – NP
   Martín Serrano Torrico, 57, manual labourer, pneumonia, Pozoblanco

Pascual Garrido Calé, 55, manual labourer, malaria, Granada
May 30 – NP
Ángel Marina Gamero, 26, woodsman, enterocolitis, Madrid
Antonio Montero Alguacil, 28 manual labourer, enterocolitis, Fuenteobejuna
May 31 – NP
Andrés García Garrido, 25, manual labourer, enterocolitis, Castro del Río

June 1
    NP – José A. Aranda Monterroso, 57, baker, enterocolitis, La Granjuela
        Francisco Romero Díaz, 53, miner, enterocolitis, Pozoblanco
    PP - José Vila López, salesman, emaciation, Malaga
June 2
    NP – Cristóbal Roríguez López, 30, manual labourer, nephritis, Malaga
        Juan Gil Torres, 27, manual labourer, enterocolitis, Huelva
    PP - Juan Cabezas Molina, 52, manual labourer, uraemia, Seville
June 5 – NP
Manuel Armada Lozano, 67, baker, pneumonia, Posadas
June 6 – NP
Blas Esquina Orellana, 58, manual labourer, enterocolitis, La Granjuela
June 9 NP
Juan Pérez Casstelo, 64, manual labourer, myocarditis, Hinoosa del Duque
Joaquim Frutos Vaquerom 29m manual labourer, enterocolitis, Pedroche
June 10
    NP – Manuel Arroyo García, 26, manual labourer, rheumatism, Jaén
        José M. Nogales Herrera, 20, manual labourer, enterocolitis, Peñarroya
    PP - Alfredo Gómez Lucena 48, manual labourer, emaciation, Cordoba
June 11 – NP
Pedro Reyes Ramos, 49, manual labourer, malaria, Montoro
June 13 – NP
Manuel Castillo Borrego, 55, manual labourer, pneumonia, Villa del Río
June 16 – NP
Anastasio Cabanillas Ponce, 33, miner, enterocolitis, Brazatortas
Amador Lora García, 53, mason, anaemia, Alanís
June 17
    NP – José Muñoz Rodríguez, 36, manual labourer, tuberculosis, Valsequillo
    PP - Jorge Blanco Conde, 31, bricklayer, agranulocytosis, Santa Eufemia
June 18 – NP
Manuel Brocal Martínez, 55, manual labourer, nephritis, Montoro
Ángel Gallardo Perales, 26, office clerk, gastric ulcer, Ciudad Real
June 19 – NP
Luís Núñez, 28, farm worker, enterocolitis, Cádiz
Ignacio Ríos Cornejo, 29, manual labourer, malaria, Valsequillo
Juan A. Peláez Santiago, 57, farmer, nephritis, Montoro
June 21 – PP
Antonio Fimia Espino, 34, bricklayer, enterocolitis, Cordoba
June 22 – PP
Manuel Bolaños Veria, 61, dealer, asystole, Cordoba
June 23
    NP – Nicolás Mareo Rodríguez, 33, manual labourer, anaemia, Badajoz
        José García Cachinero, 60, road worker, bronchitis, Villanueva Cordoba
    PP - Alfonso Cejudo Redondo, 22, manual labourer, enterocolitis, Villanueva C.
June 24 – PP
Gabriel Vázquez García, 24, coal merchant, enterocolitis. Alcaudete
June 25 – PP
Juan Molina Gómez, 58, manual labourer, septicaemia, Malaga
José Valverde Ruíz, 77, manual labourer, pneumonia, Pedroche
Manuel Sánchez Alamillo, 54, manual labourer, enterocolitis, Alcaracejos
Daniel Rodríguez Guisado, 58, manual labourer, enterocolitis, Fuente Palmera

June 27 – NP
   Pedro Moreno Martínez, 55, manual labourer, enterocolitis, Bujalance
   Luís Romero Jiménez, 23, barber, tuberculosis, Cádiz
   Melitón Tejada Cabanillas, 39, butcher, enterocolitis, Peñarroya
June 28 – PP
   Juan Manuel Sánchez Moreno, 21, manual labourer, tuberculosis, Hinojosa
   Antonio Roda Delgado, 37, manual labourer, anaemia, Úbeda
   Juan A. Romero Horrillo, 26, manual labourer, enterocolitis La Cardenchosa
June 29
   NP – Juan Barba Jiménez, 56, manual labourer, anaemia, Doña Mencia
            Vicene Fernández Cabrera, 60, manual labourer, emaciation, Villanueva del Rey
   PP - José Ramírez Martín, 23, bricklayer, enterocolitis, Malaga
            José Luna Díos, 37, manual labourer, tuberculosis, Baena
June 30 – NP
   Luís Fernández Médez, 29, manual labourer, anaemia, Palma del Río.
July 1
   NP – Alfonso Ruíz Tabas, 36, manual labourer, enterocolitis, Hinojosa
   PP – Antonio Acosta Romero, 40, manual labourer, enterocolitis, Seville
            Pedro Cuadrado Cerezo, 37, bricklayer, enterocolitis, Adamuz
July 2 – NP
   Juan Quero Bazán, 61, manual labourer, enterocolitis, Montoro
July 3
   NP – José Trujillo Badillo, 27, manual labourer, enterocolitis, Ciudad Real
   PP – Francisco Herrera Esinosa, 41, manual labourer, asystole, Cádiz
July 4 – PP
   Serápio Ortuño Delgado, 40, enterocolitis, Zamoranos
July 5 – PP
   José Bautista Losada, 60, cook, enterocolitis, Jerez de la Frontera
   José Ruíz Priego, 51, manual labourer, enterocolitis, Carcabuey
   Antonio Romero Moreno, 20, manual labourer, tuberculosis Villanueva de Córdoba
July 7 – PP
   Agustín Góngora Serrano, 19, mechanic, pellagra, Lucena
   Rafael Cádiz Miranda, 57, manual labourer, gastroenteritis, Ciudad Real
   Miguel Zarnoza Pérez, 50, market gardener, enterocolitis, Torrecampo
July 8
   NP – Antonio Jurado García, 31, bricklayer, anaemia, Belmez
            Feliciano Rubio López, 31, road worker, tuberculosis, Hinojosa
   PP – José SotoMayor Palma, 43, baker, tuberculosis, Aguilar
July 9 – NP
   Juan Boquizo Martínez, 43, manual labourer, gastroenteritis, Villanueva del Duque
July 10 – PP
   Manuel Jurado Alamillos, 28, manual labourer, gastroenteritis, Villanueva del Duque
July 11 – NP
   José López Checa, 37, miner, mitral insufficiency, Linares
July 13 – NP
   Antmonio García Torrico, 40, manual labourer, enterocolitis, El Guijo
July 14 – PP
   Florencio Moreno Barcala, 35, metalworker, uraemia, Madrid
   José Rodriguez Fernández, 59, charcoal worker, emaciation, Cordoba
July 18 – NP
   Antonio Murillo Méndez, 54, manual labourer, enterocolitis, Badajoz
July 21 – NP
   Santiago Rubio González, 23, manual labourer, tuberculosis, Espiel
July 25 – NP
   Antonio Nogales Trenado, 50, manual labourer, enterocolitis, Badajaoz
July 29
   NP – Marco Tejero Jiménez, 50, manual labourer, enterocolitis, Fernán Núñez
            Cristóbal Pacheco Lopera, 29, manual labourer, pneumonia, Iznájar
   PP – Juan Criado Orihuela, 40, designer, tuberculosis, Cabra

July 31 – NP
Fidel García Cáceres, 60, manual labourer, enterocolitis, Villaviciosa

August 2 – NP
Federico Rodríguez Mata, 31, manual labourer, tuberculosis, Jaén
August 3
NP – Antonio Bocero Rodríguez, 32, manual labourer, enterocolitis, Fuente Carreteros
PP – Dolores Jiménez Fernández, 40, housewife, pellagra, La Victoria
August 4 – NP
Antonio Salinas Pozuelo, 49, manual labourer, mitral insufficiency, Pedro Abad
Francisco Mayrga Hurtado, 60, manual labourer, enterocolitis, Lucena
August 6
NP – Francisco Gómez Ramos, 51, farm worker, enteritis, Belmez
PP – Miguel Cortés Alhaja, 22, marble worker, peritonitis, Mérida
Manuel Gordillo Jurado, 17, manual labourer, heart attack, La Cardenchosa
August 7 – NP
Juan Madrid Ramos, 50, shoemaker, enteritis, Bujalance
José Peláez García, 44, manual labourer, myocarditis, Pueblonuevo
August 8 – NP
Miguel Ortíz Luna, 26, tradesman, enteritis, Villafranca
August 9
NP – Francisco Gómez Márquez, 28, manual labourer, tuberculosis, Montilla
PP – Matias Fresno de la Orden, 43, mechanic, uraemia, Almadén
August 12 – PP
Manuel Ropero Calleja, 65, miner, enterocolitis, Fuenteobejuna
August 14
NP – Manuel Cortés Cerezo, 54, barber, mitral insufficiency, Adamuz
PP – Juan J. Jiménez Caballero, 49, salesman, pneumonia, Cabra
August 15 – NP
Manuel González Rodríguez, 44, manual labourer, tuberculosis, Hornachuelos
August 16 – NP
Antonio Romero León, 49, manual labourer, emaciation, Baena
Pedro Robles Rosauro, 43, manual labourer, mitral insufficiency, Bujalance
August 19 – PP
Pedro Garcia Carretero, 22, manual labourer, tuberculosis, Cardeña
Modesto Prats Aguilar, 49, barber, heart attack, Villanueva del Duque
August 20 – PP
Manuel Márquez Peinado, 29, manual labourer, tuberculosis, Dos Torres
Francisco Lucena Márquez, 52, employee, heart attack, Montilla
Francisco Priego Parrado, 35, manual labourer, skull fracture (possible suicide)
August 21 – NP
Manuel Gómez Torrico, 51, manual labourer, pneumonia, Villaralto
August 25 – NP
Agustín Arévalo Marta, 38, manual labourer, pneumonia, Pedroche
August 26 – NP
Conrado López Aláez, 31, manual labourer, enterocolitis, Dos Torres
August 28 – NP
Atanasio García Ramos, 50, manual labourer, enterocolitis, Espiel
Juan Rodríguez Luque, 60, manual labourer, septicaemia, Fernán Núñez
Ventura Peinado Madueño, 52, manual labourer, pneumonia, Dos Torres

September 6 – NP
Severo Gómez Gozalbo, 40, farmer, enterocolitis, Pozoblanco
Aniceto López Romera, 24, manual labourer, tuberculosis, Villaviciosa
Valentín Aragonés Hidalgo, 47, miner, septicaemia, Castuera
September 9 – PP
Demetrio Herruzo Rayo, 51, manual labourer, tuberculosis, Hinojosa
September 10 – NP
Amador Loano Heras, 49, manual labourer, bronchopneumonia, Baeza

September 11 – NP
  Francisco Pérez Cerezo, 30, manual labourer, tuberculosis, Cáceres
September 17 – NP
  Antonio Moya Torrico, 27, manual labourer, tuberculosis, Hinojosa
  Vicente Triviño Agenjo, 54, farm worker, arteriosclerosis, Los Blázquez
September 22 – NP
  Ángel López Redondo, 34, entrepreneur, tuberculosis, Santa Eufemia
September 24 – NP
  Modesto Cidoncha Cidoncha, 64, manual labourer, pneumonia, Frenteobejuna
September 26 – PP
  Enrique García Álvarez, 28, sailor, nephritis, Alcalá del Río

October 2 – NP
  Antonio Calvo García, 44, manual labourer, heart failure, Belmez
October 3 – NP
  Manuel Cotán Mejias, 42, manual labourer, enterocolitis, Seville
  José Villalba Benavides, 43, manual labourer, enteritis, Posadas
October 4 – NP
  Joaquín Mateo Ramos, 26, manual labourer, tuberculosis, Jerez de la Frontera
  Alfonso Ojeda Fernández, 60, manual labourer, erysipelas, Posadas
October 5 – NP
  José Torralbo Carbonero, 29, weaver, nephritis, Villanueva de Córdoba
  Francisco Ballestero Martínez, 44, manual labourer, tuberculosis, Almodóvar
October 6 – NP
  Andrés Luque Regaló, 27, manual labourer, tuberculosis, Adamuz
October 7 – NP
  Francisco Castillejo Molero, 35, office clerk, tuberculosis, Peñarroya
October 8 – NP
  Santiago Mantas Pontes, 60, manual labourer, pneumonia, El Carpio
  Pablo Aranda Algar, 36, manual labourer, tuberculosis, El Carpio
  Gregorio Cabrera Mansilla, 34, cattle breeder, influenza, Villanueva del Duque
October 9 – NP
  Rafael Alberca Segura, 34, manual labourer, septicaemia, Adamuz
October 16 – NP
  Juan M. Nacarino Molina, 63, schoolteacher, uraemia, Palma del Río
  Antonio Ferrero López, 56, manual labourer, rheumatism, Dos Torres
October 17 – NP
  Benito Pedregosa Uclés, 46, shoemaker, nephritis, Valenzuela
  Aurelio Martín Infantes, 26, manual labourer, tuberculosis, La Granjuela
October 18 – NP
  Benito Cejudo Moreno, 44, manual worker, enterocolitis, Villanueva de Córdoba
October 20 – NP
  Juan Martín Perea, 55, bricklayer, colitis, Belalcázar
October 22 – NP
  Ángel Morales Peña, 38, manual labourer, meningitis, Hinojosa del Duque
October 23 – NP
  Ángel Barancho Cano, 55, manual labourer, septicaemia, Fuente la Lancha
  Francisco Ferrero Lavrador, 46, manual labourer, bronchitis, Torrecampo
  Pedro Martín Infantes, 37, miner, enterocolitis, La Ganjuela
October 24 – NP
  Fransico Mejias Cerero, 58, manual labourer, bronchopneumonia, Adamuz
  Antonio Nevado Expósito, 63, manual labourer, mitral insufficiency Villaviciosa
  Fernando Ruíz Galán, 39, manual labourer, tuberculosis, Almodóvar del Río
October 25 – NP
  Antonio Álvarez Cobos, 45, manual labourer, enterocolitis, Villaviciosa
October 26 – NP
  Elias Jiménez Blanco, 49, manual labourer, pneumonia, Santa Eufemia
October 27 – NP
  Rafael Martínez Castilla, 29, manual labourer, influenza, Adamuz

Dionisio Díaz Serrano, 41, manual labourer, rheumatism, Dos Torres
October 28 – NP
Miguel Padilla Torrecilla, 41, manual labourer, arthritis, Almeria
Antonio Escamilla García, 59, manual labourer, arteriosclerosis, Villanueva de Córdoba
October 29 – NP
Federico Moral Otero, 29, manual labourer, tuberculosis, La Carlota
October 30 – NP
Manuel Romero Rodriguez, 40, miner, influenza, Belmez

November 1 – NP
Antonio Casán Molleja, 34, manual labourer, tuberculosis, Pedro Abad
November 2
NP – José Moreno González, 61, manual labourer, enterocolitis Alcalá la Real
    Emiliano Vioque Trialdo, 65, farm worker, enterocolitis, Dos Torres
PP – Juan Pérez Martínez, 60, manual labourer, emaciation, Villanueva de Córdoba
November 3 – NP
José Muñoz Muñoz, 34, manual labourer, enteritis, Cardeña
Antonio Rodríguez Rodríguez, 48, manual labourer, nephritis, Granada
November 4 – NP
Mario Pérez Fernández, 47, shoemaker, pneumonia, Espiel
Elias Ruíz González, 40, farmer, asystole, Hinojosa del Duque
November 5 – NP
Antonio Ballesteros Cañete, 42, manual labourer, tuberculosis, Luque
José Español Blasco, 40, farm worker, pneumonia, Villanueva del Duque
November 6 – NP
José Tanay Pérez, 39, bricklayer, enterocolitis, Utrera
November 10 – PP
Ángel Novembras Ortíz, 48, manual labourer, laryngeal epithelioma, Villaviciosa
November 11 – NP
Diego Ortega Pérez, 87, manual labourer, rheumatism, Adamuz
November 14 – NP
Manuel Sánchez García, 43, manual labourer, enterocolitis, Castuera
November 15 – NP
Pablo Santamarta Expósito, 50, coal merchant, enterocolitis, Priego
November 17 – NP
Santiago Benítez del Rey, 43, manual labourer, pneumonia, Espiel
November 18 – NP
Manuel Melendo Carrillo, 45, manual labourer, arteriosclerosis, Villa del Río
November 19 – NP
Martín Muñoz Frutos, 41, manual labourer, uraemia, Cabeza del Buey
November 20 – NP
Juan Romero Góez, 47, shepherd, tuberculosis, Villaralto
November 22 – NP
José Sámchez Gómez, 47, manual labourer, enterocolitis, Villanueva de Córdoba
Urbano Martínez Martínez, 37, manual labourer, tuberculosis, Cuenca
November 24 – PP
Bibiana Romero Gómez, 70, unemployed, enterocolitis, Villanueva de Córdoba
November 27 – PP
Joaquín Mendoza Carpio, 46, manual labourer, heart attack, Castro del Río
November 28 – NP
Pedro Gómez Moreno, 28, manual labourer, bronchopneumonia, Montoro
Antonio Fernández Gómez, 56, manual labourer, influenza, Villaralto
November 29 – NP
David Garacía Toril, 46, manual labourer, enterocolitis, El Viso
Francisco Murillo González, 49, manual labourer, rheumatism, Villanueva del Rey
November 30 – PP
Mª Josefa Hidalgo Rodríguez, 35, housewife, enterocolitis, Montoro

December 1 – NP
    José Plaza Cuevas, 42, manual labourer, influenza, Fuente Palmera
    Agustín Murillo Sánchez, 60, blacksmith, enterocolitis, Hinojosa del Duque
December 2 – NP
    José Escamilla García, 66, bricklayer, enterocolitis, Villanueva de Córdoba
    Felipe Molina Herruzo, 46, manual labourer, bronchopneumonia, El Guijo
December 3 – NP
    Julián Madueño Muñoz, 37, manual labourer, avitaminosis, El Viso
    Pedro Hidalgo Caballero, 66, manual labourer, enterocolitis, Montoro
December 5 – NP
    Francisco Santíago Gordillo, 42, manual labourer, enterocolitis, La Granja
December 6
    NP - Diego Valiño Trejo, 53, farm worker, septicaemia, Badajoz
    PP – Juliana Almena Moreno, 54, myocarditis, Belalcázar
December 8 – NP
    Ángel Díaz Morales, 41, miner, enterocolitis, Fuente La Lancha
    Juan Olarte Gutiérrez, 43, railway worker, enterocolitis, Cerro Muriano
December 11
    NP – Joaquín Ruíz Moreno, 28, tin worker, influenza, Guadalcanal
    PP – Juan Sánchez Vela, 55, manual labourer, enterocolitis, Jaén
December 12 – NP
    Ramón Cano Moya, 38 manual labourer, septicaemia, Baena
December 15 – NP
    Juan Moyano Platero, 56, manual labourer, septicaemia, Villanueva del Río
December 16 – NP
    Francisco Romero Horrillo, 60, shepherd, endocarditis, Ojuelos Altos
December 17 – NP
    Francisco Benítez Molero, 47, manual labourer, enterocolitis, Bujalance
December 18 – NP
    Juan Romera Moya, 46, manual labourer, pneumonia, Morente (Bujalance)
    Antonio Cuevas Éstrada, 56, manual labourer, tuberculosis, Posadas
December 19 – NP
    José López Abad, 28, manual labourer, septicaemia, Carmona
December 20
    NP – Juan José García López, 45, manual labourer, typhus, Valenzuela
    PP – Baroncio Zarza Nieto, 40, manual labourer, cerebral haemorrhage, Ávila
December 21 – NP
    Bartolomé Morales Fernández, 62, manual labourer, gastric haemorrhage,
    Hinojosa del Duque
December 22 – NP
    Luciano Tapia Morales, 60, manual labourer, nephritis, Baena
    Antonio Briones Molina, 47, manual labourer, typhus, Fuente Tójar
    Manuel López González, 25, manual labourer, valvular lesion, Torrecampo
December 23
    NP - José Serrano Fernández, 45, manual labourer, arteriosclerosis, Dos Torres
    Pedro Moya Sánchez, 28, manual labourer, typhus, Pedroche
    Manuel Pérez Romero, 50, manual labourer, cerebral embolism, Seville
    PP – Antonio López Bejarano, 21, manual labourer, tuberculosis, Añora
December 25 – NP
    Francisco Medina Pérez, 38, manual labourer, typhus, Montoro
December 26 – NP
    Rafael Écija Casas, 35, manual labourer, septicaemia, Cabra
    Ángel García Gallego, 52, entrepreneur, enterocolitis, Cabeza del Buey
    Francisco Romero Velasco, 37, manual labourer, septicaemia, Cañete
December 27
    NP – Francisco Perales Lozano, 41, manual labourer, pneumonia, Marmolejo
    PP – Victoriano Torrico Sánchez, 23, manual labourer, tuberculosis, Alcaracejos
December 28 – NH
    Víctor Ruíz González, 38, manual labourer, uraemia, Hinojosa del Duque

Juan Sámchez Álvarez, 22, manual labourer, tuberculosis, Hinojosa del Duque
Casimiro Padillo Izquierdo, 61, manual labourer, rheumatism, Castillo Locubín
Manuel Cerdán Roca, 25, manual labourer, tubercular enteritis, Teruel
December 29 – NP
    Juan Jurado Jiménez, 23, office worker, tuberculosis, Bujalance
    Pablo Segovia Morales, 32, manual labourer, typhus, Belmez
December 30 – NP
    Rafael Toledano Pino, 51, manual labourer, enterocolitis, Adamuz

1942
January 2 – PP
    Antonio Doblas Muñoz, 59, manual labourer, tuberculosis, Moriles
January 5 – NP
    Cirilo Romero Rísquez, 38, manual labourer, rheumatism, Torre
    Salvador Luque Valverde, 36, manual labourer, tubercular enteritis, Adamuz
    Pedro Hernández Vicente, 54, manual labourer, enterocolitis, Almería
January 6
    NP – Ignacio Millán Fernández, 22, manual labourer, typhus, Peñarroya
    Adalberto Serrano Rodas, 52, schoolteacher, encephalitis, Cuenca
    PP – Alejandro Márquez Rodríguez, 47, miner, emaciation, Fuenteobejuna
January 7 – NP
    Manuel Figueroa Via, 54, schoolteacher, hypertension, Puente Genil
January 8 – NP
    Ramón Roso Gomater, 24, manual labourer, tuberculosis, Cádiz
January 9
    NP – Antonio Ortega Ponce, 17, manual labourer, tuberculosis, Huelva
    PP – Isidoro Valle Jiménez, 60, manual labourer, tuberculosis, Lucena
January 10 – NP
    Acarías Muñoz Fernández, 60, entrepreneur, typhus, Conquista
    Antonio Vigara García, 55, manual labourer, cerebral haemorrhage, Belalcázar
January 12 – NP
    Andrés Paredes Tapía, 56, manual labourer, tuberculosis, Valsequillo
    Antonio Martínez Nieto, 59, manual labourer, enterocolitis, Adamuz
    Esteban Cejudo Montes, 57, barber arteriosclerosis, Doña Mencía
January 13 – NP
    Jacinto Cano Asencio, 57, manual labourer, arteriosclerosis, Montoro
    Antonio Rojas Martínez, 43, bricklayer, anaemia, Cañete de las Torres
    Manuel Carmona Leiva, 35, chauffeur, typhus, Montoro
    José Barbero Tena, 34, manual labourer, arteriosclerosis, Villaviciosa
    Lorenzo Guillén Buenosvinos, 47, manual labourer, enterocolitis, Bujalance
    Cándido Toril López, 58, blacksmith, pneumonia, Fuenteobrejuna
January 14 – NP
    Calixto Santos Cabezas, 47, salesman, typhus, Villanueva de Córdoba
January 15 – NP
    Cristóbal Serrano Castro, 37, manual labourer, nephritis, Luque
    Sebastián Blanco, 55, manual labourer, typhus, Torrecampo
January 16 – NP
    Antonio Mariscal Jurado, 36, manual labourer, tuberculosis, Villaviciosa
January 19 – NP
    Alejandro Delgado Moyano, 26, manual labourer, myocarditis, Torrecampo
January 20 – NP
    Juan Fernánde Navarrete, 31, manual labourer, typhus, Villaviciosa
Januarya 21 – NP
    José Aranda Ávila, 57, manual labourer, typhus, Fuente La Lancha
    Miguel Cejudo Fernández, 57, farmer, heart attack, Santa Eufemia
    Isidro Díaz Guerrero, 44, manual labourer, enterocolitis, Obejo
    Manuel Murillo Martínez, 54, miner, arteriosclerosis, El Hoyo
    Francisco Barbero Tena, 38, manual labourer, typhus, Villaviciosa

January 23 – NP
  Juan López Flores, 37, manual labourer, endocarditis, Fuente Palmera
January 26 – NP
  José González Nevado, 32, manual labourer, tuberculosis, Villaviciosa
  Francisco Gallego Toledano, 35, baker, typhus, Montoro
January 28 – NP
  Manuel Caballero Alcalde, 53, manual labourer, arteriosclerosis, El Viso
January 31 – NP
  Juan Jiménez Gómez Calero, 51, bricklayer, tuberculosis, La Granja

February 4 – NP
  José Peña Valverde, 51, manual labourer, mitral insufficiency, Adamuz
February 6 – NP
  Francisco Vázquez Gallego, 23, miller, typhus, Badajoz
February 10 – NP
  Antonio López Pescuezo, 57, baker, pneumonia, Villanueva del Río
February 15 – NP
  Luís Núñez Alonso, 56m market gardener nephritis, Castuera
February 21 – NP
  Manuel Rivera López, 30, miner, tuberculosis, Pueblonuevo
  Antonio Serrano Castro, 55, muleteer, pneumonia, Fernán Núñez
February 24 – PP
  Lucía García Valverde, 55, housewife, enterocolitis, Adamuz
February 26 – NP
  Juan A. Hidalgo Medina, 37, manual labourer, pneumonia, Belalcázar

March 3 – PP
  Salvador López Rabadán, 45, manual labourer, acute nephritis, Luque
March 4 – PP
  Agustín Jurado Fernández, 20, manual labourer, tuberculosis, Hinojosa
March 5 – PP
  Diego Soler Jerez, 19, manual labourer, tubercular meningitis, Huelva
March 8 – PP
  Antonio Carmona Relaño, 51, manual labourer, pneumonia, Adamuz
  Ignacio Martínez Reguez, 51, boiler maker, angina, Bujalance
March 17 – PP
  Domingo Aguayo Caballero, 18, bricklayer, tuberculosis, Baena
March 21 – PP
  Juan Carracedo Culeabra, 33, manual labourer, tuberculosis, Obejo
  Carmen González Zarza, 48, housewife, laryngeal epithelioma, Estepa
March 26 – PP
  José Alfaro Puerto, 56, mçiner, stroke, El Viso

April 9 – PP Womens Prison
  Malio Giraldo Molero, 7 months, enterocolitis, Badajoz
April 11 – PP
  Juan Hernández Peña, 30, stoker, bronchopneumonia, Espiel
  Sinforiano Paredes Dominguez, 33, railway worker, mitral insufficiency, Peñarroya
April 13 – PP
  Antonio Chacón Benítez, 35, muleteer, tuberculosis, Villaviciosa
April 14 – PP
  Tomás Gómez Marín, 40, manual labourer, tuberculosis, Castro del Río
April 17 – PP
  Antonio Pantoja Pérez, 56, manual labourer, stroke, Córdoba
April 20 – PP
  Cristóbal Medina Muñoz, 42, manual labourer, endocarditis, Rute
April 27 – PP
  Antonio Caballero González, 47, miner, uraemia, Alcaracejos

April 30 – PP
   José Lázaro de Diego, 59, manual labourer, cirrhosis of the liver, Bujalance

May 7 – PP
   Galo Adamuz Montilla, 45, schoolteacher, pneumonia, Córdoba
May 8 – PP
   José Sepúlveda Arjona, 38, miner, tuberculosis, La Carolina
May 13 – PP
   Jacinto Sánchez Campillo, 34, manual labourer, cerebral haemorrhage, Iznájar
May 14 – PP
   Rafael Expósito Leal, 48, manual labourer, pneumonia, Adamuz
May 21 – PP
   Antonio Jaut Castilla, 19, manual labourer, tuberculosis, Pozoblanco
May 27 – PP
   Adriano Valencia Quero, 47, manual labourer, myocarditis, Alcaudere
   Manuel Puerto Nieto, 51, manual labourer, heart attack, Villanueva de la Serena
May 28 – PP
   Pedro Morillo Pinto, 29, shoemaker, pneumonia, Badajoz

June 14 – PP
   Francisco Santofimia Carrillo, 33, manual labourer, tuberculosis, Córdoba
June 20 – PP
   Aquilino Sáchez Navarro, 53, foundry worker, mitral insufficiency, Peñarroya

July 2 – PP
   José Trujillo Muñoz, 55, painter, heart attack, Málaga

August 26 – PP
   Eulalio Medrán Gañán, 26, manual labourer, tuberculosis, Dos Torres
August 129 – PP
   Eufemio Cabello Merlo, 30, manual labourer, tuberculosis, Villaviciosa

September 8 – PP
   Miguel Oviedo Rodríguez, 18, manual labourer, tuberculosis, Añora

October 7 – PP
   Julián Vergara Ventura, 56, manual labourer, enterocolitis, Fuenteobejuna
October 9 – PP
   Antonio Patiño Suárez, 28, baker, enterocolitis, Santa María

November 8 – PP
   Manuel Ruíz Romero, 40, manual labourer, enterocolitis, Pozoblanco
November 21 – PP
   María Sanz Cisneros, 63, housewife, gastric epithelioma, El Viso

December 16 – PP
   Francisco Carrillo Nevado, 44, manual labourer, tuberculosis, Pedroche
December 20 – PP
   Eusebio Castro Robles, 58, manual labourer, enterocolitis, Granada

1943
January 15 – PP
   Rafael Pedraza, manual labourer, pylori ulcer, Villanueva de Córdoba
   Manuel Jiménez Jurado, 4, pneumonia, Bujalance (women's prison)

February 2 – PP
   Manuel Mendoza García, 24, manual labourer, enterocolitis, Granada
February 15 – PP
   Antonio Medina Romero, 40, manual labourer, pneumonia, Pedroche

665

February 17 – PP
> Pablo García Rubio, 52, manual labourer, heart failure, Villaralto

March 25 – PP
> Cándido Olor Munich, 52, farmer, septicaemia, Lérida

April 1 – PP
> Bernardino López Morales, 30, road worker, skull fracture, Villanueva del Duque (possible suicide)

April 8 - PP
> José Belmonte Blanco, 48, manual labourer, pulmonary gangrene, El Viso

April 16 – PP
> Gabriel Rodríguez Ordóñez, 25, manual labourer, heart attack, Luque
> Antonio Cañero Llamas, 66, tradesman, gastric ulcer, Fernán Núñez

April 19 – PP
> José Fernández Castillejo, 52, miner, asystole, Peñarroya

June 28 – PP
> Baldomero Corredera Ávila, 40, manual labourer, uraemia, La Carlota

July 12 – PP
> Juan Pablo Caballro Romero, 31, farm worker, tuberculosis, Hinojosa del Duque

July 20 – PP
> Antonio Morenas Polaina, 38, butcher, tuberculosis, Jaén

August 7 – PP
> Miguel Bolaños Cabrera, 69, manual labourer, asystole, Montoro

August 11 – PP
> Roque Serrano Fernández, 43, manual labourer, haemoptysis, Dos Torres

August 29 – PP
> Victoriano Ollero Moreno, 55, manual labourer, tuberculosis, El Viso

September 16 – PP
> Manuel Carmona García, 21, manual labourer, myocarditis, Montoro

October 17 – PP (women's prison)
> José Toro Luque, 7 months, meningitis, Priego

November 24 - PP
> José Ortiz Arjona 75, gastric ulcer, Palenciana

1944
January 13 – PP
> Agustín Morilla aTejada, 62, farmer, cerebral embolus, Venta del Charco

February 1 – PP
> Bartolomé Muñoz Garacía, 38, manual labourer, Maltese fever, Seville

June 11 – PP
> Joaquín Sánchez Sánchez, 41, miner, tuberculosis, Peñarroya

July 1 – PP
> Miguel Díaz López, 22, manual labourer, tuberculosis, El Viso

1945
April 11 – PP
> Rafael García Gutiérrez, 51, baker, skull fracture, Villanueva de Córdoba (possible suicide)

June 5 – PP
    Juan Luque Rubio, 50, manual labourer, syncope, Montilla
June 28 – PP
    Juan Molina Jiménez, 21, manual labourer, heart attack, Fernán Núñez

September 23 – PP
    Ignacio García y García del Barrio, 55, tradesman, bronchopneumonia, Montilla

1946
January 20 – PP
    Francisco Manosalvas Medina, 60, tradesman, septicaemia, Pedroche

June 7 – PP
    Francisco Romero Paredes, 24, boiler worker, spinal fracture, Belmez (possible suicide)

July 3 – PP
    Florencio López Cortés, 50, cattle breeder, heart attack, Torrecampo

August 3 - PP
    Juan Díaz Martínez, 23, goatherd, tuberculosis, Montoro
August 7 – PP
    Antonio Avilés Díaz, 26, farm worker, meningitis, Pedroche
August 20 – PP
    Antonio Fabre Sánchez, 48, telegraphist, heart attack, Rota
August 31 – PP
    Juan Sánchez Cabrera, 29, manual labourer, heart attack, Almería

September 2 – PP (women's prison)
    María Gómez Aguilar, 3 days, Córdoba

October 9 – PP
    Juan Morales Rodríguez, 64, coalman, asystole, Badajoz
October 21 - PP
    José Fernández Gómez, 54, manual labourer, heart attack, Villaralto

November 8 – PP
    Mª Antonia Baena Granado, 82, housewife, cerebral haemorrhage, Rute

December 4 – PP
    José Palma Martínez, 27, basket weaver, tuberculosis, Jaén

1947
January 5 – PP
    José Murillo Alegre, 49, cattle breeder, hanging, El Viso
    (suicide, according to his son José 'Commandante Ríos')
January 27 – PP
    Isabelino Granados Moya, 51, manual labourer, pneumonia, Villanueva del Duque

TOTAL: 756

James Maley, from Glasgow, and other British prisoners captured
in Salamanca 1936. Front row, first on right.
Movietone newsreel in local cinemas exhibited while he was in prison.[4]

29-year-old James Maley, in the prison yard after his capture.
A volunteer from Glasgow with 28 Machine Gun Co, he was captured
by Moors on horseback with cutlasses.[5]

4   https://i2prod.dailyrecord.co.uk/incoming/article6472468.ece/ALTERNATES/s1200/JS72351236.jpg
5   https://www.dailyrecord.co.uk/news/real-life/play-tells-tale-glaswegians-who-6472382

APPENDIX VII

WRITTEN TESTIMONIALS FROM PRISONERS REGARDING THE FIRST CONCENTRATION CAMPS AT THE END OF THE WAR AND AFTERWARDS. PRISON DIARIES AND LETTERS.

Descriptions of the journeys to the concentration camps and the terrible days that followed: Castuera, Badajoz

<u>José María Carnise Casas</u> a Catalan from Reus, Tarragona. Written testimony sent to Moreno Gómez October 19 1987 regarding his story that begins March 26 1939, when he is taken prisoner at the end of the war.

- "We drove through Alcaracejos without problems, but when we arrived at Villanueva del Duque, we saw advance Nationalist troops entering the village. We were stopped a couple of kilometres ahead and told that we should go no further as Franco's troops had entered Hinojosa del Duque. We jumped off the truck and I went off on my own looking for our own troops, but I had not gone even one kilometre when on my right, I saw a column of Moroccan soldiers and on my left, some infantry. Obviously, I chose the latter. In other words, I surrendered to the first group I found... to members of the 13th Volunteer Infantry Regiment.

  We took to the road. We arrived at the village of El Viso. I was left in a fenced-in yard where I remained for 6 days. I only had one meal during all this time. The rest of the time I had to make do with a little bit of bread... I and the other prisoners-of-war were near the road and from there we watched truck after truck full of all kinds of armed men: Falangistas, troops, Moroccans, infantry. I was also overwhelmed by the number of priests and Guardia Civil who accompanied them. No wonder It was three years since I last saw them.

  Every day we were told that we would be fed the next day... The camp was continuously filing up and hope for food decreased with every day that passed. One day, we were told that we were to be moved to another camp. Four Guardia Civil arrived: two took their place

at the head of the column and two at the end. We began walking.
There were so many of us. That day, we got as far as Villanueva del
Duque, some 14 kilometres distant. The next day they gave us a little
bread and we continued our march. This time we went further – 31
kilometres to Peñarroya-Pueblonuevo where we were put into a large
yard. I think it was a flour factory. There were a lot of women at the
entrance to the yard threw us pieces of bread over which we fought
ferociously.

The next morning, we started walking again until we arrived in
La Granjuela. Again, that damn village. There, I first met Vicents
de la Selva. The first thing he asked me was whether I had any
tobacco. When I replied that I hadn't, he gave me a few small boxes.
What a great guy. La Granjuela was a disaster: not a single house
remained standing. Three times it was taken by the Nationalists
and three times we took it back. We remained there ten or twelve
days. We lived between four walls, with another fellow from Reus,
one from Montroig, another from Cabra del Campo, another from
Valls (called Gomis), another from Cambrils (calld Sentís), another
from the province of Lérida (called Florensa) and me. In the morning
they gave us a little bread and the occasional can of sardines for two
people; sometimes, for a change, it was a can of tuna, but always for
two people, To that, we added our own special stew The fellow from
Montrog would go out into the fields and bring some grasses we boiled
in a pot we had found; when they were well cooked, we ate them.

One afternoon, they told us that we would resume our march the
next day as we were being transferred to another camp. They gave us
our rations for the following day: a roll of bread and a can of sardines
per person, but when it had to happen, happened. We had built up
such a hunger that as soon as we returned to our 'hotel' we devoured
all that we had been given. Then to bed, to wait for the next day and
find out where we were going.

We set off very early, I cannot tell you exactly where we went. All
I can say is that we became exhausted. We crossed some towns and
villages and knocked on the doors of the houses asking for a bit of
bread, The only thing we received were words of consolation or the
oft-repeated cry *Sons of sorrow!* We marched 50 kilometres that day.
As if something was missing that day, it began to rain mid-morning.
I walked with Florensa, and I remember that whenever we stopped to

sit a moment, one of us had to keep standing so as to pull the other up because our joints were no longer working on their own. Another memory from the day: as we marched through an acorn oak forest we foraged for acorns, like so many pigs. I don't know how many of us there were, but there must have been at least a thousand. That evening, we arrived at Castuera, and that particular ordeal was over.

We were ordered to stay in the town and sleep on the streets and in the squares. It may seem impossible that we didn't all take off, but the thought never crossed our minds as the only thing we could think of was how to fill our stomachs. They told us that the next day we would be taken to the concentration camp and once we got there, we would be fed... I imagine that the fearful spectacle we must have presented the townspeople must have been quite disturbing. As you can imagine, we looked terrible. Wearing one shoe and one espadrille, washing any way we could, unable to shave, we must have looked extremely macabre.

Florensa and I squeezed back-to-back in a corner and waited for the next day. At dawn, we left for the concentration camp. It could not have been far away because I remember that we got there relatively quickly. We were each told the number of the hut that we were to occupy. With a small door and large windows, there was little living space per person as each hut was occupied by 80 men. We had to beg pardon to turn around. Also, we were unable to change clothes or shave during the entire time that we remained in the camp....

I am now going to talk about our infamous concentration camp. A large yard. In the centre, some flagpoles where the flags of Spain, the Carlistas and the Falange waved. The huts were arranged around the yard. On one side, the kitchen. On the other, the latrines, which were no more than trenches that we had to dig beforehand... If we needed to go to the toilet at night, we had to call the guard and a soldier with a fixed bayonet would accompany us... And we were not even criminals! The camp was surrounded by a double row of barbed wire, a trench 3 metres deep and 6 metres wide, then another double row of barbed wire. And finally, all around the camp, machine gun posts.

In the morning, at sunrise, reveille. We had to fall in immediately in front of our hut; if we didn't, the sergeants would enter through the windows and lash us with whips. Once we had fallen in, we were made to sing the set hymns: the hymn of the Legion, of the Falange and the *Orimendi*. We then left the camp, two by two. One side of the gate there was a mountain of pickaxes and on the other, a mountain of shovels. There we remained unit noon when we returned to the camp to eat. Depending on the day, we got either cold or hot food. When it was cold food, we were given a small can of tuna. When it was hot food, we went to get it, two by two, and were given a ladle of water and some chickpeas. In the evening, we got either cold or hot food, the opposite of what we had been given at lunch. As far as bread was concerned, we were given less than half a roll. Every hut was allotted 30 or 35 rolls, but as there were 80 of us, we each got less than half a roll.

As far as personal hygiene was concerned, I can say that the whole time that I spent in Castuera we were never given any water in which to wash. As most of us had no razors, we just let our beards grow. As we couldn't wash our clothes, we were a mess. As far as bugs were concerned, in addition to being abundant, they were all kinds and colours One of our pastimes was racing body lice.

When we first arrived at the camp, the guards were Falangistas. When we went out to work, there was a guard for every 8 prisoners. In addition to making sure that we did not escape, another one of their duties was making sure that we did not sleep on the job. These guards always carried rifles. A month or a month and a half later, the Falangistas were replaced by soldiers, and we all benefited from that."

José María Carnicer was able to leave Castuera well into the month of June 1939 thanks to a good behaviour reference his father and other persons were able to obtain for him. He had been interned for three months.

Miguel Cruz, from Villanueva de Córdoba. Interviewed by Moreno Gómez in Villanueva de Córdoba Summer 1980.

▪ "The arrival of the Nationalist troops in Villanueva marked the beginning of mass detentions in the town, led by individuals such as Diego *El Cunga* Cachinero, Vicenti *Salado* Muñoz, Bartolomé *Berenguer* Cepas, Emilio *El del Lunar* and others.

April 20 in the afternoon, I left for Castuera in the convoy that had been organized in the town for this purpose. The previous convoys had been sent to Los Blázquez and Valsequillo. There were some 200 of us who were first taken to the bullring in Puertollano. From there, we were sent to Castuera by train. Castuera camp was bursting with prisoners, so much so that the commander had to be forced to accept us. Our names were recorded and then we began to suffer all kinds of punishment, privations and hunger. We were made to work digging an enormous 3-metre-wide trench enclosed on either side with barbed wire. Improvised showers and building huts, each of which contained some 80 prisoners.

Juan *El de la Luna*, one of those who managed to escape from town earlier when he heard they were looking for him, returned to Villanueva where, unfortunately, he was caught. Diego Ranchal, brother of the Mayor of Villanueva del Duque, was also there with me. Another good companion was Pablo Agenjo, who was shot near Cardeña by the Guardia Civil, under the Law of Escapes.

To be released from the camp one needed good behaviour references from people of good standing in our hometown. I was finally released after two months in the camp, with no more food than the bread roll and can of sardines I was given when I left."

Francisco Romero. Interviewed by Moreno Gómez in Villanueva de Córdoba August 1983.

▪ "An edict was published in Villanueva according to which everyone who had served in the Rojo Army had to present himself, without exception, excuse or pretext, for transportation to Castuera concentration camp. Several hundred of us did so and we were taken from Villanueva Station to Puertollano. There, we slept under the stars with frequent threats from the guards in case anyone should think of escaping.

We left for Castuera and there we were received by the commander, who greeted us with an aggressive and insulting speech. The repression began in the camp. Many died there, either from the beatings they received from the Falangistas, from hunger or the cold, as there were practically no sheds. We managed to build ourselves a kind of hut with brush from the nearby hills."

From Valsequillo to Castuera via Los Blásquez

José Martí Prats.
▪ "After two or three days in Valsequillo, we were taken (actually, we took ourselves) to Los Blásquez, to something that looked like a concentration camp. In effect, the village had been surrounded by

rows of barbed wire, on either side of which there were two 7 x 8 metres dirt trenches that showed the footprints of anyone who attempted to cross them.

One evening we were told that the next day we would be leaving for Castuera. The march was no joke: it was a 50-kilometre walk. In brief, 800 of us left, escorted by or under the custody of – or whatever you wish to all it – some 40 soldiers. Small groups of us were left behind along the way; they just could go no further as we were not given anything to eat and those who had managed to keep something, ate it all at the beginning of the walk.

Valsequillo and La Granjuela, Cordoba

Mariano Martín Sierra. Unpublished memoires.

- "Valsequillo and La Granjuela were two villages totally destroyed by the bombing. They were deserted by every inhabitant except the prisoners. The parts of the houses that had not totally collapsed served as refuges and we huddled in them like rats. I shared an attic with 12 men from Madrid, all of whom also were bakers as I was. The stairs to the attic were completely destroyed so getting up there was a circus act.

The commander of Valsequillo was a 'good Falangista' who would frequently order the prisoners to fall in even though many could not even stand because they were so weakened by hunger. Two or three Falangistas, wearing their blue shirts, would walk between the ranks and start looking at us... The unlucky men who were chosen were called out of the ranks, taken away, and we never saw them again...

When I could get up early in the morning, I would go to the place where they threw some garbage. I ate whatever I could find or whatever others who had been there before me had left behind. Potato peelings, banana skins, the occasional mouldy acorn were, in that order, my first, second and third course.

[There was a changing of the guard in Valsequillo May 6 1939 and the Falangista commander was replaced by a militiaman. Martín Sierra was sent to the workshop.]

"As letters arrived from everywhere, our work consisted, among other things, of opening them and sending them to the office. These frequently consisted of allegations and accusations against prisoners from various local authorities in each prisoner's hometown, all of which put that prisoner's life at risk. This is where my 'job' began.

In the absence of the commander, we would gather the accusations that we considered to be the most damaging and, very carefully, I would go outside and tear them up or burn them wherever I could, returning to my post as if nothing had happened."

Rafael Bedmar Guerrero from Puente Genil but interned in Higuera de Calatrava, Jaén, described his experience in his diary *Memorias de uma guerra*.

- "I presented myself to the Military Headquarters in Linares. They had a file on me that contained a good behaviour reference from persons of recognized standing, but they only referred to when I had lived in Linares. I was sent to a room that was set aside for suspect individuals and was full of people. An hour later, our hands were tied together with wire, and we were taken outside. When we left the building, we met with a convoy of trucks full of prisoners collected from other centres. We were put onto one of the trucks, crushed so tightly against each other than we could not move. The convoy took off down the middle of an empty road, flanked all around by armed soldiers.

We had no idea whatsoever of where we were going. Continuing along bad roads, we rode a hundred and fifty kilometres until we arrived at a town that had been destroyed during the war. Higuera de Calatrava, located between two lines of fire. Only the church remained standing. We were shoved off the back of the truck and as soon as we landed, several officers beat us on our backs with their riding crops, urging us to move faster.

As best we could, we removed the wires from our hands. We were crushed together in a building as if we were objects. We could not move and there was little air to breathe. Any toilet functions had to be made where we stood, stuck to each other. Several prisoners who were unable to survive the cruel treatment died. The next day we were taken out and out names were taken down at some tables they had set up on the street. As we left the table, some soldiers gave us a roll and a can with three sardines. Other soldiers led us to ruined houses that remained more or less standing.

There was an attic in the prison that served as a torture chamber. Eight days after I was imprisoned, I was taken to this attic. An official sat at a table, a typewriter in front of him, and he was flanked by four other officials, riding whips in hand. There was a chair in front of the official who was writing and in the centre of the roof, under a metal beam, a wooden stool. Above the stool, hung some ropes and some wires. I was seated in front of the scribe and told: 'All right, we think this is going to end soon. You are going to tell us what we want to know. A friend of yours has made a statement. He says that you accompanied the military patrol from Malaga and that you went to his house to arrest him. That they wanted to kill him on his doorstep, but you interfered, and you saved his life. Therefore, we believe that you are a person of some standing among the Rojos and so you must know who it was who killed Tom, Dick, Harry. Many of those responsible are here, imprisoned in the cells and you are going to tell us who they are. So, begin singing and we'll finish quickly.'

I knew what to expect. I absolutely refused everything. If I were to turn in just one of my comrades, they would make me talk, and talk, and later shoot me with them. When they saw that no amount of soft talk could get me to talk, the guard who was interrogating me punched me in the nose and knocked me off the stool onto the floor, blood pouring from my nose. Two other guards then picked me up. The one who was interrogating me asked: 'Have you had enough, or do you want some more?' I continued to deny everything.

They then tied the rope around my neck with a loose knot and tied my hands with wire, made me stand on the stool and, when they were ready, asked: 'Make up your mind or you'll leave this room for the cemetery.' I knew that one way or another, I was going to die, and I preferred that it be this way. At least, I would take nobody with me. I continued to deny everything they asked. My tormentors got angrier and angrier, and they pulled the noose and wires tighter and tighter. I was almost hanging from my neck. A single kick at the stool and my body would be left hanging.

One of the tormentors took off one of my shoes and socks, got hold of my big toe with some pliers and pressed down on my toenail. He crushed it so hard that I felt my toe go numb. When my toe turned black, they removed the piers. They then took some tweezers and, getting hold of my nail, pulled it so hard that all I felt was a strong electric shock that reached right into my brain. They did the same thing with my other foot, and I almost went into shock. I couldn't feel my legs any longer. I felt pins and needles all over my body. I closed my eyes so I couldn't see their faces. They kicked my stool, and I was left hanging like a sack of sand. One guard swung me back and forth. When my body swung towards the guard in front of him, the latter would give me a lash on my back with his riding crop.

They did this forty or more times, back and forth, whip lash after whip lash.... I stopped counting, I only felt as if my feet were being wrenched from my body.

They stopped once or twice and sat me on the stool. Always the same questions, but I could no longer hear a thing. I almost fainted. Furious with my resistance, the guard who was in front of me kicked me in the stomach. I felt a large knot rising from my stomach and vomit coming into my mouth. My tormentor stopped and almost fell on the bloody garbage that littered the floor. A few seconds later, I lost consciousness.

When I opened my eyes, the jailor, with a bucket and cloth was wiping my body with a watery vinegar solution. I was in a small cell, apart from the others. I don't know how I was taken to that place. My body shivered convulsively; I had an extremely high fever. I struggled between life and death for a week. Little by little, I got better. My physical strength once again enabled me to resist. No sooner had I recovered from the torture that I was taken from the small cell and put in amongst a hundred or more other prisoners. My comrades were horrified when they saw the scars that covered my body. They shook with fear at the thought that they, too, would be called to be interrogated.

When the time came to give me the food my mother had brought for me, the jailer threw it down the toilet to make the others think I had eaten it. They still discovered nothing. I hid the undershirt that I wore during the torture: pieces of skin from my back stuck to the cloth. I scooped out the crumbs from a loaf of bread that my mother sent me and returned it to her with the shirt hidden inside, so that the jailer did not see what I had done.

When my family discovered the mistreatment that I had been subjected to, my mother took that shirt and ran through the streets shrieking like a madwoman: 'Criminals! Murderers! Look what they have done to my son!'

We were divided into companies, and a leader was appointed for each one. Once the operation was over a bugle called us to go to the square and listen to a speech by a colonel. For more than an hour, we were harangued by the colonel as he ranted about the wonders of the Glorious Movement and the marvellous Caudillo that God had miraculously sent us. We then had to fill out forms with our names, telling where we were caught by the Movement, political affiliation, etc. Naturally, I wrote down what I saw fit.

Several days passed and we were put to digging a ditch, which I soon understood would serve as a mass grave. As the information arrived from the various hometowns, they called out those who had the worse references, shoved them into the basement at the command post (it filled up every evening), and the next morning, took them out and shot them."

Excerpts from Prison Diaries and testimonials regarding conditions in penitentiaries of all kinds, and especially the *sacas*.

Antonio Baena Moreno. Pozoblanco. Teacher. Secretary of the Provincial Socialist Federation of Cordoba; Commissar General for the Army of Andalusia and Extremadura. Arrested March 30 1939 and detailed in Pozoblanco, where he was sentenced to death by garrotte April 22 1940. Unpublished personal Diary of his experience in the local jail July 19 – September 12 1940 before being sent to Cordoba, then to Burgos, then back to Cordoba where he was executed November 17 1941.

- "19 July 1940. Last night we entertained ourselves for a few hours playing cards and smoking cigarettes. I fell asleep at midnight and, as usual, woke up at the ill-fated hour. It appears that all is quiet and no saca is expected. The spyhole to the cell remains open... The danger is past!

21 July 1940. Day of euphoria and optimism. As there are no sacas on Sundays, we can enjoy the luxury of a long siesta and at night, sleep without worrying.

22 July 1940. Today marks 3 months since I was tried. Three short months! How much more human it would be for the sentences to be carried out quickly. Families would stop having to spend what they don't have and once it was all over, they might begin to accept

the situation. Unfortunately, it appears that what they want to do is to cause the most possible suffering.

26 July 1940. We were apprehensive all night. After we were locked down for the night, a companion assured us that a large saca was planned. We stayed awake all night, waiting for dawn. When I awoke, the fatal hours had almost passed. If I had been taken in that saca, I think I would have reacted as if it were a simple *paseo*. I was very tense and felt an Olympian hatred for my enemies. I have reached such a state that I am indifferent to whatever is going to happen to me. With every passing day, I am more and more convinced of the utter irrationality of these procedures.

27 July 1940. Saturday. Two days of safety. Everyone is sleeping in total relaxation.

6 August 1940. It seems that there are rumours of a tragedy. This morning we all thought there was going to be a saca. A truck stopped by the gate at the fatal hour and many of us thought that our time had come. We were pessimistic all day. We have heard that several death sentences were commuted to several years in prison and when this happens, this is always followed by executions.

4 September 1940. I woke up in the middle of the night when the light in the corridor went out. I thought that the dance had begun. I waited, patiently smoking a few cigarettes, until daybreak. The fatal hour had passed.

12 September 1940. The blow that some of us had been expecting fell. This morning at dawn, three companions of ours were shot. It has hit us like a bomb... It was about four in the morning, almost two hours earlier tan the usual fatal hour, when they closed the spyhole in the cell door, a sign that the saca was about to begin. The fact that it started so early led us to believe that it would be larger than usual and that many of us would be executed. We all thought that we would be one to have this terrible bad luck. Of the numerous sacas we have lived through, none caused such fear. Regarding two of the comrades who were shot, there were no serious accusations against one, and the other was a kindly person by nature, incapable of harming a single soul.

When I awoke, the spyhole was already closed. I asked my cellmates, who were already sitting on six baskets, how long it had been closed – only a few minutes, they replied. Because of that, I endured a good couple of hours expecting it to open and to hear them call my name. You can never appreciate how relative time is in such situations. If it were not because we are used to suffering... two hours like this would be enough to age you a dozen years. All the same, when I went into the patio, the ravages of those two hours were etched on our comrades' faces. Everyone looked as if he was just recovering from a serious illness.

Arthur Koestler. Hungarian-born British war correspondent, captured and imprisoned by Nationalists after the fall of Malaga February 5 1937. Several monographs and extracts from his diary entitled *Dialogue with Death. A Spanish Testimony*.

- "I was taken to a large empty room. In one corner, there was a stool on which I was made to sit. Two Guardia Civiles sat down in front of me, next to the door, with their rifles on their knees. We remained that way for some time.

I then heard screams coming from the patio and a young man whose chest was bare and covered in blood was brought into the room. His face was battered covered with cuts; for a moment I thought that he had been run over by an engine. Holding him up by the armpits, they dragged him to the other side of the room, screaming and moaning. The Falangistas who were dragging him spoke to him softly: 'Quiet man, we are not going to beat you anymore'. They took him out through another door, closed it, and a few second later we heard the noise of beatings, slaps and kicks. The man screamed at regular intervals.

Then there were a few seconds of silence. I could only hear rapid, panting breaths. I don't know what they did to him during those seconds. He then screamed again, just once, an abnormally sharp cry; and finally, he was quiet. A little later the door opened, and they dragged him across the room in which I was seated, to the patio. I could not tell whether he was dead or simply unconscious. I did not have the guts to look at him any longer. Later, a second victim passed through this room and was subjected to the same treatment; and then another."

- "That night, when I opened my eyes, it was still not daylight; I had been awakened by a noise. I listened: someone was singing. He seemed to be very nearby. The man who was singing appeared to be in one of the solitary confinement cells across from mine. I sat up, and my heart stopped: he was singing the *Internationale.*

He was singing off-key; he was hoarse. Obviously, he was hoping that the other condemned men would join in, but nobody did. He sang alone in his cell, in the jail, and at night... I was hearing the Internationale being sung by a man who knew that he was about to die... He repeated the chorus two or three times, stretching it out so that the song would last longer. I stood up, went close to the door and, teeth chattering, raised my fist... and I got the feeling that everyone in all the cells around me were standing next to their doors, like me, solemnly raising their fists in a gesture of goodbye.

He sang. I could see him before me: unshaven, broken nose, tortured eyes... but none of us sang along with him; we were too afraid.

...About four in the morning, there was a noise in the corridor. A saccharine voice was reading out a list of forty or fifty names; some doors opened, then were slammed shut. The sound of steps, whispering, mysterious noises.

I then put my ear instead of my eye next to the spyhole. I could only discern the sound of a long line of men shuffling along the corridor, slowly, hesitantly, as if they were walking against their will. The sound of steps slowly disappeared. Forty or fifty men were walking to their death.

I lay down on my cot and asked myself if the singer was one of them, and whether they would be shot one by one or in groups, with rifles or machine guns.

I must have just fallen asleep when I was woken by the same saccharine voice that I had heard earlier that morning. This time it came through the bars of the window in the cell, from one of the patios that I had crossed when I was looking for Sir Peter. He read out twenty or thirty names. I couldn't count them exactly; the length and complexity of Spanish surnames confused me. On this occasion, all those who heard their names called had to answer 'present' and when someone didn't reply quickly enough, the saccharine voice burst out swearing. The guard then called out:

'Everyone in cell number 17'.

'Everyone in cell number 23'.

Koestler was transferred to Seville prison, where he lived in constant panic. There, in April, he befriended a young farmer who had been recently arrested.

- "He was taken prisoner ten days ago on the Almería front and was sentenced three days later. He, like all prisoners of war, is accused of 'military rebellion'. As we walked in the patio, Nicolás told us about his trial before the Seville court martial. It lasted three minutes. The presiding magistrate read out the name of the prisoner, his place of birth, and the name of the place where he had been captured. The prosecutor asked for the death sentence and added: 'I am only sorry that I cannot put this pinko in a cage and send him to Geneva before we shoot him, to show the League of Nations the kind of miserable individuals are its supposed defenders of justice and democracy.'

... I offered to lend Nicolás a book, but he said he couldn't read. He said he had hoped to learn how to read after the end of the war."

April 14, in the morning, Koestler looked for his friend in the Seville prison patio:

"Nicolás has disappeared. Rest in peace, Nicolás. I hope that it all happened very quickly and that they did not make you suffer too much. How little you were, small Andalusian farmer with slightly bulging, soft blue eyes, the eyes of the poor, of the meek."

Mid-April, Koestler again witnessed one of many sacas in Seville prison:

- "I had fallen asleep and woke up a little before midnight. In the dark silence of the cell, seeped in the nightmare of three hundred sleeping men, I heard a priest praying softly and the tinkling of a hand bell.

A cell door opened, the third one on the left, on the same side of the corridor as mine, and someone called out a name. A sleepy voice asked: '*Que? Que pasa? What's up?*' The priest raised his voice and rang the bell more loudly.

In his cell, the sleepy man understood. At first, he just groaned then, in a muffled voiced, begged: '*Socorro! Socorro! Help!*' They took him away. I could hear him yelling outside. Still, the distant sound of shooting was some time coming.

Meanwhile, the priest and the prison guard had opened the door to the next cell: cell 42, the second one on my right. Again: *'Que pasa? What's up?'* Again, the praying and the hand bell. This prisoner was sobbing like a child. He called for his mother: *'Madre! Madre!'* Again: *'Mother! Mother!'*

They then went to the next cell. They called my neighbour, but he said nothing. He was probably awake and, like me, prepared. But when the priest finished praying, he asked softly, as if he were speaking to himself: 'Why must I die?' The priest answered him with a few words, spoken solemnly but quite rapidly: 'Have faith. Death is deliverance.'

They went on to the next cell. He was also prepared. He said nothing. Whilst the priest prayed, he began to sing the *Marseillaise* softly, but after a couple of stanzas, his voice choked and he, too, began to sob. They took him away..."

▪ "Our hearing became miraculously acute. We heard everything. The nights on which there were executions, we heard the telephone ring at 10 o'clock. We would hear the guard answer. We heard him say: 'noted... noted... noted...' at regular intervals. We knew that the guard checked a name with each 'noted'. We did not know whose names were checked, not if one of us was on the list.

The telephone always rang at 10. That left us to lie on our cots, waiting until midnight or 1 o'clock. At midnight or 1 o'clock, we could hear the strident ring of the night doorbell. The priest and the firing squad had arrived. They always arrived together.

The sound of opening doors, the tinkling of the hand bell, the priest's praying, the shrieking calls for mercy, cries for mothers...

The steps came closer down the corridor, they moved away, they came closer, they moved away. They stopped at the cell on the one side, they went to another cell in another wing, they returned. The priest's voice always rang out clearly: 'Lord, have pity on this man; Lord, forgive him his sins. Amen.' We lay on our cots, our teeth chattering like castanets. Every night, we placed our lives on the scales and every night, somebody fell out."

19 April "These past three days all the noises from afar were blanketed by the voice of a man who sobbed and called for his mother non-stop. His cell must have been near mine. Every time that I put

my ear to the door, I could hear him. I asked Ángel who was the man who did not cease crying. He is a militiaman, he said, who was sharing the cell with his brother but who had been alone since Friday night."

<u>22 April</u> "I get this burning desire to murder that small, brown, fat priest who rings the hand bell every night."

<u>25 April</u> "Sunday afternoon they brought another prisoner in and put him into my old cell, number 41. I watched as they brought him. He was incredibly young, only about fifteen or sixteen years old. I heard a guard say that they would be coming for him that night. At that moment, the young man in cell 41 began banging on the door; he must have heard everything. 'I don't want to die!' he screamed. '*Madre! Madre!* Mother! Help! I don't want to die! Help! Help!'

His cries echoed all along the corridor. The entire prison became uneasy. Confused, but clear, sounds came from all the cells. The young man kept yelling and they took him out of that cell and put him in some special solitary confinement cell. They had to call some more guards to help them.

A little later – before 10 o'clock – the priest came down the corridor, probably to take the young man's confession. A guard called out for some brandy.

At ten, the phone rang. I heard the guard answer 'noted' three times... It was not yet 11 o'clock. I fell asleep. The next day I discovered that they had shot three prisoners before midnight. The young man had not shouted. They may have used the brandy to get him drunk."

Pablo Uriel Diéz, graduated in Medicine in July 1936, is conscripted as a Doctor Soldier in Zaragoza to an engineering battalion. On leave at home, his sisters told him that his other to brothers had been arrested. One of them, Antonio, was condemned as an active member of the Republican Left Party and executed in Soria.

- "There was no singing at 6 p.m. At that time, every conversation flagged. Nobody paid attention to what was said because everyone was tense, trying to hear sounds from the workshop. At once, the entire prison acquired a single ear whose hearing was sharpened and particularly acute. If one thought that he had understood the noises that came from the workshop and he wanted to confirm his doubts, all he had to do was bang on the door to his cell. A soldier would soon appear: What do you want?' 'I want to go for a pee.' 'Not now. Nobody can leave his cell.'

This meant that the Falangistas were in the workshop, bringing with them their hateful lists, and that in the new few minutes five, six, maybe ten of us would begin their final journey. The silence continued until one herd the car engines speed away from the prison.

The four of us in our cell were feeling the usual anguish. In the silence of the prison, we heard steps coming down the corridor. Immediately, someone touched the lock on our door. When the corporal opened the door, I could see four men standing there. The corporal was holding a sheet of paper, but he was not looking at it. He looked at sergeant Sangrós: 'You're on the list; please come with me.'

Sangrós did not appear surprised, and he did what we least expected: moving automatically, from many years' habit, he shook each of our hands. As he walked to the door, he turned around: 'If any of you manage to get out of here, don't forget to go and see my mother. Tell her that I am only sorry for her.'

We sat down, despondent and silent. The same question went round and round our heads. Who decided who would die? Why? Sangrós died without knowing why. As we put the sergeant's papers in order, we took note of his mother's address in case one of us could visit her."

Another day, another of his cellmates was taken: Leonardo Navarro, a university student who was alarmed because he had seen on of his classmates, a rabid Falangista, arrive at the prison. His cellmates attempted to calm him down but:

- "The lock turned in the door and the corporal called out: 'Leonardo Navarro, please come with me.'

Hearing this, Leonardo suddenly calmed down. Calmly, he looked at us with reproach, and pronounced the unexpected: 'See how I was right? I knew it, but you didn't believe me' The, quite simply and calmly, he removed his watch, gave it to us and walked out impassively and determinedly."

Yet another day, a Nationalist sergeant was put into Paulo's cell:

- "He was from the Siguenza front. He had fought bravely, and he swore that he had been cited in the Order of the Day for his valour. Two of his brothers had been shot in his hometown earlier and he had enlisted to avoid being also eliminated. He was ordered to present himself urgently to the Zaragoza barracks.

At noon, there was some unusual coming and going in the prison: 'Everyone get into your cells! Hurry up!' As I returned to my cell from the patio, I saw the shape of an automobile in the shade and in the corridor, next to the workshop, some blue-clothed Falangistas... The door to our cell opened and the corporal told the Sergeant to go with him. The latter, not having any idea of what was awaiting him, got up cheerfully and followed the corporal without saying a word.

The corporal later returned to get the sergeant's cape and he spoke frankly to those of us in the cell, appearing that he had to get something off his chest: 'It was a terrible scene. At the moment that the Sergeant was coming out in handcuffs, about to get into the Falangista's car, his wife arrived with a lunchbox of food. She went crazy and we had to calm her down in the workshop."

Bullet holes in old wall from Francoist firing squad executions. Also, some shrapnel from fighting. Sant Felip Neri Square, Barcelona[6]

---

[6]   https://www.gettyimages.pt/detail/foto/pla%C3%A7a-de-sant-felipe-neri-in-barcelonas
      -gothic-imagem-royalty-free/1162956022

APPENDIX VIII

SELECTED EYEWITNESS ACCOUNTS OF VIOLENCE AND TORTURE IN THE PROVINCIAL TOWNSHIPS

Santiago Cepas Romero. Released from Castuera concentration camp and rearrested as soon as he returned May 20 1939. Letter from Valencia to Moreno Gómez April 20, 1984.

- "I was working in the forge at Romero's home. May 20 1939, I was arrested in my own home by two Falangistas armed with rifles, the so-called *El del Lunar* and the other, *El de los Dientes*. They said: 'Santiago, you have to come with us to answer some questions, then you can come back here.' That is what they told everyone.

I was taken to La Preturilla. There I found quite a few Falangistas, a SIPM lieutenant, and a Guardia Civil called Medina. The lieutenant told me to sit on a chair without a back and to unbutton my shirt collar. Next, the soldiers who were on either side of me, rained beatings on me until they got tired. Meanwhile Medina typed away without asking me anything. At the end he told me: 'Sign here, and if you don't, you'll get another beating.'

Some days earlier they had killed several prisoners by beating them with rods, one of whom was Gabino. Rafael Diéguez Montes, another prisoner, unsuccessfully tried to kill himself by throwing himself from the veranda onto the patio of the jail. He was taken to Burgos and later executed by firing squad in Cordoba.

The evening on which I was arrested, another ten or twelve fell: Isidro Díaz Luna, Miguel *El Merino*, Juan José Ventijera, Miguel Silva Jurado. We had all been brutally beaten. They smashed Miguel Silva's toes by stomping on them. They tied Calixto Santos up because he stood up to them, and almost killed him. From La Preturilla, we were taken to the City Hall jail and from there, some days later, to Romo's house and, finally, to the Fuente Vieja Schools, from where we were taken to be court-martialled.

In Fuente Vieja, many evenings Falangistas came to beat us. One of us who was most often beaten was Lope Ibáñez. The poor fellow was left half dead and we had to move him by dragging him along the floor on a blanket; he was covered with wounds. The guards threw buckets of water onto him to revive him. He was later executed by firing squad… The Villanueva beatings were our daily bread."

Miguel Regalón Molina. Describing his personal experience in 1939 before the end of the war. Letter from Valencia to Moreno Gómez, April 1986.

- "In Villanueva, a so-called Don Juan, the military judge, and a SIPM lieutenant, were two of the most sadistic individuals one can imagine. I never dreamt that there could be people who remained impassive as men fell beaten and bloodied before them, with nothing with which to treat their wounds or stop the bleeding.

I didn't think that my body could resist the punching, beating, kicks, being hit with rifle butts and beatings with a fire tong. I was not allowed to speak during the interrogations. The only ones who spoke were the tormentors, on my right and on my left, two monsters dressed in military uniforms, all of which took place in front of a large crucifix and a group of Señoritos and Falangistas.

When I fell to the floor unconscious, they would throw me into the patio. When I came to, I crawled on my hands and knees into the warehouse to join my comrades. In my case, I urinated blood. I lost consciousness three times from the beatings; my whole body was one swollen haematoma, and my genitals were swollen from their kicks.

At daybreak, several drunken men would come in with the sergeant of the guard, revolver in hand, and they would begin running on top of the men who were lying down, stomping on their heads and stomachs, under the pretext that they were taking a count of who was there. They then brushed us with a toilet brush covered in shit to show that we had been counted. Not content with such savagery, they took Eugenio Palmera and me and others out to the patio, slapped us and stood us up against a wall, pointed their revolvers at our chests and asked us whether we wanted to die facing them or shot in the back. The whole time they passed bottles of alcohol from one to the other until they left… until the next night.

The little food and clean clothes that my family sent me were kept by the soldiers, until my family ran out of clothes to send. My life as an inmate, like that of so many others, was saved by miracle, as my file was marked 'dangerous Rojo'. When I was transferred from one prison to another my handcuffs were so tight that they cut my circulation and my hands turned black. On one occasion I begged them, please, to loosen them a bit and they just tightened them even more. I spent several days without being able to move my wrists. I began to believe that I was alive by a miracle and even that they might still kill me.

Before the interrogations, during which only the tormentors spoke, they made me pass through what they called the 'tunnel of laughter', which consisted in sending the prisoner between two rows of soldiers who beat him, slapped him, kicked him in the butt and more, so that when he got to the interrogation room he could not stand up. I remember two bastards there: one called Higuera and another Vigorra. This happened in the Fuente Vieja Schools.

In La Preturilla it was the Guardia Civil who took charge of the first interrogations, using the same procedures. There Pedro Juan *El Chunga* read out the charges, promising the prisoner that he would not be executed Then they killed him.

In La Preturilla barracks, in the patio, on the left, there was a room or cellar with wooden beams and a pulley with a rope. The floor was covered with sawdust, stained red with blood that was also splashed on the walls. I was put in there, but I was not hung. During my interrogation, when only the tormentors spoke, a group of individuals roared with laughter whenever we fell to the ground."

Francisco Romero Cachinero. Describes his passage through the Fuente Vieja Schools jail. Letter to Moreno Gómez from Villanueva de Córdoba August 1983.

▪ "The Fuente Vieja group of schools was the most horrendous jail that was ever known in Villanueva, where hundreds of persons from Valle de los Pedroche, from Adamuz, from other townships, were tortured. A frequent major contributor to the torture was Matías Malaleche Pedraza, accompanied by a group of local political bosses who treated the inmates brutally.

The inmate's depositions were taken at the Guardia Civil headquarters, always in the presence of several Falangistas who told the Guardia Civil of the prisoner's importance and determined the extent of the beating given by a short, grubby soldier from Balicia who would stand behind the inmate, who was seated on a chair without a back, and beat him without pity until he fell unconscious.

These statements were then sent to Judge Calero and the inmates were forced to sign everything that he wanted, or else... more beatings. The Judge's principal assistants were Pepe Higuera and Pepe Delgado. The scribe was Luís *El Plancha*. We also had to put up with the beatings from the Falangistas themselves. They would come to the prisons, having drunk a bit too much, take the inmates they wanted without any opposition from the prison guard, and beat them until they were exhausted. They threw buckets of water over the inmates to revive them. This is what happened to Crisóstomo Marcilina Romero Badia and to José *El Papel*, two of several soldiers who crawled back to the cell where they remained many days in bed without being able to move. The one who did the most beating was Pedro *El Chicorro* Muñoz Ruíz and who was known for his violence."

Sebastián Gómez. Speaking of the Fuente Vieja Schools jail and the notorious Judge Juan Calero, during an interview with Moreno Gómez in Villanueva de Córdoba August 7 1982.
- "It happened in Fuente Vieja jail, I believe in Summer 1939. Matías Pedraza and others, including *El Tiraor* arrived, and the latter shouted: 'Pay attention, those who names I shall now call out: Francisco Sánchez Muñoz, Antonio *El Piñón*, Juan *El Ramo*, Antonio Casado, Pedro Torralbo, José Maria Sánchez and Francisco Illescas.'

First, he slapped Piñón, then Beatas, but the latter threw a can of olives at El Tiraor, and they spilled all over his head whilst Matías Pedraza bust his guts with laughter. El Tiraor continued now beating Illescas and saying: 'I am going to leave you with a souvenir for the rest of your life'. He tore a piece off Illescas' ear with his teeth and stomped on it.

It was a dissolute, cruel and uninhibited environment that the victors created around the defeated. One day, Matías Pedraza violently slapped José Jurado, ex-Mayor for the Radical Party during the Republic. This was an act of vengeance y Matíaz Pedraa who one day had been fined by the ex-Mayor because of some scandal involving prostitutes. Many of the worst civilian acts of violence during the post-war period were acts of revenge for personal reasons."
- "October 19 1939, Juan Calero came to Fuente Vieja to interrogate Francisco *El Villaralteño* Rubio Gómez, who had been arrested. When Francisco appeared before him, the Judge turned to

the right-winger Pedro *El Ché* Benito and handed him a lash saying: 'Come on! Avenge your father's death!'

Although Pedro Benito was reluctant to do so and as the Judge insisted, he hit him only twice with the lash. Juan Calero stood up, furious, tore the lash from his hands and launched himself against the prisoner with a rain of blows, knocking him to the ground. Calero then stood over him and began to kick him. Two soldiers came up to help the Judge; one stood on Francisco's head and another on his stomach."

Pedro Molinero.

- "Emilio *El del Lunar* came to arrest me, and I was taken to the Guardia Civil barracks in La Preturilla, which also housed the SIPM headquarters. When I arrived, there were more than 40 right-wingers present – rich men, Falangistas and accusers.

Some of them had been arrested in Jaén or Totona during the war and others were accusers because of the death of some relative. Nevertheless, when I came in, most of them left, probably because I was not considered to be all that important a prisoner.

After I made my deposition before the SIPM lieutenant, Guardia Civil Corporal Galiánaiz immediately came and stood next to me, and it was he who beat me the most. Also present to punch me and wield their whips were the Muñoz Fernández Salado brothers, two soldiers and some others.

The whole business of my making a deposition was to force me to sign a declaration that I had belonged to the local War Committee, but as I denied it, they beat me. One rich guy added a codicil to the supposed declaration: 'This is the person who handed our farms over to the workers', because I had been a member of a collective of small landowners. Generally, the great names of the oligarchy did not show their faces there, although they were in direct contact with Judge Juan Calero.

After spending some time in City Hall jail and in the house on Calle Conquista, I had to attend a second inquiry by the Military Court in Lauriano House. There they read the declaration they had previously written and continued to insist that I had belonged to the War Committee. Roughly speaking, the accusations were all the same for everyone: they had participated in the taking of the town, had belonged to the Committee, had spoken against right-wingers, and so forth. Assisting Judge Calero with the declarations were Luís *El Plancha* as recorder and Francisco Madueño. The beatings were given by two officers, a 2nd Lieutenant of Artillery and another 2nd Lieutenant from a Regular regiment. The latter, *El del Gorro* Colarao later became tragically famous as *Teniente Pepinillo* who sowed terror in Espiel and ended up committing suicide.

The principal figure was the examining magistrate., Judge Juan Calero Rubio, whose greatest explosions of rage appeared when he felt the inmate was not receiving enough of a beating.

In Summer 1939, the inmates were transferred to the Fuente Vieja Schools. That Judge went there from time to time, accompanied by the terrible Lieutenants, to take declarations. I cannot forget the horrendous beating of José Escribano whom they wanted to sign an admission that he had participated in shooting 21 Nationalists. Fortunately, Escribano was exceptionally able to get them to remove that accusation, and he survived. Miguel Carabinas Cabezas, Juan Lorenzo Cucharas Cantador and many others were also brutally beaten."

Antonio Chaparro. Interviewed by Moreno Gómez in Villanueva de Córdoba 1981. He had served as a prison guard in the Preturilla barracks.

▪ "One of the worst torturers was Bartolomé Berenguer Cepas who had been appointed head of the municipal police and who reached the rank of sergeant during the war. One day I saw him in action with corporal Caiániz, under the orders of Leopoldo Mena the SIPM Lieutenant. They brought in a prisoner when I was on guard duty and made him kneel with his hands tied behind his back. They accused him of having taken a pot-shot at Torrico and whenever he corrected them or denied something, the lieutenant gave a signal and they beat him non-stop.

At one moment, he begged for some water from a pitcher that was nearby, but in the pitcher, there was nothing else except dirty water and cigarette butts. They left him in a coma. When they took him to the hospital, they dressed him in a new shirt because the one he had been wearing was so bloody it had stuck to his back and they had to cut it in 1936 as a guard over right-wingers who had been arrested, among them El Tiraor, who was getting his revenge."

Antonio Pedraza García. Denounced by Pedro *Sargento Chicorro* Muñoz because the two had fought each other when they were boys. His son Bartolomé Pedraza recalls a moment in his childhood. Revenge ran deep.

- "No sooner was my father taken to jail, that he received terrible beatings that broke three ribs. The ones who beat him were Juan Lucio, Pepe Delgado, Miguel Higuera's sons, the Falangista Lara and *El Chicorro*. One day my mother tried to talk to my father through a window that gave onto the street but at that moment a lieutenant came out of the building and ordered my mother to strip naked. On another occasion, when I went up to the top floor of the house on Calle Conquistador to see him, Falangistas who were there kicked me down the stairs."

Francisco *Curro Beatas* Sánche Muñoz. Member of the Izquierda Republicana executed May 17 1940. Report from his daughter **Antónia Sánchez Cerezo**, Villanueva de Córdoba 1983. (Francisco Sánchez' farewell letter from prison to his family is recorded herein.)

- "At the end of the war, my father tried to escape to the hills to join Julián Caballero, but whilst at the El Minguillo farm, he was betrayed by his nephew, Mateo Sánchez, and he was arrested then and there. My father was taken for interrogation many times, always getting the same beatings. He had been a member of the local War Committee, but he also saved many right-wingers, such as Alfonso Fernández, and he was able to prevent others from setting fire to Dionisio Pedraza's house, whose son Gregorio was Mayor at the time. The one who beat him the hardest was Miguel Higuera's son, Pepe.

The first beating he received was given him by the scandalous Berenguer, who beat him until he lost consciousness. When the Falangistas heard that my mother, Agustina Cerezo, was publicly complaining about the way they were beating my father, they ordered her to go to La Preturilla and, in her presence, gave my father another savage beating. They sent her away with the warning: 'Now go round telling people that we are beating your husband'.

When my brother Francisco tried to say goodbye to Father, they told him to take 50 pesetas and they kept the money, made him wait all day, but in the end, he was still unable to bid him farewell. After he was executed, my mother asked for my father to be buried in a niche in the cemetery. She was told to take 60 pesetas to pay for the niche. She had to sell her sewing machine to get the money, but they kept it, and my father was buried in the mass grave."

Pedro *Cuadrado* Torralbo Gómez. High-ranking member of the Communist Party, served as a Captain in the famous Garcés Battalion for which he was executed after the end of the war. His son **José Torralbo Rico**'s story of visits to his father in jail is harrowing.

- "We went to the square because we knew they were interrogating my father, hoping to see him. There, a member of the public took only me by the hand and led me into the City Hall. The jail was on the left as you go in. There was a thick wooden door and behind it, another one of wire netting. I was just a child, but they opened the doors and let me in so I could see my father. It was horrible: a pile of men, leaning against each other, semi-conscious, vomiting and bleeding.

Among them – I can't forget it – I saw my father sitting on the floor, his back against the wall, legs stretched out before him and his head fallen on his chest, bleeding from his mouth, from his ears, from all over his body. Today I understand what that man must have felt when he saw me, if he even saw me, because I remember that they took me out immediately... His tormentors were Francisco *El Tiraor*, assisted by Berenger, Pedro *El Barbero* Serrano, and others."

Juan Cantador Zamora. A totally innocent man caught in the Francoist circle of torture and death. This is his daughter **Catalina Cantador Romero**'s story.

- "From the day after the war was over until April 17, my father had to present himself daily at the Court by order of the Judge. On that date, finding that he had not committed any crime, the Judge suspended the order for his arrest and set him free.

May 10 1939 at midnight, Vicente Salado and a soldier arrived at his house and took him to La Preturilla, where he was tortured and afterwards informed that there was an allegation against him. He was tortured so savagely that strips of skin from his back stuck to the short he was wearing. He was allowed home.

Two days later, he was again taken to La Preturilla where he was again beaten in such a barbaric manner that, because his screams could be heard from the street, they closed the road to traffic. When I took his breakfast to the jail the next morning, I could confirm that he was unable to move because of the beating he had received. He went home. The next day, he was back at La Preturilla and again suffered a new beating.

I, his daughter, who from the street could hear my father's terrible screams, am telling you this. How I saw how Bartolomé Capas (head of the municipal police) come outside sweating, with a whiplash in his hand, and go back in with a case of what appeared to be beer. I also saw the following individuals, among others, leave the building: Vicente Salado, Francisco Tiraor, Diego *El Chunga*, Pepe Liñán, Miguel Fernández and Fructuoso *El de los Dientes*. The son of Rojas, from the tavern in the square, stood guard at the door of the building.

From where I stood at the door of Morales' home, I could see how the priests who were in the church poked their heads out through the vestry door each time they heard a scream, closed the doors and windows tight, never saying a word or making the slightest gesture of disapproval.

May 24 1940 my father was tired by the military court in this town and sentenced to death. At the court martial, the prosecutor asked Luísa Doctor, the accuser, to point out which of the accused men sitting on the bench had killed her husband. The good lady said she did not recognize any of them and that she had heard nothing said about my father.

The Judge then asked the accuser if she had ever spoken to the accused, to which she answered that she might have spoken to him as she might have spoken to anyone. He then asked her why she said he had killed her husband. 'No sir', she replied. 'Judge Juan Calero must not have heard me correctly. What I said was that it was someone from outside one of Juan Elías' agents.

The prosecutor asked my father why he had signed a confession and my father replied that it was because he was beaten until he signed. The prosecutor then asked him whether he was ever beaten. Again, my father replied that yes, he had and that it would be impossible for him to name all those who beat him as he often lost consciousness. The examining magistrate himself had beaten him.

When the court martial was over, my father was taken to Fuente Vieja prison. No sooner did he arrive there, that a 2nd Lieutenant, the Judge and José Higuera entered the patio and, in the presence of the other prisoners, beat my father until he lost consciousness. I could hear his screams from outside the building, on the street. From the day on which he was condemned to death until September 26 1940 when he was transferred to Cordoba prison, we never saw him again."

Antonio Ramos Palomares. From Almodóvar del Río. Soldier returning after the defeat.

▪ "After we reported in 1939, at first it appeared that our tormentors had some consideration for the defeated, but when a little time had passed, when it appeared that no-one else was turning himself in, they [Francoists] began with their interrogations. As we swore that each one of us would be responsible for his acts and that no one would accuse anybody of anything, they began with the mistreatment and the punishments.

There was the case of Ángel Plazuelo Lozano, who had been a guerrilla sergeant and who they wanted to give details of the services he had rendered in the Rojo zone and the names of the companions who had fought with him, but he remained faithful and was tortured to death. He was immediately thrown down a well to simulate a suicide and the next day they took his body to the cemetery in a garbage truck.

From then on, all of us were continuously beaten. They divided us into two groups: one in the Santo Hermitage (the larger one) and another in the Trades Union building. In November, they thought that one of us had escaped, something they could not prove, and because of that they held us incommunicado. We were beaten and tortured at all hours. In my case, my comrades had to feed me for more than fifteen days because they had destroyed my arms during the sessions of torture."

Spanish refugees, mainly republicans and members of the International Brigades, guarded by French troops at a camp on Argeles beach in 1939.[7]

APPENDIX IX

TESTIMONIALS OF THE CLERICAL REPRESSION IN SPAIN AND OF THE SITUATION OF CHILDREN BEHIND BARS

Manuel Espejo. DOS TORRES. *Regarding the pressure to confess ritual in Córdoba prison.* Interviewed by Moreno Gómez, Madrid, January 1984.

- "At night, a Guardia Civil truck would arrive at the prison. They called the names of those who were to be executed and took them to a room to confess. There was a different priest at each saca, monks or Jesuits from Córdoba. The prison chaplain was Rev. García, S.J. and he took charge of confessing the most reluctant. He took them to his room and afterwards, whether they had confessed or not, said they had. Because of that, many also refused to go into his room. Approximately half in each saca confessed, many of these convinced by Rev. García to whom they owed favours at hard times or when they were hungry.

Flor Cernuda. QUINTANAR DE LA ORDEN. *Testimonial recorded by Tomasa Cuevas, regarding Ocaña prison.*

- "In Ocaña, we were taken to Mass. During Mass, another priest and minister of God would give us a sermon on how evil the Rojos were, but they, as they were so good, were willing to forgive... to forgive our souls, but they would have to kill our bodies because it was our bodies that had sinned."

Agustina Sánchez. Regarding VENTAS PRISON in Madrid. Testimonial recorded by Tomasa Cuevas. *Regarding the day they executed her mother-in-law, Josefa Perpiñán, July 24 1939.*

- "She came out looking like a ghost and was taken to the chapel. There were another six women with her; they executed all seven. They had been taken to the church the night before and from there, to the chapel. The priest told them to entrust their souls to God because they were going to die. The next morning, at 6 a.m., at the last moment, they were given a crucifix to kiss. My mother-in-law grabbed the crucifix and threw it at the priest's head. She did not kill him because he ducked...

As they took the women away, I was made to leave the church so that I could not go into the gallery and see in what condition they left. I was told that the women's lips were dark purple, almost black. I was so upset that the only reason they were killing my mother-in-law was because they could not capture her son."

Regarding Agustina's experience in Amorebieta prison from which she was released.

- "I was called for my examination and the priest told me that I had religion by a thread and that I would leave here when he wanted to let me go. 'Do you see this finger?' he asked. 'With it I have often pulled the trigger against Rojos. I have killed a lot. My only regret is that I have not killed them all.'"

Francisco Gómez Herencia. NUEVA CARTEYA, CORDOBA. *Regarding his being examined on the catechism before he could be released from jail.*

- "One day after I had spent four years in prison, the Director told me that if I could pass the catechism examination and answer the priest's questions satisfactorily, I could be released. The Director himself, Don Faustino, gave me a catechism to study. A few days later he called me: 'Herencia, how are you getting along with what I gave you?' 'The same, Don Faustino', I replied. 'I just cannot get this into my head'. The Director, seeing that it was pretty much the same after a few days, called an old village priest to come in and examine me and the Director himself helped me answer the questions. This is how I was able to go free."

The situation of the children behind bars

Tomasa Cuevas. DURANGO prison. *Description of the problem of children in prisons and the support given by the townspeople of Durango to their mothers.*

- "We were more than two thousand women in that prison. There were a great many children ranging from a few months of age

– some were born in jail – up to three and even four years of age. The Government then issued an Order by which children older than three years could not accompany their mothers in prison. Where did these children go? To a children's home. There were some problems: in some cases, it was friends or relatives who asked to be allowed to take the children. Unfortunately, we were in the North of the country, and we were all from the Centre and even some, Andalusia. The Government had set a deadline by which the older children had to leave the jail and their mothers were getting desperate. What would happen to their children? The townspeople of Durango were brilliant. They went to the Director and told him that they would take these children until their families could get them. The little ones who stayed with their mothers had a hard time of it; they only had the same food as the inmates to eat, no more milk or anything else. Two died not long after. When the townspeople heard of this and that others were not faring very well, they sent in jugs of milk to be shared by the mothers."

Nieves Waldemar. Imprisoned in the Convent of the French Nuns in GUADALAJARA, told Tomasa the following:

▪ "There were just fourteen square metres with a toilet just for us, where at night a mass of children and women slept. What one child had, the others would catch: pimples, scabies, all those illnesses that are infectious just because of the overcrowding... They executed two sisters-in-law in Auñón, Guadalajara, and when they left to be shot, with their fists raised, one of them said: 'Companions, do not salute, because you will be punished, but think of my children, I am leaving my children!' This is what the married one said. Her husband, a member of the elite police force, had been executed. This woman left her children to be taken are of by whoever wanted to because all of her relatives had been executed."

Carmen Machado. From Madrid, reporting on some serious figures for DURANGO prison.

▪ "The tragedy of these mothers was horrible. The children were not given any special food, they had to eat the same as their mothers. Many women were not able to receive any food packages because their homes had been destroyed and all their relatives had disappeared. There was an epidemic of encephalitis lethargica: the same children whom we had seen happily playing the day before, began to fall asleep and most of these never woke up. I particularly remember a little boy from the province of Toledo, an amusing little chap, one of the children who died from the encephalitis. I imagine that I see him, during the vigil of his body, wrapped in a blanket until they came to get him the next day."

Carmen Riera. Is extremely blunt as she gives an interview to the *Prohibido recordar* [Forbidden to Remember[*documentary. She describes her horrible exxperience in SANTURRARÁN prison.*

▪ "Regarding the children, more than thirty died in ten days, including my daughter. When she died, they moved me to a room where they took her little casket. The nuns picked some flowers for her and began lamenting 'Oh how exciting, another little angel who loves God!' I could not stand it anymore and kicked them out of the room. I spent the whole night sitting on the floor, next to my daughter's casket."

MOTHERS!
Protect your children from the fascist barbarians
EVACUATE MADRID
You will be given housing, food and peace
(Republican Government poster)

APPENDIX X

TESTIMONIALS REGARDING OTHER ASPECTS OF THE MULTI-REPRESSION AND THE SPANISH HOLOCAUST. CONDITIONAL FREEDOM AND SLAVE LABOUR

<u>Matías Romero Badía</u>. VILANUEVA DE CÓRDOBA
- "At about 11 p.m. on the night of 9 May of the Year of Victory, a short 2nd Lieutenant [probably the notorious *Teniente Pepinillo*] accompanied by several drunken Falangistas appeared in the room on the first floor of the building where the prisoners were held [Juan Herrerro's house on Calle Conquista], and who, revolvers in hand, had come to 'say warm goodnights to the inmates': "Let's see. Where is the bastard who said that if he had a hoe, he could make a hole in the roof of this house and get away? You have five minutes to tell me who he is and if you don't, I have a truck outside the door to cart you all to the cemetery."

About half an hour later they returned, revolvers in hand and he repeated the question. As the inmates again refused to comment, he pulled out a list with the following names: Bartolomé *El Floro* Pozuelo, a sheet shearer, and Francisco *Floro el Manco* Pozuelo, his son, and took them away. About an hour later, they returned, barely able to walk. The Lieutenant, pistol in hand again, repeated the same question and getting the same reply, read out another name: Alfonso *El Papel* Ibáñez Tamaral, whom they to headquarters where a group of tormentors, including Pedro *El Barbero*, were waiting.

And so on, back and forth, the same question and the same response. The Lieutenant kept on repeating "If anybody present still lives at the end of the night, he shall never forget the night of 9 May of the Year of Victory" …

CORDOBA. *Regarding his transfer to the Provincial Prison from the Alcázar, Winter 1942.*

- "We were transferred to the Provincial Prison in two groups. The first group marched along the banks of the River Guadalquivir to their new destination. In the second group we were formed into sets of two and three, some of us shackled to each other, others tied to the other with wires. Each prisoner had to carry his backpack over his shoulder. Once we were lined up Indian file, usually in groups of some two hundred, each one was tied to another with a long rope, head to tail, to prevent any attempted escapes. Closely guarded, we began walking along the riverside until we arrived at the New Prison, which was still under construction."

Antonio _El Cano_ Álvarez. POSADAS. _Report to the local Historic Memory Association._

- "We were sleeping one night in the anthill that was the local jail, when there was a banging at the door. 'Who is it?' A guard asked. 'Come on out... We're going to have some fun!' someone replied. They took my father and another person from this town out to Plaza Mártires, where a ring of many Señoritos and their sons were amusing themselves with castor oil, some were pouring gasoline, another was wielding a lash, another... and so on. They beat my father so badly, we had to use pincers to remove his shirt."

Miguel Hernández. ORIHUELA. _Diary and letters written in Orihuela jail._

- "I feel much worse here than I did in Madrid. There, nobody, neither those who received nothing, suffered such hunger as we do here, and we did not see the suffering and the diseases that are everywhere in this building. My fellow townspeople are especially interested in showing the evil that lies in their hearts and I have been seeing this first-hand every sine I fell into their hands.

The Señoritos will never forgive me for having put the little intelligence and learning that I have at the service of the people. These are the most brutish individuals you can imagine. But they will not f... with me or anybody else. Every moment they are making me waste now, each beating, they will have to work for.

I do not want to leave whilst they still are scoundrels and blackguards in the world. Tell me how you get along in this labyrinth of hunger and misery in which we have been placed..."

Testimonials from inmates in Córdoba Provincial Prison

José Sáenz Jiménez. Letter to Moreno Gomez from Caracas, Venezuela, March 24 1986. *Regarding his father in Córdoba Provincial Prison 1941.*

- "Every morning, they took a new victim from the cells, swollen like a balloon, to add to the list of those who died from some supposed disease that the imprisoned doctors were forced to certify. Faced with such a situation, my father denounced it to the authorities, regardless of the threats he received and that almost cost him his life."

Carlos Menéndez. DOS TORRES. *Regarding his treatment in Córdoba Prison.*

- "We ate rotten beans, mouldy turnips, squash and more, without olive oil. I was present when we left our cells as they inspected the several individuals who died each day. The death certificates, complete with diagnosis, had already been signed. The only thing missing was each dead man's personal details."

Manuel Espojo Blanco. DOS TORRES. *Driver for prison Director Escobar.*

- "One of the dormitories was occupied by those who were dying of hunger. It was a dismal scene because they were no more than skeletons who could not even stand up. They were grouped in the two dormitories above the offices in the centre of the prison. When they died, they put several bodies in the same casket and there still was room left for more. On the other hand, the inmates suffering from typhus were kept in the infirmary. The sleeping mats had to be burnt to get rid of the bed bugs and the lice."

Adriano Romero. VILLANUEVA DE CÓRDOBA and Member of Parliament for PONTEVEDRA. *Describes the dormitories in the old Córdoba prison, under the Alcázar.*

- "The dormitories in which we were packed, like a herd of sheep, had so little sleeping space that it was often less than 20 centimetres wide per person. They were located in an underground area of the old Alcázar that had at one time served as stables, under three corners of the patio, some 20 metres long and some 7 or 8 metres wide, to serve a prison population of more than 1,000 inmates... there just wasn't enough room for everybody in the patio.

In the 350-inmate dormitories, there was a single open-air toilet that we could not use during the day. An epidemic of typhus broke out amidst this appalling misery, with no sanitary conditions, with a food ration that did not supply even one fourth of the calories that a human needs to live. If you add those who died from hunger to those who died from typhus, they liquidated a third of the prison population. Some six hundred persons who had been turned into living skeletons survived... This Dantesque scenario of dead and men turned into skeletons from hunger and typhus, was repeated in all the most important prisons in Spain. All those who had nobody to help them was condemned to starve to death."

Francisco Gómez Herencia. NUEVA CARTEYA. *Regarding hunger and parasites.*

- "It is almost impossible to describe those days of horror, of anguish, of unrelenting hunger. During the day we would drag our miserable bodies from one patio to another, not even able to sit on the ground because the parasites that covered it, the logical result of the total lack of sanitation, would crawl over our bodies... We were treated worse than animals and I still have several scars to remind me of that unforgettable stage of my life. If we were treated in a humiliating and savage manner, our despair grew even more at mealtimes. The only food we were given was a watery soup made from boiled cauliflower stalks in which, to give it an appearance of containing something nourishing, they threw cubes of fat that they used to grease the cartwheels.

For twelve months, our lives slowly slipped away. The few of us who survived did so thanks to the proximity of our families who, with a superhuman effort, sent us some food when they could. We would have all died if it had not been for a young officer who, horrified with the sadism with which we were treated, informed Madrid. An inspector was sent and when he asked to see where the food was stored, all they could show him was a shack containing some rotten turnips."

Miguel Regalón. VILLANUEVA DE CÓRDOBA. *Regarding medical treatment.*

- "Every afternoon, when we were sent to our cells, several inmates who had died from hunger remained on the patio floor. We had to brush the lice off those who could not move from their mats because they were dying. They were covered with infected, foul-smelling sores. Dr. Sama, who lives in Córdoba and was also imprisoned, can attest to this, as he wrapped cotton around his ankles to keep the lice from infecting him.

In Summer, with temperatures above 40°C in the patio, we were given no water and whenever some relative brought a water bottle, the guards would break it. One inmate from Adamuz caught a little bird that had fallen from the roof and ate it raw. The guards beat him and that was the last we saw of him. The lack of vitamins (avitaminosis) began to manifest itself with a swelling of the face, the legs and the skin on our testicles. One day I was lucky in that my family brought me a canteen of olive oil and some oranges. I went into the patio, to share a spoonful of oil and a bit of orange peel with those who could no longer stand. One inmate, nicknamed *El Panolo*, died with some olive oil and a bit of orange in his mouth."

Released on Probation – the so-called Conditional Freedom

Miguel Regalón. In a letter to Moreno Gómez from VALENCIA April 1986. *Regarding the 'delights' of living as a parolee.*
- "I was released on parole and was constantly under supervision, followed by informers from all social classes, and the police would send me Communist propaganda in the mail to see if I would fall into their trap by distributing it. I was provoked, to see how I would react, but as I had a wife and daughter, grinned and bore it and humiliated myself. This is why I often say that I have more than nine lives."

Ernesto Caballero Castillo. VILLANUEVA DE CÓRDOBA
- "Something else were the daily signing ins at the Guardia Civil barracks that many men and women were required to attend, usually at the end of the day. At times there was a huge crowd of people in front of the gates to the barracks, waiting for some guards to come out with their lists and check off their names so that they could go home... until the next day. If they felt like it, the guards would stop

this or that person, interrogate them harshly and give them a beating or two, for no reason other than they were working men.

In addition to the constant fear that one might be kept back and interrogated, the required appearances made it impossible for one to work in the countryside as it forced the parolees to remain in town all day, because of which, many families went hungry.

My mother was forced to attend these signing ins for a long time. Also, with my brother and me.... every afternoon. I was about 9 years old at the time and my brother about 11. Both of us had been working in the fields, taking care of farm animals, but we were forced to leave those jobs and stay in town doing nothing.

- "Every afternoon, the Falangistas would go to the bus stop to see who was coming back into town, and when they felt like it, they arrested them again, together with any of their family who were waiting for them. At a minimum, they kept them closed up in the country jails for a few weeks, when they did not shave their heads or beat them. The people who waited at the bus stop were terrified of the uniformed and armed Falangistas... One day, my mother arrived on the bus..."

Forced labour and the Workers' Battalions

Pedro Gómez González. VILLARALTO. Letter to Moreno Gómez from Córdoba December 16 1986. *Description of he conditions in Workers Battalion 28, Company 4, La Bacolla, near Santiago de Compostela, where he was put to work building an airfield.*

- "The most rotten tricks were the rule here. The boss was a Commander in the Engineers, the greatest bastard I ever met. Each Battalion worked eight hours, one in the morning and the other in the afternoon.

The morning Battalion had to rise at 5 a.m. We were given a mug of coffee and we were marched five by five, holding each other by the hand, escorted on either side by a soldier who carried a rifle and a whip. It was a 3km march to the job. The other Battalion began working from 1 p.m. to 9 p.m. When dinner was served, boiled cabbage, it was already 11 p.m.

It was exhausting work. We had to dig and fill 8 or 10 metre-and-a-half wagons with earth and take them along a path to level a hilly area. We were put into groups, each under the command of one of the most bloodthirsty sergeants or corporals, who lashed us and made us work without stopping. We were given little clothing and paid nothing, despite the fact that the work had been given to a private contractor.

When one of us escaped, the rest of us were punished by having to attend instruction sessions after work. Two friends of mine from the same town managed to escape, but we later heard that the Guardia Civil had caught them in León railway station. They were surely killed because we never heard of them again. Another friend of mine from Villaralto, Alfonso Luna, was beaten with a sharp pointed stick they jammed into his arm. When they finally took him to Santiago Hospital it was too late; he died the following day from gangrene.

There was a great deal of hunger. The poor chap who received nothing from his family, was condemned to death. Many starved to death. When we arrived, we were only fed some broth and cattle greens. We looked like starving beggars, eaten up with misery. We were housed in an ancient tannery. At night, shivering with cold, we watched the stars through the holes in the roof.

The commander laughed at us. He called us the 'sons of the Pasionaria'. Many companions could not work because they were too weak to walk, and they fainted. When we returned to the camp from work, those who were able to stand went to look for food in the garbage cans for anything edible – fish guts and other rubbish. You had to see it to believe it."

Excerpts of testimonials from the award-winning 2009 Spanish television documentary on the Dos Hermanas, Seville, labour camp: *Prisoners of Silence.*

- "It was very, very difficult for the inmates' families. My mother had reached the end of her tether because of so much suffering. Four of her sons had died. There was no end to her pain, and she had run out of tears.
- In addition to the hunger, it was the humiliation.

- Families could visit on Sundays and Holy Days, but it cost an enormous amount and we had to travel long distances. We came from Jaén and after we got to Dos Hermanas, we had to walk 8kms to get to the prison camp.

  Once at the camp we could see my father from a distance, behind a wire fence. We had to shout to speak to him. Instead of making him happy to see us, it made him very sad. My mother cried non-stop and so did the children.

- One day, four or five drivers escaped – one from Córdoba, another from Gibraleón, the others from Morón. They made the mistake of not going far away and a few days later they were caught and taken to La Corchuela, which began with 3,500 prisoners and they added another 3,500 from El Arenoso. The fugitives were shot in front of everybody, and we were made to march past the bodies. 'Eyes right! Eyes right! It was a day of silence."

APPENDIX XI

POST-WAR FIRING SQUAD
EXECUTIONS IN CORDOBA CAPITAL
2 JUNE 1939 – 21 FEBRUARY 1945

**Individual, small group and mass executions compiled.
and annotated for the Democratic Memorial of Spain
investigation of the Francoist genocide in Cordoba**

[Francoists refused to recognize the rank prisoners-of-war held with
the Republican army when they were captured and they only recorded
the names of those they executed, contrary to the Geneva Convention,
and their lowest pre-war rank (*given in italics in brackets*). The highest
rank each held in the Republican army when they were captured
is nevertheless indicated here following their name. Details of the
military record for each is given when available.)]

1939
2 June **(1)**
    Juan B. Ruano Muñoz, 43, bricklayer, Porcuna
11 July **(1)**
    Alfonso Hernández Carrillo, 32, carpenter, Beas de Segura
26 August **(1)**
    Rafael Guerrero Juárez, 51, electrician, Espiel
1 September at 6:15 a.m. **(4)**
    Antonio Juárez Guerrero, 26, mechanic, Alora (Málaga)
    Francisco Bermúdez Canales, 54, farm worker, Baena
    José Bermúdez Expósito, 36, farm worker, Baena
    Isidoro Povedano Muñoz, 37, farm worker, Priego
28 October at 2 a.m. **(2)**
    José Navajas Espejo, 38, manual labourer, Posadas
    José Mesa Frías, 31, farm worker, Acalá la Real
6 November **(1)**
    Benjamín Gil Bona, 20, manual labourer, Aljuna (Valencia)
8 November at 6 a.m. **(11)**
    José Maras Jiménez, 39, farmer, Puente Genil
    José Luna Granados, 26, farm worker, Puente Genil
    Miguel Rey Balaguer, 25, manual labourer, Puente Genil
    Diego Cornejo Morales, 36, farm worker, Puente Genil
    Feliciano García Haba, 23, farm worker, Peraleda de Zaucejo
    Antonio Zurita Montero, 26, farm worker, Cañere
    Ricardo Rubio Calero, 27, Pozoblanco[8]

---

[8] Prisoner of war captured 1938 in Vinaroz (Castellón), Political commissar in the army
     and son of *El Calor*, the famous Socialist.

711

Manuel Moya Vivo, 28, bricklayer, Villacarrillo (Jaén)
Juan Durán Lorenzo, 34, farm worker, Azuaga
José Cuevas Simonís, 44, farm worker, Posadas
José Pérez Pozo, 49, farm worker, Gilena
9 November (3)
San Rafael cemetery
Rafael Martínez Arenales, 55, businessman, Villaviciosa
Justo Deza Montero, 39, farm worker, Puente Genil[9]
6 December at 6:30 a.m. (6)
La Salud cemetery
Antonio Sánchez Cabello, 19, bricklayer, Puente Genil
Juan B. Sánchez Aguilar, 50, farm worker, Puente Genil
San Rafael cemetery
Manuel Martín Balsera, 26, farm worker, Badajoz
Felipe Espinosa Casablanca, 31, manual labourer, Badajoz
José Jiménez Vargas, 24, barber, Lora del Río
Salvador Sánchez Fernández, 29, basket weaver
15 December (2)
Domingo Muñoz Mérida, 32, farmer, Almodóvar del Río
Juan Muñoz Torres, 29, farm worker, Posadas

1940
7 January at 7:15 a.m (1)
El Alcázar Provincial Prison
Francisco Cruz González, 40, manual labourer, Dos Torres
17 January at 7 a.m. (2) Casillas artillery range
Narciso Sánchez Aparício, 50, Lieutenant Colonel (*Artillery Major*)
from Santa Clara, Cuba[10]
Esteban Rodríguez Domingo, 45, Major (*Artillery Lieutenant*), Valencia[11]
20 January at 7 a.m (3)
Luís Giménez Romero, 29, farm worker, Espejo
Bartolomé Higuera Caballero, 31, farm worker, Villafranca
Andrés Caballero García, 27, farm worker, Villafranca
27 January at 6:30 a.m. (2) Casillas artillery range
José Bueno Quejo, 43, Commander (*Infantry Captain*), Vitoria[12]

---

[9] Famous Cordoban Socialist, Chairman of the Puente Genil War Committee, Alderman as of 1931. During the war, lived in Pozoblanco where he presided over a farmers' cooperative. Went valiantly to his death exhorting his cellmates to make sure that so much bloodshed should not go unpunished.

[10] Infantry major in Segovia, ordered to organize a battalion of volunteers 17 July 1936. Promoted Lt. Colonel, served as General Staff Officer 1938, Chief of Staff of the XVIIth Army Corps and then of the XXIII Army Corps. Source: 5 October 1987 letter to Moreno Gómez from Carlos Engel.

[11] At the end of the war was serving with Light Artillery Regiment 6 in Murcia. Ordered to suppress the rebellion in Albacete, Promoted to Captain in 1936 and to Major in 1938. Source: Carlos Engel.

[12] When war broke out, was in command of the garrison in Santoña and refused to support the coup. He commanded a unit in defence of the Port of Escudo, which later became the 2nd Division of the Republican Santander Army Corps. Promoted to Major in 1937. When Santander fell to the Nationalists, he went to France and from there, to the Centre of Spain where he was appointed Chief of Staff of the 22nd Division of the Andalusia Army Corps. Was tried by court martial in Cordoba 5 September 1939. Posthumously fined 5,000 pesetas under the Law of Political Responsibilities. Source: Carlos Engel *Op. Cit.* And Jesús Gutiérrez Flores, undated letter to Moreno Gómez from Santander.

Luís Soler Espiauba-Cánova, Major (*Infantry Lieutenant*), Cartagena[13]
29 January (1)
  Juan Aranda Martos, 41, carpenter, Cordoba
28 February at 6 a.m. (3)
  Eugenio Giménez Blanco, 30, manual labourer, Posadas
  Juan Sánchez Muñoz, 37, manual labourer, Posadas
  Juan Codines Galvez, 27, farmer, La Rambla
18 March at 6 a.m. (19)
  Francisco Mejías Sánchez, 28, farm worker, Hornachuelos
  José García Rodríguez, 46, miner, Hornachuelos
  Emilio Ramos Cantador, 60, farm worker, Hornachuelos
  José Linares Pérez, 34, farm worker, Hornachuelos
  Antonio Núñoz Pulido, 36, farm worker, Hornachuelos
  Hilario Expósito Flores, 32, farm worker, Obejo
  Higinio Morales Moraza, 35, businessman, Obejo
  Tomás Flores Puerto, 39, manual labourer, Obejo
  Francisco González Gavilán, 38, bricklayer, Almodóvar del Río
  Manuel Berjillo Rodríguez, 29, farm worker, Almodóvar del Río
  Francisco Gómez Galindo, 30, farm worker, Benamejí
  Manuel López Moya, 30, farm worker, Morente
  Gabriel Serrano González, 28, farm worker, Villa del Río
  Cristóbal Sáinz Marín, 35, farm worker, Villa del Río
  Rafael Casado Castro, 27, bricklayer, Villafranca
  Gonzalo Obrero Duque, 30, farmer, Villafranca
  Rafael Muñoz Navarrte, 25, farm worker, La Rambla
  Francisco Haro Manzano, 33, farm worker, Bujalance
  Isidoro Martínez Trucharte, 56, farm worker, Baena
5 April at 6 a.m. (6) Casillas artillery range
  Lorenzo Almaraz de Pedro, 52, Captain (*Infantry Lieutenant*), Badajoz[14]
  Damián Contreras Moreno, 39, Major (*Infantry Lieutenant*), Torredonjimeno[15]
  Felipe Gallardo Linares, 53, Major (*Infantry Lieutenant*), Linares[16]
  Eugenio Muñoz Hoyuela, 34, Major (*Infantry Lieutenant*), Palencia[17]
  Antonio Fernández Sánchez, 44, Captain (*Artillery Lieutenant*), Murcia[18]
  Enrique Medina Vega, 50, Major (*Captain*), Chafarinas[19]
8 April at 6 a.m. (24)

---

[13] Had retired to Almería in 1936. With the war, re-enlisted. Promoted to Captain, then to Major in 1938. Fought in the Granada sector with 54MB of the 23rd Division of the XXII Andalusia Army Corps.

[14] Had retired to Badajoz in 1936. When this city fell, fled to Portugal where he embarked on the ship Nyassa that took many refugees to Tarragona. Re-enlisted and rose to the rank of Captain. Fought in Catalonia, then with the Military Headquarters of Almería, then with the XXIII Army Corps and lastly, in June 1938, the Levante Army Corps.

[15] Enlisted as Lieutenant with the Jaén Volunteer Battalion No. 4 and promoted Captain November 1936. Commanded the 148MB of the 37th Division of VII Army Corps of Extremadura. Promoted Major June 1938.

[16] Retired 2nd Lieutenant, re-enlisted, promoted Lieutenant October 1936, then Captain and Major (June 1938). Always fought with the Army of Andalusia.

[17] Was a Guarda de Asalto posted in Linares in 1936. Rose to Captain December 1936. 1938 transferred to the Army of the Centre and in July promoted Major.

[18] Serving in Cartagena when war broke out. Promoted Lieutenant October 1936 then Captain March 1938. Remained the entire war with the 3rd Coastal Artillery Regiment of Cartagena. It is now known how he came to be executed in Cordoba.

[19] Had retired to Almeria as Captain. Re-enlisted October 1936 and promoted Major. Commander of Almeria Machine Gunners. Ended the war with the Almeria Recruiting Office. Source: Carlos Engel.

Rafael Pérez Alcaóde, 25, office worker, Cordoba
Bermabé Menor Molleja, 28, Villa del Río
Bernardino Escabia Molleja, 29, Villa del Río
Miguel Mantas Cantero, 27, Villa del Río
Lorenzo Moyano Morales, 25, Villa del Río
Francisco Moreno Rojas, 27, Villa del Río
Ildefonso Platero Rojano, 66, Villa del Río
Ignacio Pino Gutiérrez, 25, Aguilar
José Redondo Mata, 33, Navas de la Concepción (Seville)
Antonio López Rodríguez, 44, Hornachuelos
Pedro Mangas López, 25, Hornachuelos
Manuel García Palomares, 43, Hornachuelos
Rafael Guerra Morales, 31, Guadalcázar
Antonio Nieto Romero, 37, Puente Genil
Juan M. Giménez Rodríguez, 45, Castro del Río
Francisco Sánchez García, 48, Villanueva de Córdoba
Alfonso Ayuso Tinahones, 33, Adamuz
Francisco Caballero García, 28, Villafranca
José Arrabal Muñoz, 35, Arenas del Rey (Granada)
Vicente Rubio Molero, 33, Hinojosa del Duque
Amalio Molina Macías, 30, Villamayor de Calatrava (Ciudad Real)
Juan A. Muñoz Barba, 42, Cabra

20 April at 6 a.m. (5)
El Alcácer Provincial Prison
Ángel Gómez Ortega, 32, market gardener, Posadas
Rafael Torronteras Zafra, 36, ironmonger, Posadas
Alfonso Sánchez González, 22, marble worker, Posadas
Félix Cuenca Cruz, 39, farm worker, Villa del Río
Juan Barrera Reyes, 48, farm worker, Adamuz

4 June at 6 a.m. (16)
Juan Barrozasa Castro, 24, chauffeur, Palma del Río
Juan Fuentes Sánchez, 38, manual labourer, Palma del Río
Antonio Garcia Garcia, 36, market gardener, Palma del Río
Antonio Benítez González, 32, manual labourer, Palma del Río
José Castillo Sánchz, 38, ironmonger, Hornachuelos
Juan Días Martinez, 52, manual labourer, Posadas
Luís González Navajas, 63, farmer, Posadas
Antonio Sánchez Olmo, 25, farmer, Espiel
Antonio Franco Muñoz, 25, manual labourer, Espíel
Daniel Arévalo León, 34, farm worker, Villaviciosa
Francisco Álvarez Izquierdo, 33, farm worker, Villafranca
Virgilio Ferri Vidal, 41, hat maker, Valencia
José Alcaíde Manso, 49, farm worker, La Rambla
Francisco Caballero Bueno, 33, salesman, Rute
Ramón Romero Fernánmdez, 48, tradesman, Montoro
Ildefonso Ruíz Santiago, 38, farm worker, Valenzuela

8 June at 5 a.m. (13)
Juan Fuentes Ruíz, 41, manual labourer, Montoro
Bartolomé Mazuelas Sánchez, 34, farm worker, Montoro
Francisco Rodriguez Torres, 38, farm worker, Montoro
Antonio Navarro Aguilar, 29, manual labourer, Palma del Río
Ruperto Muñoz Martínez, 24, manual labourer, Palma del Río
José Pérez Reyes, 34, farm worker, Palma del Río
Rafael Polonio Delgado, 32, farm worker, Palma del Río
Manuel Díaz Sánchez, 41, manual labourer, Palma del Río
Juan Felipe Martínez Murillo, 34, manual labourer, Hornachuelos
Francisco Palacios Bernal, 56, farm worker, Almodóvar
Bartolomé Torralba Pastilla, 35, mechanic, Villa del Río

Cristóbal Navajas Manchado, 34, farm worker, Villa del Río
Pedro Berenguer Díaz, 55, manual labourer, Posadas

22 June at 5 a.m. (18)
Juan González Guirado, 37, bricklayer, Palma del Río
Miguel Pavón Fernández, 37, manual labourer, Posadas
Manuel Girona Rodríguez, 48, farmer, Posadas
Antonio Fructuoso García, 36, manual labourer, Posadas
José García García, 35, manual labourer, Hornachuelos
Manuel López Rodríguez, 37, manual labourer, Hornachuelos
Francisco Cardo Camacho, 27, manual labourer, Hornachuelos
Rafael Ruíz Moya, 38, farmer, Adamuz
Antonio Gómez Torres, businessman, Villafranca
Manuel García Castro, 43, transportation worker, Villa del Río
Francisco Palma Prieto, 34, farm worker, Aguilar
Francisco Díos Muñoz, 29, bricklayer, Villafranca (Bujalance)[20]
José Ranchal Tartajo, 38, manual labourer, Espiel
Antonio Zamora Espinar, 32, farm worker, Iznájar
Daniel de la Torre Nevado, 52, farm worker, Villaviciosa
Manuel Noguera Guisado, 36, manual labourer, Silillos (Fuente Palmera)
Antonio Redondo Heras, 47, manual labourer, Fuenteobejuna
Antonio Medina Pedregosa, 23, farm worker, Cordoba
10 July at 4 a.m. (1)
Antonio Ruíz Martinez, 45, manual labourer, Palma del Rio
20 July at 5 a.m. (6)
Antonio Sánchez Caler, 44, manual labourer, Montoro
Bartolomé Porras Ruano, 43, farmer, Montoro
José Lara Olmo, 57, manual labourer, Montoro
Juan A. Villaverde Vega, 32, manual labourer, Montoro
Francisco García Guilarte, 34, farmer, Montoro
José Quero Izquierdo, 26, farmer, Lopera
30 July at 5 a.m. (2)
Manuel Sáchez Pérez, 28, manual labourer, Montoro
Francisco Jurado Gutiérrez, 36, butcher, Villanueva de Córdoba
12 September at 6 a.m. (11)
Tomás de la Torre Barbero, 42, farm worker, Villaviciosa
Antonio Aranda Pulido, 29, farmer, Villaviciosa
José López Arribas, 24, farm worker, Villaviciosa
Juan Muñoz Arribas, 36, farmer, Villaviciosa
José Cabello Calvo, 25, farmer, Villaviciosa
Alberto Merino Pérez, 23, farm worker, Castro del Río
Antonio Francos Ruíz, 39, manual labourer, Palma del Río
José Deña Velasco, 31, farmer, Constantina (Seville)
Francisco Cuesta Gutiérrez, 66, farmer, Almodóvar del Río
Manuel Tapiero Cáceres, 47, manual labourer, Fuente Palmera
Santiago Mantas Cuencam 34, bricklayer, Villa del Río
20 September at 6 a.m. (7)
Rafael Cuevas Alcaíde, 27, potter, Villaviciosa
Tomás Lopera González, 36, manual labourer, Villaviciosa
Diego González Mísa, 28, farm worker, Villafranca
Manuel Lucena Padilla, 36, farm worker, Espejo
Antonnio García Urraco, 29m farm worker, Hornachuelos
Manuel Colomina Benítez, 30 baker, Montoro
Antonio Almenara Muñoz, 35, farmer, Palma del Río

---

[20] Famous Capitán Paco of the Villafranca Battalion. The poet Pedro Garfias sang his heroism in his book *Héroes del Sur* and by Joe Monks, an Irish member of the International Brigade, in his book *Con los rojos en Andalucía*, Renacimiento, Seville, 2012.

30 September at 6 a.m. (**6**)
  José Dios Criado, 61, farm worker, Castro del Río
  Miguel Porcel Redondom 49, tradesman, Castro del Río
  Francisco López Morales, 45, farm worker, Villa del Río
  Juan Solas Hernández, 41, manual labourer, Hornachuelos
  Emilio Santiago Lara, 45, runner, Valenzuela
  Francisco Cerezo Requena, 42, farmer, Cordoba
11 November at 6:30 a.m. (**1**)
  Manuel Díaz López, 26, barber, Baños de la Encina (Jaén)
12 November at 6:30 a.m. (**1**)
  Antonio Mármol Alonso, 34, manual labourer, Adamuz
  Juan Rojas Arenas, 34, manual labourer, Pedro Abad
27 November at 6:30 a.m. (**1**)
  Ernesto López Vidal, 60 hospital administrator, Agudos, Lugo
18 December at 7 a.m. (**4**)
San Rafael cemetery
  Francisco Alcaíde Cruz, 27, Santa Eufemia
  Vicente Lillo Morente, 32, Palma del Río
  Manuel Andújar Rosa, 48, Palma del Río
  Manuel Qesada López, 43, Montoro
20 December at 7 a.m (**3**)
  Juan José Velasco Mateos, 39, farm worker, Santiago de Calatrava (Jaén)
  Manuel Portero Romero, 51, manual labourer, Santiago de Calatrava (Jaén)
  Sebastián Romero Urbano, 30, manual labourer, Santiago de Calatrava(Jaén)
27 December at 7 a.m. (**34**)
  Santiago Fernández González, 61, manual labourer, Belalcázar
  José Paredes Ruíz, 30, farmer, Belalcázar
  Manuel Pizarro Bravo, 55, farm worker, Belalcázar
  Juan Valverde Castro, 25, farm worker, Torrecampo
  Tomás Romero Enríquez, 29, bricklayer, Villanueva de Córdoba
  Juan J. Serrano Cepas, 33, shoemaker, Villanueva de Córdoba
  Leopoldo Lucena de la Rubia, 33, farm worker, Porcuna
  Emiliano Hidalgo Fernández, 61, businessman, Dos Torres
  Valeriano Domenech Martínez, 32, farm worker, Porcuna
  Mateo García Gómez, 51, cattle breeder, Montoro
  Francisco Díaz Tintor, 44, manual labourer, Montoro
  Juan García Carrasco, 41, farm worker, Montoro
  Francisco avarro Majuelos, 39, farm worker, ontoro
  Manuel Carmona Rastroyo, 25, farmer, Puente Genil
  Fernando Molina Molina, 24, businessman, Puente Genil
  Francisco Caballero Díaz, 40, farmer, Obejo
  Martín Sánchez Bravo, 35, farmer, Alcaracejos
  José Fernández Galán, 30, manual labourer, Alcaracejos
  Lucas Centella Aranda, 40, farm worker, Castro del Río
  Rafael Moreno Bello, 28, farm worker, Castro del Río
  Rafael Rodríguez Gálvez, 34, bricklayer, El Carpio
  José Román Romero, 29, farm worker, El Carpio
  Lorenzo Gaitán Román, 27, chauffeur, El Carpio
  Domingo Villalba Calvo, 32, manual labourer, Cañete de las Torres
  Antonio Vázquez Navas, 50, farm worker, Fuenteobejuna
  Rafael Pérez Román, 36, manual labourer, Adamuz
  Juan Manuel Noguero Leal, 41, moulder, Peñarroya
  Juan Gutiérrez Cachinero, 37, farm worker, Azuel
  Juan Calixto López Santiago, 40, farm worker, Valenzuela
  Francisco Luque Morales, 32, farm worker, Espejo
  Rafael Limones Caro, 33, bricklayer, Palma del Río
  Juan Abril Pardiñeiro, 61, miller, Espiel
  Antonio Pvón Fernández, 46, manual labourer, Posadas
  Juan Moya Carrillo, 56, manual labourer, Pedroche

1941
13 January at 6 a.m. (**1**)
 Manuel Fuillerat Lora, 34, shoemaker, Villa del Río
20 January at 6 a.m. (**1**)
 José Suescun Moreno, 43, ironmonger, Villanueva del Rey
31 January at 6 a.m. (**25**)
 Féliz Chaves Caballero, 53, manual labourer, Alcalde de Fuente la Lancha
 Antonio Vigara Regidor *El Sabio*, 64, farmer, Belalcázar[21]
 Manuel Ruíz Fernández, 45, shoemaker, El Viso
 Vicente Rodríguez Ruíz, 44, manual labourer, Belalcázar
 Rafael Gómez Fernández, 53, cattle breeder, Belalcázar
 Emiio Ruíz Rodríguez, 23, manual labourer, Belalcázar
 Lorenzo Rodríguez Tapias, 41, manual labourer, Belalcázar
 Manuel Hidalgo Gómez, 29, manual labourer, Belalcázar
 Isidro Morales Cañamaque, 70, farm worker, Montoro
 Diego Valverde Ranchal, 26, manual labourer, Pedroche
 Matías González Calzadillo, 45, farm worker, Hinojosa del Duque
 Nicolás Vioque Caballerom, 29, carrier, Hinojosa del Duque
 José Navarro Moreno, 24, manual labourer, Hinojosa del Duque
 Victoriano Murillo Platero, 28, farmer, Hinojosa del Duque
 Teófilo Morales Arellano, 50, manual labourer, Hinojosa del Duque
 Domingo Acedo González, 65, shoemaker, Hinojosa del Duque
 Vicente Ureña Villaseca, 24, cattle breeder Hinojosa del Duque
 José Muñoz González, 35, muleteer, Palma del Río
 Marcos Luna Garacía, 44, farmer, Villaralto
 Antonio Sánchez Puerto, 40, manual labourer, Villaralto
 Maximiliano Toril Fernández, 44, landowner, Villaralto
 Florencio Luna Fernández, 45, shepherd, Villaralto
 Luís Cabanillas Agredano, 27, manual labourer, Fuenteobejuna
 Manuel Epósito Gómez, 35, chauffeur, Fuenteobejuna
 Bernardo Fernández Santos, 20, manual labourer, Alcaracejos
1 May at 6 a.m. (**34**)
La Salud cemetery
 Manuel Sánchez Ruíz *El Perla*, 33, farmer, Mayor of Montilla
 Blas Gómez Medina, 39, manual labourer, Villanueva de Córdoba
 José T. Torralbo Expósito, 32, manual labourer, Villanueva de Córdoba
 José Maria Sánchez Jurado, 65, clerk, Villanueva de Córdoba
 Pedro Padilla Moreno, 36, manual labourer, Villanueva de Córdoba
 Agustín Vigara García, 24, manual labourer, Belalcázar
 Fernando Cano Yébenes, 47, farm worker, Morente
 Manuel Ruano Borrego, 46, manual labourer, Cañete de las Torres
 Gaspar Manrique Bejarano, 40, carpenter, Cañete de las Torres
 Cristóbal Muñoz López, 45, farm worker, La Rambla
 Rafael Garrido Marin, 51, farm worker, Valenzuela
 Juan Rivas Montilla, 26, farm worker, Valenzuela
 Juan Cid Contreras, 22 baker, Bujalance
 Bartolomé Parrado Serrano, 54, manual labourer, Bujalance
 Mateo Castillo Rubio, 30, farm worker, Bujalance
 Moisés López Sánchez, 34, cattle breeder, Villaralto
 Augusto P. Martín Fernández, 26, cattle breeder, Villaralto
 Manuel Rojas Lara, 41, manual labourer, Montoro
 Francisco Mora González, 39, manual labourer, Montoro
 Vicebte Kereba Uñiguez, 25, farm worker, Villa del Río
 Andrés García Moreno, 23, farm worker, Castro del Río

---

[21] An elderly Socialist and very peaceful, President of the *Casa del Pueblo,* the community social centre, hid in his house for a year and a half until he was turned into the authorities. His son Agustín was executed soon afterwards.

Arcadio Gordillo Monje, 69, farmer, Hinojosa del Duque
José Perea Cortés, 33, porter, Hinojosa del Duque
Cándido García Arellano, 25, farm worker, Hinojosa del Duque
José Sánchez Salguero, 41, miner, Puertollano
Alfonso Regalón Román, 32, farm worker, Adamuz
Juan Pato Velázquez, 27, railway worker, Almodóvar del Río
Antonio Guerra Almodóvar, 35, farm worker, Pedro Abad
Manuel Mena Molina, 38, manual labourer, Pedroche
Francisco Casado Pedrajas, 51, farmer, Pozoblanco
Florentino Moyano Fernández, 41 manual labourer, Fuente la Lancha
Rafael Alijo Torquemada, 36, manufacturing, Puente Genil
Francisco Ramírez Molina, 28, farm worker, Luque
Leoncio Gómez Fernández, 37, manual labourer, Villaralto

3 May at 6 a.m. (**34**)

San Rafael cemetery
Fernando López Muñoz, 3, farm worker, Bujalance
Francisco Garcíz Pajuelo, 22, manual labourer, Alcaracejos
Francisco Muñoz Gutiérrez, 25, bricklayer, Montoro
Pedro Jurado Benavides, 47, tradesman, Montoro
Tomás Pizarro Rodríguez, 39, cattle breeder Belalcázar
Francisco Nieto Zamorano, 37, farm worker, Pedro Abad
Gregorio Artero Rojas, 37, farm worker, Pedro Abad
Juan Morales Millán, 43, farm worker, Pedro Abad
Mariano Morrugares. Garrido, 38, farm worker, Pedro Abad
Fernando Lara Cuadrado, 23, farm worker, Pedro Abad
Manuel Cuenca Pelado, 29, railway worker, Villa del Río
Rafael Piedrahita Juradom 56, farm worker, Santa Eufemia
Bernardino Muñoz Castillo, 46, farm worker, Santa Eufemia
Florentino Redondo Serena, 36, farm worker, Santa Eufemia
Aquilino Fernández Ruíz, 35 farm worker, Santa Eufemia
Manuel Daza Jurado, 45, farm worker, Santa Eufemia
Alberto Carrasco García, 32, chauffeur, Hinojosa del Duque
José Fernández Luque, 48, farm worker, Hinojosa del Duque
Juan Barbancho Delgado, 36, bricklayer, Hinojosa del Duque
Alejandro ima Fernández, 41, manual labourer, Villaralto
Antonio Fernández Sánchez, 35, farmer, Villaralto
José Ortigoso Moreno, 40, boiler maker, Belmez
Antonio Carisimo Prieto, 32, miner, Belmez
Vicente Blanco García, 28, tradesman, Belmez[22]
Antonio Pizarro Valero, 41, manual labourer, Villaharta
Manuel Larias Fajardo, 30, railway worker, Gadalcanal
José Díaz Morales, 31, manual labourer, Fuente la Lancha
Antonio Gómez Alcántara, 24, farm worker, Castro del Río
Alfonso Gómez Gutiérrez, 20, farm worker, Castro del Río
Francisco Sánchez Mellado, 21, farm worker, Castro del Río
Manuel Alejandre González, 34, manual labourer, Fuenteobejuna
Rafael Franco Anguita, 27, manual labourer, Palma del Río
Manuel Ocón Fleitas, 31, farm worker, Adamuz
Antonio José Calero Tirado, 37, manual labourer, Pedroche

19 May at 5 a.m. (**4**)

Bernardo Rojas Navarrom, 38, manual labourer, Hinojosa del Duque
Miguel Pérez Muñoz, 30, El Carpio
Antonio Castellano Guerrero, 29, manual labourer, Belalcázar
Diego Ruíz Medina, 24, manual labourer, Belalcázar

3 June (**28**)

---

[22] Attempted suicide in prison by cutting his wrists, because of the torture he had suffered. He was healed until he was well enough to be shot.

San Rafael cemetery
Eduardo Bujalance López, 33, manual labourer, Hornachuelos[23]
Pedro Torralbo Gómez *Cuadrado*, 41, teacher, Villanueva de Córdoba[24]
Avelino Nevado Asencio, 23, cattle breeder, Villanueva de Córdoba
José Sánchez Torralbo, 39, manual labourer, Villanueva de Córdoba
Juan ACasado Muñoz *El Ramo*, 56, manual labourer, Villanueva de Córdoba
Francisco López Cejudo, 55, manual labourer, Villanueva de Córdoba
Gaspar Coleto Cabezas, 44, manual labourer, Villanueva de Córdoba
Alfonso Bujalance Gallego, 33, manual labourer, Villanueva de Córdoba[25]
Ramón Javega Pozo, 40, manual labourer, Villanueva de Córdoba
Manuel Orellana Gómez, 28, cattle breeder, Villanueva de Córdoba
Francisco Illescas Palomo, 45, manual labourer, Villanueva de Córdoba[26]
Bartolomé Viveros Torralbo, 29m, farmer, Villanueva de Córdoba
Rafael Diéguez Montes, 32, mechanic, Cordoba
José Alvarado Borrull, 38, railway worker, Cordoba
Manuel Herrero Huertos, 31, manual labourer, Posadas
Antonio Luján Valenzuela, 29, manual labourer, Posadas
Francisco Villafranca Muñoz, 36, chauffeur, Bujalance
Juan Díaz Borrego, 50, farm manual labourer, Valenzuela
Juan Rodríguez Cobijar, 35, farm worker, Villanueva de la Serena
Leonardo Ramos Perales, 34, manual labourer, Puente Genil
Francisco Pérez Cabello, 30, farmer, Puente Genil
Juán M. Cabrera de la Torre, 49, manual labourer, Montoro
Manuel Madueño Navarro, 29, bricklayer, Montoro
Ángel Trujillo Medina, 48, manual labourer, Villanueva del Duque
Miguel Domínguez Flores, 33, manual labourer, Palma del Río
Lucas Fortado Cañete, 39, chauffeur, El Pedroso (Seville)
9 June at 5 a.m. (**9**)
Blas Gajete López, 26, schoolmaster, Villafranca
Manuel Bonilla Castillo, 25, farm worker, Baena
Mariano Alcaíde Aguilar, 23, bricklayer, La Carlota
Antonio Fernández Cruz, 29, farm worker, Torredonjimeno (Jaén)
Santiago Raos Bazán, 27, mechanic, Villa del Río
Rafael Rojas Navarro, 35, farm worker, Pedro Abad
Francisco Olanda Garrido, 28, barber, Pedro Abad
Francisco Delgado Fernández, 22, baker, Montoro
Joaquín Antequera Gámiz, 22, farmer, Palenciana, near Pedro Abad
28 June at 5 a.m. (**9**)
San Rafael cemetery
Manuel Valle Expósito, 46, manual labourer, Montoro
Gumersindo Loro Expósito, 37, miner, Pueblonuevo
Nicolás D. Marchante Luengo, 37, miner, Pueblonuevo
Francisco Moya Gómez, 55, manual labourer, Pedroche
Teófilo Sánchez Delgado, 28, manual labourer, La Parilla (Cordoba)
Ricardo Gómez Rivera, 30, miner, Villanueva del Duque

---

[23] Brother of Antonio Bujalance López, Frente Popular Member of Parliament, assassinated in Cordoba the first days of the coup. This was one of the most eminent Socialist families in Hornachuelos, all of the members of which were blacklisted for extermination, regardless of whether they were guilty of any crimes or not, as was Eduardo's case.

[24] Leading member and founder of the PCE in Villanueva de Córdoba in 1921. Alderman 1931. Provincial Member of Parliament 1936. Served as Captain in the Garcés Battalion during the war.

[25] Served as Major in the Republican army.

[26] One of the great martyrs of Fascism in Villanueva, cruelly tortured in prison where a guard tore an ear off with his teeth. His wife and three sons died in 1939 and after his death, only one member of the family, a son named Francisco, survived.

Manuel Gómez Ruíz, 43, manual labourer, Villaralto
José Caballero Expósito, 39, miner, Belmez
Carlos García Herrador, 28, fishmonger, Montilla
<u>12 July at 5 a.m.</u> (**3**)
   Andrés González Cano, 52, manual labourer, Villanueva de Córdoba
   Diego García Castellón 46, contractor, Villaviciosa
   Antonio Jiménez Rojano, 45, farm worker, Aguilar
<u>15 July at 5 a.m.</u> (**15**)
   San Rafael cemetery
   Gumersindo Pérez Capitán, 25, farmer, Villanueva de Córdoba
   Alfonso Ruíz Piedrahita, 33, stonemason, El Viso
   Juan García Arriaza, 24, manual labourer, Fuente Carretero
   Francisco Pedregosa Velasco, 39, farm worker, Valenzuela
   Eduardo Murillo Murillo, 29, farm worker, Villanueva del Rey
   Andrés Márquez Sillero, 26, farm worker, Villanueva del Rey
   Sebastián Cantador Redondo, 39, farm worker, Adamuz
   Rafael Pozo Marin, 33, manual labourer, Adamuz
   Eulogio Carracedo Culebra, 23, farm worker, Obejo
   Francisco Doncel Navajas, 35, farm worker, Castro del Río
   Alfonso Castilla Jiménez, 57, manual labourer, El Carpio
   Emilio García Lara, 37, shearer, Villa del Río
   Domingo Gil Iquierdo, 30, bricklayer, Villa del Río
   Benito Cantarero Ramírez, 39, boiler maker, Villa del Río
<u>8 August at 5 a.m.</u> (**2**)
   Pedro Roberto Benitez, 60, manual labourer, Bujalance
   Manuel Elías Sánchez, 24, manual labourer, Castro del Río
<u>20 August at 5:30 a.m.</u> (**8**)
San Rafael cemetery
   Rafael Deza Montero, 44, farmer, Puente Genil[27]
   Benito Ceballos León, 24, manual labourer, Adamuz
   Francisco Jiménez Amil, 59, farm worker, Adamuz
   Francisco García González, 22, porter, Hinojosa del Duque
   Manuel Aljarilla Montilla, 24, farm worker, Valenzuela
   Fidelio Gálvez Sánchez, 48, railway worker, El Hoyo (Belmez)
   Antonio Morilla Torres, 45, manual labourer, Montoro
   Antonio Zafra González, 40, farm worker, Villanueva del Rey
<u>23 August at 5:30 a.m.</u> (**2**)
   Nicolás Sánchez Ramirez, 28, La Parrilla (Cordoba)
   Antonio Acosta Navarrete, 25, La Rambla
<u>25 August at 5 a.m.</u> (**4**)
San Rafael cemetery
   Manuel Lora Tejada, 26, cattle breeder, Pueblonuevo
   Juan Blanco Luna, 33, farmer, Villaralto
   Alfonso Ruíz Cachinero, 47, farm worker, Cardeña
   José Rodríguez Rodríguez, 41, farm worker, Castro del Ríos
<u>30 August at 5:30 a.m.</u> (**4**)
   Martín Redondo Torralbo, 37, manual labourer, Villanueva de Córdoba
   Antonio Rísquez Medinam 27, farm worker, Añora
   Juan José Castillo López, 49, farm worker, Valenzuela
   José González Rodríguez, 22, manual labourer, Almodóvar del Río
<u>12 September</u> (**11**)
San Rafael cemetery
   Juan Cantador Zamora, 50, manual labourer, Villanueva de Córdoba
   Juan Lorenzo Cantador *Cucharas*, 30, manual labourer, Villanueva Cordoba
   Francisco Muñoz Cabezas, 32, manual labourer, Villanueva de Córdoba

---

[27] The Deza Montero family, Socialists, were another of the most martyred families in Puente Genil. Three brothers were executed by Francoists: Justo, Marcos and Rafael.

Juan Santofimia Muñoz, 70, Villanueva de Córdoba
Miguel Campos Toledo, 30, from Torrecampo, resident Villanueva Cordoba
Faustino García Calero *El Peno*, 36, manual labourer, Pozoblanco[28]
Antonio Merino Guerrrero, 44, from Valverde, resident Pozoblanco
Antonio Balbino Culeabra, 27, Alcaracejos
Juan Sánchez Culebra, 28, Alcaracejos
Emiliano Ayala Navarrete, 28, Alcaracejos
Sergio Fernández Tenor, 30, Dos Torres
29 September at 6 a.m. (**3**)
San Rafael Cemetery
    José Palomo Huertas, 59, farmer, Villanueva de Córdoba[29]
    Juan Flores López, 51, businessman, Espiel
    Manuel López Toro, 39, farm worker, Bujalance
11 October at 6:30 a.m. (**7**)
San Rafael cemetery
    Antonnio Machuca Molina, 46, Luque
    Pedro Ramos Navarro, 47, farmer, Hinojosa del Duque
    José Monje Tejero, 30, farm worker, Belalcázar
    Luís Martinez González, 26, manual labourer, Posadas
    Nicomedes de la Fuente Lanza, 22, farm worker, Villaviciosa
    Miguel Amor Jordán, 39, manual labourer, Adamuz
    Tomás Cuadrado Ruíz, 61, farm worker, Adamuz
25 October (**2**)
San Rafael cemetery
    Juan Fernández Utrero, 43, manual labourer, Badajoz
    Francisco Fernández Calderón, 35, manual labourer, Belalcázar
6 November at 7 a.m. (**8**)
San Rafael cemetery
    Juan Escoriza Segura, 44, manual labourer, Villanueva de Córdoba
    José Capitán Pozuelo, 47, manual labourer, Villanueva de Córdoba
    Juan Francisco Chuán Soto, 38, manual labourer, Villanueva de Córdoba
    Francisco Pamo Susin, 26, manual labourer, Valenzuela
    Leandro Benavente Murillom 24, manual labourer, Vilanueva del Rey
    José Domenech Martinez, 49, manual labourer, Montoro
    Francisco López Sánchez, 25, shoemaker, Alcaracejos
    Alfonso Castro Cruz, 37, manual labourer, Dos Torres
17 November at 7 a.m. (**3**)
San Rafael cemetery
    Antonio Baena Moreno, 37, schoolmaster, Pozoblanco[30]
    Jerónimo Jurado Carrillo, 28, electrician, Pozoblanco
    Andrés Márquez Tamajón, 47, farm worker, Castro del Río
24 November at 7:15 a.m. (**1**)
San Rafael cemetery
    José Cabello Béjar, 49, carpenter, Adamuz

---

[28] Served as Lieutenant in the Republican army.
[29] Together with his wife, Francisca Gómez Cuevas, leading Communist in Villanueva. The Palomo family was one of those blacklisted for elimination. His brother José Antonio was killed under the Law of Fugitives 8 June 1948 because he was listening to a clandestine radio station.
[30] One of the most peaceful people who avoided strife at all costs, even acting in favour of right-wingers. He had three strikes against him: he was one of the most influential Socialists in Pozoblanco, he had chaired the War Committee and had served as a political commissar with the VIII Army Corps and then as Commissar General for the Army of Andalusia-Extremadura.

25 November at 7 a.m. (**1**)
San Rafael cemetery
 Miguel Lindo Serrano, 31, manual labourer, Adamuz[31]
10 December at 7:30 a.m. (**10**)
San Rafael cemetery
 Luís Romero García, 29, farmer, Posadas
 Manuel Martín Bovi, 35, manual labourer, Posadas
 Miguel Fernández Navea, 25, peasant, Villa del Río
 Juan Cívico Rincón, 47, peasant, Castro del Río
 Farnando Sánchez Medina, 26, farm worker, Castro del Río
 José Delgado Medina 26, manual labourer, Villa del Río
 Antonio Prieto Asensio, 27, metalworker, Villa del Río
 Julián Claudio Carrillo, 35, farm worker, Pedroche
 Francisco Carrillo Cobos, 39, farm worker, Pedroche
 Juan Castilla Riveram 26, farm worker, Pedro Abad
20 December at 7:30 a.m. (**6**)
San Rafael cemetery
 Manuel Gómez Maratín, 42, manual labourer, Villaralto
 Manuel Villegas Gómez, 29, manual labourer, Villaralto
 Pedro Rachal Alcaíde, 28, manual labourer, Torrecampo
 José Expósito Flores, 28, manual labourer, Obejo
 Antonio Olmedo Molina, 30, manual labourer, Luque
 Manuel Cepas Piedra, 40, manual labourer, Bujalance

1942
10 January at 7:30 a.m. (**9**)
San Rafael cemetery
 Francisco Rojas Moreno, 24, manual labourer, Villanueva de Córdoba
 Antonio Santofimia Ranchal, 23, farm worker, Pozoblanco
 Juan Cazorla Muñoz, 30, hairdresser, Dos Torres
 Juan Expósito Murillo, 35, farm worker Hinojosa del Duque
 Antonio Romero Alcudia, 30, bricklayer, Torrecampo
 Rafael Aragonés Alcaíde, 29, hairdresser, La Carlota
 José Sabaquebas Martínez, 32, railway worker, Posadas
 Manuel Aguilar Arance, 53, cordwainer, Bailén
 José Rodríguez Calero, 22, manual labourer, La Granja de Torrehermosa
24 January at 7:30 a.m. (**1**)
San Rafael cemetery
 Juan Valentín Escribano, 31, manual labourer, Belalcázar
10 February at 7:30 a.m. (**2**)
San Rafael cemetery
 José Gómez Flores, 29, manual labourer, Doña Mencía
 Marcos Torres Cabanillas, 40, railway worker, Cabeza del Buey
25 February at 7:30 a.m. (4)
San Rafael cemetery
 Bartolomé Martínez Peña, 57, manual labourer, Adamuz
 Manuel Fernández Gómez, 54, manual labourer, Villaralto
 Francisco José Fernández Ruíz, 37, manual labourer, Villaralto
 Juan José López Casado, 43, metalworker, Hornachuelos
10 March at 7 a.m. (**2**)
San Rafael cemetery
 Juan Amor Arvás, 30, manual labourer, Montoro
 Joaquín Moreno Muñoz, 43, manual labourer, Baena

---

[31] Belonged to the famous Lindo family of Adamuz, some of whom, Rafael and Diego Luque Lindo, fled to the hills to fight with the Romera guerrillas, surviving until 1949 in the mountains.

24 March at 7 a.m. (**1**)
San Rafael cemetery
    Francisco Molina Toledano, 27, manual labourer, Posadas
9 April at 6:15 a.m. (**1**)
San Rafael cemetery
    Juan Manuel de San Silvestre 57, manual labourer, Posadas
11 May at 6:30 a.m. (**1**)
San Rafael cemetery
    Antonio Ruíz Bejarano, 32, manual labourer, Pozoblanco
23 May at dawn (**3**)
San Rafael cemetery
    José León Gómez, 45, manual labourer, Villaralto
    Juan Liñan Suárez, 42, manual labourer, Adamuz
    Manuel Mesa Rodríguez, 36, manual labourer, Montoro
25 June at 6:15 a.m. (**11**)
San Rafael cemetery
    Ambrosio Alcalde Bejarano, 30, manual labourer, Villanueva de Córdoba
    José Márquez Morales, 34, manual labourer, Hinojosa del Duque
    José Luna Aranda, 30, manual labourer, Hinojosa del Duque
    Antonio Pérez Murillo, 33, Hinojosa del Duque
    Alejandro Barbero Capilla, 25, debt collector, Esiel
    Bernabé Navado Plaza, 46, manual labourer, Adamuz
    Antonio Izquiano Barea, 44, Adamuz
    Gerardo Muñoz Gómez, 31, electrician, Villaralto
    José Girón Villamón, 46, manual labourer, Pedroche
    Francisco Sáchez Castillo, 46, manual labourer, Pedroche
    Juan Ruíz Fernández, 49, Cabezarrubia
3 July at 6:15 a.m. (**1**)
San Rafael cemetery
    Francisco Rosales Rojano, manual labourer, Baena
8 August at 6 a.m. (**7**)
San Rafael cemetery
    Pedro Barea Higura, 39, Villanueva de Córdoba
    Blas Fernández Díaz, 28, Dos Torres
    Juan Gutiérrez Peralvo, 28, Dos Torres
    Dámaso del Rey Cáceres, 38, Villanueva del Duque
    Andrés Lindo Pérez, 35, Adamuz
    Carlos Murillo Copé, 63, Belalcázar
    Juan Fernández Fernández, 33, Alcaracejos
22 September at 6 a.m. (**2**)
San Rafael cemetery
    Francisco Ceballos Cano, 43, shoemaker, Adamuz
    Agustín Sánchez Sánchez, 37, electrician, Alcaracejos
2 October (**2**)
    Ramón Lleida Gómez, 45, retired soldier, Valencia
    Manuel Galván Valdivia, 27, draftsman, Málaga
7 November at 7 a.m. (**8**)
    Rafael Zamorano Pella, 52, peasant, Espejo
    Juan Manuel Cáceres Oliva, 31, shoemaker, Palma del Río
    Saturnino Mulloz Luna, 31, cattle breeder, Villaralto
    Bartolomé Ayllón Quesada, 45, peasant, Adamuz
    Francisco Osgaz Pedregosa, 36, peasant, Adamuz
    José Hidalgo Acedo, 29, manual labourer, Dos Torres
    Antono López Moreno, 30, manual labourer, El Viso
    Manuel Navarro Fernández, 41, nurse, Seville
9 November at 7 a.m. (**1**)
    Juan Aguilera Ruíz, 40, manual labourer, Iznájar
24 November at 7 a.m. (**4**)
    Bernabé Canno Delgado, 43, chauffeur, Adamuz

Pedro Crespo Alcaíde, 36, farmer, Torrecampo
Antonio Montoro Ávila, 23, manual labourer, Montefrío (Granada)
Francisco Montoro Ávila, 26, manual labourer, Montefrío (Granada)
10 December at 7 a.m. (1)
Francisco Amor Cuadrado, 37, peasant, Adamuz
23 December (1)
Antonio Rmá Villamón, 25, baker, Pedroche

1943
23 January at 7 a.m. (3)
San Rafael cemetery
Rafael Pérez Expósito, 32, manual labourer, Montoro
Luís Lucena Plata, 29, manual labourer, Espejo
José Arévalo Villas, 43, welder, Belmez
10 March at 7 a.m. (3)
Francisco Ruíz Olalla, 38, Montoro
José Rodríguez García, 37, Montoro
Avelino Pedrajas Jaut, 32, Pozoblanco
25 March at 7 a.m. (5)
Juan Sánchez Pozuelo *El de la Loma*, 27, manual labourer, Villanueva de Córdoba
Joaquín Albacete Fernández, 29, manual labourer, Adamuz
Gregorio Sánchez Mulloz, 28, manual labourer, Cardella
Manuel Arroyo Rodríguez, 29, manual labourer, Baena
José Cuesta Meero, 28, manual labourer, Linares (Jaén)
9 April at 6:30 a.m. (2)
Feliciano García Castillo, 49, manual labourer, Belalcázar
Ceferino Lamelas Armesto, 44, sharpener, Orense
26 April at 6:30 a.m. (1)
Agapito Fernández Sánchez, 34, farmer, Villaralto
24 May at 6:30 a.m. (2)
Andrés Mulloz Sánchez, 43, Villaralto
Francisco Valcarrers Cambronero, 27, Pedro Abad
June at 6:30 a.m. (1)
Pedro Marchna Molina, 41, Luque
9 July at 6:30 a.m (1)
Leopldo Mulloz Jurado, 41, Espejo
9 August at 7 a.m. (2)
Bartolomé Juan Garcia Dueñas, 36, cattle breeder, Pozoblanco
Juan Zurita Solar, 51, businessman, Montoro

1944
25 January at 7:45 a.m. (2)
Alfonso Ruíz Díaz, 46, peasant, Montoro
Francisco Cabezas, 40, farmer, Puente Genil

21 June at 5:45 a.m. (1)
Antonio José Castillo Benavides, 35, fishmonger, Montoro
19 October at 6:45 a.m. (13)
Manuel Álvarez Agudo, 20, bank employee, Madrid
Juan Vélez López, 32, plumber, Posadas
Celestino Lara Ruíz, 39, cook, Cazorla (Jaén)
Francisco Medina Rodríguez, 34, metalworker, Cordoba
Antonio Fernández Cuenca, 29, tradesman, Brazil
Alfonso Cerezo Regalón, 40, bricklayer, Adamuz
Emilio Jiménez Rascón, 20, clerk, Cordoba
Eusebio Gómez Gusmán, 32, manual labourer, Iznájar
José Molero Berlanga, 32, telegraph operator, Espiel
Rafael Medrán Navarrete, 33, miner, Alcaracejos
Antonio Cobos León, 25, Army corporal, El Carpio

Pedro Roldán Ortiz, 26, mechanic, Lucena
Sebastián Caravaca Martínez *El Niño del Dinero*, 20, peasant, Bujalance[32]
9 November at 7:30 a.m. (8)
    Antonio Sánchez Serrano *El Chepa*, 42, railway worker, Llerena[33]
    Francisco García Rael, 22, manual labourer, Cordoba
    Pedro Gómez Machado, 25, manual labourer, Villa del Río
    Sebastián Rivera Moyanom 25, manual labourer, Villa del Río
    Bartolomé Canales aGonzález, 24, manual labourer, Villa del Río
    Enrique Ramos Bazán, 34, bricklayer, Villa del Río
    Juan Carmona García, 27, manual labourer, Montoro
    Diego Bautista Caparrós, 36, manual labourer, Almeria
30 November at 7:30 a.m. (1)
    Vicente Resina Camino, 36, manual labourer, Montoro

1945
1 February 1945 at 7:30 a.m. (1)
    Justino Dominguez Mellado, 40, farmer, Espejo
21 February 1945 at 7 a.m. (1)
    Bartolomé Jurado Barrera, 39, manual labourer, Adamuz

TOTAL EXECUTED: 584

---

[32] Arrested at the end of 1943 when he came down from the mountains where he had been fighting as one of the Los Jubiles guerrillas, to visit his family in Bujalance. Not known if he was captured or whether he turned himself in; beaten to force him to betray his comrades, they then shot him.

[33] Another member of the Los Jubiles guerrillas, arrested in September 1940 in Alcaracejos. For some unknown reason, they waited four years to kill him.

Mass grave in Estépar, Burgos
These victims were exhumed 2014 by the local branch of the
Association for the Recovery of Historical Memory [34]

According to historians and Historic Memory Associations, an estimated
130,000 people are buried, unidentified, in mass graves throughout
Spain, 90,000 of whom were executed during the civil war and 40,000
in the post-war period, especially during the Triunnium of Terror.

---

[34] https://en.wikipedia.org/wiki/File:Spanish_Civil_War_-_Mass_grave_-_Est%C3%A9
par,_Burgos.jpg

APPENDIX XII

CIVILIANS ASSASSINATED BY THE GUARDIA CIVIL UNDER THE LAW OF FUGITIVES JUST BEFORE AND DURING THE 1947-1949 TRIENNIAL OF TERROR

Date	Name & Age	Place of Birth	Where killed
1941			
July 19	Antonio Pizarro Illescas, 38	Vn de Córdoba	Torrecampo
	Juan Fernández García, 20	Pozoblanco	Espiel
	Bernardino Mansilla Villarreal, 39	Pozoblanco	"
	Eusebio Vioque Sánchez, 38	Espiel	"
	Máximo Peralbo Caballero, 18	Espiel	"
	Teodoro Sánchez Luna, 18 *brother*	Villaralto	"
	Restituto Sanchez Luna, 22 *brother*	Villaralto	"
	Manuel Gómez Valverde, 33	Espiel	"
	Francisco Marabé Campos, 31	Pueblonuevo	"
	Baudilio Muñoz Márquez, 39	Pozoblanco	"
	Andrés Espinosa Martínez, 54	Pozoblanco	"
	Honorato Sánchez Gómez, 20	Pozoblanco	"
	Ángel Egea Risco, 28	Pozoblanco	"
	Eladio Rubio González, 27	Espiel	"
	Antonio Arévalo Bajo, 20	Pozoblanco	"
	Adrián Arévalo Bajo, 20	Pozoblanco	"
July 27	Antonio Jurado Muñoz, 59	Dos Torres	Dos Torres
	Genato Gazorla Muñoz, 59	Dos Torres	"
July 30	Pedro Romero Fernández, 23	Dos Torres	"
	José Talero Tapia, 60	Dos Torres	"
	Sebastián Lunar Rubio, 51	Dos Torres	"
July 31	Florencio Rísquez Andújar, 29	Torrecampo	Torrecampo
	Sebastián Pastor Romero, 41	Torrecampo	"
August 1	José Romero Iglesias, 20	Dos Torres	Dos Torres
September 2	Rafael Parra	Córdoba	Vn de Córdoba
November 22	Alfonso Alharilla Morales, 19	Bujalance	Bujalance
	José Gallardo Gómez, 19	Bujalance	"
	Francisco Nieves Galiano, 18	Bujalance	"
1943			
May 18	Andrés Cepas Luna, 45	Vn de Córdoba	Vn de Córdoba
1946			
August 3	Diego García Cachinero, 43 (*hung*)	Obejo, Vn de Córdoba	La Candelera
October 26	Fernando Chacón Benito, 29	Villaviciosa	Villaviciosa
November 11	Miguel Esquina Carrión, 69 *father*	Hinojosa	Pueblonuevo
	Julián Esquina Barbarroja, 42 *son*	Hinojosa	Hinojosa
	Andrés Esquina Barbarroja, 29 *son*	Hinojosa	"

Date	Name & Age	Place of Birth	Where killed
December 27	Manuel Sánchez Nocetto, 51 *father*	Fuente Tojar	Fuente Tojar
	Francisco Sánchez Moral, 31 *son*	Fuente Tojar	"
	José Mª Leiva Pimentel, 36	Fuente Tojar	"
	Josefa Briones Molina, 58	Fuente Tojar	"
1946	Francisco Sánchez López (*beaten*)	Vn de Córdoba	Córdoba
1947			
January 18	Juan J. Ortíz Castillejo, 57	La Cardenchosa	Hornachuelos
	Diego Zújar Monterroso, 35	Cañada de Gamo	"
March 26	Daniel Gallardo Algaba, 49	Los Sánchez	Fuenteobejuna
	Santiago Benavente Pérez, 58	Argallón	"
March	Ramón Oriego Salamanca	-	Baena
April 12	Antonio Capitán Pizarro, 33	Dos Torres	Cardeña
	Antonio Vioque Alcalde, 33	Dos Torres	Cardeña
April 18	Miguel Fabios Amor, 19 (*beaten*)	Pozoblanco	Hospital/Córdoba
June 17	Francisco Perea Gallardo	Fuenteobejuna	Fuenteobejuna
July 28	José Mª Jurado Zarnoza, 30	Vn de Córdoba	Vn de Córdoba
August 23	Pedro Molero Izquierdo, 24	Vn de Córdoba	Montoro
August 26	Rafael Gómez Rivera, 18	Villaharta	Villaharta
1948			
January 25	Rafael Muñoz Sánchez 32, *brother*	Navas Concepción	Km 14, Navas
	Manuel Muñoz Sánchez, *brother*	Navas Concepción	"
	Santos M. Muñoz Sanchez, *brother*	Navas Concepción	"
	El Conejo's father	-	Espiel
January 28	Antonio Caballero Fernández, 68	Villanueva del Rey	Vn. Del Rey
March 8	Pedro Moya Tejada, 32	Pozoblanco	Pozoblanco
	Juan Mejias Cerezo, 50	Adamuz	"
March 11	Enrique Muñoz Agudo, 33	Navas Concepión	Hornachuelos
March 17	Antonio Salado Alonso, 39	Hornachuelos	"
March 27	Antonio Camacho Invernó, 38	Hornachuelos	"
April 18	Juan Ruíz Calero	Pozoblanco	Pozoblanco
	Lucas Rodríguez Fernández, 34	Pozoblanco	"
April 24	Epifanio Delgado Hidalgo, 27	Obejo	Obejo
May 12	Pedro Gómez Jurado, 25	Rute	Hornachuelos
May	*Unknown man*	Villanueva d Duque	Vn del Duque
June 2	Andrés Cano Ruíz, 40	Vn de Córdoba	Vn de Córdoba
June 8	Manuel Torralbo Cantador	Vn de Córdoba	"
	José A. Palomo Huertas, 48	Vn de Córdoba	"
	Juan Romero Cachinero, 39	Vn de Córdoba	"
	Isidoro Calero Pozo, 45	Vn de Córdoba	"
	Andrés Díaz Gutiérrez, 58	Vn de Córdoba	"
	Catalina Coleto Muñoz, 52	Vn de Córdoba	"
June 17	Juan García Serrano, 26	Vn de Córdoba	"
	Padro Coleto Díaz, 45	Vn de Córdoba	"
	Genaro Ruíz Zamora, 27	Vn de Córdoba	"
July 5	Andrés Molero Redondo, 51	Villafranca	Obejo
	Francisco Romero Huertas, 46	Vn de Córdoba	"
July 7	Pedro Rojas Serrano, 51	Montoro	Cardeña
July 8	Pablo Agenjo Rodríguez	Vn de Córdoba	"
August 4	Juan Moyano Márquez, 23	Pozoblanco	Pozoblanco
	Juan A. Fuentes Cardador, 25	Pozoblanco	"

Date	Name & Age	Place of Birth	Where killed
August 31	Felipe González Torrico, 18	Hinojosa	Hinojosa
September 3	Antonio Gómez Soto, 43	Albuñán, Granada	Adamuz
	Rafael Quesada Carvajal, 35	Villafranca	Adamuz
September 10	Amelia Rodríguez Lopes, 49 *mother*	Pozoblanco	Pozoblanco
	Amelia García Rodrig., 18 *daughter*	Pozoblanco	"
	Isabel Tejada López, 60	Pozoblanco	"
	Antonio Cabanillas Rodríguaez, 34	Pozoblanco	"
September 12	Juana Cabello Moreno, 52	Vn de Córdoba	Conquista
	Andrés Gañán Calventos, 30	Conquista	"
September 13	Sixto Fernández Gómez, 46	Villaralto/Cardeña	Cardeña
September 14	Cipriano Redondo Moreno 63 *father*	Obejo	"
	Brígida Muñoz Díaz, 60 *wife/mother*	Obejo	"
	Juan Redondo Muñoz, 27 *son*	Obejo	"
September 24	Bernabé Sánchez Torralbo, 52	Adamuz	Adamuz
September 25	Fernando Gallego Pontes, 19	Vn del Duque	Vn del Duque
September 28	Pedro Gómez Calero, 36	Vn de Córdoba	Vn de Córdoba
	Miguel Fabios Dueñas, 59	Pozoblanco	"
September 29	Matías Valro Aranda	Hinojosa	Villanueva del Rey
	Fernando Litón Cano, 54 *father*	Villanueva del Rey	"
	Jacinto Litón Cano, 27 *son*	Villanueva del Rey	"
October 17	Ángel Sojo Llamas, 54	Fuentes Andalucia	Hornachuelos
	Fernando Antínuez Bajo, 27	Navas Concepción	"
October 26	Pedro Torrecilla Alias, 46	Adamuz	Adamuz
October 28	Pedro Márquez Rodríguez, 41	Pozoblanco	Pozoblanco
	Juan Arévalo Calero, 39	Pozoblanco	"
	Clemente Márquez Galán, 42	Ciudad Real	"
	Manuel Fernández Fernández, 35	Alcaracejos	"
November 10	Andrés González Fernández, 62 *bro.*	Villafranca	Villafranca
	Diego Gozález Fernández 6o *brother*	Villafranca	"
November 11	Juan A. Redondo Monteagudo, 57	Adamuz	Adamuz
December 2	Pedro Herruzo García, 25	Pozoblanco	Pozoblanco
	Pedro Caballero Olmo, 38	Añora	"
	Antonio Olmo Caballero, 34	Añora	Vn de Córdoba
	Eufrasio Madero Expósito, 38	Vn de Córdoba	"
December 7	Joaquín Heredia Giménez, 39	Vn de Córdoba	"
	Gaspar Martín Valverde, 45	Vn de Córdoba	"
December 19	Rafael Fernández Muõz, 36	El Guijo	Pedroche
	Juan Aperador García, 42	El Guijo	"
	Pedro Castillo Fuente, 65	Pedroche	"
1948	Joaquim Chamizo Zoilo	Córdoba	Almodóvar
April 48	Salado Alonso		Montoro*
May 48	José Sánchez Cambrón	Hinojosa	Hinojosa*
	Maximiliano Ruíz		Hinojosa*
June 48	Manuel Gutiérrez		Rute*
	Antonio Roldán		Cabra*
September 48	Francisco Revilla Martín		Montilla*
Undated	So-called Baltasar		Villaviciosa§
Undated	Enrique de la Fuente Arribas		Villaviciosa§

*Incomplete information from official records
§ Information from oral sources

729

Date	Name & Age	Place of Birth	Where killed
1949			
February 12	Francisco Moreno Castro, 43	Hornachuelos	Hornachuelos
February 26	Manuel Zurita Cuadrado, 62	Fuenteobejuna	Fuenteobejuna
February 27	Higino Diéguez Carcía, 43, *brother*	Fuenteobejuna	Belmez
	José Diéguez García, 26 *bro./husband*	Castillo Guardas	Belmez
	Teresa Molina Sánchez, 26, *wife*	Espiel	Belmez
	Antonio Medina Moreno, 59	Belalcázar	Belmez
March 5	Rafael Santacruz Rojano, 42 *cousin*	Obejo	Hornachuelos
	Féliz Rubio Rojano, 46 *cousin*	Obejo	"
April 6	Francisco Guijo Redondo, 22	Pozoblanco	Pozoblanco
April 10	Amador Cabanillas Castillejo, 69	Hornachuelos	Villaviciosa
April 12	Diego García Vázquez, 25	Fuenteobejna	"
	Juan Calero de los Río, 40	Villaviciosa	"
July 17	Manuel Vigara Regidor (*tortured*)	Baelalcázar	Belalzázar
July 27	Antonio Muñoz Fernández, 38	Villanueva del Rey	Vn. del Rey
	Pedro Manuel Cano Ruíz, 44	Villanueva del Rey	"
	Antonio Sánchez Jódar, 46	Villanueva del Rey	"
September 10	Françisco Cebrián Fernández, 43	Adamuz	Adamuz
September 17	Rafael Rodríguez Carmona	Villafranca	Villafranca
September 27	Pedro Gómez Caro, 32	Valsequillo	Los Blázquez
	Juan Menjíbar Murillo, 64	Castuera	"
	Leoncio Rubio Sánchez, 54	Hinojosa	"
	Lorenzo Gutiérrez Pérez, 23	Hinojosa	"
	Isidoro Rodríguez Rubio, 29	Pozoblanco	Pozoblanco
1950			
January 13	Miguel Lira Cano, 48	Granja de T.	Belalacázar
	Ángel Paredes Mansilla, 23	Belalcázar	"
	Pedro Benitez Medina, 27	Belalcázar	"
June 27	Diego Porras Piedra, 39	Rute	Rute
	Gumersindo Bueno Reina, 44	Rute	Rute

TOTAL CIVILIANS EXECUTED: 160

Missing from the above, Antonio Vargas Montes, born in Seville, a member of the Regional Committee of the PCE, murdered by the Guardia Civil in Belmez and buried in the local mass grave. Source: an email from Francisco Espinosa to Moreno Gómez, containing the deceased's widow's, María Luísa, testimony dated 24 February 2007.

SOME AUDIO-VISUAL SOURCES

Newspapers
ABC Córdoba
ABC Sevilla
Amanecer, Zaragoza

Córdoba, Cordoba
Diário Azul, Cordoba
> Founded in 1936 as the official organ of the FET-JONS and published until 1941. Predecessor of today's *Diário de Córdoba*.

El Defensor de Córdoba
Heraldo de Aragón, Zaragoza
Villanueva, Villanueva de Córdoba

Documentaries

Albino Garrido's oral testimony can be heard as part of a 2011 documentary on Spanish TVE2 by Juan Sella ad Rafael Robledo, *El pesadillo de Castuera Badajoz* (The Castuera Badajoz Nightmare), which includes multiple visual recordings of unnamed oral testimonies from survivors and relatives of disappeared prisoners at this concentration camp, Uploaded to YouTube January 25 2014 by CGT Barcelona:
https://youtube/MArWunQbbQM.

Carolina Gil Fonce. Interview with Judge Baltazar Garzón May 12 2008, Radio Argentina. Can be downloaded at: http://www.informam.nl .

Guerra de Gila. Caustic monologue by the comedian Miguel Gila Cuestas, where a soldier appears to have a phone conversation with 'the enemy'. *Is that the enemy speaking?* 2014. YouTube August 17 2009. Available at: https://www.youtube.com/watch?v=R7d4Aj4tFA4

José Martínez and Txsaber Larreategi. *Prohibido recordar* (Forbidden to Remember). Basque television (ETB) documentary, Tentazioa Rec 2010.

Juan Caunedo Domínguez. *Sombra, niebla y tiempo*. (Shadow, fog and time). Freelance producer. Documentary for the Madrid Forum for Memory.
Mariano Agudo and Eduardo Montero, Directors. *Presos del Silencio* (Prisoners of Silence). Documentary produced by LaZanfoña Producciones, Canal Sur TV, March 2009. Uploaded by Intermedia productions to: https://www.kaosenlared.net/noticia/presos-silencio-documental-fortaleza-historica

Marisa Paredes and José Luís Peñafuerte. *Los caminos de la memoria*. (The paths of memory). Documentary 2009. Available on the Internet at: https://cinepeliculasflv.com/21524-los-caminos-de-la-memoria-online-peliculas-gratis-hd-espanol.html

Martín Jönsson and Carl Pontus Hjorthén. Documentary. *Maria Carmen España – The end of Silence*. 2011 Swedish TV Broadcast. Available on the Internet at: https://www.youtube.com/watch?v=wmlgo8uSXcg.

Montse Armengpou & Ricard Belis. *Los niños perdidos del Francoism*. (Franco's lost children). Award-winning television documentary by Televisó de Catalunya, Barcelona, 2002.

Patxi Eguilaz. Documentary. *Desafectos. Esclavos de Franco en el Pirineo*. (The disaffected. Franco's Slaves in the Pyrenees). Regarding a Workers' Battalion in El Roncal-Salzar, employed in building a road between these two towns, 1939-1941. Un-named survivors' oral testimonies. 2007. Uploaded to YouTube at:
https://www.youtube.com/watch?v=OH89dek8hPk

Patzi Eguilaz. Documentary. *Nos quitaron todo*. (They took everything from us). Produced in collaboration with Professors César Laiana, Mirta Núñez, Emilio Majuelo, Miguel A. Rodriguez Arias, and José Miguel Castón. and directed by Patzi Eguilaz. 85 min. Creative Commons, 2011. Distributed by Eguzki at Bideoak.geronimouztariz.com.

Guillermo Carnero Rosell and Carlos Ceacero, Producers and Directors. Documentary. *Una inmensa prisión. Imágines contra el olvido*. (An immense prison. Images against oblivion). Sub-titled in English. By Impulso Records, 2005. Recording available to download from the Internet at: https://www.vimeo.com/111282336 .

The Internationale. Hymn of the International Communist Party. Listen to in English, with substitles, at:
https://www.youtube.com/watch?v=3sh4kz_zhyo.

APPENDIX XIII

ESTIMATED BALANCE OF VICTIMS OF FRANCOIST GENOCIDE IN CÓRDOBA PROVINCE AND CAPITAL

LOCATION	EXECUTED BY FIRING SQUAD — TOWNSHIPS WAR-TIME	TOWNSHIPS POST-WAR	CAPITAL WAR-TIME	CAPITAL POST-WAR	NAZI CAMPS	CÓRDOBA PRISON 1939-1946	LAW OF FUGITIVES 1939-1950	MAQUIS KILLED	TOTALS	TOTAL CÓRDOBA CAPITAL	WAR-TIME RIGHT-WING VICTIMS
	1	2	3	4	5	6	7	8	9	10	11
ADAMUZ		3	(5)	(29)	0	22	6	10	41	(+34)	61 (+48)
AGUILAR	140		(16)	(3)	2	2			145	(+19)	0
ALCARACEJOS		1	(2)	(11)	4	4		1	10	(+13)	36
ALMEDINILLA	20		(0)	(0)	0	1			21	(+0)	1
ALMODOVAR DEL RÍO	30	6	(9)	(8)	2	5		1	44	(+17)	14
AÑORA			(1)	(1)	3	3	3	2	11	(+2)	30
BAENA	700	32	(6)	(8)	14	13	1		760	(+14)	110
BELALCÁZAR	1	30	(1)	(18)	2	9	6	23	71	(+19)	170
BELMEZ	40	36	(25)	(6)	8	16	2	3	105	(+31)	41
BENAMEJÍ	45		(10)	(1)	1	0			46	(+11)	0
LOS BLAZQUEZ	16		(0)	(0)	0	2			18	(+0)	0
BUJALANCE	3	55	(15)	(13)	3	13	3	22	99	(+28)	112
CABRA	100		(7)	(1)	1	7	1		109	(+8)	0
CAÑETE DE LAS TORRES		3	(6)	(4)	1	4			8	(+10)	25
CARCABUEY	20		(2)	(0)	1	2			23	(+2)	0
CARDEÑA	3	5	(0)	(0)	0	6	1	4	19	(+0)	4
LA CARLOTA	110		(17)	(2)	2	4		1	117	(+19)	0
EL CARPIO	3		(8)	(6)	3	2			8	(+14)	25
CASTRO DEL RÍO	40	167	(16)	(18)	6	7			220	(+34)	79
CONQUISTA		1	(0)	(0)	0	1	1		3	(+0)	0
CÓRDOBA CAPITAL			4,000	584	15	26	2	2	4,629		1
DOÑA MENCÍA	20		(15)	(1)	1	4			25	(+16)	0

LOCATION	TOWNSHIPS WAR-TIME	TOWNSHIPS POST-WAR	CAPITAL WAR-TIME	CAPITAL POST-WAR	NAZI CAMPS	CÓRDOBA PRISON 1939-1946	LAW OF FUGITIVES 1939-1950	MAQUIS KILLED	TOTALS	TOTAL CÓRDOBA CAPITAL	WAR-TIME RIGHT-WING VICTIMS
	1	2	3	4	5	6	7	8	9	10	11
DOS TORRES		6	(3)	(8)	1	13	7	3	30	(+11)	78
ENCINAS REALES	10	6	(0)	(0)	0	0			10	(+0)	0
ESPEJO	80	6	(9)	(7)	6	2			94	(+16)	64
ESPIEL	?	16	(3)	(11)	2	9	5	1	33	(+14)	23
FERNÁN NÚÑEZ	150	4	(13)	(0)	1	5			160	(+13)	11
FUENTE LA LANCHA			(0)	(3)	0	3		2	5	(+3)	0
FUENTEOBEJUNA	400	40	(5)	(5)	14	19	8	10	491	(+10)	57
FUENTE PALMERA	15		(2)	(4)	7	6			28	(+6)	0
FUENTE TOJAR	10		(1)	(0)	0	2	4	4	20	(+1)	2
LA GRANJUELA	10		(8)	(0)	2	5		1	18	(+8)	14
GUADALCÁZAR	?		(4)	(1)	0	0			0	(+5)	0
EL GUIJO			(0)	(0)	0	2	2		4	(+0)	0
HINOJOSA DEL DUQUE	16	67	(3)	(20)	8	23	9	31	154	(+23)	124 (+34)
HORNACHUELOS	40	8	(11)	(19)	7	2	4	1	62	(+30)	18
IZNÁJAR	80		(3)	(3)	1	3			84	(+6)	0
LUCENA	127		(12)	(1)	2	5			134	(+13)	0
LUQUE	31		(12)	(4)	3	5			39	(+16)	13
MANTALBÁN	17	1	(2)	(0)	1	0			19	(+2)	0
MONTEMAYOR	25	22	(6)	(0)	2	0			27	(+6)	0
MONTILLA	200		(8)	(2)	7	7	1		237	(+10)	10 (+1)
MONTORO	40	21	(16)	(40)	7	25	2	5	100	(+56)	118
MONTURQUE	15		(1)	(0)	0	1			16	(+1)	0
MORILES	10		(1)	(0)	0	1			11	(+1)	0
NUEVA CARTEYA	70		(10)	(0)	1	2			72	(+10)	0
OBEJO	?		(2)	(7)	1	2	6	2	11	(+9)	0

LOCATION	EXECUTED BY FIRING SQUAD — TOWNSHIPS WAR-TIME	TOWNSHIPS POST-WAR	CAPITAL WAR-TIME	CAPITAL POST-WAR	NAZI CAMPS	CÓRDOBA PRISON 1939-1946	LAW OF FUGITIVES 1939-1950	MAQUIS KILLED	TOTALS	TOTAL CÓRDOBA CAPITAL	WAR-TIME RIGHT-WING VICTIMS
	1	2	3	4	5	6	7	8	9	10	11
PALENCIANA	20		(4)	(0)	0	2			22	(+4)	0
PALMA DEL RÍO	300	40	(10)	(22)	14	7		1	361	(+32)	42
PEDRO ABAD	20	40	(11)	(11)	3	6	1	1	70	(+22)	26
PEDROCHE			(3)	(11)	0	11	1		13	(+14)	69 (+30)
PEÑARROYA-PUEBLONUEVO	40	88	(11)	(4)	10	24	1	5	168	(+15)	20
POSADAS	20	19	(16)	(24)	17	12			68	(+40)	77
POZOBLANCO	1	209	(10)	(10)	11	8	25	7	261	(+20)	55 (+126)
PRIEGO	80	2	(4)	(1)	0	8			90	(+5)	0
PUENTE GENIL	900	29	(32)	(17)	5	10			944	(+49)	115
LA RAMBLA	60		(2)	(6)	8	0			68	(+8)	8
RUTE	55		(2)	(11)	2	3	4	2	66	(+13)	0
S. SEBASTIÁN DE LOS B.	23		(2)	(0)	0	0			23	(+2)	0
SANTAELLA	36		(1)	(0)	1	0			37	(+1)	0
SANTA EUFENIA		21	(1)	(6)	0	6		5	32	(+7)	37
TORRECAMPO		3	(0)	(4)	3	8	2		16	(+4)	33
VALENZUELA	25		(1)	(11)	1	9			35	(+12)	19
VALSEQUILLO			(0)	(0)	0	3	1		4	(+0)	0
LA VICTORIA	?		(6)	(0)	1	2			3	(+6)	0
VILLA DEL RÍO	75		(3)	(28)	3	8			86	(+31)	30
VILLAFRANCA	51	8	(9)	(11)	1	2	5	2	69	(+20)	13
VILLAHARTA	60		(4)	(1)	9	2	1	1	73	(+5)	32

LOCATION	EXECUTED BY FIRING SQUAD				NAZI CAMPS	CÓRDOBA PRISON 1939-1946	LAW OF FUGITIVES 1939-1950	MAQUIS KILLED	TOTALS	TOTAL CÓRDOBA CAPITAL	WAR-TIME RIGHT-WING VICTIMS
	TOWNSHIPS		CAPITAL								
	WAR-TIME	POST-WAR	WAR-TIME	POST-WAR							
	1	2	3	4	5	6	7	8	9	10	11
VILLANUEVA DE CÓRDOBA		102	(2)	(35)	4	23	22	20	171	(+37)	98
VILLANUEVA DEL DUQUE	?		(1)	(4)	2	7	2	3	14	(+5)	4
VILLANUEVA DEL REY	15		(2)	(5)	1	14	7	3	40	(+7)	12
VILLARALTO	4	1	(0)	(20)	1	7	4	2	19	(+20)	32
VILLAVICIOSA	40	3	(13)	(13)	6	22	4	24	99	(+26)	33
EL VISO		6	(1)	(3)	0	10		7	23	(+4)	33
ZUHEROS	10		(2)	(0)	1	0			11	(+2)	0
Elsewhere in Spain 1947-1950						150	6	51	207		
POST-WAR						98			98		
VILLAFRANCA	51	8	(9)	(11)	1	2	5	2	69	(+20)	78
VILLAHARTA	60		(4)	(1)	9	2	1	1	73	(+5)	13
											32
TOTALS	4,472	1,102	4,000	584	246	756	160	262	11,582	4,629	2,107 (+239)

- Column 1: Victims of the Francoist repression in the towns and villages of Córdoba province during the war (4,472 executed by firing squad).
- Column 2: Victims of the Francoist repression in the towns and villages of Córdoba province during the post-war (1,102 executed by firing squad). Some towns are left blank or with few victims; this may be due to the fact that the executions were carried in larger neighbouring towns, in party headquarters elsewhere or there is no available data.
- Columns 3 and 4: These figures are in brackets because they are not computed in the totals as they are included in the data for Córdoba capital. They serve solely to indicate the number of residents of towns and villages executed in Córdoba capital. These total 4,000 shot during the war and 584 during the post-war.
- Column 5: Cordovans who died in Nazi concentration camps (246).
- Column 6: Inmates who were starved to death or who died from the sub-human conditions in the Provincial Prison in Córdoba capital. The victims are listed by towns until 1946. From 1947-1950 accounted for jointly (98 victims). Total number exterminated in this fashion for the decade 756.
- Column 7: Victims of the Law of Fugitives from 1941-1950, as a result of summary executions, on suspicion of collaborating with the guerrillas, because related to someone who had fled to the hills, os simply to extend the terror (160 assasinated in this manner).
- Column 8: Estimated number of *maquis* shot by the Guardia Civil in the Córdoba mountains (262 victims).
- Columns 9 and 10: Estimated number of republicans killed by the Francoist genocide in Córdoba province: total: 11,582 victims. By locality: in Córdoba capital: 4,629 victims; in towns and villages, 6,953 victims.
- Columna 11: Right-wing victims, by location: General total 2,107 victims (2,029 during the war and 78 in the post-war) 239 victims are given in brackets as they were not killed in Córdoba province but elsewhere in Spain, especially in Valencia through the actions of the People's Court.

Note: Until the Spanish government offers free access to all sorts of registers and archives to confirm the exact number of arrested, dead, etc., historians will only be able to present incomplete figures.

9 781960 861283